Space Sciences Series of ISSI

Volume 51

Andre Balogh · Andrei Bykov ·
Jonathan Eastwood · Jelle Kaastra

Editors

Multi-scale Structure Formation and Dynamics in Cosmic Plasmas

Previously published in *Space Science Reviews* Volume 188,
Issues 1–4, 2015

 Springer

Editors
Andre Balogh
Department of Physics Blackett Laboratory
Imperial College London
London, UK

Jonathan Eastwood
Department of Physics Blackett Laboratory
Imperial College London
London, UK

Jelle Kaastra
SRON Netherlands Institute for Space
   Research
Utrecht, The Netherlands

Andrei Bykov
Ioffe Institute
St. Petersburg, Russia

ISSN 1385-7525   Space Sciences Series of ISSI
ISBN 978-1-4939-3546-8   ISBN 978-1-4939-3547-5 (eBook)
DOI 10.1007/978-1-4939-3547-5

Library of Congress Control Number: 2016931277

Springer New York Heidelberg Dordrecht London
© Springer Science+Business Media New York 2016

Cover Image: X-ray image of the cluster of galaxies 1E0657-56 (the Bullet cluster) made with Chandra X-ray observatory (NASA). The Bullet cluster is a large cosmological plasma structure of a scale size about a megaparsec, which was formed after the collision of two large clusters of galaxies dominated by dark matter. For more details see the chapter Structures and Components in Galaxy Clusters in the book. Credit: M. Markevitch (NASA GSFC)

Printed on acid-free paper

Springer is part of Springer Science+Business Media (www.springer.com)

Contents

DOI 10.1007/978-1-4939-3547-5_1
Reprinted from *Space Science Reviews* Journal, DOI 10.1007/s11214-015-0140-4

Multi-scale Structure Formation and Dynamics in Cosmic Plasmas

A. Balogh[1] · A. Bykov[2,3,4] · J. Eastwood[5] · J. Kaastra[6]

Received: 12 February 2015 / Accepted: 13 February 2015 / Published online: 3 March 2015
© Springer Science+Business Media Dordrecht 2015

1 Plasma Structures: From the Earth Magnetosphere to Clusters of Galaxies

Being statistically homogeneous on cosmological scales the Universe is demonstrating a very rich variety of structures and components at smaller scales. Gravity is the source of energy and the driving force for formation of the large scale structure (LSS), clusters of galaxies, galaxies and stars, while the electromagnetic fields are drastically important for the microphysics of the momentum and energy transfer in the baryonic matter and radiation. The Workshop "Multi-scale structure formation and dynamics in cosmic plasmas" which was held at International Space Science Institute in April 2013 was devoted to a broad discussion of different aspects of formation, dynamics, and observational appearance

✉ A. Bykov
byk@astro.ioffe.ru

A. Balogh
a.balogh@imperial.ac.uk

J. Eastwood
jonathan.eastwood@imperial.ac.uk

J. Kaastra
j.kaastra@sron.nl

[1] Imperial College London, Prince Consort Road, London SW7 2AZ, UK

[2] A.F. Ioffe Institute for Physics and Technology, 194021 St. Petersburg, Russia

[3] St. Petersburg State Politecnical University, St. Petersburg, Russia

[4] International Space Science Institute, Bern, Switzerland

[5] The Blackett Laboratory, Imperial College London, South Kensington Campus, London SW7 2AZ, UK

[6] SRON Netherlands Institute for Space Research, Sorbonnelaan 2, 3584 CA Utrecht, The Netherlands

 Springer

of plasma structures at different scales ranging from LSS to the Earth's magnetosphere. The present book is composed from the reviews, which are based on the discussions at the ISSI Workshop. It contains review papers on the basic processes of structure formation in cosmic plasmas starting from electric currents, which produce magnetic structures in planet magnetospheres, stellar winds, and relativistic plasma outflows like pulsar wind nebulae and Active Galactic Nuclei jets. The important role of the helicity concept on the structure formation and evolution of the large scale magnetic fields in highly conductive cosmic plasmas is emphasized in the book. Microscopic dynamics of plasma flows and magnetic fields was discussed in depth in Space Sciences Series of ISSI, Volume 47 "Microphysics of Cosmic Plasmas", originally published as Space Science Reviews (Balogh et al. 2013), and thus, it is highly recommended to refer to that volume while reading the present volume. Cosmological aspects of plasma structures are reviewed within a discussion of large-scale structure formation from the first non-linear objects to massive galaxy clusters, which is followed by a review of observations and current models of structures and components in galaxy clusters. Supernova remnants interacting with molecular clouds are among the most important ingredients of the global galactic ecology with a profound effect on the phenomena related to star formation. The multi-wavelength view from the radio to gamma-rays with modern high resolution telescopes revealed a beautiful and highly informative picture of both coherent and chaotic plasma structures tightly connected by strong mutual influence. The same plasma processes are likely to control the structure and dynamics of Earth's magnetosphere where detailed direct satellite observations are available. The properties of magnetic field fluctuations and structures in the outer solar atmosphere and Earth's magneto-tail, which have direct implications for the general problem of structure formation in hot plasmas, are discussed in depth in the volume.

This volume is aimed at graduate students and researchers active in the areas of astrophysics and space science.

Acknowledgements The Editors are greatly indebted to all the participants of the Workshop held in ISSI Bern on 15 to 19 April 2013 who brought their broad range of expertise and interest in the astrophysics of plasmas to discuss the vast range of scales of plasma structures in the Universe and how the study of their formation and dynamics can illuminate processes at the different scales. We thank the staff of ISSI for their dedicated support: Prof. Rafael Rodrigo, Executive Director, and his colleagues Prof. Rudolf von Steiger, Jennifer Fankhauser, Andrea Fischer, Saliba Saliba and Sylvia Wenger. The resulting collection of review papers was the outcome of the exchanges and fruitful collaboration among the participants; we thank them for their successful efforts to integrate the lessons learned in the different topics, as the reviews in the volume testify. Thanks are also due to the reviewers of the papers; in all cases the reviews were thorough and constructive and the volume bears witness to their contribution. Finally the Editors thank the staff of Space Science Reviews, as well as the production staff for their patience on occasion and for an excellently produced volume.

References

A. Balogh, A. Bykov, P. Cargill, R. Dendy, T. Dudok de Wit, J. Raymond, Microphysics of cosmic plasmas: background, motivation and objectives. Space Sci. Rev. **178**, 77–80 (2013). doi:10.1007/s11214-013-0027-1

DOI 10.1007/978-1-4939-3547-5_2
Reprinted from *Space Science Reviews* Journal, DOI 10.1007/s11214-014-0041-y

Electric Current Circuits in Astrophysics

Jan Kuijpers · Harald U. Frey · Lyndsay Fletcher

Received: 27 October 2013 / Accepted: 26 February 2014 / Published online: 5 June 2014
© Springer Science+Business Media Dordrecht 2014

Abstract Cosmic magnetic structures have in common that they are anchored in a dynamo, that an external driver converts kinetic energy into internal magnetic energy, that this magnetic energy is transported as Poynting flux across the magnetically dominated structure, and that the magnetic energy is released in the form of particle acceleration, heating, bulk motion, MHD waves, and radiation. The investigation of the electric current system is particularly illuminating as to the course of events and the physics involved. We demonstrate this for the radio pulsar wind, the solar flare, and terrestrial magnetic storms.

Keywords Cosmic magnetism · Electric circuits · Radio pulsar winds · Solar flares · Terrestrial magnetic storms

1 Introduction

Magnetic field structures in the cosmos occur on many scales, spanning a range of over 15 decades in spatial dimension, from extragalactic winds and jets down to the terrestrial magnetosphere. Yet in all these objects the properties of magnetic fields lead to a very similar, multi-scale, spatial and temporal structure. Magnetic fields originate in electric currents, as described by Maxwell's equations. They have energy, pressure, and tension, as quantified by their energy-momentum (stress) tensor. They exert a force on ionized matter, as

J. Kuijpers (✉)
Department of Astrophysics, IMAPP, Radboud University Nijmegen, P.O. Box 9010,
6500 GL Nijmegen, The Netherlands
e-mail: kuijpers@astro.ru.nl

H.U. Frey
Space Sciences Laboratory, 7 Gauss Way, Berkeley, CA 94720-7450, USA
e-mail: hfrey@ssl.berkeley.edu

L. Fletcher
SUPA School of Physics and Astronomy, University of Glasgow, Glasgow G12 8QQ, UK
e-mail: lyndsay.fletcher@glasgow.ac.uk

 Springer

expressed by the Lorentz force. Finally, their equivalent mass is small when compared to the mass-energy of ambient matter.

These basic properties lead to a common appearance in a variety of objects which can be understood as follows. Since there are no magnetic charges, magnetic fields are not neutralized and they extend over large stretches of space and time. The Lorentz force which keeps ionized matter and magnetic fields together allows gravitation to anchor a magnetic field in a condensed ionized object. Since there gas and dynamic pressures dominate over the magnetic pressure, magnetic fields tend to be amplified by differential motions such as occurs in a stellar convection zone, a differentially rotating accretion disk, binary motion, or a stellar wind around a planet. Expressed in circuit language, a voltage source is set up by fluid motions which drives a current and forms a dynamo in which magnetic field is amplified at the expense of kinetic energy. Next, the small equivalent mass of the magnetic fields makes them buoyant, and as ionized matter slides easily along magnetic fields, they pop up out of the dense dynamo into their dilute environs. Whereas this central domain is dominated by fluid pressure the environs are dilute so that there the magnetic pressure dominates. The tension of the magnetic field then allows for transport of angular momentum from the central body along the field outward into a magnetized atmosphere such as a corona or magnetosphere. As more and more magnetic flux rises into the corona, and/or differential motions at the foot-points of magnetic structures continue to send electric currents and associated Poynting flux into the corona the geometries of these nearly force-free structures adapt, expand and generally lead to the appearance of thin sheets of concentrated electric currents. The same process occurs in the terrestrial magnetosphere but now the Poynting flux is going inward toward the central body. Finally, the magnetic structure in the envelope becomes unstable and the deposited energy is released in a process equivalent to an electromotor or Joule heating but now in a multitude of small 'non-force-free' regions created by the currents themselves, often together in explosions, such as storms, (nano-)flares, ejected plasmoids, jets, but also in a more steady fashion, such as super-fast winds.

The description of the formation and dynamics of complicated magnetic structures in terms of simplified electric current circuits in a variety of objects elucidates the fundamental physics by demanding consistency, and distinguishing cause and effect. Also, it allows for a unified answer to a number of pertinent questions:

– *Current Distribution*: what is the voltage source, how does the current close, which domains can be considered frozen-in, where (and when) is the effective resistivity located?
– *Angular Momentum Transport*: where are the balancing (decelerating and accelerating) torques?
– *Energy Transport*: what is the relative importance of Poynting flux versus kinetic energy flux?
– *Energy Conversion*: what is the nature of the effective resistivity (Lorentz force, reconnection, shocks), and what is the energy partitioning, i.e. the relative importance of gas heating, particle acceleration, bulk flow, and MHD waves?

For this review we have chosen to zoom in on the magnetosphere of the *radio pulsar*, on the *solar flare*, and on *terrestrial aurorae and magnetic storms*. We will point out parallels and similarities in the dynamics of the multi-scale magnetic structures by considering the relevant electric circuits. Many of the same questions (and answers) that are addressed below are relevant to other objects as well, such as extragalactic jets, gamma ray bursts, spinning black holes, and planetary magnetospheres.

Equations will be given in Gaussian units throughout to allow simple comparisons to be made while numerical values are in a variety of units reflecting their usage in the fields they come from.

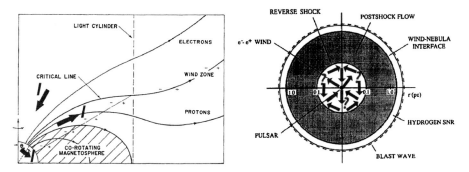

Fig. 1 *Left*: The Goldreich-Julian picture of the pulsar magnetosphere for an aligned magnetic rotator (Goldreich and Julian 1969). The potential difference across the polar cap drives a current I, which enters the star along polar field lines, crosses the stellar surface layers, leaves the star along field lines on the other side of the 'critical' field line, and is supposed to close somewhere in the wind zone. Note that the incoming/outgoing current is carried here by respectively outgoing electrons/outgoing protons, but could equally well be carried by a surplus of outgoing electrons/outgoing positrons. Adaptation from Goldreich and Julian (1969), reproduced by permission of the AAS. *Right*: Sketch of the closure of the current system in the Crab wind, which has to take place before the reverse shock at about 0.1 pc

2 Electric Circuit of the Pulsar Wind

The pulsar wind presents an important class of Poynting flux-dominated outflows. These are dominated by electromagnetic rather than kinetic energy. Here we want to study the closed electric current system, which involves regions both very near to the pulsar and very far away in the pulsar wind, a problem for which no general agreement exists as to its solution. The relative importance of the magnetic and kinetic energy flows is conveniently written as the magnetization σ:

$$\sigma \equiv \frac{\text{Poynting flux}}{\text{kinetic energy flux}} = \left\{ \frac{B_{tor}^2/4\pi}{\rho v_{pol}^2/2}, \frac{B_{tor}^2/4\pi}{\Gamma \rho c^2} \right\} > 1, \tag{1}$$

where we have included the contribution only from the toroidal (i.e. transverse to the radial direction) magnetic field B_{tor} and neglected the poloidal (i.e. in the meridional plane) magnetic field B_{pol} since the latter falls off faster with distance than the former in a steady wind. v_{pol} is the poloidal component of the wind velocity, ρ the wind matter density in the observer frame, and Γ the Lorentz factor of the wind. The first term inside curly brackets applies to non-relativistic and the second to relativistic winds. Further, it is assumed that either the ideal MHD condition applies in the wind:

$$\mathbf{E} = -\mathbf{v} \times \mathbf{B}/c, \tag{2}$$

or that the wind is nearly a vacuum outflow in which case the condition

$$\mathbf{E} = -\mathbf{c} \times \mathbf{B}/c \tag{3}$$

takes over smoothly from the ideal MHD condition. Here $\mathbf{B}$ is the magnetic induction (magnetic field), $\mathbf{v}$ the fluid velocity, and c the speed of light.

The rotating magnetized star which forms the pulsar is a so-called 'unipolar inductor' (Goldreich and Julian 1969). The rotation of the magnetized conductor creates a potential drop across the moving field lines from the magnetic pole towards the equator (see Fig. 1, Left). This voltage drop appears along the field lines between the star and infinity. The reaction of the star to this strong field-aligned potential drop is that, under certain conditions, an electric current is set up (see Fig. 1, Left). Charged particles are drawn out of the crust and accelerated to such high energies that a dense wind of electron-positron pairs leaves the star and provides for the electric currents. The magnitude of the electric current density near the stellar surface is expected to be of order

$$j_{GJ}(r_\star) \equiv \tau_{GJ}(r_\star)c, \tag{4}$$

where the Goldreich-Julian (GJ) charge density is defined as

$$\tau_{GJ}(r) \equiv -\frac{\mathbf{\Omega}_\star \cdot \mathbf{B}_0(\mathbf{r})}{2\pi c}. \tag{5}$$

$\mathbf{B}_0(\mathbf{r})$ is the background magnetic field at position $\mathbf{r}$, $\mathbf{\Omega}_\star$ is the stellar rotation vector, and $r_\star$ the stellar radius. The GJ charge density just provides for a purely transverse electric field and a corresponding $\mathbf{E} \times \mathbf{B}$-drift which causes the (ideal) plasma to co-rotate with the star at the angular speed $\mathbf{\Omega}_\star$ (2). As a result, when the charge density is equal to the local GJ density everywhere, the parallel electric field vanishes. This is the situation on the 'closed' field lines which are located near the star.

On the open field lines the speed of the charges is assumed to be the speed of light since the wind is expected to be relativistic from the beginning (4). Things are complicated here because a strictly steady state pair creation is not possible. It is clear that the *parallel* electric field momentarily vanishes as soon as the charge density reaches the value τ_{GJ}. However, to produce this GJ-density one needs a very strong parallel electric field to exist. Actually, the strong time-variations within a single radio pulse are believed to mirror the temporal process of pair creation. For our purpose we assume a steady-state relativistic wind to exist in the average sense. This is the reason for the appearance of the GJ density in (4). The current (and the much denser wind) only exist on the so-called 'open' field lines, the field lines which connect to infinity (see Fig. 1, Left). For a steady current the incoming current (poleward for the aligned rotator of Fig. 1, Left) must be balanced by the outgoing current (equatorward). This defines a critical field line in the wind (Fig. 1, Left) separating the two parts of the current. Note, however, that in contrast to Fig. 1, Left the incoming current may consist of outgoing electrons while the outgoing current may be composed of outgoing positrons. This current brakes the rotation of the star through the torque created by the Lorentz force density $\mathbf{j} \times \mathbf{B}/c$ in the stellar atmosphere. Stellar angular momentum is then transported by the electric current (which twists the magnetic field) in the wind, and dumped somewhere far out in the wind. Where, is the big question.

The main problem with the pulsar wind is that apparently the value of σ in (1) changes from $\sigma \gg 1$ near the star to $\sigma \ll 1$ somewhere in the wind. The first value follows from the assumption that the radio pulsar is a strongly magnetized neutron star with surface fields $B_\star \sim 10^9$–10^{13} G whereas observations of the Crab nebula demonstrate that $\sigma \sim 2 \cdot 10^{-3}$ in the un-shocked wind (Fig. 1, Right). To solve this problem one needs to know where and how the electric current closes in the wind so as to dump the Poynting flux in the form of kinetic energy of the wind.

Our approach to the pulsar wind problem starts with the 1D, non-relativistic, stellar wind description in terms of an electric circuit in Sect. 2.1. We then include 2D effects in Sect. 2.2, consider the aligned magnetic rotator in vacuo in Sect. 2.3, discuss numerical results for the electric circuit in the aligned rotator in Sect. 2.4, consider the importance of current starvation in Sect. 2.5, consider the multiple effects of obliquity on the electric circuits in Sect. 2.6, critically review the nature of current sheets in Sect. 2.7, mention the effect of a 3D instability in Sect. 2.8, and conclude the first of the three parts of this review by summarizing our findings about the pulsar electric current circuit in Sect. 2.9.

2.1 Stellar Winds: 1D MHD, Partial Current Closure

Already in the early days of stellar MHD, Schatzman (1959) and Mestel (1968) realized that magnetic fields anchored in a star and extending into its atmosphere force an ionized stellar wind to co-rotate with the star to distances much larger than the photosphere. As a result the specific angular momentum carried off by a magnetized wind is much larger than its specific *kinetic* angular momentum at the stellar surface because of the increased lever arm provided by the magnetic field. Our starting point is the 1D MHD description of a magnetized stellar wind by Weber and Davis (1967). Their aim is to find a steady-state smooth wind. They assume the validity of ideal (i.e. non-resistive) MHD and consider an axially symmetric magnetic structure of a special kind—a so-called split magnetic monopole—which is obtained by reversing the magnetic field of a magnetic monopole in one half-sphere. This structure rotates around the axis of symmetry (*aligned* rotator), and it is assumed that the poloidal structure is not changed when rotation sets in. Finally, they consider the outflow just above the equatorial plane. They define an *Alfvén* radius r_A in the equatorial plane where the ram pressure of the radial flow part equals the radial magnetic tension. More generally, the Alfvén radius is determined by the *poloidal* components (Mestel 1999, Chap. 7):

$$\rho(r_A)v_{pol}^2(r_A) = \frac{B_{pol}^2(r_A)}{4\pi},\tag{6}$$

or equivalently, where the poloidal flow speed equals the poloidal Alfvén speed (i.e. an Alfvénic Mach number unity). They find that the angular momentum loss per unit mass equals

$$r v_\phi - \frac{r B_r B_\phi}{4\pi\rho v_r} = \Omega_\star r_A^2.\tag{7}$$

Here r, ϕ are, formally, spherical radius, and azimuth respectively, but note that their derivation is valid just above the equatorial plane ($\theta = \pi/2$) so that r really is the cylindrical distance. Again the mass density ρ and the other quantities are measured in the observer (lab) frame. Their main finding then is that the total torque of a magnetized wind is equal to producing effective co-rotation of the wind out to the Alfvén radius. Nice and simple though this result is, it is also deceptive in that the current does not close at the Alfvén surface. Actually, most of the current goes off to infinity. The authors find that the distribution of angular momentum over electromagnetic and kinetic parts with distance follows Fig. 2, Left which shows that only a part of the Poynting torque is converted into a kinetic torque, and that asymptotically at large distance the Poynting contribution becomes constant and dominates. For the current closure in the wind then (which necessarily requires more than one dimension to describe) we can conclude that there is only *partial current closure*, and that most of the current goes out into the wind along the field lines in a force-free manner, i.e. asymptotically the remaining current flows along the magnetic field lines (Fig. 3, Right). Indeed, Weber and Davis (1967) find that the current within a magnetic surface satisfies

$$I(r) = \frac{c}{2} B_r r^2 \frac{v_\phi/r - \Omega_\star}{v_r},\tag{8}$$

and becomes constant at large distances. Therefore, the result of this study—which turns out to be applicable to the wind from a *'slow magnetic rotator'*, here defined as emitting a *wind mainly driven by thermal gas pressure*—is that at large distances the magnetic angular momentum and the Poynting energy flux dominate, so that $\sigma_\infty > 1$. In the next sections, we will investigate how the introduction of more realistic conditions and the transition to the pulsar wind open up the possibility of a small magnetization at infinity.

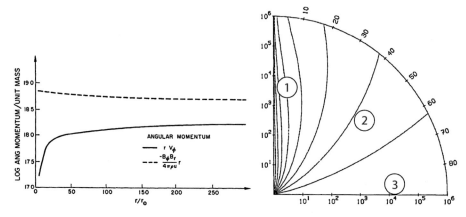

Fig. 2 *Left*: Partition of wind specific angular momentum over kinetic and magnetic contributions with distance, expressed in solar radii, for the Sun in 1D approximation; *dashed line* is magnetic torque and *drawn line* kinetic angular momentum. The Alfvén radius is between 30–50 solar radii. It follows that the torque remains ultimately electromagnetic, and that only a smaller part of the current closes at a finite distance. From Weber and Davis (1967), ©AAS. Reproduced with permission. *Right*: Focussing of a stellar wind in 2D approximation. As a result, there is subsequent complete conversion of Poynting into kinetic angular momentum. The computation is for an aligned rotator, an initially split monopole magnetic field, and an ideal MHD plasma. The bending back of the field lines towards the axis is not real and due to the logarithmic plotting of distance of a field line as a function of polar angle. Distance is in Alfvén radii. Credit: Sakurai (1985), reproduced with permission ©ESO

2.2 Stellar Winds: 2D, Complete Current Closure

The next major step in our understanding of a magnetized wind was taken by Sakurai (1985) who investigated the angular dependence of a steady, axially symmetric, stellar wind of an aligned rotator, again under the assumption of smooth, ideal MHD, and for an initially split monopole. Now, it is not sufficient to solve the 1D equation of motion (the *Bernoulli equation*) but at the same time the equation describing force balance in the meridional plane (the *trans-field equation*) is required. An important result of his study is that the entire Poynting flux is converted into kinetic energy. This result can be understood from the requirement of force-balance across the poloidal field. Figure 3, Right demonstrates the focussing of a magnetic wind towards the rotation axis. Three regimes can be distinguished in the wind solution:

1. In the polar region, a dense jet exists where gas pressure $-\nabla p$ is balanced by the contributions from the Lorentz force $-\nabla \frac{B_\phi^2}{8\pi}$ and $\frac{(\mathbf{B}_\phi \cdot \nabla)\mathbf{B}_\phi}{4\pi}$;
2. At intermediate latitudes, the magnetic field is force-free $\mathbf{j} \times \mathbf{B} = 0$, i.e. the magnetic pressure $-\nabla \frac{B_\phi^2}{8\pi}$ and the hoop stress $\frac{(\mathbf{B}_\phi \cdot \nabla)\mathbf{B}_\phi}{4\pi}$ are in balance;
3. Finally, in equatorial regions, the magnetic pressure $-\nabla \frac{B_\phi^2}{8\pi}$ balances the inertial force $-\rho(\mathbf{v} \cdot \nabla)\mathbf{v}$.

The result of magnetic focussing away from the equatorial plane is an inertial current density, given by the MHD momentum equation:

$$\mathbf{j} = c\frac{\mathbf{f} \times \mathbf{B}}{B^2}, \tag{9}$$

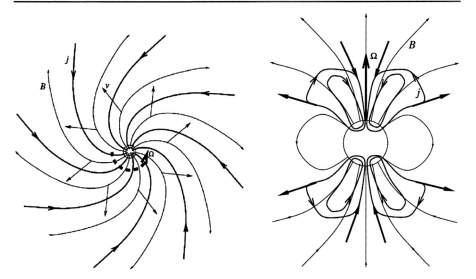

Fig. 3 *Left*: 2D sketch of the magnetic wind of an axially symmetric, aligned rotator just above the equatorial plane as seen by a poleward observer. *B* is the magnetic field, *v* the wind speed, *j* the electric current density, and *Ω* the rotation rate. *Right*: Sketch of the same structure as projected onto the meridional plane. Note that part—but not all—of the current is closing and thereby accelerating the wind close to the star. Adapted from Kuijpers (2001), ©Cambridge University Press, reprinted with permission

driven by the inertial force density

$$\mathbf{f} = -\rho(\mathbf{v} \cdot \nabla)\mathbf{v}. \tag{10}$$

Since the magnetic field has both a poloidal and a toroidal component the Lorentz force associated with the inertial current density not only focusses the wind in the poloidal plane toward the rotation axis but also accelerates the stellar wind radially outward, thereby converting magnetic into kinetic energy.

MHD Integrals for Axially Symmetric Cold Winds   The axially symmetric case of an MHD plasma is especially illuminating because of the four integrals of motion allowed by the (2D) MHD equations (Mestel 1968). Here we consider cold gas, neglect gravity, but allow for relativistic motion (Lorentz factor Γ, axial distance r). In terms of quantities in the laboratory frame, these conserved quantities can be written as (Mestel 1999, Chap. 7):

$$\frac{B_\phi}{B_{pol}} = \frac{v_\phi - \Omega(\psi)r}{v_{pol}} \tag{11}$$

This is the *frozen-in field condition* which derives from the requirement that the gas exerts no stresses on the magnetic field. The parameter ψ labels the magnetic flux surfaces. Expressed in the co-rotating frame in which the field pattern is static, it amounts to requiring the gas to be moving along magnetic field lines.

$$\frac{B_{pol}}{4\pi\rho v_{pol}} = cF(\psi) \tag{12}$$

This equation is tantamount to a *constant mass flux* along a unit outgoing poloidal flux tube.

$$r\Gamma v_\phi - cF(\psi)rB_\phi = L(\psi) \tag{13}$$

If one multiplies this equation with the constant mass flux per unit poloidal flux tube one obtains the *conservation of total angular momentum flux* per unit poloidal flux tube as the matter moves out. This angular momentum is made up out of kinetic specific angular momentum and electromagnetic angular momentum.

$$\Gamma c^2 - cF(\psi)r\Omega(\psi)B_\phi = c^2 W(\psi) \tag{14}$$

Similarly, from this equation follows a generalization of Bernoulli's equation, the *conservation of total energy flux* per unit poloidal flux tube (for the cold gas) as the gas moves out. This total energy flux resides in kinetic (mass) energy flow and in Poynting flux.

2.3 Pulsar: Aligned Rotator; Vacuum Versus Plasma

Before we are in a position to study the degree of magnetic focussing in a pulsar wind we have to explain why an *aligned* magnetic rotator is relevant at all to the radio pulsar wind. Of course, a radio pulsar only exists if the magnetic rotator is *oblique*, or at least if deviations from axial symmetry exist. However, there are separate important electromagnetic effects which come in already for an aligned relativistic wind apart from the effects of obliquity, and we will try to disentangle these. Two basic consequences from Maxwell's theory require our attention:

– In vacuo, an axially symmetric magnetic rotator is—in a steady state—surrounded by an axially symmetric magnetic field. Since the stellar environs are assumed to be a vacuum no electric current flows. Any external electric fields are axially symmetric and therefore time-independent. Therefore, there is no displacement current. The absence of both material and displacement electric currents imply that the surrounding magnetic field is purely poloidal. Since the magnetic field is time-independent any electric field has to be poloidal as well. A *circulating* Poynting flux does exist in the toroidal direction but there is *no outgoing* Poynting flux. A fortiori then, there are no radiative electromagnetic losses.
– In the presence of an ionized wind, the situation is different. Now, poloidal electric currents do exist, both as a result of the unipolar induction by the rotating magnet, and because of the drag on the wind from the rotating field. As a result, the external magnetic field has a toroidal component. Also, because of the ideal MHD condition (2), an electric field appears in the lab frame which has a poloidal component. These electric and magnetic field components imply an outgoing Poynting flux (as in a common stellar wind). The magnetic field and the gas motion are, however, again time-independent, and as a result the electric field is time-independent as well. Therefore, there is no displacement current, and again no radiative losses (such as will be the case for an oblique rotator).

Thus, by confining ourselves to the aligned rotator first, we postpone the discussion about the importance of displacement electric fields, and isolate the Poynting flux which appears in the ideal MHD approximation. Such a pulsar wind is the (relativistic) extension of the above magnetized stellar wind. The electric circuit is expected to close along the path of least resistance, which means across the magnetic field inside the star and along the magnetic field in the wind. However, an important difference with an ordinary star appears. The neutron star may not be able to provide sufficient gas to short out the electric field component in the wind along the magnetic field. This happens, when the local charge density is not everywhere equal to the GJ charge density (4). There and then parallel potential drops develop.

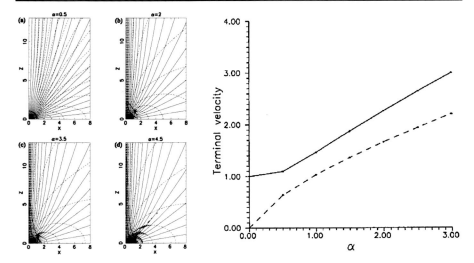

Fig. 4 *Left*: Sequence of shapes of the poloidal field lines (drawn) with increasing magnetic rotator parameter α from $\alpha = 0.5$ (solar-wind-type slow magnetic rotator) to $\alpha = 4.5$ (fast magnetic rotator). The initial non-rotating monopole magnetic field has a spherical Alfvén surface located at one Alfvén radius (r_A). Distances are in units of r_A with the base located at $x = 0.5$. *Dotted lines* indicate poloidal currents. *Thick lines* indicate Alfvén and fast critical surfaces. From Bogovalov and Tsinganos (1999), Fig. 2. *Right*: The terminal velocity as a function of α (*solid line*). For comparison, the corresponding terminal speed in Michel's minimum energy solution is also plotted (*dashed line*). From Bogovalov and Tsinganos (1999), Fig. 5

2.4 Aligned Rotator: Numerical Results on Collimation and Acceleration

Does the relativistic nature of the pulsar wind promote collimation and reduction of σ? The effects of fast rotation on collimation and acceleration of both non-relativistic and relativistic stellar winds have been investigated numerically by Bogovalov and Tsinganos (1999). They find both the collimation and the acceleration of the wind to increase with the *magnetic rotator parameter* (Bogovalov 1999)

$$\alpha \equiv \frac{\Omega r_A}{\Gamma_0 v_0}. \tag{15}$$

Here, Ω is the wind angular rotation rate (at most equal to the stellar rotation rate), v_0 the outflow speed at the photosphere, Γ_0 is the corresponding Lorentz factor, and r_A the Alfvén radius. The behavior of the non-relativistic outflow is shown in Fig. 4, Left. Since relativistic outflows such as in pulsars effectively have $\Gamma_0 \sim 100$ they belong to the domain of slow rotators with $\alpha \ll 1$, and collimation becomes ineffective (Fig. 4, Right; Fig. 5, Left).

The precise role of collimation for relativistic wind acceleration has been elucidated by Komissarov (2011). Consistent with his numerical results, he shows that *differential* collimation as sketched in Fig. 5, Right, is required to obtain high Lorentz factors. Such differential collimation is associated with increasing conversion efficiency of Poynting into kinetic energy flux. Tentatively, we conclude that aligned relativistic rotators lack sufficient differential collimation to establish substantial conversion of Poynting flux if the ideal MHD limit—including inertia—applies throughout the wind. Clearly, the relativistic wind motion does not help to reduce the wind magnetization.

The Effect of a Parallel Potential Gap Contopoulos (2005) shows that a constant potential gap at the basis of the open field lines allows the (otherwise ideal) wind to sub-rotate with

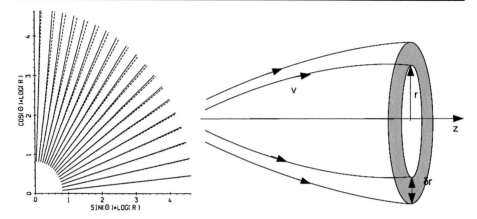

Fig. 5 *Left*: Inefficient collimation and conversion of Poynting angular momentum in the far zone of a rotating magnetic rotator ejecting relativistic plasma. The shape of the poloidal magnetic field lines is given by *solid lines*. For comparison are drawn lines of pure radial outflow (*dashed*). From Bogovalov and Tsinganos (1999), Fig. 13. *Right*: Continuous relativistic acceleration requires differential collimation which increases towards the axis. From Komissarov (2011), ©SAIt, reproduced by permission

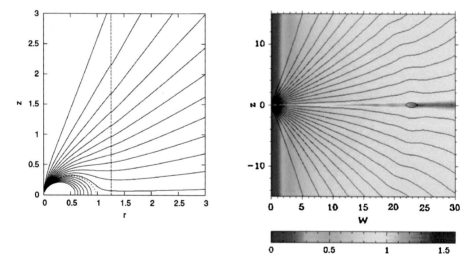

Fig. 6 *Left*: Poloidal field lines (drawn) for an aligned rotator with a potential gap at the base, and for otherwise ideal MHD. The wind relative rotation rate is 0.8. The poloidal flux increases between field lines in steps of 0.1, counting from zero on the vertical axis. The *dotted line* shows the separatrix (boundary between open and closed field lines) at a flux value of 1.23. Distance is expressed in light cylinder radii. Credit: Contopoulos (2005), reproduced with permission ©ESO. *Right*: The wind zone structure of a dipolar magnetosphere at much larger distance than in the *Figure on the left* is again practically radial. The contours show the field lines of poloidal magnetic field. The *color image* shows the distribution of the logarithm of the Lorentz factor. From Komissarov (2006), Fig. 7

respect to the stellar rotation, assuming an *aligned*, initially split monopole for the poloidal field (Fig. 6, Left). However, two remarks are in place. First, both Contopoulos (2005) and Timokhin (2007) neglect the influence of the outgoing/incoming current on the drift speed.

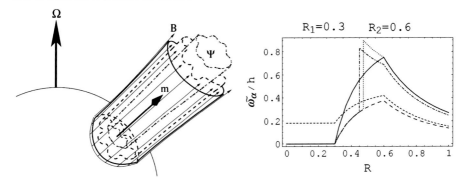

Fig. 7 *Left*: In an oblique magnetic rotator a potential gap at the base of the open field lines does not lead to an average rotation of plasma different from stellar rotation but causes differential rotation over flux surfaces around the magnetic axis of the polar cap. Credit: Fung et al. (2006), reproduced with permission ©ESO. *Right*: Examples of differential rotation in charged, outflowing beams of mixed electrons and positrons around the magnetic axis (z-direction) above the polar cap of an aligned magnetic rotator and depending on the electric current density and charge density distributions. Computed are the equilibrium angular velocities *with respect to the lab frame* as a function of axial distance R for *hollow* beams of mixed electrons and positrons, flowing out at approximately the same (relativistic) speed, with or without *core*, and with or without return current, for cylindrically symmetric beams and a uniform background magnetic field. Dimensionless angular velocity is plotted as a function of dimensionless axial distance. The core region extends from the rotation axis to $R_1 = 0.3$, the hollow beam(s) from $R_1 = 0.3$ to $R_2 = 0.6$, and the pulsar magnetosphere (which co-rotates with the star) starts at $R_3 = 1$. Dimensionless angular velocity is given by $\tilde{\omega}_\alpha(R) \equiv \omega_\alpha(R)/\Omega_\star$. Dimensionless positron/electron charge excess in the hollow beam(s) is given by $h \equiv (n_+ - n_-)/n_{GJ}$ where the GJ density is given by $n_{GJ} \equiv \tau_{GJ}/|e|$ from (5). Defining the parameter $Q \equiv (-\beta_z^0 j_z)/(j_{GJ})$, where $\beta_z^0 \equiv v_z/c$ and j_{GJ} is defined in (4), the various curves are for: *solid*: a hollow beam with $Q = 0$; *long dash*: a hollow beam with $Q = 0.5$; *short dash*: a beam with $Q = 0.5$ and with a core $h_I = 0.2$, $Q_I = 0.1$; *dotted*: a beam with $Q = 0.5$ surrounded by a neutralizing return current; *dash-dotted*: a beam with $Q = 0.5$ and a surrounding return current of the same radial extent. Credit: Fung et al. (2006), reproduced with permission ©ESO

In case of a current, the general expression for the drift speed with respect to the lab frame is given by (cylindrical coordinates z, R, ϕ) (Fung et al. 2006)

$$v_{drift} = -c\frac{E_R}{B_0} + v_z\frac{B_\phi}{B_0}, \qquad (16)$$

where B_ϕ is generated by the current itself. For a relativistic plasma as in the pulsar wind, the final term cannot be neglected, and, in fact can lead to a near-cancellation of the drift so as to cause the gas not to co-rotate with the pulsar at all (Fig. 7, Right). Secondly, as can be seen in Fung et al. (2006) differential rotation of the open field lines takes place around the magnetic axis, not around the rotation axis. Indeed, in an oblique rotator, a potential gap leads to circulation of the open field lines around the oblique magnetic axis, superposed on a pattern which still is rotationally locked to the star (Fig. 7, Left). It is interesting to note that in the terrestrial magnetosphere (where rotation is unimportant) differential rotation is also oriented around the dipole axis, not around the rotation axis.

More Numerical MHD Results Komissarov (2006) models the aligned split monopole with relativistic ideal MHD, including inertia, artificial resistivity, and artificially resetting gas pressures and densities. His spatial domain extends out to a much larger distance than in Contopoulos (2005) (Fig. 6, Right). Again, his computations do not show any collimation of the field lines towards the rotation axis. Neither do they solve the issue of current closure nor the conversion of Poynting flux as required by the observations of σ.

2.5 Current Starvation and Generalized Magnetic Reconnection

The solution to the problem of current closure in an aligned rotator may be given by what is called current starvation. If one insists on a force-free wind the total electric current is conserved if one follows the circuit outward since the current density is everywhere parallel to the open magnetic field lines. Since these field lines become increasingly toroidal in the outward direction the same happens with the electric current density. A conserved current I then implies the following dependence of current density with distance r to the pulsar:

$$j \approx \frac{I_{GJ}}{r \pi r_{LC} \delta},$$
(17)

where δ is the asymptotic opening angle between two selected magnetic flux surfaces. However, the available current density falls off as (quantities in the co-moving frame are dashed)

$$j_{\max} = j'_{\max} = n'ec = \frac{nec}{\Gamma} \propto \frac{1}{r^2}.$$
(18)

Here we have assumed that the (force-free) current density is mainly toroidal (as is the magnetic field), and therefore invariant under a Lorentz transformation to the co-moving frame, while the density is not and obtains a Lorentz factor upon transformation back to the observer frame. Apparently, since the drift speed runs up against the speed of light a current starvation problem arises at a radius well within the termination shock (Kuijpers 2001)

$$\frac{R_{\max}}{r_{LC}} \approx \frac{M}{2\Gamma}.$$
(19)

Here $r_{LC} \equiv c/\Omega_\star$ is the light cylinder radius, Γ is the wind Lorentz factor (before current closure), and $M \equiv n/n_{GJ}$ is the multiplicity of the pair plasma in the wind expressed in terms of the GJ density at the base $n_{GJ} \equiv \Omega_\star \cdot \mathbf{B}(2\pi ec)^{-1}$. Such a shortage of charges leads to strong electric fields parallel to the ambient magnetic fields which try to maintain the current. The effect of the parallel electric fields can be described as *Generalized Magnetic Reconnection*, a term introduced by Schindler et al. (1988) and Hesse and Schindler (1988). In this region ideal MHD must break down and particles are accelerated (heated) along the magnetic field. At the same time, the parallel electric field causes the external magnetic field pattern to slip over the inner fast rotating pattern at a much smaller rate. This implies a much smaller external toroidal magnetic field, and this implies a much smaller current. In other words, the current closes across the field, and the associated Lorentz force accelerates the wind in both radial and toroidal directions. As a net result the current dissipates in this layer and causes both heating and bulk acceleration of the wind. A sketch of such a dissipative shell is given in Fig. 8. This current closure due to current starvation is a natural candidate for the main conversion of the Poynting flux in the MHD wind of a pulsar into kinetic energy (Kuijpers 2001). Finally, note that this current starvation occurs throughout the wind and already for the aligned pulsar, which is different from the current starvation in Melatos (1997) who considers the singular current layer which he postulates to occur at the site of the displacement current, and which exists for an oblique rotator only.

2.6 Oblique Rotator: What to Expect?

From the study of the aligned magnetic rotator we come to the conclusion of the applicability of

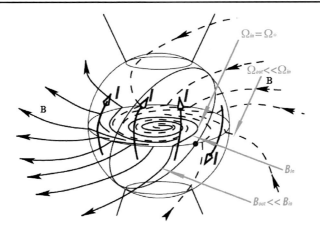

Fig. 8 Sketch of the magnetic field lines (*black arrows*) and of the electric current lines (*open arrows*) far out in the wind where current starvation occurs. Here strong electric fields are generated parallel to the ambient magnetic fields, the electric current closes and dissipates, thereby converting most of the Poynting angular momentum into kinetic, and the wind is strongly heated and accelerated. As a result, the toroidal magnetic field component drops sharply. Outside the dissipative layer the field pattern rotates much slower and slips over the inside field structure. From Kuijpers (2001), ©Cambridge University Press, reprinted with permission

- ideal and force-free MHD in the main part of the pulsar wind,
- ideal, non-force-free MHD inside the neutron star,
- effectively resistive MHD in the accelerating gaps above the magnetic poles,
- and a far-out current starvation region of current dissipation.

Since the formation of an MHD wind does not require a magnetic rotator to be aligned, the above considerations remain applicable to the oblique rotator, be it in a modified form. Further, in an oblique rotator also new effects appear because of the change of magnetic geometry. In particular the time-dependence causes the appearance of a displacement current which is absent in the aligned case. The overall magnetic geometry is sketched in Fig. 9. Although axial symmetry is lost on a small scale, still the rotation together with magnetic (hoop) stresses are expected to lead to some global axial symmetry in two polar jets each of one (opposite) polarity only. In between, there is an equatorial wind consisting of alternating 'stripes' of magnetic field combining both polarities (Coroniti 1990). All three domains contain electric currents. However, whereas the magnetic field inside a jet varies smoothly, the magnetic field in the equatorial wind has an alternating 'striped' spiral structure, and is therefore strongly time-dependent. As a result, a new phenomenon appears in the equatorial wind, the displacement current which usually is absent in a non-relativistic MHD approximation.

Displacement Current in an Oblique Rotator Wind Melatos and Melrose (1996) point out that, in an oblique rotator, the displacement current density $\partial \mathbf{E}/(c\partial t)$ increases with respect to the 'conduction' current density $\mathbf{j}$ with distance from the star:

$$\left| \frac{\frac{\partial \mathbf{E}}{c\partial t}}{\mathbf{j}} \right| > \frac{e E \Omega_\star}{m c \omega_0^2} \{1, \Gamma\} \propto r, \tag{20}$$

where it is assumed that $E \propto 1/r$ and the total plasma frequency is given by

$$\omega_0^2 \equiv \frac{4\pi (n^+ + n^-) e^2}{m_e} \propto \frac{1}{r^2}. \tag{21}$$

Fig. 9 Sketch of a relativistic oblique magnetic rotator. Three magnetic domains can be distinguished: two polar jets, each with magnetic field of one polarity, and an equatorial wind with alternating flux bundles originating at the respective magnetic poles. From Kuijpers (2001), ©Cambridge University Press, reprinted with permission

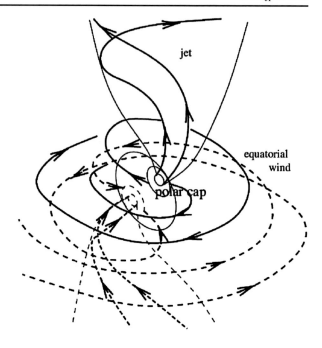

Here the first term inside curly brackets applies to a current transverse to the relativistic outflow and the second term to a parallel current. For the Crab Nebula with $r_{shock} = 2 \times 10^9 \, r_{LC}$ the displacement current density becomes larger than the charged current density at an estimated critical radius of $r_{cr} = 1 \times 10^5 \, r_{LC}$ in case of a transverse current. We note that this critical radius is larger than our estimate for current starvation for Crab values $M \sim 10^5$ and $\Gamma \sim 10^2$ so that current starvation remains important. Their point, however, that the displacement current cannot be neglected is well made.

Force-Free Wind of an Oblique Rotator Contains a Displacement Current Sheet Appealing though the concept of a striped wind (Coroniti 1990; Bogovalov 1999) is, an inconsistency exists in its description since the assumption of the ideal MHD condition (2) to be valid everywhere leads to a contradiction. The wind speed is continuous over each stripe while the magnetic field reverses sign (Fig. 9), and therefore the electric field as given by (2) is a step-function in time for a static observer. It then follows immediately from Ampère's law that there is a singular sheet of displacement current. In a realistic situation, the width would be finite and represents a strong radiative pulse which is not discussed at all. Instead, the striped wind concept studies the energy liberated in this sheet by reconnection.

In a follow-up Bogovalov (1999) presents the transition between two stripes as a tangential discontinuity with a charge current (Fig. 18) instead of a displacement current. He constructs this equatorial current sheet by picking out field lines starting in the equatorial magnetic plane from an initial oblique split-monopole (Fig. 10).

Simple Proof for the Existence of a Displacement Current Here we would like to point out a simple proof for the relative importance of a displacement current versus a charge current in the wind of an oblique rotator. Let us start by assuming that the electric field is determined by the ideal MHD condition everywhere (2). Now, substitute this electric field into Ampère's law. For a static observer $|\partial \mathbf{B}/\partial t| \sim Bv/(\pi r_{LC})$ and $|\nabla \times \mathbf{B}| \sim B/(\pi r_{LC})$.

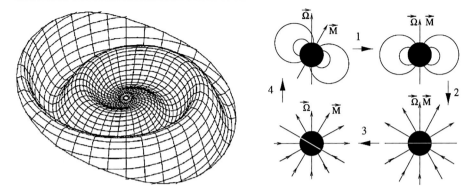

Fig. 10 *Left*: Sketch of the 'ballerina' curtain separating the polarities of the magnetic field lines of a rotating oblique split monopole. The field lines constituting this separatrix surface are selected from the 2D wind model by Sakurai (1985) for the magnetic monopole-like basic configuration by inserting a plane tilted at 10° with respect to the (rotational) equator near the star and following the individual field lines starting at the cross-section of plane and stellar surface. The magnetic field lines below this surface are then reversed. Credit: Sakurai (1985), reproduced with permission ©ESO. *Right*: The same procedure has been used by Bogovalov (1999) to obtain the model wind for an (initial) split-monopole: the author first considers the aligned case, then computes the wind from a corresponding magnetic monopole, and finally reverses the field line directions within a tilted stellar hemisphere. Essentially, the procedure is allowed since a monopole has no (magnetic) equator. See also Fig. 17, Left. Credit: Bogovalov (1999), reproduced with permission ©ESO

One then immediately finds, contrary to the initial assumption, that the time-varying part of the wind is associated with a displacement current of relative importance (for the extreme case of a perpendicular rotator)

$$\frac{|\frac{1}{4\pi}\frac{\partial \mathbf{E}}{\partial t}|}{|\frac{c}{4\pi}\nabla \times \mathbf{B}|} \sim 1 - \frac{1}{\Gamma^2}. \tag{22}$$

Further this result implies that the stationary part of the wind—which satisfies the ideal MHD condition, and which may be subject to charge starvation, satisfies

$$\frac{|\mathbf{j}|}{|\frac{c}{4\pi}\nabla \times \mathbf{B}|} \sim \frac{1}{\Gamma^2}. \tag{23}$$

Clearly, since the pulsar wind is believed to have $\Gamma \sim 100$–200 already at its base, a displacement current cannot be neglected, and in fact dominates over the charge current in the equatorial wind. This turns out to be important since a displacement current has a different way of dissipating (i.e. 'Co-moving Poynting Flux Acceleration') than the charge current.

Vacuum Versus Plasma: Numerical Results Li et al. (2012) model the oblique pulsar wind for various inclinations while neglecting inertia and assuming a uniform conductivity. The power in their wind solutions varies depending on conductivity, which ranges from infinity (the force-free case) to zero (the vacuum), and is given in Fig. 11, Left. Of course, these results do not represent a real pulsar wind, mainly since the actual conductivity is highly varying in space and time. Although an infinite conductivity is probably a good approximation in most of space the nature of the wind depends critically on many small, self-consistently determined, non-ideal domains, e.g. the conductivity of vacuum in polar and outer wind gaps, and the anomalous resistivity in regions of pair creation and radio emission. Nevertheless, the results are instructive (Fig. 11, Right), in particular as to the current circuit and the

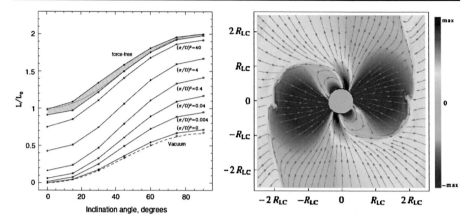

Fig. 11 *Left*: Dimensionless energy losses as a function of inclination angle of a rotating magnetic dipolar star, and as a function of (uniform) resistivity in a resistive MHD computation. Conduction and displacement currents weaken with decreasing conductivity. From Li et al. (2012), reproduced by permission of the AAS. *Right*: Force-free magnetic field lines in the plane containing both the rotation and the magnetic axes for a 60° inclined dipole. Color represents out-of-plane magnetic field into (*red*) and out of (*blue*) the page. The *color table* shows only values up to 30 % of the maximum of the out-of-plane magnetic field, and a square root stretching has been applied to its magnitude. Ideal force-free case. From Li et al. (2012), reproduced by permission of the AAS

relative role of Poynting flux associated with the MHD current on one hand versus displacement current on the other hand. For the case of infinite conductivity, the (wind) power of an aligned rotator—where the displacement current is absent—is found to be only half that of the perpendicular rotator where the displacement current reaches its maximum.

Wind from an Oblique Rotator as a TEM Wave   Does the displacement current in the wind dissipate, and reduce σ before the shock? Skjæraasen et al. (2005) model the pulsar wind as a large-amplitude, superluminal, nearly Transverse Electromagnetic (TEM) wave in a relativistically streaming (Lorentz factor γ_d) electron-positron plasma with dispersion relation

$$\left(\frac{ck}{\omega}\right)^2 = 1 - \frac{2\omega_p^2}{\omega^2 \gamma_d \sqrt{1 + \hat{E}^2}},\tag{24}$$

where the plasma frequency is given by

$$\omega_p^2 = \frac{4\pi n^{\pm} e^2}{m_e},\tag{25}$$

the density of electron-positron pairs is $n^{\pm}$, the dimensionless wave electric field is given by

$$\hat{E} = \frac{eE}{mc\omega},\tag{26}$$

and ω is the wave frequency. They then study the evolution of this TEM wave when a shock confines the TEM, and find that as the distance in the wind increases, the wave amplitude decays and its frequency width broadens, while the thermal background spectrum grows (Fig. 12).

Crab Wind Energy Carried by a TEM Wave   Melatos (1998) solves for exact transverse wave solutions and finds two circularly polarized waves, one subluminal wave in which the

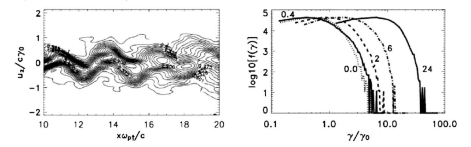

Fig. 12 *Left*: Phase space diagram of shock precursor: a nonlinear TEM wave decays, and broadens with distance, while the thermal background spectrum grows. Initially, the particles are strongly phase-coherent with the wave; thermal broadening can be seen at positions beyond $x > 11$. u is velocity, γ Lorentz factor, and ω_p plasma frequency. From Skjæraasen et al. (2005), ©AAS. Reproduced with permission. *Right*: Energy spectra of particles with position. As the distance increases particles are accelerated to higher energies by the decaying TEM. From Skjæraasen et al. (2005), ©AAS. Reproduced with permission

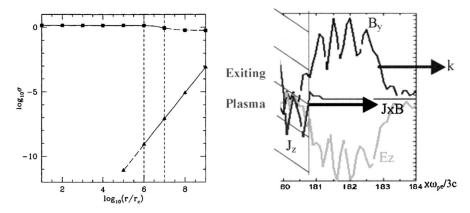

Fig. 13 *Left*: The σ-parameter for the two exact transverse wave solutions in a pulsar wind as function of distance. Initially, the pulsar wind can be fitted to the subluminal wave (*top*) where the particles are magnetized in the radial magnetic field. How the transition is from a wind which is predominantly a subluminal wave to a superluminal wave (*bottom*), where $\sigma \ll 1$ and the particles are unmagnetized, is unclear. From Melatos (1998), ©SAIt, reproduced by permission. *Right*: Co-moving Poynting Flux Acceleration: A finite-width electromagnetic pulse leads to an accelerating force (3D sketch). From Liang and Noguchi (2009), reproduced by permission of the AAS

particles are magnetized in the radial magnetic field, and consequently do not oscillate in the wave field so that $\gamma^{\pm} \approx \gamma_d$, and one superluminal wave in which the particles are unmagnetized and oscillate with the wave field so that $\gamma^{\pm} = \gamma_d(1 + \hat{E}^2)^{0.5} \approx \gamma_d \hat{E}$. Clearly, since $\hat{E} \sim 10^{11}$ at the light cylinder for the Crab, the superluminal wave has a small magnetization as required by the observations (see Fig. 13, Left). The problem, however, is that the properties of the pulsar wind at the base more resemble those of the subluminal wave. How the transformation from one wave into the other takes place is not clear.

Co-moving Poynting Flux Acceleration (CPFA) What then is the main dissipation mechanism for strong electromagnetic pulses with large magnetization? Transverse EM pulses with moderate electric fields in a plasma cause particles to move at the steady drift speed $c\mathbf{E} \times \mathbf{B}/B^2$. However, for near-vacuum waves with $E \sim B$, particles can be accelerated im-

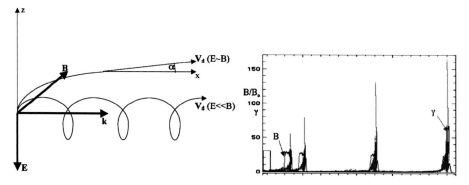

Fig. 14 *Left*: When the electric field amplitude in an electromagnetic pulse becomes comparable to its magnetic field amplitude, the drift speed approaches the speed of light, and particles are accelerated by the ponderomotive force in the front of the pulse. This happens in the 'impulsive plasma gun'. From Liang and Noguchi (2009), reproduced by permission of the AAS. *Right*: Computation of impulsive particle acceleration (Lorentz factor in *red*) by a discrete magnetic pulse (*black*) as a function of time. From Liang and Noguchi (2009), reproduced by permission of the AAS

pulsively by the ponderomotive force in the wave front, catching up with and retarding the wave slightly (Fig. 13, Right). This process has been proposed by Contopoulos (1995) to operate in oblique pulsar winds ('impulsive plasma gun', Fig. 14, Left). Figure 14, Right shows impulsive acceleration as computed by Liang and Noguchi (2009) for a strong EM pulse in the absence of a guiding magnetic field. Clearly, the process is effective although the Lorentz factors obtained in this study are still far below what is required in a pulsar wind.

2.7 Current Sheets

Within the pulsar context one often encounters the assumption of—infinitesimally thin— current sheets. As long as these are invoked as approximate circuit elements required for current closure, there is no problem. However, when these current sheets become ingredients of dissipation it is relevant to ask how they have developed out of space-filling 'body' currents, as for example one is used to do in the solar corona flare context. Below, we will discuss the nature of two main classes of current sheets invoked in pulsar studies: singular return currents at the boundary between open and closed field lines, and current sheets between the stripes in the striped wind model.

Current Sheets on the Boundary Between Open and Closed Field? Models of the magnetosphere of a magnetic rotator show distributed currents along the open magnetic field lines above the pulsar polar caps but often return currents in the form of sheets, such as indicated by arrows in Fig. 15, Left. The current sheet then is along the last open magnetic field line, separating zones of open and closed magnetic field lines. However, as we see it, this is just a simplification. By its very origin the return current is not a singular layer but a body current, just as the polar cap current. Actually, this was discussed already in the early seminal paper by Goldreich and Julian (1969) (Fig. 15, Right). Also, numerical studies find that the electric (polarization) currents to and from the pulsar in the polar cap region are quite broad as can be seen in the representative example of Fig. 11, Right, which is computed for an oblique 60° dipole and under the force-free assumption. Another recent numerical study (Timokhin and Arons 2013) shows a distributed return current with dimensions comparable to the polar cap width (Fig. 16, Left). In particular, the perpendicular rotator presents an interesting

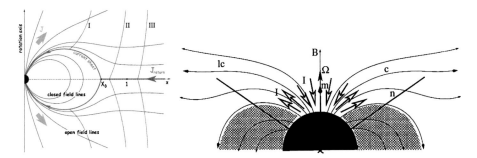

Fig. 15 *Left*: Sheet currents in a typical theoretical model (*blue*). From Timokhin (2007), Fig. 1. *Right*: Finite-width currents in original Goldreich-Julian model (polar cap is exaggerated). Credit: Fung et al. (2006), reproduced with permission ©ESO

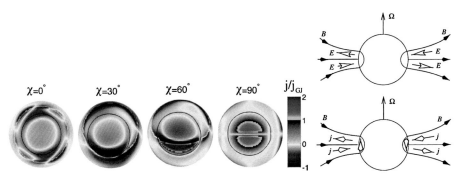

Fig. 16 *Left*: Finite width return current in actual computation. From Timokhin and Arons (2013), Fig. 1. *Right*: Perpendicular rotator with electric fields (*top*) and electric currents (*bottom*). Current and return current have similar widths. From Kuijpers (2001), ©Cambridge University Press, reprinted with permission

example where the symmetry causes currents and return currents to have the same finite width as can be seen in Fig. 16, Right. Note that there is a puzzling difference between the two papers in the directions of the currents for the perpendicular rotator: in Fig. 16, Left, the currents in the polar cap above the equator have the same directions as those below the equator, while in Fig. 16, Right, the currents are opposite, and form part of one circuit. In our opinion, the latter picture is correct as it is based on the sign of the surface charges for a perpendicular vacuum rotator which is given by the sign of $\mathbf{\Omega}_\star \cdot \mathbf{B}$, and which is opposite on both sides of the equator.

Current Sheets Between Equatorial Stripes? The other class of singular currents occurs at the tangential discontinuities between field bundles of different polarities when these meet in the equatorial region. The most simple case is that of the aligned rotator (Fig. 15), Left, which shows the meridional projection of the equatorial tangential discontinuity. In 3D this current spirals inward in the equatorial plane. Of course, in the oblique case this flat current sheet becomes undulating (Fig. 17, Left). Many authors concentrate on this tangential discontinuity in the striped equatorial wind as a source of converting magnetic into kinetic energy (Coroniti 1990; Bogovalov 1999). There are, however, two objections to be made to these studies. One is that this wrinkled current sheet is in reality made up out of a combination of the tangential discontinuity current plus the displacement current we have discussed

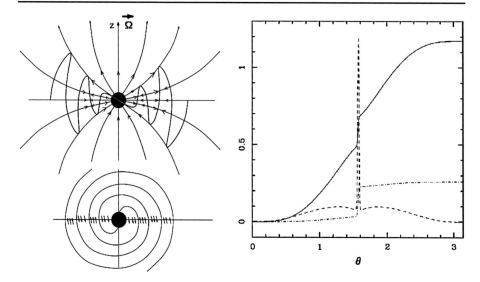

Fig. 17 *Left*: Meridional (*top*) and equatorial (*bottom*) cross-section of the oblique split-monopole wind. Credit: Bogovalov (1999), reproduced with permission ©ESO. *Right*: The angular distribution of the relative luminosities of a pulsar wind from an axially symmetric computation with numerical resistivity only, at $r/r_{LC} = 10$: total luminosity (*solid line*), hydrodynamic kinetic (*dot-dashed line*) within the polar cone of angle θ. The *dashed line* shows the total flux density in the radial direction. From Komissarov (2006), Fig. 9, Right

above. Surely, dissipation of this current sheet then cannot be handled purely from a resistive MHD point of view. The second objection is that, in confining the study to the singular layer, one neglects the 'body' currents within the stripe which connect to the magnetic poles and which carry both angular momentum and energy from the star. How such body currents would condense into a singular sheet is not made clear.

Towards a Realistic Description of the Equatorial Wind of an Oblique Rotator   Figure 18 shows sketches of the meridional cross-section of part of the equatorial pulsar wind, according to existing lore (Right) and our view (Left). As we see it, the equatorial region of the wind of an oblique rotator consists of mainly toroidal magnetic flux tubes which come from both poles and join up, arranged in an alternating pattern in radial direction. Each flux tube carries a body current, with opposite signs above and below the equator. These currents correspond to the internal twist of the flux tubes caused by the polar in-coming and outgoing current system. We have constructed this sketch based on the evolutionary scenario of the events when a finite flux tube is drawn out from a star when it starts rotating (Fig. 19). Apart from these currents there is a current which separates two flux tubes. This current consists to a negligible amount (of order Γ^{-2}) out of a charge current, and mainly out of a displacement current. Altogether, the Poynting flux of the oblique rotator is supported by two electric current systems: a charge body current, which may be subject to current starvation and consequent dissipation at some distance from the star, and a displacement current, which probably is distributed instead of singular as well, and which dissipates by the CPA process. The third component of a postulated singular charge current separating flux tubes from both hemispheres—on which several papers focus—can be completely neglected as to its contribution to the total Poynting energy flux. In our opinion this is corroborated by the results from Komissarov (2006) who computes the amount of Poynting flux dissipated

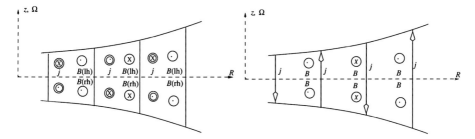

Fig. 18 *Left*: Meridional cross-section of the equatorial wind. The field in each segment connects to the corresponding polar cap on the star. The horizontal width of a flux tube is roughly constant in the force-free cold flow, and equal to πr_{LC}. The force-free nature of the current implies that the magnetic field in each flux tube is helical as indicated in brackets (*lh* for left-handed; *rh* for right-handed). The pitch of the field on the northern hemisphere is left-handed. *Right*: Conventional current picture which focuses on the singular current sheets associated with the tangential discontinuity and leaves out any body currents. From Kuijpers (2001), ©Cambridge University Press, reprinted with permission

Fig. 19 Sketch of electric currents (*open arrows*) which are set up as a flux tube connecting both magnetic poles is drawn out from a perpendicular magnetic rotator so that it brakes the stellar rotation. For clarity, only two magnetic field lines (*black arrows*) are shown. The inclined plane on the right is the cut presented in Fig. 18 in the asymptotic regime when the top of the loop has vanished to infinity. From Kuijpers (2001), ©Cambridge University Press, reprinted with permission

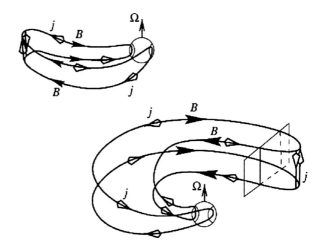

by numerical resistivity in MHD approximation of a pulsar wind constructed according to the recipe by Bogovalov (1999) (Fig. 17, Right). As can be seen the contribution from the equatorial current sheet is minor with respect to the total energy flux.

2.8 3D-Instability of a Toroidal Wind

An intrinsically different solution to the sigma problem has been proposed by Begelman (1998). He points out that the strong toroidal field amplified by the reverse shock of the wind is Rayleigh-Taylor unstable, and has the tendency to rise out of the equatorial plane. The shocked magnetic field and the electric current system are then expected to acquire a turbulent shape, and reconnection of this magnetic turbulence provides the required conversion of magnetic into kinetic energy. Numerical computations of the field evolution by Porth et al. (2013) indeed demonstrate that the shocked toroidal field is (kink-)unstable, and acquires a turbulent aspect (see Fig. 20). It should, however, be mentioned that these authors use a large numerical resistivity (cell-size $2 \cdot 10^{16}$ cm!) to dissipate kinks of mixed polarities, and thereby reduce σ.

Fig. 20 Kink instability of toroidal field behind the shock. 3D rendering of the magnetic field structure in a model with $\sigma_0 = 3$ at $t = 70$ yr after the start of the simulation. Magnetic field lines are integrated from sample points starting at $r = 3 \cdot 10^{17}$ cm. *Colors* indicate the dominating field component, *blue* for toroidal and *red* for poloidal. From Porth et al. (2013), Fig. 2, Left

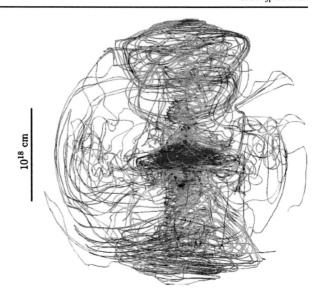

10^{18} cm

2.9 Conclusions for Radio Pulsar Winds

The key problem in pulsars—and in cosmic magnetic structures in general—is to explain how magnetic energy is transported away from a dynamo to a conversion/dissipation region where energetic particles, radiation and outflows are observed. We have described our present understanding of pulsar winds by focussing on their electric current systems.

We have pointed out, on the basis of simple arguments, that an oblique magnetic rotator emits both an MHD wind and an embedded TEM, both of which are initially dominated by the magnetic field so that $\sigma \gg 1$. Both the wind and the TEM wave carry off angular momentum and energy away from the star in the form of Poynting flux. The electric current system has three domains: at the base it runs through the surface layers of the star across its magnetic field, thereby exerting a decelerating torque on the rotating star; next there is a largely force-free domain of the superfast wind where the currents follow the magnetic field lines; finally, the Poynting flux is converted into kinetic energy, so that $\sigma \ll 1$. This can happen by a combination of effects:

- partly by inertial focussing towards the rotation axis and subsequent bulk acceleration in polar jets. However, since the pulsar is a slow rotator this does not happen unless important differences exist between the geometries of the polar winds of an oblique versus an aligned rotator;
- partly by current starvation, the subsequent evolution of parallel electric fields, direct particle acceleration, current closure and bulk acceleration by the Lorentz force in a process which has been termed Generalized Magnetic Reconnection;
- and partly by (CPFA) particle acceleration in the strong TEM.

To what degree 'common' magnetic reconnection (Parker 1957) and tearing play a role is not clear. We expect it to be confined to equatorial regions, and it may well be that its role has been severely overestimated. We like to point out that similar singular current layers, which exist in the solar corona between regions of different polarity, are not per se characteristic sites of solar flares.

We wish to emphasize the twofold effects of parallel electric fields: First, they develop naturally in a current circuit with local or temporal shortages of charge carriers, where they lead to direct particle acceleration. An important example is the polar gap in a pulsar, which leads to abundant pair creation. Secondly, they also act as regions of 'rupture' on both sides of which ideal, force-free magnetic structures slide along each other ('slippage'), and where electric currents cross the magnetic field, thus leading to bulk acceleration by the Lorentz force. The example here is the region of current starvation far out in the pulsar wind.

Further, we have argued that singular current layers are often assumed to be present for convenience whereas consideration of realistic distributed ('body'-)currents leads to a better understanding of the physics of the current circuit and its dissipation.

In the numerical computations it remains unclear whether the computed macrostructures are insensitive to details of resistivity and viscosity, in particular the effects of numerical resistivity. This is a difficult question to answer because of the large range of length scales involved. For the radio pulsar the length scale varies over 11 decades, from the stellar radius of order 10 km, via the distance of the light cylinder $r_{LC} \equiv c/\Omega_* \sim 1.58 \cdot 10^3 P_{0.033}$ km ($P_{0.033}$ is the pulsar period in units of 0.033 sec, the period of the Crab pulsar) to the distance of the reverse shock 0.1 pc. For jets in Active Galaxies it varies over 9 decades, from a few Schwarzschild radii $R_S = (2GM_{BH}/c^2)^{0.5} \sim 3 \cdot 10^{14} M_{BH}/M_\odot$ cm to the distances of knots and shocks at 10^{25} cm (M_{BH} is the black hole mass).

3 Current Circuits in Solar Flares

In a solar flare, energy previously stored in a coronal current system is suddenly released and converted into heat, the kinetic energy of large numbers of non-thermal particles, and the mechanical energy of magnetized plasma set into motion. Broadly speaking, the heat and the excess radiation resulting from the non-thermal particles and their interactions with their environment constitutes the *flare*, whereas the motion of the magnetized plasma constitutes a *coronal mass ejection* (CME) which is often, but not always associated with a flare. The overall duration of the primary energy release phase (the impulsive phase) is on the order of 10 minutes, with significant variations in the energy release rate indicated by X-ray 'elementary flare bursts' on ~ 10 s timescales (and also lower-amplitude variations on shorter scales). The broadly accepted view of a solar flare is that the magnetic field is permitted to reconfigure by reconnection of the magnetic field—a highly localized process which nonetheless allows the global field/current system to relax to a very different energy state. The magnetic field changes dramatically in the impulsive phase, and steady-state models are certainly inappropriate. It was for a long time assumed that the photospheric magnetic field did not change during a solar flare meaning that the photospheric current also did not change, but this is not correct. Substantial non-reversing changes in the photospheric line-of-sight and vector fields are observed to coincide with, and occur over, the few minutes of the impulsive phase of strong flares (Sudol and Harvey 2005; Petrie 2012).

The usual approach in solar physics is to consider the corona to be an ideal MHD plasma (see (2)) in which the primary variables are plasma velocity and magnetic induction (magnetic field), $\mathbf{v}$ and $\mathbf{B}$, with current density $\mathbf{j}$ and electric field $\mathbf{E}$ being derived quantities. Of course, a flare can only happen when the ideal approximation breaks down, allowing field dissipation, and reconfiguration. The approach based on $\mathbf{v}$ and $\mathbf{B}$ as primary variables is in part because what is most readily *observed* or at least inferred (e.g. from looking at EUV images of the solar corona to give an idea of projected field direction). One or more components of the magnetic field can be deduced from spectropolarimetric observations, while plasma flows can also be deduced spectro-scopically or from feature tracking. Explaining field and flow observations has driven very successful developments in large-scale solar MHD models. However, one result of this is that the current circuit and how it changes is not much discussed at present. Simple current circuit models for flares have long existed (e.g. Alfvén and Carlqvist 1967; Colgate 1978), but we do not often grapple with the reality of the time-dependent driving, re-routing and dissipation of currents in an intrinsically complex magnetic structure.

This section overviews the several ways that electrical currents are considered in the context of solar flares, primarily as a means for storing energy in the magnetic field, and several models for the onset of a flare involve the classical instabilities of a current-carrying loop. The release of energy in a single current-carrying loop or a pair of interacting loops illustrates the transformation of non-potential magnetic energy to kinetic energy of accelerated particles, and to a mechanical and Poynting flux as the twist redistributes. However currents are also associated with *singular magnetic structures* in a more complex coronal field geometry, such as *X-lines*, *current sheets* and *separatrix surfaces*, and when reconnection occurs here fast particles can be accelerated forming a non-thermal particle beam, also constituting a current. More generally, the transport of the energy of a flare is often attributed to a beam of electrons which flows together with a counter-streaming population in a beam-return current system.

3.1 Coronal Currents and Energy Storage

The energy for a solar flare—i.e. the current system—is thought to arrive in the corona primarily by emerging through the photosphere already embedded in the twisted and stretched field of a magnetic active region (e.g. Leka et al. 1996; Jeong and Chae 2007). Small-scale photospheric stressing following emergence may also occur and be important in destabilization of a current-carrying coronal structure but it is not expected to be energetically significant. Magnetic active regions are composed of a small number of dominant magnetic bipoles (quadrupolar fields are commonly observed in complex, flare-productive regions) and the strongest magnetic poles are associated with sunspots. The best data available at present demonstrate significant non-neutralized currents in strong field regions, meaning that at photospheric heights there are net currents entering the corona on one side of the polarity inversion line, and re-entering the photosphere on the other side. In particular, this is the case close to the active region magnetic polarity inversion line (see e.g. Wheatland 2000; Georgoulis et al. 2012). The sign of these currents is strongly correlated with the sign of their host magnetic element, suggesting a dominant sense of *current helicity*. Overall the current is balanced within observational errors, in the sense that currents emerging on one side of the polarity inversion line have, somewhere, an oppositely-directed counterpart submerging on the other side. This non-neutralized current flow is an observed property of the active region on scales of tens of arcseconds, or thousands to tens of thousands of kilometers, i.e. both the separation and the size of the main current and magnetic sources in the region. Figure 21 shows the magnetogram and net electrical currents calculated by Georgoulis et al. (2012) flowing in individual magnetic fragments of the active region that hosted the last large (X-class) flare of Solar Cycle 23. The size of the current sources, as illustrated in this figure may at least conceptually be linked to a scale of sub-photospheric generation or aggregation of ropes or bundles of magnetic flux carrying like (i.e. similarly directed) currents. The character of the sub-photospheric magnetic dynamo is of course not yet known, but both turbulent convective flows and ordered shear and rotational flows are likely to play a role (e.g. Tobias and Cattaneo 2013) and thus a bundle with both a strong net current flow direction, and also

26

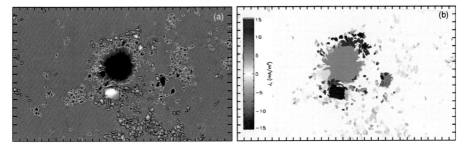

Fig. 21 Vertical magnetic field in magnetic active region (NOAA AR 10930) partitioned into flux elements (*LH panel*; *black/white* for opposite polarities) and showing calculated current density (*RH panel*). Non-neutralized currents tend to flow in strong field close to magnetic polarity inversion lines. The axes are solar x and y, and tickmarks are $10''$ apart—see Georgoulis et al. (2012) for detail; reproduced by permission of the AAS

a fragmentary character are to be expected. Oppositely-flowing currents already present in the sub-photospheric structures, and maintained during rise into the photosphere are hinted at by the interspersed red and blue sources in this figure, on the scale of a few arcseconds, though some of these current sources are weak and are not considered as being significantly non-neutral by the authors. The net currents in the largest negative and positive polarities in Fig. 21 have values of 45×10^{11} A and -35×10^{11} A respectively. The imbalance indicates that the main negative polarity must also provide current to sources other than the main positive polarity, as would be revealed by field reconstructions. Sub-resolution structures are also to be expected, and these may—indeed should—include coronal return currents in thin surface sheets, providing partial neutralization of currents flowing through the photospheric boundary.

A magnetic active region is generally more complicated than just a single twisted 'rope' of magnetic flux carrying current from one sunspot to another, and magnetic field extrapolations based on the observed photospheric vector magnetograms are used to determine the 3D structure of the coronal magnetic field **B**. Coronal electrical currents are obtained from the curl of this field.

$$\mathbf{j} = \frac{c}{4\pi} \nabla \times \mathbf{B}. \tag{27}$$

An underlying assumption is that the magnetic field is force-free, which means that electrical currents are aligned with the magnetic field, i.e. $\mathbf{j} \times \mathbf{B}/c = 0$. By definition, this assumption must break down locally in a flare when the current has to re-route and therefore at some location must cross from its pre-flare magnetic field to its post-flare magnetic field. The parameter quantifying the current is the α-parameter which is constant along a given field line but may vary from magnetic field line to field line, i.e. from point to point across the magnetic source surface:

$$\mathbf{j} = \alpha \mathbf{B}. \tag{28}$$

The case of $\alpha = 0$ corresponds to no current flowing and no free energy in the system: this minimum energy state is called the potential state. If $\alpha = const$ the field is called a linear force-free field, and if α varies from field line to field line the field is a non-linear force-free field (NLFFF). The α parameter is calculated at the photosphere, and each photospheric location paired up with another location having the same α, defining the *magnetic connectivity* in a region. Magnetic field extrapolations are still somewhat of a 'dark art', beset with

Fig. 22 The results of a non-linear force-free field extrapolation from the magnetic field observed around 5 hours before a major flare on 15 February 2012. The *lines* are extrapolated field lines color coded by the calculated current; highest currents are concentrated in highly sheared and twisted magnetic field lying close to parallel to the polarity inversion line. From Sun et al. (2012), reproduced by permission of the AAS

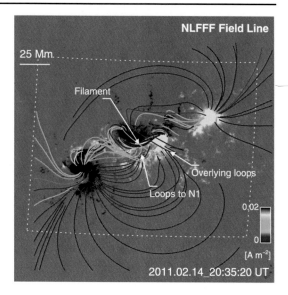

difficulties (in particular in resolving the 180° ambiguity in the measured in-the-plane component of **B**) but good agreement is found between analytic non-linear force-free fields and their reconstructed counterparts particularly in strong field, strong current locations (Schrijver et al. 2006). The mathematical formulation of magnetic field extrapolations, with field lines having at least one end in a photospheric source, means that it is not possible to find field-aligned currents closing completely within the corona. However, one could envisage energy-storing force-free 'tokamak' structures entirely disconnected from the photosphere yet prevented from leaving the corona by overlying field, or (perhaps rather implausibly) by being threaded by an anchored magnetic loop. We note also that non-force-free structures in which $\mathbf{j} \times \mathbf{B}/c$ is balanced by a gas pressure gradient are thought not likely to be significant in flares, because coronal gas pressure gradients are weak, except perhaps in the neighbourhood of a dense coronal filament.

An example of the coronal currents deduced before a major flare on 15th February 2011 is shown in Fig. 22. It indicates the variety of different interconnected current systems that can arise and that can be resolved with NLFFF extrapolation techniques. If extrapolations are carried out before and after a flare, generally speaking the calculated currents in the flare region have changed. Depending on the time resolution available, observations do not always show that the magnetic energy represented by the currents has decreased (as must nonetheless happen). However the flare may be a relatively small and transient perturbation on a system in which the overall magnetic energy is increasing.

Magnetic extrapolation, and MHD modeling of evolving active regions, calculates the force-free coronal current but does not address the current closure; in these models, current emerges from one source on the photosphere and disappears into a sink. This is adequate for most purposes, but in reality currents must close. This is presumed to happen somewhere below the photospheric layer (at least in conditions of slow evolution) in a layer where flux tube identity is established, and yet the flux tubes themselves are small enough that they can be stressed by plasma flow (McClymont and Fisher 1989). This implies a plasma dense enough that it can provide the mechanical stresses on the field-carrying plasma to twist it up. Melrose (1995) argues that this may happen deep in the solar convection zone. However, during a solar flare the energy release and therefore the coronal current evolution is much faster than the timescales on which signals about changing magnetic fields can propagate (Alfvénically) out of or into the region below the photosphere. The evolving current system in a flare requires current re-routing across the field at higher layers, in processes possibly analogous to magnetospheric substorms.

3.2 Instability of Current-Carrying Flux Ropes and Flare Onset

Prior to a flare and CME a coronal current system exists in a stable or metastable equilibrium. There are many theoretical models for the processes leading to the loss of equilibrium (see e.g. a recent review by Aulanier 2013); we focus our discussion here around the role of the current. Roughly speaking instability is to be expected when the axial current becomes strong, such that the toroidal component of the field exceeds the poloidal component. This can be seen as consistent with the result by Aly (1984) that the maximum energy above the potential field that can be stored in a 3D stressed force-free field with the same normal field at the boundary is of the same order as that of the potential field itself.

In one of the earliest models for an electrical current in the corona by Kuperus and Raadu (1974), an over-dense (straight) current-carrying filament in the corona is kept in stable equilibrium by a balance between three forces: the downward Lorentz force from the current acting on the background field of the overlying magnetic arch and the (relatively small) force of gravity are balanced by the upward Lorentz forces due to screening currents induced above the photosphere by the filament's electrical current (also described mathematically in terms of a sub-photospheric 'mirror current'). It was demonstrated by van Tend and Kuperus (1978) that an initially low-altitude current filament would become unstable and that the filament field structure and its entrained plasma would accelerate upwards if its current exceeded some critical value, or if it were given a sufficiently large upward displacement. If the filament is allowed to bend around to become a twisted loop, or 'flux rope', anchored in the photosphere a semi-toroidal configuration arises and the destabilizing role of the curvature (or hoop) force of the loop on itself must be considered. An overlying dipolar field can keep this system in equilibrium. The resulting magnetic field of even a simple construction of this kind shows features—separatrix surfaces and concave-up field lines tangential to the photosphere (known as bald patches)—indicating the structural complexity present in realistic force-free fields (e.g. Titov and Démoulin 1999). An example of the field and currents arising in an MHD model of this configuration is shown in Fig. 23, from Gibson et al. (2004). On the left panel the current-carrying flux-rope emerges from one magnetic polarity (indicated by color contours on the surface) and enters another. Color contours running parallel to the rope axis show the feet of the overlying potential magnetic arcade. A selection of field lines are plotted for the twisted flux rope, and the concave-up bald-patch field. The bald-patch field lines are also plotted in the RH panel, and overlaid with current isocontours showing the current sheets set up at the interfaces between flux rope and arcade field.

Numerical simulations can reveal details of the instability of current-carrying configurations, and how magnetic reconnection in the evolving field allows the instability to develop into a CME and flare. These simulations can exhibit quite baffling complexity; the results must be analyzed carefully to unravel the dynamic features that appear and disappear as instability and reconnection occurs. Reconnection takes place at the current sheets around, above and also below the flux rope, and this allows the kind of dramatic field reconfiguration necessary to extract energy from the field and launch a CME.

3.3 Fixed Circuit (Single Loop) Flare Models

Flare models can be much simpler than the elaborate MHD model shown above, their purpose being to explain the energy release, transport and conversion rather than dynamics.

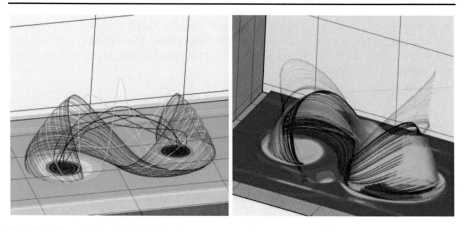

Fig. 23 *Left panel*: Sample field lines from an MHD simulation of a magnetic flux rope emerging into an overlying coronal arcade. Magnetic field contours are shown on the photospheric surface, with opposite polarities in *magenta* and *blue* colors. Field lines are shown for the flux rope and for the bald-patch separatrix surface. *Right panel*: associated current density shown by *green* photospheric color contours and coronal isocontours. From Gibson et al. (2004), reproduced by permission of the AAS

A basic flare structure is a simple, single loop (as is often seen in soft X-ray or extreme UV images) and models for flares invoking the enhanced dissipation of field-aligned electrical currents within such a loop have a long heritage, starting with Alfvén and Carlqvist (1967). In this type of model an electrical current I encounters a region of enhanced resistivity R. By analogy with various laboratory examples, Alfvén & Carlqvist argue that—if the current cannot divert around it—the entire magnetic energy stored in the current system will tend to dissipate in the region of enhanced resistivity. The enhanced resistivity in the Alfvén & Carlqvist model starts with a small local plasma density depletion. Persistence of the current across this region means that the (more mobile) electrons must be accelerated into the depletion and decelerated as they exit it. The local parallel electrostatic electric field thus generated drives ions in the opposite direction, further enhancing the depletion of charge-carriers and creating an electrostatic double-layer. This requires that the current-carrying electrons move through the double layer faster than the electron thermal speed (otherwise the potential can be shorted by redistributing thermal electrons). It was noted by Smith and Priest (1972) that the situation where electrons drift faster than the local thermal speed is equivalent to the condition for the current-driven Buneman instability. The suggestion follows that the microinstability resulting from the Buneman instability provides the enhanced resistivity in the Alfvén & Carlqvist model. The ion-acoustic instability has an even lower threshold for onset, occurring when the electrons and ions have a relative drift that is greater than the ion sound speed but requires the electrons to be much hotter than the ions. Both of these instabilities provide an anomalous resistivity by setting up waves which scatter electrons and ions.

The power I^2R dissipated in the resistive region produces flare heating, and the potential drop across the resistance may be able to accelerate charged particles. The typical power needed for a flare is $\sim 10^{27}$–10^{28} ergs s^{-1} (10^{20}–10^{21} W), and with a current of $\sim 10^{20}$–10^{21} statA ($3.3 \cdot 10^{10}$–$3.3 \cdot 10^{11}$ A) this means that the resistance is around 10^{-13}–10^{-12} s cm^{-1} ($9 \cdot 10^{-1}$–$9 \cdot 10^{-2}$ Ω). This gives rise to a potential drop of around $3 \cdot 10^{10}$–$3 \cdot 10^9$ V, producing electrons and protons up to about 10^{10} eV. This is very much larger than the kinetic energy of a few tens of keV typical of flare electrons, so a single potential drop in this system is not a good model for flare particle acceleration.

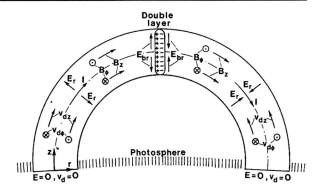

Fig. 24 The radial electrostatic field E_{br} formed at the boundary of a double layer crossed into the magnetic field $\mathbf{B} = (0, B_\phi, B_z)$ gives rise to a plasma drift $\mathbf{v} = c\mathbf{E} \times \mathbf{B}/E^2 = (0, v_{d\phi}, v_{dx})$, and a Poynting flux $S = c(\mathbf{E} \times \mathbf{B})/4\pi$. From Carlqvist (1979), ©1979, D. Reidel Publishing Co

In a later iteration of this model, Carlqvist (1979) demonstrated how part of the energy, previously stored in the twisted field structure, arrives at the double layer as a Poynting flux. Figure 24 illustrates an electrostatic double layer introduced into a (twisted) coronal loop, leading to a radial electrostatic field on its boundary (due to its finite radial extent). When crossed into magnetic field of the current-carrying loop this electrostatic field gives rise to a Poynting vector $\mathbf{S} = c(\mathbf{E} \times \mathbf{B})/4\pi$, and a plasma drift with the same direction $\mathbf{v}_d = c(\mathbf{E} \times \mathbf{B})/B^2$, and therefore with both an axial and a toroidal component. The azimuthal component of the drift represents the unwinding of the twisted flux tube across the resistive region at the top of the loop. This idea, forming an interesting bridge between Alfvénic energy transport and particle acceleration, was subsequently generalized by Melrose (1992) to include an arbitrary source of resistance, which could in principle be placed anywhere in the current-carrying loop.

A potential drop can of course accelerate particles, but it is not straightforward in this case. For a flare, we require a large number of non-thermal electrons with energy of a few tens of keV. The single potential drop implied by the current and flare parameters would instead produce extremely high energy particles, but a small number of them (preserving the current $n_e q v$ across the resistance). A more sophisticated view is necessary.

3.4 Current Diversion Flare Models

There is ample observational evidence for the involvement of multiple magnetic structures in a flare. The presence of more than two bright chromospheric (or occasionally photospheric) foot points during a flare demonstrates that multiple flux systems are present. Quadrupolar systems, with at least 4 groups of foot point, are rather common. The spreading of those foot points or ribbons in the chromosphere that is particularly evident in Hα and UV emission indicates that from instant to instant the identity of the bundle of magnetic flux that is involved in the flare is changing—in a process very similar in appearance to the racing of auroral curtains across the sky (see Sect. 4.4 and Fig. 36). The appearance of previously invisible flare loops indicates that the topology has changed. This happens during the impulsive and the gradual phases of the flare; a particularly (perhaps the only) compelling solar observation of this was recently presented by Su et al. (2013), but the ample circumstantial evidence for reconnection means that its existence in the flare corona is not generally in doubt. A large fraction of flares thus involve the formation of new flux and new current systems as magnetic connections are broken and reformed.

Melrose (1997) associates the flare energy with the change in energy stored in coronal currents as the current path shortens, and this process was explored in a model of two interacting current-carrying loops, as sketched in Fig. 25. Reconnection was involved only

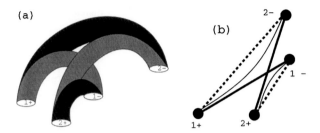

Fig. 25 Two current-carrying loops (*grey*) which reconnect to give the (*black*) post-flare configuration. Their coronal current paths are represented by *thick solid* and *dashed lines* respectively (with the *thin solid line* representing an intermediate stage during which the field-lines/current paths are shortening). From Melrose (1997), reproduced by permission of the AAS

in so far as it allowed the field to reconfigure and the currents to re-route and the field lines to shorten, corresponding to the release of energy. More recent work by Melrose (2012a, 2012b) demonstrates how this process happens; it is argued that the cross-field current enabling the re-routing of previously field-aligned currents in the corona is a result of the intrinsically time-varying character of the flare magnetic field, which is quite appropriate for the flare impulsive phase. The resulting inductive electric field plays a critical role in the overall problem. The discussion below summarizes Melrose's arguments.

The inductive electric field is given by

$$\nabla \times E_{ind} = -\frac{1}{c}\frac{\partial \mathbf{B}}{\partial t}. \tag{29}$$

A time-varying $\mathbf{E}_{ind}$ sets up a polarization drift—in this way the individual responses of the plasma particles to the changing magnetic field become important. This polarization drift is

$$\mathbf{v}_{drift} = \frac{mc^2}{q|\mathbf{B}|^2}\frac{d\mathbf{E}_{ind}}{dt}. \tag{30}$$

The changing magnetic field thus leads to a drift current (since the drift velocity depends on particle charge). It is usually assumed that the component of the inductive electric field along the field is neutralized by the flow of mobile electrons. The perpendicular component of this current is the polarization current carried mainly by the ions (see (30)) and flowing across the field, exactly as is required for current re-routing. It is related to the inductive electric field by

$$\mathbf{j}_\perp = \mathbf{j}_{pol} = \frac{c^2}{v_A^2}\frac{1}{4\pi}\frac{\partial \mathbf{E}_\perp}{\partial t} \tag{31}$$

As can be seen from (30) and (31), the polarization current is proportional to the displacement current, but the latter is negligible in the solar case (unlike in the pulsar case, see Sect. 2.6, (22) and (23)). Melrose also argues that in the flare as a whole the $\mathbf{j}_{pol} \times \mathbf{B}/c$ force that develops, once the change in magnetic field has been triggered by some external process, can drive field into the magnetic reconnection region, sustaining the process without any mechanical forces (i.e. pressure gradients) being necessary.

Within an individual current-carrying (twisted) flux tube the $\mathbf{j}_{pol} \times \mathbf{B}/c$ force generates a rotational motion of the plasma, i.e. a propagating twist or torsional Alfvén wave that carries energy away (see previous section). The idea that the energy in a flare can be transported by Alfvénic perturbations has also been explored by, e.g., Haerendel (2006) and Fletcher and Hudson (2008). The reconnection region, in which the rapid field variations at the root of the process begin, can be arbitrarily small (with arbitrarily small energy dissipation locally); its role instead is to allow the transfer of current from pre-reconnection field to post-reconnection field, and in doing so it launches Alfvénic Poynting flux along just-reconnected field lines, redistributing energy throughout the corona with the current.

The discussion above indicates how the currents in the corona may be rerouted and in the process energy transported away from the reconnection site and a new current profile established throughout the corona. In the magnetosphere/ionosphere system the current restructuring implied by substorms (flare analogues) involves current closure in the partially ionized layers of the ionosphere, supported by the Pedersen resistivity (Sect. 4.4). The deep chromosphere is also a partially ionized plasma, however the high density and resulting strong collisional coupling between ions and neutrals means that the collisional friction that provides the Pedersen resistivity is too small to support cross-field currents (e.g. Haerendel 2006). That is not to say that ion-neutral collisions have no role to play in the flare energy dissipation; if the period of the Alfvénic pulse (moving the ions) is short compared to the ion-neutral collisional timescale then frictional damping can take place. The ion-neutral collisional timescale is longest around the temperature-minimum region of the solar chromosphere where the number density of protons is also at a minimum, and significant wave damping can occur here for wave pulses of duration a second or less (Russell and Fletcher 2013).

An Alfvén wave entering the chromosphere passes through a region in which the Alfvén speed is varying rapidly; the reflection and transmission and damping of the wave needs to be carefully calculated (e.g. De Pontieu et al. 2001). Moreover, if the wave enters at an angle to the density gradient, mode conversion can also occur (e.g. Khomenko and Collados 2006). Therefore, the wave energy is unlikely to be dissipated straightforwardly, in a "single pass" through the chromosphere or photosphere, but multiple reflections leading to counterstreaming waves, generation of plasma turbulence, and other complexities, are to be expected. However, at some level the photospheric currents are fixed, and somehow the new and old currents must match, which implies current dissipation or current diversion into a singular layer. At some point in the atmosphere the field becomes line-tied, and the high inertia of the material means that the Alfvén wave front reaching the photosphere cannot continue. Its energy and current must be dissipated in full or in part. Dissipation at this layer, as well as the damping occurring elsewhere in the atmosphere, must be what provides the intense heating and particle acceleration that characterize a solar flare. In broad terms, the stressed magnetic field in the wave pulse leads to field-aligned electric fields which can be a source of both heating and acceleration in the plasmas of the lower atmosphere. This is described further in Sect. 3.7. Another effect of the high inertia of the photospheric material is that any excess photospheric current can be diverted across the magnetic field into a thin envelope around the coronal flaring region. It can do so since the photospheric material can take up the stresses resulting from the Lorentz force.

3.5 Singular Current-Carrying Structures in a Complex Corona

As indicated in Sect. 3.2 the magnetic field and the current structures arising from something even as simple as a current-carrying flux rope embedded in an overlying potential field arcade can be quite complex. The extrapolated magnetic field shown in Fig. 22 suggests multiple magnetic domains (regions with the same magnetic connectivity).

Coronal field configurations can be analyzed to identify sheet-like magnetic discontinuities—regions of rapid change of field-line connectivity—known as separatrices. These are regions where, in non-potential configurations, current sheets can develop (e.g. Low and

Fig. 26 A rendering of the separate coronal magnetic domains over the entire Sun (*semi-transparent colored surfaces*), and a few selected field lines (in *cyan*). The semi-transparent surfaces are in fact separatrix surfaces, separating regions of different magnetic connectivity. Part of the heliospheric current sheet (in *semi-transparent yellow*) is also shown. From van Driel-Gesztelyi et al. (2012), ©2012, Springer Science+Business Media B.V.

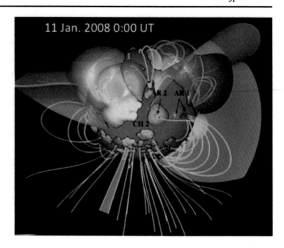

Wolfson 1988). In fields determined from realistic magnetic boundaries singular separatrices are not found; the relevant structures are instead quasi-separatrix layers (QSLs). An example of this for the whole Sun field is shown in Fig. 26, where the semi-transparent separatrix domes indicate the interfaces between fields which are closed on the scale of an active region, from those which open to the heliosphere.

It is now well-known that, in flaring regions, the intersection of QSLs with the photosphere corresponds closely with sites of strong flare radiation, i.e. the flare ribbons (e.g. Demoulin et al. 1997). The flare ribbons—or at least the subset of them defined by the nonthermal X-ray and optical foot points—are where the majority of the flare energy arrives at the lower atmosphere. However, the field extrapolations suggest that the bulk currents are stored in and released from large-scale flux ropes. It is not clear how these pictures match up—whether in fact sufficient current can be stored in QSLs as opposed to flux ropes, or whether the process of flux-rope destabilization involves reconnection with and current transfer onto the QSL field.

A further type of current system often discussed in the solar flare context is that which exists in the dissipation region of a reconnecting structure, where the magnetic field is sufficiently small and thus the particle Larmor radii are large compared to the scale length on which the magnetic field varies. Reconnection—even in the steady state—involves the advection of magnetized plasma into a current sheet or X-line. This results in a $-\mathbf{v} \times \mathbf{B}/c$ electric field, and electrons which are demagnetized in a small region around the X-point can be accelerated in the direction of this electric field. Ions, due to their larger gyro radii, are demagnetized over a larger region and are accelerated in the opposite direction. Where this current closes will depend on the configuration of the magnetic field and on the orbits of the particles; these exhibit properties of dynamical chaos meaning that they can be deflected out along the magnetic separatrices into higher field regions where they become re-magnetized and may leave the system, or mirror and return for further acceleration (Hannah and Fletcher 2006). In 2.5D the X-line or current sheet extends indefinitely and currents run uniformly along it, but of course in 3D these currents must close; the X-line is generalized as a magnetic separator which meets the photosphere at either end—the model of Somov (1986) shown in Fig. 27 indicates one possible geometry for this.

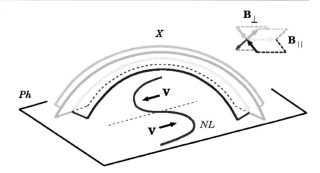

Fig. 27 The X-line, indicated by the field vectors in the *top right*, is the separator between reconnecting magnetic domains. In this cartoon it is projected from one photospheric endpoint to the other, offering a current path. In a localized region such as extracted in the *top right*, the reconnection electric field flows parallel to the X-line. Credit: Somov (1986), reproduced with permission ©ESO

3.6 Beams and Return Currents

Non-thermal particles are a defining property of the majority of solar flares, and explaining where and how they are accelerated is a problem of long standing. The standard model of solar flares separates the region where particles are accelerated from the region where the flare energy is dissipated as radiation, placing the former in the corona close to (or perhaps in) the magnetic reconnection region and the latter in the lower atmosphere—the chromosphere. The chromosphere is where the flare radiation is primarily generated and where the signatures of accelerated particles are primarily seen. If they are accelerated in the corona the particles must then travel from the corona to the chromosphere transporting energy, and must therefore have a significant component of velocity directed along the magnetic field, in a beam-like distribution.

The properties of the accelerated particles can be deduced from the radiation that they produce—hard X-rays (at typically a few 10s of keV) for electrons, and γ-rays for ions. Most of the work on flare non-thermal particles focuses on electrons, simply because HXRs are more readily observed than are γ-rays, and our knowledge of ions is thus rather poor. In particular, we do not know whether electrons and ions are always accelerated together in flares, but electron acceleration is a central feature. Energy transported to the chromosphere by a beam of electrons must be dissipated there, which means that the electrons give up their energy and stop; at the minimum this involves Coulomb collisions in the dense chromospheric plasma. Under this assumption, the so-called '*collisional thick-target model*', the rate of incident electrons required to explain the observed HXRs can be calculated (Brown 1971). The deduced rate of $\sim 10^{35}$–10^{36} electrons s^{-1} would, if not neutralised, constitute a colossal current, of 10^{14}–10^{15} A ($3 \cdot 10^{23}$–$3 \cdot 10^{24}$ statA), orders of magnitude greater than is present in the pre-flare corona (and with a correspondingly large associated magnetic field). However, this current will not exist un-neutralized. Driven by the inductive electric field of the beam and the (transient) electrostatic field at its head, a return current of electrons will flow (e.g. van den Oord 1990), drawn presumably from the chromosphere where there is a large charge reservoir. It has frequently been proposed that this return current could replenish the accelerator, thus neatly solving the problem of how to replenish the acceleration region (which operates in a tenuous corona, and cannot pump out 10^{35} electrons per second for long). It remains to be seen what the overall electron current circuit joining a chromospheric foot point and the coronal acceleration region would look like. In the case of a free-streaming beam, a return-current could flow co-spatially with the beam. Such a return current would not be able to re-enter an accelerator that involves a large-scale direct electric field; in other DC field accelerators such as around a reconnecting X-line the return current could re-enter the accelerator if particles could drift across the field (similarly in accelerators with many small-scale and possible transient current sheets). However, there are doubts

Fig. 28 Schematic of ion and electron currents produced by a cross-field driving current in the corona, and closing across the field in the lower atmosphere. From Winglee et al. (1991), reproduced by permission of the AAS. In the *upper panel*, the electrons stream down the central channel into the chromosphere, and chromospheric electrons are drawn back up in external channels. In the *lower panel* the ions, which are less strongly magnetized than the electrons, drift across the field to provide partial current closure in the beam channel, and once there are accelerated by the same electric field that accelerated the electrons

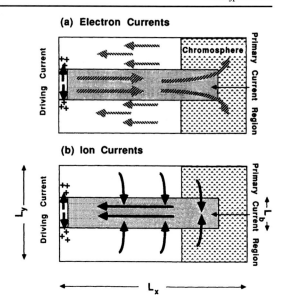

about whether the intense electron beam demanded by observations (e.g. Hoyng et al. 1976; Krucker et al. 2011) can co-exist stably with its co-spatial counter-streaming return current, or instead have its energy substantially dissipated as heat via an instability. Other configurations with a large-scale DC electric field, and thus without a co-spatial return current, have also been discussed. For example in the sketch from Winglee et al. (1991) shown in Fig. 28 the return currents flow over the surface of the channel through which the electron beam is accelerated and flows, and ions are drawn into the channel across the magnetic field, and are themselves accelerated (in the opposite direction to the electrons).

An alternative is that a co-spatial beam of electrons and ions flows together through the corona in a neutral beam (Martens and Young 1990) carrying zero net current. The ions would carry the bulk of the energy, and it is not entirely clear that such a beam arriving at the chromosphere can readily produce the non-thermal HXR spectrum observed though it is plausible under certain conditions (Karlický et al. 2000). It would be premature to discount this model, though there is strong evidence for ions in only a few flares.

It should be emphasized that the main challenge in the electron-beam model is meeting the requirement of a high electron flux from the corona implied by HXR observations interpreted in the collisional thick-target model. Electrons decelerate collisionally in the chromosphere and only have a finite radiating lifetime (thus a finite and well-specified amount of radiation they can produce) before they join the thermal population. Electron flux requirements can be alleviated by allowing the electrons to be re-accelerated once they have entered the chromosphere (Brown et al. 2009), e.g. in multiple current sheets or turbulence. However, the flare *energy* must still reach the chromosphere; the quiescent magnetic energy density of the chromosphere is not sufficient to explain a flare (e.g., compare the energy flux required to heat the chromosphere, of $\sim 10^3$–10^4 W m^{-2} (10^6–10^7 erg cm^{-2} s^{-1}), with the flare requirement of $\sim 10^8$ W m^{-2} (10^{11} erg cm^{-2} s^{-1})). Flare energy may also be transported through the atmosphere by conduction (slow) or by MHD waves. The latter is an integral part of the holistic view of coronal currents discussed in Sect. 3.4.

3.7 Current Circuits and Flare Particle Acceleration

A large fraction—estimates suggest up to 50 %—of the flare energy emerges in the kinetic energy of mildly relativistic electrons and ions. Electrons, which are more straightforward to detect via their bremsstrahlung radiation, typically have non-thermal energies of a few tens of keV. Spectra suggest that a true non-thermal tail can be distinguished from a core thermal distribution at energies exceeding ~25 keV, and the spectral distribution is a power-law in energy $F(E) \sim E^{-\delta}$ with spectral index δ typically between 3 and 8. Imaging demonstrates that hard X-ray radiation, which is collisional bremsstrahlung from the non-thermal electrons, usually originates primarily in the dense chromosphere, except in rare 'dense-loop flares' where the coronal density is high enough to lead to significant coronal non-thermal emission. In the flare standard model electrons are accelerated in the corona and propagate approximately collisionlessly to the chromosphere where they radiate. In this way the flare energy is also carried to the chromosphere, powering the optical and UV emission which is the flare's main radiation loss. As mentioned in Sect. 3.6 the electron beam required may not be able to propagate stably together with its return current; this condition arises because the beam current density required by observations (e.g. Krucker et al. 2011) is so high that it flows relative to its return current at speeds well in excess of the ion or even electron thermal speed, a condition that should lead to the ion-acoustic or Buneman instabilities respectively (Brown and Melrose 1977). Other ideas for flare particle acceleration, including placing the accelerator in or near the chromosphere, are being investigated (Fletcher and Hudson 2008; Brown et al. 2009), which may considerably reduce the number of accelerated electrons required (though not their total energy).

There are numerous models for particle acceleration in solar flares, and again we will be selective here, mentioning only those closely associated with the existence or disruption of electrical currents and thus providing a very incomplete view of flare particle acceleration. A recent overview of the many acceleration processes that may be operating in the solar atmosphere can be found in Zharkova et al. (2011). As mentioned in Sect. 3.5 reconnecting structures such as current sheets or X-lines are a location where particle acceleration may take place, in the zone close to the singular region where one or more components of the magnetic field goes to zero and the particles become demagnetized and can accelerate freely in the electric field of the current sheet. Outside these regions particles are magnetized and the population as a whole experiences various kinds of drift, rather than the acceleration of a small tail to high energies. It is hard to see how a single macroscopic current sheet in the corona can fulfill the number requirement of 10^{36} electrons s^{-1} in a large flare; if the electrons are advected into the current sheet by the reconnection inflow at the external Alfvén speed $v_A = B/(4\pi n_e m_i)^{1/2}$ then the total electron flux that can pass through the sheet of area A is $2n_e v_A A$ electrons s$^{-1} = 1.4 \times 10^{22} n_9 B_2 A$ for n_9 the electron number density in units of 10^9 cm^{-3} and B_2 the magnetic field strength in units of 100 G $= 10^{-2}$ T. For typical coronal values of $n_9 = 1$ and $B_2 = 1$ the current area required is on the order of 10,000 km on a side (while being of the order of an ion skin-depth thick). Such a structure will not be stable and will fragment into many smaller structures. Fragmented current sheets distributed throughout a coronal volume may offer a better prospect. The need to continually supply the accelerating volume with fresh electrons remains the same though. For this reason, moving the current sheet accelerators into the low atmosphere—for example the transition region or upper chromosphere—is an attractive prospect since both the number density of electrons and the magnetic field strength are significantly increased; for example if $n_9 = 100$ and $B_2 = 10$, which might be found in the upper chromosphere, then A is a more manageable 1000 km on a side, comparable in height to the thickness of the chromosphere. For acceleration, the electric field in the current sheet must exceed the Dreicer field, E_D to allow significant electron runaway;

$$E_D = \frac{e \ln \Lambda}{\lambda_D^2} \tag{32}$$

where $\ln \Lambda$ is the Coulomb logarithm and λ_D^2 is the plasma Debye length. If we let the temperature in the upper chromosphere/transition region be 10^5–10^6 K then $E_D = 2.3 \cdot 10^{-5}$–$2.3 \cdot 10^{-4}$ statvolt cm^{-1} = 0.7–7 V m^{-1}. If a parallel potential drop at least this large can be generated in a chromospheric/transition region current sheet, over a distance of a couple of thousand km, this could potentially supply the required electron flux.

Field-aligned acceleration by parallel electric fields is much more explored in pulsar and terrestrial magnetospheres than in solar flares; the literature dealing the obvious solar analogies is small. Field-aligned potential drops were part of the early current-interruption models of Alfvén and Carlqvist (1967) and Carlqvist (1979) mentioned in Sect. 3.3, but the basic problem with these is that for the inferred currents the parallel potential drops generated are of the order 10^{10} V, much higher than the typical electron energy. A more complex geometry is needed, and current fragmentation may provide the solution. Holman (1985) suggested a very large number n of filamented counter-streaming currents. If these currents provide the total power radiated by the flare, then they can do so at a lower potential drop per current channel. If the potential drop along each current channel is reduced by a factor n then the current per channel is increased to give the same power. Potential drops of ~100 keV are obtained with $n \sim 10^5$ small currents, implying filamentation of the large-scale magnetic field into the same number of bundles (Holman 1985). A model with many counter-streaming currents is in conflict with the observations of large net currents flowing in flare regions, and the current channels would be very narrow. We know that most of the energy emerges in the hard X-ray part of the spectrum, and it is very hard to determine the size of HXR footpoints, but in well-observed flares scales on the order of a couple of arcseconds are found, and are consistent with optical source sizes on the order of 1–2 $\times$ 10^{16} cm^2 (Krucker et al. 2011). Each of the 10^5 current channels would then have an area of 1–2 $\times$ 10^{11} cm^2 at the footpoint. The coronal current channels could be larger, allowing for expansion of the field from the footpoint into the corona, but if the expansion of the flux tube is large then one must also incorporate the effect on particles of magnetic trapping.

In the magnetospheric case the existence of parallel electric fields is not in doubt. The exact altitude distribution of these fields is still debated (Birn et al. 2012). The electric fields can be quite extended along the magnetic field in a quasi-static fashion, or they can be concentrated in multiple thin, so called double layers as for instance envisaged by Alfvén and Carlqvist (1967). In the case of the magnetosphere, there seems to be no general agreement on the process generating the $E_{\parallel}$ that is observed as also dispersive Alfvén waves have been observed and speculated as the source of small-scale structures in the aurora (Birn et al. 2012). The solar case however, is much less well explored.

Perhaps of most relevance here is that parallel electric fields in general play a role in dissipating magnetic stresses, as is required when the large-scale magnetic field relaxes. The general argument made by Haerendel (1994) and Melrose (2012b) is that a parallel electric field allows the decoupling of two regions of magnetized plasma experiencing a shear flow relative to one another and perpendicular to the magnetic field threading both. In the solar case these would correspond to the photosphere where the magnetic field is frozen into a high-inertia fluid, and the corona through which an Alfvén wave pulse is traveling. Figure 29 sketches this process; in the left-hand panel, an initially vertical field (indicated by the solid lines with arrowheads), frozen in to the lower plasma, is distorted by a mechanical stress

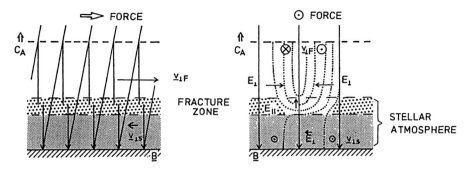

Fig. 29 View of a magnetic fracture zone, face-on (*LH panel*) and in cross-section (*RH panel*). See text for full description. From Haerendel (1994), reproduced by permission of the AAS

on the magnetized plasma at some (distant) top boundary. In a zone termed the 'fracture zone' by Haerendel, field reconfiguration takes place to release the stress, and the result is a plasma flow indicated in the Figure by $v_{\perp F}$. The 'fracture' allows energy to be dissipated. Examining this scenario in cross-section, as in the right-hand panel of Fig. 29, the $v_{\perp F}$ in the region above the fracture results in an electric field $E_{\perp}$ perpendicular to the magnetic field (RH side of this panel). The $E_{\perp}$ is an electrostatic field, existing for much longer than the transit of the magnetic disturbance through the fracture zone, and is balanced by an oppositely-directed electrostatic field in the adjacent plasma (LH side of this panel), which sets that plasma into motion. The electrostatic equipotentials must be continuous, as sketched in the dashed lines, meaning that somewhere in the fracture zone there is an electric field parallel to the magnetic field, where acceleration may take place. The physical scales of the structure can be estimated based on the fact that in the region of parallel electric field in the fracture zone, the potential drop should be up to 10 MeV, to allow the acceleration of ions to γ-ray producing energies. The electric field should be in excess of the Dreicer field, which was earlier calculated as 0.7–7 V m^{-1} in the upper chromosphere, but—varying as it does with the plasma density—can easily reach 100 V m^{-1} in the mid-chromosphere. If we take 100 V m^{-1} as a plausible value, this means that to obtain MeV energies would require an accelerating length of around 100 km. The width of the equipotential structure shown in the right-hand panel of Fig. 29 would then be on the order of 200 km. This is not dissimilar to the arcsecond-scale of flare ribbon widths.

The question of what provides the resistivity to support a parallel current requires an examination of the microphysics, however in the presence of an external driver, the microphysics will have to find a way to keep up. So in principle the propagation of a magnetic shear (or twist) disturbance into a line-tied region can result in particle acceleration, in a region somewhere between the inertially line-tied photosphere and the frozen-in low-β field of the corona. Initial evaluations by Haerendel (1994) concluded that for reasonable chromospheric parameters this process was capable of generating megavolt potential drops. Its application to the magnetosphere, the environment for which it was developed, is discussed in Sect. 4.4.

We remind the reader that a very similar conversion of energy has been proposed for the pulsar stellar wind in Sect. 2.5. Also there, parallel electric fields develop but now due to current starvation, particles are accelerated by these parallel electric fields, the outer wind slips over the inner fast rotating wind, the electric current closes, and leads to bulk acceleration (Generalized Magnetic Reconnection).

There are other possibilities for producing a parallel electric field, for example in a propagating Alfvén wave pulse traversing a low-β plasma, such that the kinetic ($\beta \sim (m_e/m_p)^{1/2}$) or inertial ($\beta \sim (m_e/m_p)$) regimes apply. In these cases a parallel field is sustained because the wave is traveling so fast that the finite thermal speed, and the finite inertia, of electrons cannot be ignored. Thus the electrons cannot instantaneously short out the $E_{\parallel}$ that develops because of the cross-field polarization current arising from the induced electric field (see Sect. 3.4). The value of $E_{\parallel}$ depends linearly on the parallel and perpendicular wavenumber of the wave, and in general needs small scales. This process has frequently been discussed in the magnetosphere/ionosphere context (e.g. see review by Stasiewicz et al. 2000). It was examined by Fletcher and Hudson (2008) for the solar analogue and found to be a viable process (i.e. generating an electric field in excess of the Dreicer field) in coronal densities and temperatures for waves with perpendicular scales of less than 5 km or so. Supplying adequate electrons from the tenuous corona to meet the HXR requirements remains problematic, but it is also possible that as the chromosphere heats during the flare onset and the Dreicer field decreases the effective acceleration in the $E_{\parallel}$ of an inertial Alfvén wave becomes possible here too.

3.8 Conclusions for Solar Flares

The complicated, current-bearing magnetic fields in solar active regions hosting flares must relax and release their energy in such a way as to inject it into the kinetic energy of nonthermal particles, which carry the bulk of a flare's energy which is then dissipated as heat. This may happen directly in the solar corona resulting in an intense beam of electrons—a current—propagating through the corona and balanced by a return flow of ambient electrons in a beam-return current system. Or, in processes more analogous with the magnetosphere, the relaxation might proceed in the form of Alfvénic perturbations launched by the introduction of an interruption to the current flow in a single loop, or by reconnection in a system of two or more interacting loops. In the case of two interacting loops it may be shown how the inductive electric field leads to cross-field currents which allow the currents to reroute during reconnection, and drives a Poynting flux along the reconnected field. However we know that coronal magnetic topology is often more complicated than two interacting loops and this must be included in our thinking. Electrical currents with high individual particle energies can also be formed in the diffusion region of a reconnecting coronal structure, where individual particles become decoupled from the reconnecting components of the field and are accelerated in the $-\mathbf{v} \times \mathbf{B}/c$ field. However we note that Longcope and Tarr (2012), who treat the redistribution of energy stored in a current sheet in 2D, demonstrate that the vast majority of the stored energy is carried away from the sheet by fast magnetosonic waves launched by unbalanced Lorentz forces in the sheet, leaving little to be dissipated locally. They also comment that in 3D Alfvénic waves running along the separators will instead result. It remains to be seen how these different views of current-carrying structures can be reconciled, also with the complex current sheets that arise and that are observed to be strongly associated with the chromospheric locations of energy deposition.

An energy flow carried by magnetic perturbations must be dissipated somewhere, and there are several avenues through which it could lead to both heating and particle acceleration (though observations dictate that eventually non-thermal particle kinetic energy must dominate). Assuming that it is large enough to overcome Coulomb collisional friction, the parallel electric field set up by the (inertial) Alfvén wave propagating through the very low β environment of the flare corona and upper transition region may accelerate or heat particles, depending on the ambient plasma conditions. The release of shear stresses in the wave as it travels towards the deep atmosphere where it is line-tied also generates a parallel field. More complex processes such as partial wave reflection in a converging field and the development of a turbulent cascade may also provide a source of first heating and then, as the plasma heats and becomes less collisional, particle acceleration in the lower atmosphere. The idea of waves rather than particles carrying flare energy through the corona to the chromosphere finds close analogy with the terrestrial magnetosphere-ionosphere system described in the next section, and deserves further examination.

4 Electric Currents in the Earth's Magnetosphere and Ionosphere

The main difference between electric currents around pulsars or in solar flares and currents in the Earth's magnetosphere and ionosphere is that the former are dominated by the effects coming from inside the pulsar or the Sun while the latter are generated by the interaction of the Earth's magnetosphere with the solar wind. At Earth the plasma of the solar wind travels with about 400–600 km/s, is comprised of electrons and mainly protons of 10^5–10^6 K, carries a magnetic field of 1–10 nT (10^{-5}–10^{-4} G) with variable orientation, and has an average density of 1–10 particles/cm^3. The solar wind plasma interacts with the originally dipolar magnetic field of the Earth, with a terrestrial surface magnetic field of 25–65 μT (0.25–0.65 G), (and all other magnetized bodies in the solar system like Jupiter, Saturn, and Uranus (Russell 2003)) whereby the Earth's magnetosphere is formed into a tear-drop shape with a compressed side facing the sun and a long tail extending anti-sunward (Roelof and Sibeck 1993).

4.1 Magnetopause Current

Large-scale currents in space are the source of magnetic fields in distant regions and deform planetary dipole magnetic fields to produce planetary magnetospheres (Parks 2004). The prime example for this interaction is the generation of the magnetopause current, also called *Chapman-Ferraro current*, and the subsequent deformation of the Earth's dipole field and generation of the magnetosphere (Fig. 30). The magnetopause current is a typical example of a boundary current at the interface that separates the internal Earth's magnetic field and strongly magnetized, low density plasma (plasma $\beta < 0.1$) from the weakly magnetized, compressed and thus high-density ($\beta \approx 1$) plasma of the solar wind in the magnetosheath. It is generated by the solar wind protons and electrons that partially penetrate the geomagnetic field and spiral back out into the solar wind after half a gyration. The gyro radii of protons and electrons are very different and positive and negative particles gyrate in opposite directions which leads to the generation of the surface current. The magnetopause current responds to changes in the solar wind properties (especially pressure) and can quickly move the magnetopause which is very often seen in spacecraft data while crossing this space boundary (see e.g. Gosling et al. 1967).

The balance of external and internal forces at the magnetopause requires an equilibrium between the solar wind pressure p_{sw}, the magnetopause current $\mathbf{j}_{mp}$, and the Earth's magnetic field $\mathbf{B}$:

$$\nabla p_{sw} = \mathbf{j}_{mp} \times \mathbf{B}/c. \tag{33}$$

This can be rewritten for the magnetopause current which for Earth runs from dawn to dusk (Parks 2004):

$$\mathbf{j}_{mp} = \mathbf{B} \times \frac{\nabla p_{sw}}{B^2} c. \tag{34}$$

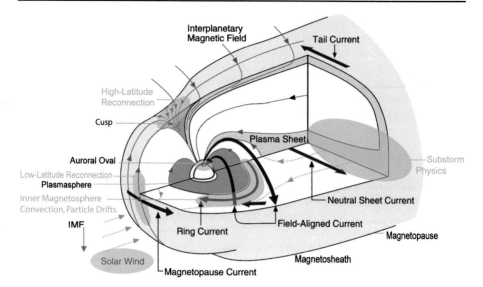

Interplanetary
Magnetic Field

Tail Current

High-Latitude
Reconnection

Cusp

Plasma Sheet

Auroral Oval

Substorm
Physics

Low-Latitude Reconnection
Plasmasphere

Inner Magnetosphere
Convection, Particle Drifts

IMF

Neutral Sheet Current

Field-Aligned Current

Ring Current

Magnetopause

Solar Wind

Magnetosheath

Magnetopause Current

Fig. 30 Schematics of the Earth's magnetosphere with its major plasma regions, physical processes shown in *green*, and primary current systems shown in *red*. Modified from an original by C.T. Russell and R. Strangeway; from Frey (2007), ©American Geophysical Union

The magnetopause current roughly doubles the local magnetic field strength just inside the magnetopause and completely cancels it just outside. Besides particle signatures (change in density, temperature, and flow velocity) this is one indication in spacecraft data that the magnetopause has been crossed (e.g. Dunlop and Balogh 2005). A similar current should flow at the boundary between the solar system and interstellar space (heliopause), and at the interface region between a neutron star environment and intergalactic space, but so far has only directly been measured at Earth and Jupiter (Kivelson 2000).

In the equatorial plane the magnetopause current flows from dawn to dusk. In the magnetotail it splits into a northern and southern current, termed tail currents. The tail currents flow from dusk to dawn across the tail magnetopause and close to Earth they continue the dayside magnetopause current. Further down the tail the magnetopause currents close the neutral sheet currents that will be covered in the following section.

4.2 Neutral Sheet Current

The interaction between the solar wind and the Earth's magnetosphere gets substantially enhanced whenever the solar wind magnetic field has a southward component (negative B_z) and can connect itself to the Earth's dipole field in the subsolar region in the process named reconnection (Parker 1957; Paschmann 2008). Reconnected field lines are anchored on one end at the Earth's surface and the other end is carried by the solar wind flow downtail across the polar caps (Fig. 31). One consequence of reconnection and plasma transport into the tail is the continuous supply of magnetic flux and fresh plasma towards the center of the magnetotail. Magnetic flux and plasma pressure would continuously increase if there were not a limiting process that converts magnetic field energy into heat and kinetic energy of particles and also removes plasma from the central magnetotail and that process is reconnection at the x-line in the tail.

Fig. 31 Flow of plasma within the magnetosphere (convection) driven by magnetic reconnection at the dayside magnetopause. The *numbered field lines* show the succession of configurations a geomagnetic field line assumes after reconnection with an IMF field line (*1*), drag across the polar cap (*2–5*), reconnection at the x-line in the tail (*6*), ejection of plasma down the tail and into the solar wind (*7′*), and the subsequent return of the field line to the dayside at lower latitudes (*7–9*). From Kivelson and Russell (1995), ©Cambridge University Press, reprinted with permission

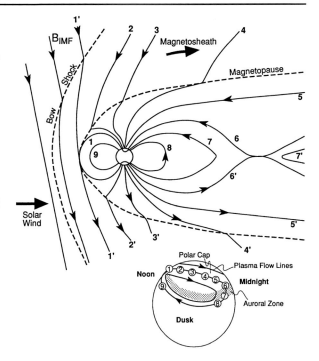

The magnetospheric tail at larger distances from Earth is separated into a northern and a southern lobe with opposite magnetic field directions, low density plasma (<0.01 cm^{-3}), and open magnetic field lines that extend well beyond the lunar orbit (see Fig. 30). The flow of plasma towards the center of the magnetic tail creates the plasma sheet. This higher-density (0.1–1 cm^{-3}), hot ($T_i \sim 2$–20 keV, $T_e \sim 0.4$–4 keV) plasma region resides on closed magnetic field lines and contains plasma originating both from the solar wind and from the ionosphere (Paschmann et al. 2003).

The center of the plasma sheet is also called the neutral sheet, because the magnetic field direction reverses and the magnitude becomes very small (<5 nT, $<5 \cdot 10^{-5}$ G). The screening of the magnetic field of different orientations is achieved by a diamagnetic current flowing across the magnetospheric tail from dawn to dusk (Baumjohann and Treumann 1996). Its cause is the gradient in plasma pressure pointing from north to south in the upper half and from south to north in the lower (southern) half. An estimate of the neutral sheet current per unit length follows from

$$J = \frac{cB_T}{2\pi} \tag{35}$$

which separates the lobes and balances the tail magnetic field $B_T \approx 20$ nT which gives $J = 30$ mA/m ($3 \cdot 10^4$ statA/cm). Particles entering the neutral sheet from the northern or southern lobes become demagnetized and are transported toward dusk (positive ions) and towards dawn (negative electrons). Current continuity of the neutral sheet current is maintained through current closure with the tail magnetopause current in the north and south. If one could look down the tail from Earth and actually see the currents they would appear to form the Greek letter Θ.

Fig. 32 Development of the Dst index during the large magnetic storm in February 1986 and the increase of the energy content of the ring current between $L = 3$–5 (expressed in R_E). The *line of Dst* refers to the *left scale*, the *plusses* refer to the *right scale*, which is inverted to follow the Dst trends. Adaptation from Hamilton et al. (1988), ©American Geophysical Union

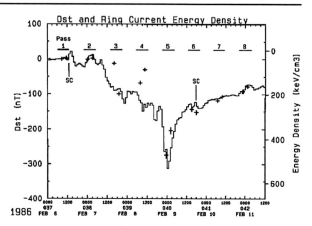

4.3 Ring Current

The ring current consists of ions trapped inside of 4–5 R_E at the equator in the same region as the outer electron radiation belt, and their energy ranges from a few tens of keV up to a few MeV (Paschmann et al. 2003). It is generated by the motion of trapped particles in an inhomogeneous planetary magnetic field as the particles undergo gradient and curvature drifts (Parks 2004). The gradient drift of a particle with charge q, mass m, perpendicular velocity $v_\perp$, and parallel velocity $v_\parallel$ is given by

$$\mathbf{v}_{\nabla B} = \frac{1}{2} \frac{m v_\perp^2}{q B} \frac{\mathbf{B} \times \nabla \mathbf{B}}{B^2} c, \tag{36}$$

while the curvature drift of the particle is given by Parks (2004)

$$\mathbf{v}_{curv} = \frac{m v_\parallel^2}{q B} \frac{\mathbf{B} \times \nabla \mathbf{B}}{B^2} c. \tag{37}$$

The action of both drifts will cause positive particles to drift westward in the Earth's dipole field, while negative particles drift eastward, creating a net westward current, the ring current. The effect of this westward current is a reduction of the local magnetic field at the surface of the Earth, especially close to the equator. The ring current becomes important during magnetic storms when the penetration of plasma from the tail drastically enhances the ring current and leads to a decrease of the magnetic field on the ground. During strong magnetic storms the depression of the equatorial magnetic field at the surface of the Earth can reach several hundred nT, or up to 1 %. Defining W as the total energy of the particles one can write the total current as

$$I = -\frac{3W}{2\pi r^2 B} c, \tag{38}$$

and the perturbation $\Delta \mathbf{B}_T$ of the Earth's magnetic field B_s at the surface by this current can be expressed by

$$\frac{\Delta \mathbf{B}_T}{B_s} = -\frac{2W}{B_s^2 R_E^3} \mathbf{z} \tag{39}$$

The unit vector $\mathbf{z}$ points northward and the negative sign accounts for the reduction of the surface magnetic field by the westward flowing ring current.

Geomagnetic storms usually start with the impact of a strong and persistent southward interplanetary magnetic field together with an increased solar wind dynamic pressure on the dayside magnetopause. The enhanced pressure pushes the magnetopause inward by several Earth radii, enhances the magnetopause current, and temporarily causes an increase of the magnetic field on the ground. This phenomenon is known as the storm sudden commencement (SSC) and can be seen as the slight increase of the Dst (Disturbance storm time index, Mayaud 1980) index in Fig. 32. During the main phase of the magnetic storm charged particles in the near-Earth plasma sheet are energized and injected deeper into the inner magnetosphere drastically enhancing the quiet time ring current and producing the so-called storm time ring current. This leads to a strong decrease of the Dst index and marks the main phase of the storm (Fig. 32). This main phase lasts generally for several hours but can last for several days during severe storms. After a turn of the IMF to northward directions the recovery phase of the storm begins during which the injection of plasma sheet material slows down or stops completely and various loss processes (primarily charge exchange and precipitation) weaken the ring current. The recovery phase can last several days up to several weeks.

During the early years of space research it was thought that the ring current consists only of protons. However, observations showed that a substantial fraction of the ions can be O^+ and N^+ (Hamilton et al. 1988). During the example in Fig. 32 it was shown that more than 50 % of the ring current ions were oxygen and nitrogen thus demonstrating the importance of the ionosphere as a potential source of ring current particles during major magnetic storms. Clues about how these ionospheric ions are accelerated and transported out into the magnetosphere come from correlated observations of electric field fluctuations and upward flowing ions. These observations indicate that low frequency (frequency < ion cyclotron frequency) electric field fluctuations with amplitudes of ≥ 0.1 V/m accompany upward flowing ions. Several researchers have shown theoretically that large-amplitude electromagnetic waves can give rise to a significant ponderomotive force that can accelerate and transport ions out of the ionosphere and favor acceleration of heavier mass species. Thus, O^+ could be accelerated more effectively than H^+ ions, thereby explaining the dominance of oxygen species in the ring current during some major storms (Parks 2004).

Besides its importance on the ground magnetic disturbance during magnetic storms, the ring current also influences the size of the auroral oval (Schulz 1997). The storm time ring current stretches the magnetospheric magnetic field outward thus increasing the amount of open magnetic flux in the polar cap. As the inside of the auroral oval marks the boundary between open and closed magnetic field lines this increase in open flux enhances the overall size of the auroral oval by about $4°$ for each -100 nT $(-10^{-3}$ G) of Dst.

4.4 Field-Aligned Currents

The electric fields that are associated with the convection process in the magnetosphere map along the magnetic field to the ionosphere where they create currents that are perpendicular to the magnetic field. In the terrestrial ionosphere both electrons and ions are scattered by collisions with neutrals and the generalized Ohm's law describing the relationship between the electric and magnetic fields and the current becomes:

$$\mathbf{j} = \sigma_\parallel \mathbf{E}_\parallel + \sigma_P \mathbf{E}_\perp - \sigma_H (\mathbf{E}_\perp \times \mathbf{B})/B \qquad (40)$$

where the ionospheric Pedersen conductivity σ_P governs the Pedersen current in the direction of that part of the electric field $\mathbf{E}_\perp$ which is transverse to the magnetic field. The Hall conductivity σ_H determines the Hall current in the direction perpendicular to both the electric and magnetic field $-\mathbf{E} \times \mathbf{B}/c$. The parallel conductivity σ_P governs the magnetic

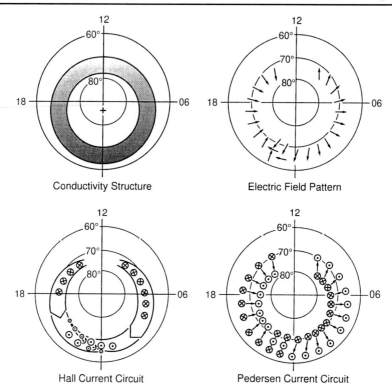

Fig. 33 Sketch of conductivity, electric field, and Hall and Pedersen currents in the auroral ionosphere. The *dots* show the upward and the *crosses* the downward field-aligned currents. From Baumjohann and Treumann (1996), ©Imperial College Press

field-aligned current that is driven by the parallel electric field component $E_{\parallel}$ (Baumjohann and Treumann 1996).

The large-scale field-aligned currents are traditionally divided into two classes: Region-1 and Region-2 currents. Region-1 currents flow at higher latitudes poleward of the Region-2 currents and close at the magnetospheric boundaries. Region-2 currents run at lower latitudes equatorward of the Region-1 currents and close in the inner magnetosphere. In the ionosphere these currents close through horizontal currents as shown in Fig. 33. Magnetospheric convection determines the orientation of the currents with Region-1 flowing out of the ionosphere in the evening sector and into the ionosphere in the morning sector (see Fig. 34). On the dayside these currents connect to the dayside magnetospheric boundaries (e.g. Janhunen and Koskinen 1997). However, the nightside currents most likely originate from the plasma sheet (Tsyganenko et al. 1993). The total current in this circuit is 1–2 MA $(3–6 \cdot 10^{15} \text{ statA})$ (Paschmann et al. 2003).

The Region-2 currents flow equatorward of the Region-1 currents and in opposite directions. They flow into the ionosphere in the evening sector, and they flow out of the ionosphere in the morning sector (see Fig. 34). At high altitudes these currents merge with the ring current in the inner magnetosphere. The total current in the Region-2 circuit is slightly smaller than in the Region-1 current circuit, typically less than 1 MA. The slight overlap of the two regions with the reversal of current directions shortly before midnight is called the Harang discontinuity (Koskinen and Pulkkinen 1995) which may play an important role

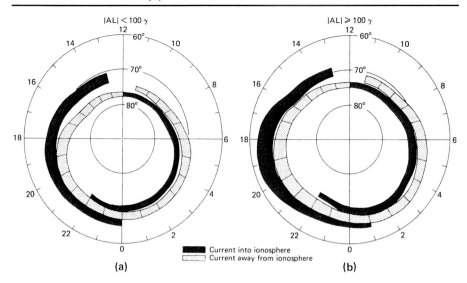

Fig. 34 Field-aligned current distribution in the polar ionosphere during low geomagnetic activity (*left*) and during disturbed conditions (*right*) ($1\gamma = 1$ nT $= 10^{-5}$ G). From Iijima and Potemra (1976), ©American Geophysical Union

in the localization of substorm onsets, even though the details are still debated (see e.g. Weygand et al. 2008).

Magnetospheric convection which is driven by reconnection at the dayside and at the tail x-line transports plasma from the tail towards Earth and develops a plasma pressure maximum in the midnight region of the near-Earth tail. The accumulated plasma has to come into equilibrium with the surrounding plasma and the dipole-like magnetic field. That sets up a pressure gradient pointing towards Earth and as the increased pressure wants to relax, it can most easily do so perpendicular to the magnetic field in the east-west direction. This plasma expansion has to overcome friction by ion-neutral collisions in the ionosphere or Ohmic dissipation of the auroral current system which is set up between the magnetosphere and the ionosphere (Fig. 35). The current system is of the Type II as originally suggested by Boström (1964) where the current flows into the ionosphere on the equator-ward side of the auroral arc all the way along the arc and flows out of the ionosphere on the poleward side. This current system is equivalent to the Region-1 and 2 current systems on the evening side and just reversed on the morning side.

Magnetic Fractures   However, the general picture in Fig. 35 is too simple as it depicts a rather static situation, balances only the magnetospheric pressure and ionospheric friction with the energy transfer from the magnetosphere into the ionosphere, and ignores the fact that even during quiet times the aurora very often appears filamented with many parallel arcs stretching over long distances east-west, but having a rather small (1–10 km) north-south extent (Fig. 36). Some additional processes must be at play and create the structures and dominate their dynamics that can be seen in the aurora.

In the downward field-aligned current region the current is carried by relatively cold up-going electrons of ionospheric origin. In the upward field-aligned current region the current is carried by hot downgoing electrons of magnetospheric origin. However, measurements by satellites over auroral arcs showed that the downgoing electrons have much higher energy (2–25 keV) than the electrons in the plasmasheet (≈ 1 keV). Therefore an additional

Fig. 35 Dynamo forces, auroral current system, and resulting convection under frictional control by the ionosphere. From Paschmann et al. (2003), ©2002, Kluwer Academic Publishers

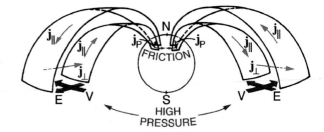

Fig. 36 Example of an aurora all-sky image mapped onto a geographic grid. The picture was taken on 2008-03-09 at 04:35:00 in Fort Simpson, Canada and shows at least 7 parallel thin auroral arcs. Each side of the mapped image is about 800 km

acceleration must be at work and early in the space age it was recognized that in the upward current region the parallel current is driven by a field-aligned potential $\Delta\Phi_\parallel$ and that it can be related to the physical parameters of the source population through a linear relationship, the Knight relation (Knight 1973):

$$j_{\parallel ion} = K\Delta\Phi_\parallel \quad \text{with } K = \frac{e^2 n_e}{\sqrt{2\pi m_e k_B T_e}} \tag{41}$$

where K is called the field-aligned or Knight conductance, T_e is the electron temperature, and n_e the density in the source region. The values for the field-aligned conductance can be deduced from satellite measurements and vary over a rather large range of 10^{-9}–10^{-11} S/m^2 ($9 \cdot 10^{-2}$–$9 \cdot 10^{-4}$ (cm s)$^{-1}$). One reason can be that it was deduced ignoring additional particle sources, like trapped electrons, secondary and backscattered electrons (Sakanoi et al. 1995), and/or wave acceleration (Frey et al. 1998). The resulting current densities are in the range of several μA/m^2 and current continuity requires that similar current densities are also found in the downward current region. The requirement for current closure combined with the low plasma density along auroral field lines are then the reasons for the generation of substantial potential drops parallel to the magnetic field in the range of a few hundred to a few thousand volts. The current-voltage relation along such magnetic field lines can be determined for simple profiles of the background ion density. For typical parameters, the current density is found to be a few times larger in the downward current region compared to currents in the upward current region for similar potential drops. Thus potential drops up to a few thousand volts and the consequent acceleration of ionospheric electrons up to keV energies, such as has been observed by the FAST satellite, are a necessary consequence of the observed current densities in the downward auroral current region (Temerin and Carlson 1998).

Looking at quiet auroral arcs in the evening sector they almost always slowly move equator-ward. As auroral arcs are the ionospheric foot point of the upward field-aligned current the question arises why they are moving and what is the context of their motion. Given the observational fact of motions of arcs and of the oval as a whole it would just be a 'logical' conclusion to assume that the whole ionosphere and with it the frozen-in auroral arcs move equator-ward in the evening sector and poleward in the morning sector following the general distorted shape of the auroral oval which extends to much lower ge-omagnetic latitudes at midnight than it extends at dawn or dusk. Coordinated campaigns of an ionospheric radar measuring the ionospheric plasma motion and an all-sky camera observing the motion of several auroral arcs however, showed that there is a speed differ-ence of 100–200 m/s, and that these arcs perform proper motion (Haerendel et al. 1993; Frey et al. 1996). Almost always this proper motion is performed into the direction towards the downward field-aligned current.

The relative arc-plasma motion provides a way to supply auroral arcs with both fresh plasma and energy. The process is driven by the 'generator' region in the magnetotail which provides the energy that is then used for the conversion into particle kinetic energy and Joule heating $\mathbf{j} \cdot \mathbf{E}$. Haerendel (1994) studied a system in which an auroral arc propagates into a region of stored magnetic energy, drawing power from a region bounded by upward and downward current sheets. The reason for their propagation is the exhaustion of energy stored in form of magnetic shear stresses in the flux tubes pervading the arc. As these shear stresses are released within time scales of the order of the Alfvén transit time, τ_A, between the generator region in the plasma sheet and the auroral acceleration region at 1–2 R_E above the ground, the maximum $(\nabla \times \mathbf{B})_\| = 4\pi j_\|/c$ moves into the stressed field region. Hence the propagation speed must be $v_n = w_{arc}/\tau_A$, where w_{arc} is the effective arc width (Paschmann et al. 2003). Some of the consequences of this propagating system are summarized in Fig. 37.

The field-aligned current carried from the generator by oblique Alfvén waves becomes more intense near the ionosphere because the flux tube converges in the dipole field. At some point this current exceeds a threshold for the generation of micro-instabilities, allowing for a non-zero field-aligned electric field and breakdown of the frozen-in condition. This magnetic 'fracture' zone is therefore a region where magnetic field lines break and reconnect again. Most of the liberated magnetic energy is then transferred into kinetic energy of field-aligned accelerated electrons. Most of the energy that is dissipated in the arc is deposited as heat in the ionosphere. A consequence of this heating is the expansion and expulsion of plasma from that region. Most of the ionospheric plasma is excavated by the front of the arc and the so-called auroral cavity is formed (Fig. 38). The fracture zone is highly structured and dynamic, launching disturbances toward the ionosphere in the form of small-scale Alfvén waves. These waves can reflect both from the ionosphere and the fracture zone. Multiple reflections lead to highly structured fields in the interference region which can modulate the auroral electron flux and energy. This interference mechanism has been proposed as a potential source of auroral fine structure and multiple arcs as seen in Fig. 36 (Haerendel 1983).

Fig. 37 Qualitative picture of a magnetic fracture region with the generator region (*top*) in motion with respect to the ionosphere (*bottom*). In the fracture region (*middle*) concentrated field-aligned currents become unstable, allowing finite potential drops and field-aligned acceleration. An interference region between the fracture zone and the ionosphere is characterized by multiple reflections. The fracture zone itself is shown in more detail in Fig. 29. From Haerendel (1994), reproduced by permission of the AAS

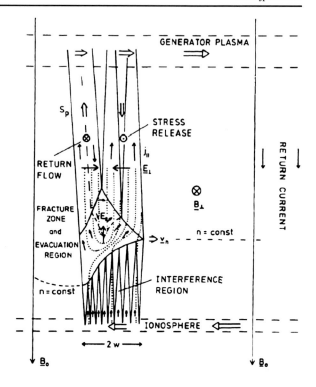

Preventing Current Starvation   The evacuation of the plasma and creation of the auroral density cavity has the consequence that in a steady state the auroral current system would undergo a similar effect as the current starvation mentioned earlier in this paper. In addition to extracting energy from the current circuit the proper motion of the auroral arc will maintain the current flow by providing fresh plasma to the acceleration process. To prevent a shortage of current carriers a local field-aligned parallel potential is generated to enlarge the loss cone for thermal magnetospheric electrons. By increasing the parallel velocity of downgoing electrons their pitch angle is decreased and more of them will not mirror in the converging dipole field but will rather penetrate deep enough to be finally lost in the ionosphere through collisions. One clear example where this process happens was demonstrated with High Latitude Dayside Aurora (HiLDA) (Frey et al. 2004).

Space-based observations revealed the occurrence of a bright aurora spot well inside the generally empty polar cap. The spot appeared with a 20–60 minutes delay after a substantial drop in solar wind density. No obvious change in solar wind properties could be related to the time of first appearance of the bright spot, which pointed to an internal magnetospheric process rather than a solar wind driven process. Two observations coincided with passes of the FAST satellite just on top of the spots where a strong upward field-aligned current and monoenergetic particles with a clear signature of parallel acceleration were determined. On field lines threading the cusp region (Fig. 30) downward field-aligned current was determined (Frey et al. 2003). Statistical analysis of solar wind conditions demonstrated a clear preference of spot occurrence during positive IMF B_y and B_z when a current circuit is set up (Richmond and Kamide 1988) with downward field-aligned current in the cusp region, Pedersen current in the ionosphere, and the upward field-aligned current in the polar cap in the center of the dayside convection cell over the HiLDA spot (Le et al. 2002).

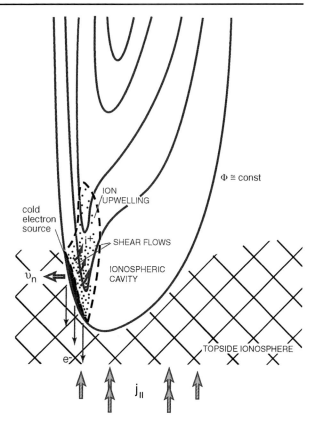

Fig. 38 Sketch of an auroral cavity. The upward field-aligned current sets up the acceleration region which evacuates the ionospheric plasma by accelerating ions up the field lines and electrons down the field lines and moves into the region with fresh cold electrons. From Paschmann et al. (2003), ©2002, Kluwer Academic Publishers

The impact of the low solar wind density on the occurrence of the aurora spot may be twofold. When the solar wind density decreases, the magnetosphere expands in order to maintain pressure balance with the external solar wind plasma medium. This expansion, combined with the reduced supplement of fresh plasma through diffusion, will reduce the number of available current carriers to maintain the upward current. In order to keep the current flowing (Siscoe et al. 2001) the system sets up a parallel potential that accelerates the few existing electrons into the ionosphere and subsequently creates the HiLDA at the footprint, thus avoiding current starvation. The other possible impact of the low solar wind density is the reduction of the Alfvén Mach number and a lower plasma β in the magnetosheath adjacent to the reconnection site. The reduced Mach number enhances the strength of the dusk convection cell (Crooker et al. 1998) and thus the strength of the upward current at the HiLDA footprint. The lower plasma β is considered favorable for the onset of reconnection and can create a stronger downward current into the cusp (Fig. 39, Frey 2007).

4.5 Conclusions for Terrestrial Magnetosphere and Ionosphere

As mentioned earlier the main difference between electric currents around pulsars or solar flares and currents in the Earth's magnetosphere and ionosphere is that the former are dominated by the effects coming from inside the pulsar or the sun while the latter are dominated

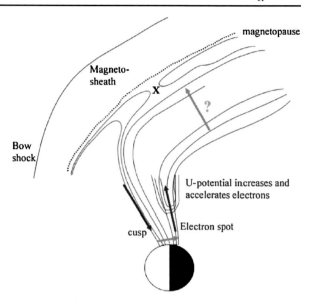

Fig. 39 Qualitative picture of the field-aligned current distribution for the case of a high latitude dayside aurora spot with the upward current over the spot, downward current connecting the cusp foot point with the high latitude magnetopause, and Pedersen current in the ionosphere. Current closure most likely happens through the magnetopause current

and generated by the interaction of the Earth's magnetosphere with the solar wind. The magnetized Earth and the plasma bound by the Earth's magnetic field represent an obstacle to the solar wind which compresses the magnetic field at the dayside and stretches it out into the long tail on the nightside. The magnetopause and neutral sheet currents are driven by this interaction.

Magnetic reconnection transfers plasma and energy from the solar wind into the magnetosphere and drives convection. Magnetic flux is transferred from the dayside magnetosphere to the tail lobes. As the system cannot accumulate energy forever, it has to employ certain methods to get rid of the accumulated energy and one way is to set up field-aligned electric fields that convert electromagnetic energy into kinetic particle energy and drive field-aligned currents. The system also has to remove accumulated plasma and that is achieved through reconnection at the x-line and the ejection of plasmoids down the tail out into the solar wind. It has been shown that the energy conversion above a system of auroral arcs can make a substantial or even larger contribution to the magnetospheric energy loss than ionospheric friction and Ohmic dissipation, and exactly for this reason the notion of an '*auroral pressure valve*' was introduced (Haerendel 2000). The optical aurora is just a small result of this energy conversion as most of the precipitating particle energy goes into heat and only $\approx 10\%$ is actually released as energy of photons.

Auroral acceleration is not just found at Earth but on all magnetized planets of the solar system (Bhardwaj and Gladstone 2000; Paschmann et al. 2003). Additionally, it has also been suggested that auroral acceleration processes happen on other astrophysical objects too, like solar flares, cataclysmic variables, and accreting neutron stars (Haerendel 2001). Intense field-aligned currents driven by forces in the outer, weak magnetic fields of an astrophysical object are the key to the generation of field-aligned potential drops. In the strong fields near the object the currents may reach a critical limit where the mirror force or current-driven anomalous resistivity require the setup of parallel potential drops for the maintenance of current continuity.

5 Conclusion

We have investigated electric current systems in three very different magnetized objects in the cosmos: radio pulsar winds, the solar corona, and the terrestrial magnetosphere. Their dimensions and physical conditions differ greatly, yet their main physical effects are very similar. Essentially, this happens because, in all three cases, an electric current system is set up inside a magnetically dominated plasma by a kinematically dominated plasma, i.e. the rotating magnetic neutron star, the dynamo below the solar surface, and the solar wind. These 'drivers' continue to try and increase the free energy of the electric circuit via transport of Poynting flux. However, ultimately, this results in electric current densities which in a number of small spatial domains are too large to be maintained. These domains are the locations of episodic conversion of electric current energy and the release of energized plasma in the form of ejected plasmoids, particle acceleration and heating, and magnetic turbulence.

These electric circuits are largely force-free inside their magnetically dominated domains, and can be described to a first approximation within the framework of MHD. Yet we underline that it is important to consider the complete electric circuit and not just the force-free part of the magnetic field. While it is well-known that a force-free structure cannot exist on its own and has to be anchored in a non force-free bounding plasma, these boundary conditions are easy to overlook in a magnetic field picture only. When one considers an electric circuit, one automatically includes the voltage source and the resistive regions, which of course are non force-free. Although their relative volumes are minor they do determine the evolution of the system.

Finally, one would like to know to what extent the end products of bulk motion, particle acceleration and heating, and the release of magnetic turbulence come about by the same physical effects in these different objects. With the proviso that we can make in situ observations inside the terrestrial magnetosphere only, while this is not possible in the corona, let alone in a pulsar wind, we conclude that possibly in all three cases one and the same process does play an important role in converting magnetic/current energy into particle acceleration. This is the occurrence of electric fields along the magnetic field whenever the local plasma conditions cannot provide for the electric current dictated by the global circuit. It is clear, however, that 'common' reconnection occurs as well next to or in combination with this 'Generalized Magnetic Reconnection'.

Acknowledgements HUF was supported through NSF grant AGS-1004736. LF gratefully acknowledges support from the UK's Science and Technology Facilities Council through grant ST/I001808/1, and from the HESPE project (FP7-2010-SPACE-1-263086) funded through the 7th Framework Programme of the European Community. JK thanks the Leids Kerkhoven-Bosscha Fonds for a travel grant. The authors acknowledge the efforts by ISSI for convening the workshop and providing support, which facilitated this joint publication.

References

H. Alfvén, P. Carlqvist, Currents in the solar atmosphere and a theory of solar flares. Sol. Phys. **1**, 220–228 (1967). doi:10.1007/BF00150857

J.J. Aly, On some properties of force-free magnetic fields in infinite regions of space. Astrophys. J. **283**, 349–362 (1984). doi:10.1086/162313

G. Aulanier, The physical mechanisms that initiate and drive solar eruptions. arXiv e-prints (2013)

W. Baumjohann, R.A. Treumann, *Basic Space Plasma Physics* (1996)

M.C. Begelman, Instability of toroidal magnetic field in jets and plerions. Astrophys. J. **493**, 291 (1998). doi:10.1086/305119

A. Bhardwaj, G.R. Gladstone, Auroral emissions of the giant planets. Rev. Geophys. **38**, 295–354 (2000). doi:10.1029/1998RG000046

J. Birn, A.V. Artemyev, D.N. Baker, M. Echim, M. Hoshino, L.M. Zelenyi, Particle acceleration in the magnetotail and aurora. Space Sci. Rev. **173**, 49–102 (2012). doi:10.1007/s11214-012-9874-4

S.V. Bogovalov, On the physics of cold MHD winds from oblique rotators. Astron. Astrophys. **349**, 1017–1026 (1999)

S. Bogovalov, K. Tsinganos, On the magnetic acceleration and collimation of astrophysical outflows. Mon. Not. R. Astron. Soc. **305**, 211–224 (1999). doi:10.1046/j.1365-8711.1999.02413.x

R. Boström, A model of the auroral electrojets. J. Geophys. Res. **69**, 4983–4999 (1964). doi:10.1029/JZ069i023p04983

J.C. Brown, Sol. Phys. **18**, 489–502 (1971). doi:10.1007/BF00149070

J.C. Brown, D.B. Melrose, Collective plasma effects and the electron number problem in solar hard X-ray bursts. Sol. Phys. **52**, 117–131 (1977). doi:10.1007/BF00935795

J.C. Brown, R. Turkmani, E.P. Kontar, A.L. MacKinnon, L. Vlahos, Astron. Astrophys. **508**, 993–1000 (2009). doi:10.1051/0004-6361/200913145

P. Carlqvist, A flare-associated mechanism for solar surges. Sol. Phys. **63**, 353–367 (1979). doi:10.1007/BF00174540

S.A. Colgate, A phenomenological model of solar flares. Astrophys. J. **221**, 1068–1083 (1978). doi:10.1086/156111

J. Contopoulos, A simple type of magnetically driven jets: an astrophysical plasma gun. Astrophys. J. **450**, 616 (1995). doi:10.1086/176170

I. Contopoulos, The coughing pulsar magnetosphere. Astron. Astrophys. **442**, 579–586 (2005). doi:10.1051/0004-6361:20053143

F.V. Coroniti, Magnetically striped relativistic magnetohydrodynamic winds—the Crab Nebula revisited. Astrophys. J. **349**, 538–545 (1990). doi:10.1086/168340

N.U. Crooker, J.G. Lyon, J.A. Fedder, MHD model merging with IMF B_y: lobe cells, sunward polar cap convection, and overdraped lobes. J. Geophys. Res. **103**, 9143–9152 (1998). doi:10.1029/97JA03393

B. De Pontieu, P.C.H. Martens, H.S. Hudson, Chromospheric damping of Alfvén waves. Astrophys. J. **558**, 859–871 (2001). doi:10.1086/322408

P. Demoulin, L.G. Bagala, C.H. Mandrini, J.C. Henoux, M.G. Rovira, Quasi-separatrix layers in solar flares. II. Observed magnetic configurations. Astron. Astrophys. **325**, 305–317 (1997)

M.W. Dunlop, A. Balogh, Magnetopause current as seen by Cluster. Ann. Geophys. **23**, 901–907 (2005). doi:10.5194/angeo-23-901-2005

L. Fletcher, H.S. Hudson, Astrophys. J. **675**, 1645–1655 (2008). doi:10.1086/527044

H.U. Frey, Localized aurora beyond the auroral oval. Rev. Geophys. **45**, 1003 (2007). doi:10.1029/2005RG000174

H.U. Frey, G. Haerendel, D. Knudsen, S. Buchert, O.H. Bauer, Optical and radar observations of the motion of auroral arcs. J. Atmos. Terr. Phys. **58**, 57–69 (1996)

H.U. Frey, G. Haerendel, J.H. Clemmons, M.H. Bochm, J. Vogt, O.H. Bauer, D.D. Wallis, L. Blomberg, H. Lühr, Freja and ground-based analysis of inverted-V events. J. Geophys. Res. **103**, 4303–4314 (1998). doi:10.1029/97JA02259

H.U. Frey, T.J. Immel, G. Lu, J. Bonnell, S.A. Fuselier, S.B. Mende, B. Hubert, N. Østgaard, G. Le, Properties of localized, high latitude, dayside aurora. J. Geophys. Res. **108**, 8008 (2003). doi:10.1029/2002JA009332

H.U. Frey, N. Østgaard, T.J. Immel, H. Korth, S.B. Mende, Seasonal dependence of localized, high-latitude dayside aurora (HiLDA). J. Geophys. Res. **109**, 4303 (2004). doi:10.1029/2003JA010293

P.K. Fung, D. Khechinashvili, J. Kuijpers, Radio pulsar drifting sub-pulses and diocotron instability. Astron. Astrophys. **445**, 779–794 (2006). doi:10.1051/0004-6361:20053040

M.K. Georgoulis, V.S. Titov, Z. Mikić, Non-neutralized electric current patterns in solar active regions: origin of the shear-generating Lorentz force. Astrophys. J. **761**, 61 (2012). doi:10.1088/0004-637X/761/1/61

S.E. Gibson, Y. Fan, C. Mandrini, G. Fisher, P. Demoulin, Observational consequences of a magnetic flux rope emerging into the corona. Astrophys. J. **617**, 600–613 (2004). doi:10.1086/425294

P. Goldreich, W.H. Julian, Pulsar electrodynamics. Astrophys. J. **157**, 869 (1969). doi:10.1086/150119

J.T. Gosling, J.R. Asbridge, S.J. Bame, I.B. Strong, Vela 2 measurements of the magnetopause and bow shock positions. J. Geophys. Res. **72**, 101–112 (1967). doi:10.1029/JZ072i001p00101

G. Haerendel, Acceleration from field-aligned potential drops. Astrophys. J. Suppl. Ser. **90**, 765–774 (1994). doi:10.1086/191901

G. Haerendel, Outstanding issues in understanding the dynamics of the inner plasma sheet and ring current during storms and substorms. Adv. Space Res. **25**, 2379–2388 (2000). doi:10.1016/S0273-1177(99)00527-X

G. Haerendel, Auroral acceleration in astrophysical plasmas. Phys. Plasmas **8**, 2365–2370 (2001). doi:10.1063/1.1342227

G. Haerendel, Commonalities between ionosphere and chromosphere. Astrophys. J. Suppl. Ser. **124**, 317–331 (2006). doi:10.1007/s11214-006-9092-z

G. Haerendel, An Alfven wave model of auroral arcs, Technical report, 1983

G. Haerendel, S. Buchert, C. La Hoz, B. Raaf, E. Rieger, On the proper motion of auroral arcs. J. Geophys. Res. **98**, 6087–6099 (1993). doi:10.1029/92JA02701

D.C. Hamilton, G. Gloeckler, F.M. Ipavich, B. Wilken, W. Stuedemann, Ring current development during the great geomagnetic storm of February 1986. J. Geophys. Res. **93**, 14343–14355 (1988). doi:10.1029/JA093iA12p14343

I.G. Hannah, L. Fletcher, Comparison of the energy spectra and number fluxes from a simple flare model to observations. Sol. Phys. **236**, 59–74 (2006). doi:10.1007/s11207-006-0139-9

M. Hesse, K. Schindler, A theoretical foundation of general magnetic reconnection. J. Geophys. Res. **93**, 5559–5567 (1988). doi:10.1029/JA093iA06p05559

G.D. Holman, Acceleration of runaway electrons and Joule heating in solar flares. Astrophys. J. **293**, 584–594 (1985). doi:10.1086/163263

P. Hoyng, H.F. van Beek, J.C. Brown, High time resolution analysis of solar hard X-ray flares observed on board the ESRO TD-1A satellite. Sol. Phys. **48**, 197–254 (1976)

T. Iijima, T.A. Potemra, The amplitude distribution of field-aligned currents at northern high latitudes observed by Triad. J. Geophys. Res. **81**, 2165–2174 (1976). doi:10.1029/JA081i013p02165

P. Janhunen, H.E.J. Koskinen, The closure of region-1 field-aligned current in MHD simulation. Geophys. Res. Lett. **24**, 1419–1422 (1997). doi:10.1029/97GL01292

H. Jeong, J. Chae, Magnetic helicity injection in active regions. Astrophys. J. **671**, 1022–1033 (2007). doi:10.1086/522666

M. Karlický, J.C. Brown, A.J. Conway, G. Penny, Flare hard X-rays from neutral beams. Astron. Astrophys. **353**, 729–740 (2000)

E. Khomenko, M. Collados, Numerical modeling of magnetohydrodynamic wave propagation and refraction in sunspots. Astrophys. J. **653**, 739–755 (2006). doi:10.1086/507760

M.G. Kivelson, Currents and flows in distant magnetospheres, in *Geophysical Monograph Series*, vol. 118 (Am. Geophys. Union, Washington, 2000), pp. 339–352. doi:10.1029/GM118p0339

M.G. Kivelson, C.T. Russell, *Introduction to Space Physics* (1995)

S. Knight, Parallel electric fields. Planet. Space Sci. **21**, 741–750 (1973). doi:10.1016/0032-0633(73)90093-7

S.S. Komissarov, Simulations of the axisymmetric magnetospheres of neutron stars. Mon. Not. R. Astron. Soc. **367**, 19–31 (2006). doi:10.1111/j.1365-2966.2005.09932.x

S.S. Komissarov, Magnetic acceleration of relativistic jets. Mem. Soc. Astron. Ital. **82**, 95 (2011)

H.E.J. Koskinen, T.I. Pulkkinen, Midnight velocity shear zone and the concept of Harang discontinuity. J. Geophys. Res. **100**, 9539–9548 (1995). doi:10.1029/95JA00228

S. Krucker, H.S. Hudson, N.L.S. Jeffrey, M. Battaglia, E.P. Kontar, A.O. Benz, A. Csillaghy, R.P. Lin, High-resolution imaging of solar flare ribbons and its implication on the thick-target beam model. Astrophys. J. **739**, 96 (2011). doi:10.1088/0004-637X/739/2/96

J. Kuijpers, Equatorial pulsar winds. Proc. Astron. Soc. Aust. **18**, 407–414 (2001). doi:10.1071/AS01048

M. Kuperus, M.A. Raadu, The support of prominences formed in neutral sheets. Astron. Astrophys. **31**, 189 (1974)

G. Le, G. Lu, R.J. Strangeway, R.F. Pfaff, Strong interplanetary magnetic field B_y-related plasma convection in the ionosphere and cusp field-aligned currents under northward interplanetary magnetic field conditions. J. Geophys. Res. **107**, 1477 (2002). doi:10.1029/2001JA007546

K.D. Leka, R.C. Canfield, A.N. McClymont, L. van Driel-Gesztelyi, Evidence for current-carrying emerging flux. Astrophys. J. **462**, 547 (1996). doi:10.1086/177171

J. Li, A. Spitkovsky, A. Tchekhovskoy, Resistive solutions for pulsar magnetospheres. Astrophys. J. **746**, 60 (2012). doi:10.1088/0004-637X/746/1/60

E. Liang, K. Noguchi, Radiation from comoving Poynting flux acceleration. Astrophys. J. **705**, 1473–1480 (2009). doi:10.1088/0004-637X/705/2/1473

D.W. Longcope, L. Tarr, The role of fast magnetosonic waves in the release and conversion via reconnection of energy stored by a current sheet. Astrophys. J. **756**, 192 (2012). doi:10.1088/0004-637X/756/2/192

B.C. Low, R. Wolfson, Spontaneous formation of electric current sheets and the origin of solar flares. Astrophys. J. **324**, 574–581 (1988). doi:10.1086/165918

P.C.H. Martens, A. Young, Neutral beams in two-ribbon flares and in the geomagnetic tail. Astrophys. J. Suppl. Ser. **73**, 333–342 (1990). doi:10.1086/191469

P.N. Mayaud, Derivation, meaning, and use of geomagnetic indices, in *Geophysical Monograph Series*, vol. 22 (Am. Geophys. Union, Washington, 1980), p. 607. doi:10.1029/GM022

A.N. McClymont, G.H. Fisher, On the mechanical energy available to drive solar flares, in *Geophysical Monograph Series*, vol. 54 (Am. Geophys. Union, Washington, 1989), pp. 219–225

A. Melatos, Spin-down of an oblique rotator with a current-starved outer magnetosphere. Mon. Not. R. Astron. Soc. **288**, 1049–1059 (1997)

A. Melatos, The ratio of Poynting flux to kinetic-energy flux in the Crab pulsar wind. Mem. Soc. Astron. Ital. **69**, 1009 (1998)

A. Melatos, D.B. Melrose, Energy transport in a rotation-modulated pulsar wind. Mon. Not. R. Astron. Soc. **279**, 1168–1190 (1996)

D.B. Melrose, Energy propagation into a flare kernel during a solar fire. Astrophys. J. **387**, 403–413 (1992). doi:10.1086/171092

D.B. Melrose, Current paths in the corona and energy release in solar flares. Astrophys. J. **451**, 391 (1995). doi:10.1086/176228

D.B. Melrose, A solar flare model based on magnetic reconnection between current-carrying loops. Astrophys. J. **486**, 521 (1997). doi:10.1086/304521

D.B. Melrose, Generic model for magnetic explosions applied to solar flares. Astrophys. J. **749**, 58 (2012a). doi:10.1088/0004-637X/749/1/58

D.B. Melrose, Magnetic explosions: role of the inductive electric field. Astrophys. J. **749**, 59 (2012b). doi:10.1088/0004-637X/749/1/59

L. Mestel, Magnetic braking by a stellar wind-I. Mon. Not. R. Astron. Soc. **138**, 359 (1968)

L. Mestel, *Stellar Magnetism*. Int. Ser. Monogr. Phys., vol. 99 (1999)

E.N. Parker, Sweet's mechanism for merging magnetic fields in conducting fluids. J. Geophys. Res. **62**, 509–520 (1957). doi:10.1029/JZ062i004p00509

G.K. Parks, *Physics of Space Plasmas: An Introduction* (2004)

G. Paschmann, Recent in-situ observations of magnetic reconnection in near-Earth space. Geophys. Res. Lett. **35**, 19109 (2008). doi:10.1029/2008GL035297

G. Paschmann, S. Haaland, R. Treumann, *Auroral Plasma Physics* (2003)

G.J.D. Petrie, The abrupt changes in the photospheric magnetic and Lorentz force vectors during six major neutral-line flares. Astrophys. J. **759**, 50 (2012). doi:10.1088/0004-637X/759/1/50

O. Porth, S.S. Komissarov, R. Keppens, Solution to the sigma problem of pulsar wind nebulae. Mon. Not. R. Astron. Soc. **431**, 48–52 (2013). doi:10.1093/mnrasl/slt006

A.D. Richmond, Y. Kamide, Mapping electrodynamic features of the high-latitude ionosphere from localized observations—technique. J. Geophys. Res. **93**, 5741–5759 (1988). doi:10.1029/JA093iA06p05741

E.C. Roelof, D.G. Sibeck, Magnetopause shape as a bivariate function of interplanetary magnetic field B_z and solar wind dynamic pressure. J. Geophys. Res. **98**, 21421 (1993). doi:10.1029/93JA02362

C.T. Russell, The structure of the magnetopause. Planet. Space Sci. **51**, 731–744 (2003). doi:10.1016/S0032-0633(03)00110-7

A.J.B. Russell, L. Fletcher, Propagation of Alfvénic waves from corona to chromosphere and consequences for solar flares. Astrophys. J. **765**, 81 (2013). doi:10.1088/0004-637X/765/2/81

T. Sakanoi, H. Fukunishi, T. Mukai, Relationship between field-aligned currents and inverted-V parallel potential drops observed at midaltitudes. J. Geophys. Res. **100**, 19343–19360 (1995). doi:10.1029/95JA01285

T. Sakurai, Magnetic stellar winds—a 2-d generalization of the Weber-Davis model. Astron. Astrophys. **152**, 121–129 (1985)

E. Schatzman, Sur la perte de masse et les processus de freinage de la rotation, in *The Hertzsprung-Russell Diagram*, ed. by J.L. Greenstein. IAU Symposium, vol. 10 (1959), p. 129

K. Schindler, M. Hesse, J. Birn, General magnetic reconnection, parallel electric fields, and helicity. J. Geophys. Res. **93**, 5547–5557 (1988). doi:10.1029/JA093iA06p05547

C.J. Schrijver, M.L. De Rosa, T.R. Metcalf, Y. Liu, J. McTiernan, S. Régnier, G. Valori, M.S. Wheatland, T. Wiegelmann, Nonlinear force-free modeling of coronal magnetic fields part I: a quantitative comparison of methods. Sol. Phys. **235**, 161–190 (2006). doi:10.1007/s11207-006-0068-7

M. Schulz, Direct influence of ring current on auroral oval diameter. J. Geophys. Res. **102**, 14149–14154 (1997). doi:10.1029/97JA00827

G.L. Siscoe, G.M. Erickson, B.U.Ö. Sonnerup, N.C. Maynard, K.D. Siebert, D.R. Weimer, W.W. White, Global role of $E_\parallel$ in magnetopause reconnection: an explicit demonstration. J. Geophys. Res. **106**, 13015–13022 (2001). doi:10.1029/2000JA000062

O. Skjæraasen, A. Melatos, A. Spitkovsky, Particle-in-cell simulations of a nonlinear transverse electromagnetic wave in a pulsar wind termination shock. Astrophys. J. **634**, 542–546 (2005). doi:10.1086/496873

D.F. Smith, E.R. Priest, Current limitation in solar flares. Astrophys. J. **176**, 487 (1972). doi:10.1086/151651

B.V. Somov, Non-neutral current sheets and solar flare energetics. Astron. Astrophys. **163**, 210–218 (1986)

K. Stasiewicz, P. Bellan, C. Chaston, C. Kletzing, R. Lysak, J. Maggs, O. Pokhotelov, C. Seyler, P. Shukla, L. Stenflo, A. Streltsov, J.-E. Wahlund, Small scale Alfvénic structure in the aurora. Astrophys. J. Suppl. Ser. **92**, 423–533 (2000)

Y. Su, A.M. Veronig, G.D. Holman, B.R. Dennis, T. Wang, M. Temmer, W. Gan, Imaging coronal magnetic-field reconnection in a solar flare. Nat. Phys. **9**, 489–493 (2013). doi:10.1038/nphys2675

J.J. Sudol, J.W. Harvey, Longitudinal magnetic field changes accompanying solar flares. Astrophys. J. **635**, 647–658 (2005). doi:10.1086/497361

X. Sun, J.T. Hoeksema, Y. Liu, T. Wiegelmann, K. Hayashi, Q. Chen, J. Thalmann, Evolution of magnetic field and energy in a major eruptive active region based on SDO/HMI observation. Astrophys. J. **748**, 77 (2012). doi:10.1088/0004-637X/748/2/77

M. Temerin, C.W. Carlson, Current-voltage relationship in the downward auroral current region. Geophys. Res. Lett. **25**, 2365–2368 (1998). doi:10.1029/98GL01865

A.N. Timokhin, The differentially rotating force-free magnetosphere of an aligned rotator: analytical solutions in the split-monopole approximation. Mon. Not. R. Astron. Soc. **379**, 605–618 (2007). doi:10.1111/j.1365-2966.2007.11864.x

A.N. Timokhin, J. Arons, Current flow and pair creation at low altitude in rotation-powered pulsars' force-free magnetospheres: space charge limited flow. Mon. Not. R. Astron. Soc. **429**, 20–54 (2013). doi:10.1093/mnras/sts298

V.S. Titov, P. Démoulin, Basic topology of twisted magnetic configurations in solar flares. Astron. Astrophys. **351**, 707–720 (1999)

S.M. Tobias, F. Cattaneo, Shear-driven dynamo waves at high magnetic Reynolds number. Nature **497**, 463–465 (2013). doi:10.1038/nature12177

N.A. Tsyganenko, D.P. Stern, Z. Kaymaz, Birkeland currents in the plasma sheet. J. Geophys. Res. **98**, 19455 (1993). doi:10.1029/93JA01922

G.H.J. van den Oord, The electrodynamics of beam/return current systems in the solar corona. Astron. Astrophys. **234**, 496–518 (1990)

L. van Driel-Gesztelyi, J.L. Culhane, D. Baker, P. Démoulin, C.H. Mandrini, M.L. DeRosa, A.P. Rouillard, A. Opitz, G. Stenborg, A. Vourlidas, D.H. Brooks, Magnetic topology of active regions and coronal holes: implications for coronal outflows and the solar wind. Sol. Phys. **281**, 237–262 (2012). doi:10.1007/s11207-012-0076-8

W. van Tend, M. Kuperus, The development of coronal electric current systems in active regions and their relation to filaments and flares. Sol. Phys. **59**, 115–127 (1978). doi:10.1007/BF00154935

E.J. Weber, L. Davis Jr., The angular momentum of the solar wind. Astrophys. J. **148**, 217–227 (1967). doi:10.1086/149138

J.M. Weygand, R.L. McPherron, H.U. Frey, O. Amm, K. Kauristie, A. Viljanen, A. Koistinen, Relation of substorm onset to Harang discontinuity. J. Geophys. Res. **113**, 4213 (2008). doi:10.1029/2007JA012537

M.S. Wheatland, Are electric currents in solar active regions neutralized? Astrophys. J. **532**, 616–621 (2000). doi:10.1086/308577

R.M. Winglee, G.A. Dulk, P.L. Bornmann, J.C. Brown, Interrelation of soft and hard X-ray emissions during solar flares. II—simulation model. Astrophys. J. **375**, 382–403 (1991). doi:10.1086/170197

V.V. Zharkova, K. Arzner, A.O. Benz, P. Browning, C. Dauphin, A.G. Emslie, L. Fletcher, E.P. Kontar, G. Mann, M. Onofri, V. Petrosian, R. Turkmani, N. Vilmer, L. Vlahos, Recent advances in understanding particle acceleration processes in solar flares. Space Sci. Rev. **159**, 357–420 (2011). doi:10.1007/s11214-011-9803-y

DOI 10.1007/978-1-4939-3547-5_3
Reprinted from *Space Science Reviews* Journal, DOI 10.1007/s11214-014-0038-6

Magnetic Helicity and Large Scale Magnetic Fields: A Primer

Eric G. Blackman

Received: 13 January 2014 / Accepted: 4 February 2014 / Published online: 16 April 2014
© Springer Science+Business Media Dordrecht 2014

Abstract Magnetic fields of laboratory, planetary, stellar, and galactic plasmas commonly exhibit significant order on large temporal or spatial scales compared to the otherwise random motions within the hosting system. Such ordered fields can be measured in the case of planets, stars, and galaxies, or inferred indirectly by the action of their dynamical influence, such as jets. Whether large scale fields are amplified in situ or a remnant from previous stages of an object's history is often debated for objects without a definitive magnetic activity cycle. Magnetic helicity, a measure of twist and linkage of magnetic field lines, is a unifying tool for understanding large scale field evolution for both mechanisms of origin. Its importance stems from its two basic properties: (1) magnetic helicity is typically better conserved than magnetic energy; and (2) the magnetic energy associated with a fixed amount of magnetic helicity is minimized when the system relaxes this helical structure to the largest scale available. Here I discuss how magnetic helicity has come to help us understand the saturation of and sustenance of large scale dynamos, the need for either local or global helicity fluxes to avoid dynamo quenching, and the associated observational consequences. I also discuss how magnetic helicity acts as a hindrance to turbulent diffusion of large scale fields, and thus a helper for fossil remnant large scale field origin models in some contexts. I briefly discuss the connection between large scale fields and accretion disk theory as well. The goal here is to provide a conceptual primer to help the reader efficiently penetrate the literature.

Keywords Magnetic fields · Galaxies: Jets · Stars: Magnetic field · Dynamo · Accretion · Accretion disks · Cosmology: Miscellaneous

1 Introduction

Planets, stars, galaxies are all examples of astrophysical rotators that reveal direct or indirect evidence for large scale ordered magnetic fields (Schrijver and Zwaan 2000; Brandenburg and Subramanian 2005; Shukurov 2005; Beck 2012; Roberts and King 2013). Here "large"

E.G. Blackman (✉)
Department of Physics and Astronomy, University of Rochester, Rochester, NY 14618, USA
e-mail: blackman@pas.rochester.edu

Influence of Rotation & Large Scale B-Fields:

	~25-30 YEARS AGO	TODAY
planetary nebulae (and pPNe)	-quasi-spherical/elliptical -fate of all low mass stars; -interacting fast+slow outflows with inertial channeling	-aspherical/multi-polar -fate of <50% (?) of low mass stars,often very collimated
young stellar objects	bipolar jets ?	bipolar jets !
active galactic nuclei	many produce jets	many produce jets
gamma-ray bursts	spherical explosions? (distance unknown)	long: jets from massive stars? short: jets from NS-NS mergers?
supernovae (?)	spherical explosions	aspherical

Fig. 1 Chart highlighting an emerging realization in astrophysics that more sources originally perceived as spherical may in fact be bipolar, and thus harborers of large scale magnetic field mediated jets

scale implies a coherent flux on scales comparable to the size of the hosting rotator and most importantly, larger than the scale of fluctuations associated with chaotic turbulent flows. In fact, all of these systems show evidence for both small and large scale magnetic fields, so the fact that large scale order persists even amidst a high degree of smaller scale disorder is a core challenge of explaining the emergence of large scale magnetic structures across such disparate classes of rotators.

Large scale magnetic fields are also likely fundamental to coronae and jets from accretion engines around young and dying stars and compact objects (e.g. Blandford and Payne 1982; Konigl 1989; Field and Rogers 1993; Blackman et al. 2001; Blackman and Pessah 2009; Lynden-Bell 2006; Pudritz et al. 2012; Penna et al. 2013). Accreting systems typically exhibit a continuum spectrum and luminosity best explained by matter accreting onto a central object falling deeper into a potential well and thereby releasing positive kinetic energy in the form of radiation or jet outflows. In fact, as Fig. 1 shows, the classes of objects likely harboring jets has increased as observations have improved, and jets likely indicate the role of large scale magnetic fields.

In young stars, pre-planetary nebulae, micro quasars, and active galactic nuclei the jets typically have too much collimated momenta to be driven by mechanisms that do not involve large scale magnetic fields (Bujarrabal et al. 2001; Pudritz et al. 2012). From the jets of AGN, Faraday rotation from ordered helical magnetic fields is directly observed (Gabuzda et al. 2008, 2012; Asada et al. 2008). Because jets are anchored in the accretion engines, they play a role in extracting the angular momentum that allows remaining disc material to accrete. The ionization fractions of accretion disks are commonly high enough, at least in regions near the very center, to be unstable to the magneto-rotational instability (MRI) (Balbus and Hawley 1991, 1998; Balbus 2003). A plethora of numerical simulations now commonly reveal that the systems evolve to a nonlinear turbulent steady state whose Maxwell stress dominates the Reynolds stress and for which large scale ordered magnetic fields emerge with cycle periods ~ 10 orbit periods (e.g. Brandenburg et al. 1995; Davis et al. 2010; Simon et al. 2011; Guan and Gammie 2011; Sorathia et al. 2012; Suzuki and Inutsuka 2013). With the caveat that angular momentum plays a comparatively subdominant role in structural support for stars, the coronae of stars and the emergence of large scale ordered fields that thread coronal holes of the sun (e.g. Schrijver and Zwaan 2000) can help shed light on related phenomena of the rising and opening up of large scale fields that form jets and coronae from accretion disks.

There are three possibilities for the origin of large scale fields in astrophysical rotators: The first is that the contemporary field is simply the result of advection and compression of the field that was present in the object before it was formed (e.g. Braithwaite and Spruit 2004; Kulsrud et al. 1997; Lovelace et al. 2009; Subramanian 2010; Widrow et al. 2012). The second is that the field is in fact dynamo produced in situ, extracting free energy from shear, rotation and turbulence in such a way as to sustain the field against turbulent diffusion (e.g. Moffatt 1978; Parker 1979; Ruzmaikin et al. 1988; Shukurov 2005; Charbonneau 2013). The third possibility is some combination of the two (e.g. Kulsrud and Zweibel 2008). In systems like the Sun and Earth where cycle periods involving field reversals are observed, the need for in situ dynamo amplification is unambiguous. For galaxies or accretion disks, the evidence for in situ amplification of large scale fields is more indirect. Regardless of whether the fields are initially frozen in and advected, amplified in situ, or a combination of the two, the evolution of magnetic helicity is very helpful for understanding the physics of magnetic field origin as we shall see.

Much of what we can observationally infer about magnetic fields of planets, stars, and accretion disks comes from information external to where the real action of magnetic field amplification and conversion of kinetic to magnetic energy occurs. Yet, most theory and simulation of astrophysical dynamos focuses on the interiors. This situation contrasts that of our own galaxy where we observe the field from within (e.g. Van Eck et al. 2011; Beck 2012), albeit on time scales too short to observe its dynamical evolution. A related point is that the hidden interiors of astrophysical rotators are typically flow dominated, with the magnetic field energy density generally weaker than that of the kinetic energy. However in the surrounding coronae of stars, accretion disks (and maybe even for galaxies) the field dominates the kinetic energy. Thus we must learn about the flow-dominated interiors from observations of the magnetically dominated exteriors and understand the coupling between the two. Laboratory plasmas of fusion devices are in fact magnetically dominated and there are many lessons learned from this context. Tracking magnetic helicity evolution has proven helpful for understanding the field evolution in both magnetically dominated and flow dominated circumstances.

In this paper I discuss basic principles of magnetic helicity evolution to guide the physical intuition of how large scale magnetic fields arise and evolve. This overview represents one path through the subject and is intended as a conceptual primer to ease immersion in the literature rather than a complete detailed review. Many relevant papers will therefore regrettably go uncited—a situation that I find increasingly difficult to avoid. A example of an earlier more detailed review is Brandenburg and Subramanian (2005).

In Sect. 2, I discuss the key physical properties of magnetic helicity that are central to all subsequent topics discussed. In Sect. 3, I summarize the conceptual progress of how these principles apply to modern developments in magnetic dynamo theory. In Sect. 4, I describe the simple two-scale model for closed systems that has come to be useful for gaining unified insight of large scale field evolution and dynamo saturation more quantitatively. I discuss several applications of the two-scale theory: dynamo saturation, the resilience of helical fields to turbulent diffusion, and dynamical relaxation. In Sect. 5, I discuss the role of helicity fluxes and their relation to the results Sect. 4. In Sect. 6, I briefly discuss the issue of gauge non-invariance of magnetic helicity. I conclude in Sect. 7, emphasizing the expectation that both signs of magnetic helicity on different scales should appear in astrophysical rotators

Fig. 2 Magnetic helicity of two linked flux equals twice the product of their magnetic fluxes

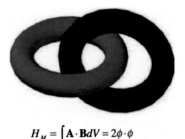

$$H_M = \int \mathbf{A} \cdot \mathbf{B} dV = 2\phi \cdot \phi$$

with coronal cycle periods, and comment on the connections between magnetic helicity dynamics and accretion disk theory.

2 Key Properties of Magnetic Helicity

Here I introduce some basic properties of magnetic helicity that underlie its role in large scale field generation and are essential for the sections that follow.

2.1 Magnetic Helicity as a Measure of Magnetic Flux Linkage, Twist, or Writhe

Magnetic helicity is defined as volume integral of the dot product of vector potential $\mathbf{A}$ and magnetic field $\mathbf{B} = \nabla \times \mathbf{A}$, namely

$$\int \mathbf{A} \cdot \mathbf{B} dV. \tag{1}$$

To see why this is a measure of magnetic linkage (e.g. Moffatt 1978; Berger and Field 1984), consider two thin linked magnetic flux tubes as shown in Fig. 2, with cross sectional area vectors $d\mathbf{S}_1$ and $d\mathbf{S}_2$ respectively. Let $\Phi_1 = \int \mathbf{B}_1 \cdot d\mathbf{S}_1$ be the magnetic flux in tube 1, where $\mathbf{B}_1$ is the magnetic field. Similarly, for flux tube 2 we have $\Phi_2 = \int \mathbf{B}_2 \cdot d\mathbf{S}_2$ where $\mathbf{B}_2$ is the magnetic field. For both tubes, we assume that the fields are of constant magnitude and parallel to $d\mathbf{S}_1$ and $d\mathbf{S}_2$ respectively. The volume integral of Eq. (1) contributes only where there is magnetic flux. Thus we can split the magnetic helicity into contributions from the two flux tube volumes to obtain

$$\int \mathbf{A} \cdot \mathbf{B} dV = \iint \mathbf{A}_1 \cdot \mathbf{B}_1 dl_1 dS_1 + \iint \mathbf{A}_2 \cdot \mathbf{B}_2 dl_2 dS_2, \tag{2}$$

where we have factored the volume integrals into products of line and surface integrals, with the line integrals taken along the direction parallel to $\mathbf{B}_1$ and $\mathbf{B}_2$. Since the magnitudes of $\mathbf{B}_1$ and $\mathbf{B}_2$ are constant in the tube, we can pull $\mathbf{B}_1$ and $\mathbf{B}_2$ out of each of the two line integrals on the right of Eq. (2) to write

$$\int \mathbf{A} \cdot \mathbf{B} dV = \int A_1 dl_1 \int B_1 dS_1 + \int A_2 dl_2 \int B_2 dS_2 = \Phi_2 \Phi_1 + \Phi_1 \Phi_2 = 2\Phi_1 \Phi_2, \tag{3}$$

where we have used Gauss' theorem to replace $\int A_1 dl_1 = \Phi_2$, the magnetic flux of tube 2 that is linked through tube 1. and similarly $\int A_2 dl_2 = \Phi_1$. Note that if the tubes are not linked, then the line integral would vanish and there would be no magnetic helicity.

Fig. 3 Picture of a "twisted horseshoe roll" element of a roller coaster at Cedar Point in Sandusky, Ohio, USA. The large scale loops are "writhe" as defined in the text and the rotation of the tracks along the direction of the coaster path is "twist". The *loop on the left* has left-handed writhe and right handed twist. The *loop on the right* has right-handed writhe and left-handed twist. The overall trip on the coaster is a closed path, the rest of which is not shown

Fig. 4 The sketch illustrates two concepts. The first is magnetic relaxation: The small scale twist of the ribbon on the left has relaxed to the largest scale available to it in the *right figure*. The right configuration is a lower energy state for the same magnetic helicity as that in the left. The second concept is that these structures can also be viewed as two strands, so that one can qualitatively appreciate some equivalence between twist of a ribbon and linkage of a pair of strands. In the text, this is made more quantitative for a different example shown in Fig. 5

Having established that the magnetic helicity measures linkage, we are poised to understand how helicity can also be equivalently characterized as a measure of magnetic twist and writhe. Examples of local twist and writhe are seen in Fig. 3, which is a roller coaster element at Cedar Point in Sandusky Ohio (USA). The large loops each correspond to writhe, and twist is measured along the track. (Note that "writhe" here is what Bellan (2000) calls "overlap" and "twist" here is what Bellan (2000) calls "writhe".) A twisted ribbon is also shown in Fig. 4, where the amount of twist is conserved between the two panels but transferred from small to large scales.

Quantitatively relating linkage to twist is nicely accomplished experimentally with strips of paper, a scissors and some tape (e.g. Bellan 2000). This is illustrated in Fig. 5. The two panels show configuations with equivalent amounts of helicity as I now describe. Start with a straight strip of paper (say 20 cm long and 2 cm wide) and give the strip a full right handed twist around its long axis (by holding the bottom with your left and twisting at the top with your right hand). Now fasten the ends so that it is a twisted closed loop. This provides a model for a flattened, twisted closed magnetic flux tube. The result is Fig. 5a. Now consider that this tube could in fact have been composed of two adjacent flattened tubes pressed together side by side. Separating these adjacent tubes is accomplished by use of a scissors. Cut along the center line of the strip all the way around and the result is two linked ribbons, each of 1/2 the width of the original, and each with one right handed

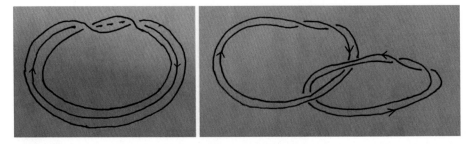

Fig. 5 The two panels show schematics of magnetic ribbon systems with equivalent amounts of magnetic helicity. The *left panel* shows a single twisted ribbon. If this were represented by a strip of paper, cutting along the centerline then results in the *right panel*. The single thick ribbon with one right handed twist has been converted to two linked ribbons of 1/2 half the width of the original, and each with a right handed twist. This figure illustrates why helicity can be represented as twist, linkage, or some combination of the two

twist. The result is shown in Fig. 5b. The use of scissors in this way has not changed the overall helicity of the system, but has transformed it into different forms as follows: If the original unseparated ribbon had a magnetic flux Φ then each of the 1/2 width ribbons have flux $\Phi/2$. From the above discussion of linkage, the linkage of these two new ribbons gives a contribution of helicity $2\Phi^2/4 = \Phi^2/2$. But the original uncut ribbon had a single right handed twist with total flux Φ. For the helicity of the initial twisted ribbon to equal that of the two linked twisted ribbons, any twisted ribbon must contribute a helicity equal to its flux squared. Thus helicity is thus conserved as follows: Φ^2 is the helicity associated with the initial twisted ribbon, and this is then equal to the sum of helicity from the linkage of the two half-thickness flux tubes $\Phi^2/2$, plus that from the right handed twists in each of these two half- thickness ribbons $2 \times (\Phi/2)^2 = \Phi^2/2$. This conveys how both linkage and twist are manifestations of magnetic helicity.

To see the relation between twist and writhe, consider again a straight paper ribbon. As above, give the ribbon one right handed twist around its long axis holding the other end fixed as in Fig. 6a. Imagine that the ends A and B are identified as the same location so it is really a closed twisted loop. Now push the ends A and B inward toward each other and the ribbon will buckle, as seen in Fig. 6b. A side view of this buckling is shown in Fig. 6c. If the ends A and B are identified, then the result is now a loop with a single unit of writhe—a large scale loop through which you can thread a rigid pole—that was derived from a single unit of twist. Thus one unit of twist helicity is equivalent to one unit of writhe helicity.

In short, one unit of twist helicity for a tube or ribbon of magnetic flux Φ is equal to one unit of writhe helicity for the same flux tube, and both are separately equal to 1/2 of the helicity resulting from the linkage of two untwisted flux tubes of flux Φ. These three different, but equivalent ways of thinking about magnetic helicity are instructive for extracting the physical ideas in what follows.

2.2 Evolution of Magnetic Helicity

The time evolution equation for magnetic is simply derived in magnetohydrodynamics: The electric field is given by

$$\mathbf{E} = -\nabla\Phi - \frac{1}{c}\partial_t\mathbf{A}, \qquad (4)$$

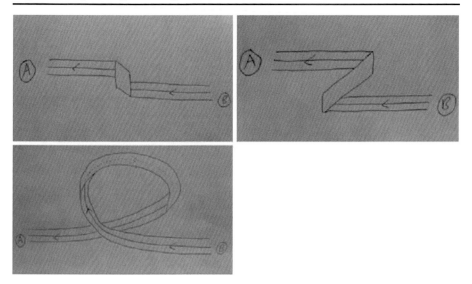

Fig. 6 The *three panels* show how to transform twist helicity into writhe helicity. (*Top left*) Start with a straight flat ribbon with one unit of right handed twist. One could imagine that the ends A and B are identified with each other and so it is a mathematically a closed ribbon loop with a twist. (*Top right*) Push the ends A and B toward each other until the ribbon buckles upward. (*Bottom*) A side view shows the writhe. The result is a writhed, closed loop

where Φ and $\mathbf{A}$ are the scalar and vector potentials. Using $\mathbf{B} \cdot \partial_t \mathbf{A} = \partial_t (\mathbf{A} \cdot \mathbf{B}) + c\mathbf{E} \cdot \mathbf{B} - c\nabla \cdot (\mathbf{A} \times \mathbf{E})$, where the latter two terms result from Maxwell's equation

$$\partial_t \mathbf{B} = -c\nabla \times \mathbf{E} \tag{5}$$

and the identity $\mathbf{A} \cdot \nabla \times \mathbf{E} = \mathbf{E} \cdot \mathbf{B} - \nabla \cdot (\mathbf{A} \times \mathbf{E})$, we take the dot product of Eq. (4) with $\mathbf{B}$ to obtain the evolution of the magnetic helicity density

$$\partial_t (\mathbf{A} \cdot \mathbf{B}) = -2c\mathbf{E} \cdot \mathbf{B} - \nabla \cdot (c\Phi\mathbf{B} + c\mathbf{E} \times \mathbf{A}). \tag{6}$$

If we average this equation over a simply connected volume that has no boundary terms, the last terms would not contribute and we obtain

$$\partial_t \langle \mathbf{A} \cdot \mathbf{B} \rangle = -2c \langle \mathbf{E} \cdot \mathbf{B} \rangle. \tag{7}$$

In MHD Ohm's law is

$$\mathbf{E} = -\mathbf{v} \times \mathbf{B}/c + \eta\mathbf{J}, \tag{8}$$

where $\mathbf{V}$ is the flow velocity and $\eta = 4\pi \nu_M / c^2$ is the resistivity for a magnetic diffusivity ν_M. For ideal MHD ($\eta = 0$) the right hand side of (7) is zero, highlighting the conservation of magnetic helicity density (and thus magnetic helicity) for a closed volume under ideal conditions and independent of the presence of velocity flows.

For comparison, the magnetic energy density evolution, obtained by dotting Eq. (5) with $\mathbf{B}$ and using Eq. (8) is given by

$$\frac{1}{8\pi} \partial_t \langle \mathbf{B}^2 \rangle = -\eta \langle \mathbf{J}^2 \rangle - \frac{1}{c} \langle \mathbf{v} \cdot (\mathbf{J} \times \mathbf{B}) \rangle, \tag{9}$$

where $\mathbf{J} \equiv \frac{c}{4\pi} \nabla \times \mathbf{B}$.

2.3 Magnetic Helicity is Typically Better Conserved than Magnetic Energy

In the absence of dissipation, magnetic helicity is exactly conserved in MHD for a closed system but magnetic energy can be exchanged with a velocity field. If we ignore the latter, when the ratio of time scale for resistive decay of magnetic energy to that of magnetic helicity is small, magnetic helicity is more strongly conserved than magnetic energy. To demonstrate when this is true, we must consider that a typical astrophysical system is often turbulent, so there is not just one scale of the field but a spectrum. The question becomes for what spectra of magnetic energy and magnetic helicity does the latter decay more slowly than the former? Following Blackman (2004) and working in the Coulomb gauge, we write the (statistically or volume) averaged magnetic energy density as

$$\langle \mathbf{B}^2 \rangle / 8\pi = M = \int_{k_0}^{k_{\nu_M}} M_k dk, \qquad (10)$$

where the brackets indicate an average, k_0 and k_{ν_M} are the minimum and maximum (resistive) wave numbers, and the one-dimensional magnetic energy density spectrum is given by

$$M_k \equiv \frac{1}{8\pi} \int |\tilde{\mathbf{B}}|^2 k^2 d\Omega_k = \int |\tilde{\mathbf{A}}|^2 k^4 d\Omega_k \propto k^{-q}. \qquad (11)$$

Here Ω_k is the solid angle in wave-number space, q is an assumed constant, and the tilde indicate Fourier transforms. The magnetic helicity density spectrum is then

$$H_k \equiv \frac{1}{16\pi} \int \left[\tilde{\mathbf{A}}(k) \tilde{\mathbf{B}}^*(k) + \tilde{\mathbf{A}}^*(k) \tilde{\mathbf{B}}(k) \right] k^2 d\Omega_k$$
$$= M_k f(k)/k, \qquad (12)$$

where $*$ indicates complex conjugate and $f(k) \propto k^{-s}$ is used to define the fraction of magnetic energy that is helical at each wave number and s is taken as a constant. We then also have correspondingly

$$\langle \mathbf{A} \cdot \mathbf{B} \rangle = \int_{k_0}^{k_{\nu_M}} f(k) M_k k^{-1} dk, \qquad (13)$$

and the current helicity density

$$\langle \mathbf{J} \cdot \mathbf{B} \rangle = \int_{k_0}^{k_{\nu_M}} f(k) k M_k dk. \qquad (14)$$

Now, using Eq. (7) for a closed system we have

$$8\pi \partial_t H = \partial_t \langle \mathbf{A} \cdot \mathbf{B} \rangle = -2\nu_M \langle \mathbf{J} \cdot \mathbf{B} \rangle, \qquad (15)$$

and Eq. (9) gives in the absence of velocity flows,

$$8\pi \partial_t M = \partial_t \langle \mathbf{B}^2 \rangle = -2\nu_M \langle (\nabla \mathbf{B})^2 \rangle. \qquad (16)$$

Then combining (13)–(16), we then obtain

$$\tau_H = \frac{-H}{\partial_t H} = \frac{\int_{k_L}^{k_{\nu_M}} f(k) M_k k^{-1} dk}{2\nu_M \int_{k_L}^{k_{\nu_M}} f(k) k M_k dk} \qquad (17)$$

and

$$\tau_M = \frac{-M}{(\partial_t M)_{res}} = \frac{\int_{k_L}^{k_{\nu M}} M_k dk}{2\nu_M \int_{k_L}^{k_{\nu M}} k^2 M_k dk}, \tag{18}$$

for the time scales of magnetic helicity and magnetic energy decay respectively. The subscript "*res*" indicates the contribution from the penultimate term in (16) only. The range of s and q for which $R \equiv \frac{\tau_H}{\tau_M} > 1$ corresponds to regime in which the magnetic helicity decays more slowly than the magnetic energy. Blackman (2004) showed that $R > 1$ for the combination of $s > 0$ and $3 > q > 0$ and that $R < 1$ for small $0 < q < 1$ and $s < 0$. The latter range would correspond to a very unusual circumstance in which all the magnetic helicity were piled up at small scales. Most commonly therefore, the range for which $R > 1$ applies and magnetic helicity is typically better conserved than magnetic energy.

2.4 Minimum Energy State of Helical Magnetic Fields

The conclusion of the previous section—that magnetic helicity usually decays more slowly than magnetic energy—justifies *a posteriori* the relevance of the question that Woltjer (1958a) considered: If the magnetic helicity is conserved for a magnetically dominated system (ignoring velocities), what magnetic field configuration minimizes the energy? Using a variational principle calculation, Woltjer (1958a) found that the answer is a configuration for which $\mathbf{J} = \frac{c}{4\pi}\nabla\times\mathbf{B} = f(\mathbf{x})\mathbf{B}$, where $f(\mathbf{x})$ is a scalar function that must therefore satisfy $\mathbf{B} \cdot \nabla f = 0$. Taylor (1974, 1986) considered the same question but assumed that magnetic helicity is approximately conserved when averaged over sufficiently large scales even in the presence of a small but finite resistive dissipation. Essentially, f becomes a measure of the inverse gradient scale of the helical magnetic field and so minimizing the energy means decreasing f as much as possible. The small amount of dissipation aids this relaxation via small scale reconnection as needed, such that the overall relaxed state of the field is one in which the small scale gradients smooth out to allow the gradient scales to reach the largest possible subject to the boundary conditions. This, in turn, uniquely determines f as a function of the specific boundary conditions. (Figure 4 shows a simple helicity conserving relaxation process where reconnection is not actually needed.) The arguments of Woltjer (1958a) and Taylor (1974) essentially assumed that $R > 1$. The previous section shows the specific spectral conditions for this to be viable, and solidifies the assumptions on which these results were based.

That Woltjer (1958a) arrived at a force-free, minimum energy state under the conditions imposed—no velocity flows, no dissipation, and a closed volume—is evident even without a variational calculation. Imagine a system with initially no pressure gradients and no velocity with a field configuration that is not force free, i.e., $\mathbf{J} \times \mathbf{B} \neq 0$. A velocity flow will swiftly develop, violating the assumption that there is no velocity. The only way that the velocity (and thus kinetic energy) could remain zero is if the field produces no acceleration, i.e. is force free. The assumption of no velocity is therefore enough to conclude that the field must be force-free in a steady state. Moreover, in a closed system, a force free field is fully helical, namely $\langle\mathbf{J} \cdot \mathbf{B}\rangle = -\langle\mathbf{B} \cdot \nabla^2\mathbf{A}\rangle$, so that the 1-D Fourier spectrum of current helicity $H_c(k)$ would be $k^2 H_M(k)$ where $H_M(k)$ is the magnetic helicity spectrum. For a fully helical system, $k H_M(k) = M(k) = H_c(k)/k$. Suppose the magnetic energy is predominately at a single wavenumber k_E, and we ask whether k_E must increase or decrease to minimize the magnetic energy: If magnetic helicity is conserved then $k_E H_M(k_E) = M(k_E)$ would remain constant as k_E changes. The magnetic energy $k_E M(k_E)$ thus decreases with decreasing k_E.

As a result, minimizing the magnetic energy for a fixed magnetic helicity would lead to as small of a k as possible. This is the essence of the "Taylor relaxed" state.

Note also that if we drop the assumption that there is no velocity flow, then the momentum equation for incompressible ($\nabla \cdot \mathbf{v} = 0$, $\rho = $ constant) flow in the absence of microphysical dissipation is

$$\frac{\partial \mathbf{v}}{\partial t} = \rho \mathbf{v} \times (\nabla \times \mathbf{v}) - \nabla\left(\mathbf{v}^2/2 + \frac{P}{\rho}\right) + \frac{\mathbf{J} \times \mathbf{B}}{\rho}. \tag{19}$$

Dotting with $\mathbf{v}$ and averaging over a closed volume gives

$$\langle \partial \mathbf{v}^2/\partial t \rangle = \langle \mathbf{v} \cdot (\mathbf{J} \times \mathbf{B}) \rangle. \tag{20}$$

In the absence of dissipation, the right hand side of Eq. (20) is also the only contributing term to the evolution of magnetic energy in Eq. (9). Thus $\langle \mathbf{v} \cdot (\mathbf{J} \times \mathbf{B}) \rangle = 0$ is the generalization to the force free state when velocities are allowed, for it is the only way for both the magnetic and kinetic energies to remain steady. Woltjer (1958b) extended Woltjer (1958a) by deriving integrals of the motion for more general hydromagnetic flows and Field (1986) focused on a static extension of Woltjer (1958a) to include pressure gradients. However, the derivation of Eq. (20) above provides a simple articulation of the generalization to the force free condition when velocity flows are allowed.

This subsection has focused on the steady state, not the dynamical relaxation to that state. In fact, magnetic relaxation is a time dependent. Even in the magnetically dominated regime of astrophysical coronae and laboratory fusion plasmas where magnetic relaxation is considered (e.g. Bellan 2000; Ji and Prager 2002); this relaxation is in fact a large scale dynamo (LSD), because large scale helical fields grow where none were present initially, and as a consequence of helical (magnetic) energy input on small scales. In the discussion of dynamos in the next sections, I will discuss how this relaxation and the more traditional flow driven large scale field growth are different flavors within a unified framework. Magnetic helicity evolution is fundamental to both.

3 Helicity and Large Scale Dynamo Saturation: Conceptual Progress

3.1 Types of Dynamos and Approaches to Study Them

Dynamos describe the growth or sustenance of magnetic fields against the otherwise competing exponential decay. They can be divided into two major classes:

Small Scale Dynamo (SSD): This corresponds to magnetic energy amplification by turbulent velocity flows for which the dominant magnetic energy growth occurs primarily at and below the velocity forcing scale (e.g. Kazanstev 1968; Schekochihin et al. 2002; Bhat and Subramanian 2013; Brandenburg and Lazarian 2013). There is a large body of work addressing the overall magnetic energy spectra of such small scale dynamos when the system is isotropically forced without kinetic helicity. We will not focus on small scale dynamos in what follows.

Large Scale Dynamo (LSD): For LSDs, the magnetic energy grows on spatial or temporal scales larger than the dominant input scale. The input energy can take different forms but in all cases, the fact that the field grows on scales large compared to those of the input energy always requires a large scale electromotive force (EMF) aligned with the large scale or mean magnetic field, such that $\overline{(\mathbf{v} \times \mathbf{b})} \cdot \overline{\mathbf{B}} \neq 0$, where the overbars indicate a spatial, temporal, or ensemble average and $\mathbf{v}$ and $\mathbf{b}$ are the velocity magnetic fields associated with turbulent fluctuations when the total velocity and magnetic field are written as sums of fluctuations plus mean values as $\mathbf{V} = \mathbf{v} + \overline{\mathbf{B}}$ and $\mathbf{B} = \mathbf{b} + \overline{\mathbf{B}}$ respectively. This implies a source of large scale magnetic helicity is involved in field amplification as we shall see more explicitly later.

Two sub-classes of LSDs can be distinguished, based on the nature of the dominant energy input:

1. *Flow Dominated*: For this subclass, the input energy is kinetic energy dominated and the EMF can be sustained, for example, by kinetic helicity, as in the classic Parker-type solar dynamo and its extensions (e.g. Moffatt 1978; Parker 1979; Krause and Rädler 1980; Pouquet et al. 1976; Blackman and Field 2002 and Blackman and Brandenburg 2002). Alternatively, there is emerging agreement that a combination of shear and fluctuating kinetic helicity can conspire to produce an LSD (Vishniac and Brandenburg 1997; Brandenburg 2005; Yousef et al. 2008; Heinemann et al. 2011; Mitra and Brandenburg 2012; Sridhar and Singh 2013). The EMF can also be sustained by magnetic helicity fluxes, either local (within sectors of a closed volume) or global (fluxes thought the system boundary) as we will later discuss.

2. *Magnetic Relaxation LSD*: This subclass of LSDs occurs when a system is initially magnetic energy dominated and the EMF is sustained by magnetically dominated quantities. Typically, injection of small scale magnetic helicity drives instabilities that facilitate the relaxation of the system and transfer of the magnetic helicity to large scales. It is a dynamo because field grows on large scales where there was little initially, and the field on large scales is sustained against decay. In plasma fusion devices where such MRD occur, helicity fluxes are the key sustainer of the EMF (e.g. Strauss 1985, 1986; Bhattacharjee and Hameiri 1986; Ortolani and Schnack 1993; Ji et al. 1995; Bellan 2000; Ji and Prager 2002). In astrophysics such MR-LSDs likely occur in coronae, where the field is injected from below on small scales relative to the corona and further relaxation occurs. Astrophysical rotators with coronae likely have flow driven dynamos in the interior, coupled to MRDs in coronae. Ironically, for the sun and stars, the observed field measured is from the base of the corona outward. Thus we can directly observe the MR-LSD processes better than the interior flow driven LSD, although the two are dynamically coupled.

In studying dynamos there are different approaches used depending on the goal. One approach is the "kitchen sink" approach, that is, try to perform quasi-realistic numerical experiments to match observational features with as realistic of conditions as possible given numerical limitations (e.g. Glatzmaier 2002). A second approach is to carry out semi-empirical model calculations based on linear theories but with dynamo transport coefficients empirically tuned to match general observational features and cycle periods (e.g. Dikpati and Gilman 2009; Charbonneau 2013), and without going after the physics of nonlinear quenching. A third approach, and the one most reviewed in this primer, is identifying basic principles of EMF sourcing and quenching. This involves pursuit of a "first principles" mean field theory and comparison to minimalist simulations to test the predictions of the theory. The goal is to develop a mean field theory that isolates the key physics from the nonlinear mess and eventually use the insights gained to inform more detailed models.

3.2 20th Century vs. 21st Century Dynamos: Physical Role of Magnetic Helicity Conservation

The 20th century textbook mean field dynamo theory (MFDT) of standard textbooks (Moffatt 1978; Parker 1979; Krause and Rädler 1980) is a practical approach to modeling LSDs

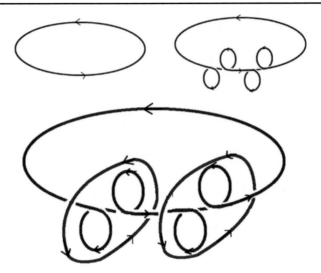

Fig. 7 Two-stage schematic for magnetic field structure in an α^2 dynamo driven by negative kinetic helicity in the conventional 20th century approach with the magnetic field represented as *lines*. The *top right panel* shows a large untwisted toroidal magnetic field ribbon. The *top right panel* shows the action of kinetic helicity on the initial ribbon. The small scale negative kinetic helicity produces each the four small scale poloidal loops. Each small loop incurs a writhe (or overlap) of positive (= right-handed) magnetic helicity. The two intermediate scale poloidal loops encircling the small scale loops in the *bottom panel* represent the resultant mean poloidal field averaged separately over each pair of loops. These intermediate scale loops are linked to the initial large scale torioidal loop. Since a single linked pair of ribbons has 2 units of magnetic helicity, we see that the two poloidal loops linking the torioidal loop have a total 4 total units of right handed magnetic helicity. This helicity in the "large scale field" has come from zero initial magnetic helicity and thus cannot be correct for MHD at large R_M which conserves magnetic helicity. The diagram does not account for the missing small scale magnetic helicity of opposite sign (compare to Fig. 8)

in turbulent rotators. These approaches focused mostly on initially globally reflection asymmetric rotators where the EMF is sustained by kinetic helicity. But for ~ 50 years, this theory lacked a saturation theory to predict how strong the large scale fields get before quenching via the back-reaction of the field on the driving flow.

In fact, the inability of the 20th century textbook theories to predict mean field dynamo saturation arises because they fail to conserve magnetic helicity. This is illustrated in a most minimalist way in the diagrammatic representations of Fig. 7 for the so called α^2 dynamo, a dynamo that just depends on small scale helical velocity motions (discussed more quantitatively later). The figure shows that from an initial toroidal loop, the four small scale helical eddies, each with the same sign of kinetic helicity $\mathbf{v} \cdot \nabla \times \mathbf{v}$ make four small poloidal loops as shown. The result is a net toroidal EMF $\overline{\mathcal{E}}_\phi = \langle v_z b_r \rangle$ which has the same sign inside the loops of all the eddies whether they moved up or down. This in turn leads to two net large scale poloidal loops (shown in blue). But recall that magnetic helicity is a measure of linkage. The first panel has only a single red loop, but seemingly evolves to a configuration with the red loop linked to the two blue loops. That is, the mean field has gone from no helicity, to two units of linkage helicity. This lack of helicity conservation highlights an unphysical feature of 20th century LSD theory.

The commonly used schematics and diagrams of more finely tuned and observationally relevant solar dynamo models, such as flux-transport models (Dikpati and Gilman 2009), also do not conserve magnetic helicity for essentially the same reason as in Fig. 7. There is typically a step where the large scale field gains helicity when a poloidal loop emerges

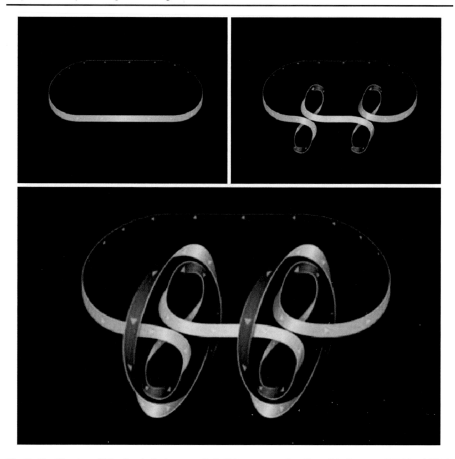

Fig. 8 Modification of Fig. 7 to include magnetic helicity conservation (from Blackman and Hubbard 2014) to include magnetic helicity conservation. The *top left panel* shows a large untwisted toroidal magnetic field ribbon. The *top right panel* shows the action of kinetic helicity on the initial ribbon. The small scale negative kinetic helicity produces each the four small scale poloidal loops. Each small loop incurs a writhe (or overlap) of positive (= right-handed) magnetic helicity. Since magnetic helicity is conserved, each of these four loops also has a negative (= left-handed) twist along the field ribbon. The two intermediate scale poloidal loops encircling the small scale loops in the *bottom panel* represent the resultant mean poloidal field averaged separately over each pair of loops. The intermediate scale poloidal loops have accumulated two units of magnetic twist, one from each of the small loops that they encircle. These intermediate scale loops are also linked to the initial large scale torioidal loop. Since a single linked pair of ribbons has 2 units of magnetic helicity we see that the two poloidal loops linking the torioidal loop have a total 4 total units of right handed magnetic helicity in linkage which exactly balances the sum of 2 + 2 left handed units of twist on these poloidial loops. In general, the small scale of the twists need not correspond to the same scale as the velocity driving the small scale writhes, though the calculations of Sect. 4 assume such for simplicity

from an initially toroidal field via the action of the Coriolis force. Flux transport dynamos are otherwise impressive in that they can be tuned to agree with large scale observations of the solar magnetic field (e.g. Wang and Sheeley 2003), but the fact that certain parameters must be tuned by hand (rather than derived from first principles) highlights that magnetic helicity dynamics have not yet been incorporated.

How do we reconcile LSDs with the conservation of magnetic helicity? The answer is shown in Fig. 8 (figure from Blackman and Hubbard 2014). There the same set of diagrams as in Fig. 7 are shown with the field lines represented by ribbons or flux bundles instead of lines. Now conservation of helicity is maintained: the writhe of the loops in the second panel is compensated for by the opposite sign of twist helicity along the loops. The linkage that results in the third panel for the large scale field is compensated for by the exact opposite amount of small scale twist helicity along the loops. This conservation of magnetic helicity of an untwisted loop subjected to writhe can be demonstrated by using an ordinary belt.

The diagrammatic solution to the missing helicity problem is also the key to understanding nonlinear quenching. In a system driven with kinetic helicity, the back reaction on the driving flow comes primarily from the buildup of the small scale twist. Field lines with ever increasing small scale twists become harder to bend. It is this small scale twist that produces a small scale current $\mathbf{j} \equiv \mathbf{J} - \overline{\mathbf{J}}$ (where $\overline{\mathbf{J}} = \nabla \times \overline{\mathbf{B}}$) and a Lorentz force when coupled to the mean field, $\mathbf{j} \times \overline{\mathbf{B}}$, that produces the back reaction against the small scale flow. The driving kinetic helicity acts as a pump, segregating the signs of the magnetic helicity between scales. but the presence of small scale twist eventually quenches which has no dynamo.

How catastrophic this quenching is, depends on the specific type of dynamo: For the α^2 dynamo (a dynamo without differential rotation), the quenching slows the growth to resistively limited rates after an initial fast growth phase, but the field does not decay (Field and Blackman 2002; Blackman and Field 2002). In contrast, for the $\alpha - \Omega$ dynamo (i.e. a dynamo for which the toroidal field growth is dominated by shear from differential rotation), the quenching is such that the field starts to grow and then decays rapidly (e.g. Shukurov et al. 2006; Sur et al. 2007). That helicity fluxes might alleviate quenching and sustain the EMF (e.g. Blackman and Field 2000a; Vishniac and Cho 2001; Brandenburg and Subramanian 2005; Shukurov et al. 2006; Ebrahimi and Bhattacharjee 2014) has emerged as the most plausible way around this decay. Because of the way our understanding of the quenching has developed, the role of magnetic helicity flux is often expressed as a solution to the quenching problem, however the helicity flux could have been the driver of the growth from the start (as in laboratory plasma dynamos, e.g. Strauss 1985, 1986; Bhattacharjee and Hameiri 1986) and thus the issue of quenching would not arise in the discussion in the first place. This is something to keep in mind in reading the literature across subfields.

The physical description described in the previous subsection is supported by detailed calculations. Mathematically coupling the dynamical evolution of magnetic helicity into the dynamo equations can largely explain the saturation seen in simulations. The connection between magnetic helicity and large scale dynamos was first evident in the spectral model of helical MHD turbulence of Pouquet et al. (1976). They demonstrated an inverse transfer growth of large scale magnetic helicity for which the driver is the difference between kinetic and current helicities. In Kleeorin and Ruzmaikin (1982) an equation that couples the small scale magnetic helicity to the mean electromotive force is present, but the time evolution was not studied. The spectral work of Pouquet et al. (1976) was reinterpreted (Field and Blackman 2002) and re-derived (Blackman and Field 2002) as a 'user-friendly' time-dependent mean field theory exemplified for the α^2 dynamo case. Indeed it was found that $\langle \mathbf{v} \cdot \nabla \times \mathbf{v} \rangle$ grows large scale mag. helicity of one sign and small scale mag. helicity of opposite sign. The latter quenches LSD growth and matches simulations of Brandenburg (2001) and subsequent papers. A similar framework shows that large scale helical fields are resilient to turbulent diffusion (Blackman and Subramanian 2013; Bhat et al. 2014) These developments will be discussed later in Sect. 4.

The study of magnetic helicity dynamics in systems unstable to the magneto rotational instability (MRI) is also emerging (e.g. Vishniac 2009; Gressel 2010; Käpylä and Korpi 2011; Ebrahimi and Bhattacharjee 2014): MRI simulations exhibit LSDs driven by an EMF sustained by something other than kinetic helicity, e.g. current helicity or helicity fluxes. Helicity fluxes may sustain EMF and alleviate premature quenching in realistic systems with global boundaries OR in sub-volume local sectors even if the system is closed (e.g. vertically periodic shearing box). This will be addressed in Sect. 5.

4 Two-Scale Approach to Helicity Dynamics Provides Physical Insight

We now summarize the derivation of the two-scale equations for magnetic helicity evolution from which the above physical discussion arises. We will then discuss two applications of these equations, one for the α^2 dynamo and the second for the resilience of helical fields to decay. A third application to dynamical relaxation is briefly mentioned as well.

4.1 Basic Equations

We follow standard procedures (Blackman and Field 2002; Brandenburg and Subramanian 2005; Blackman and Subramanian 2013) and break each variable into large scale quantities (indicated by an overbar) and fluctuating quantities (indicated by lower case). We indicate global averages by brackets. The overbar indicates a more local average than brackets, (and can also be an average over reduced dimensions) but still over a large enough scale such that both local and global averages of fluctuating quantities vanish.

The analogous procedure that led to Eq. (6) for the evolution of the total magnetic helicity, leads to separate expressions for the contributions to magnetic helicity from the large and small scale given by

$$\frac{1}{2}\partial_t \langle \mathbf{a}\cdot\mathbf{b}\rangle = -\langle \overline{\mathcal{E}}\cdot\overline{\mathbf{B}}\rangle - \nu_M \langle \overline{\mathbf{b}\cdot\nabla\times\mathbf{b}}\rangle - \frac{1}{2}\nabla\cdot(\overline{c\phi\mathbf{b}} + \overline{c\mathbf{e}\times\mathbf{a}}), \tag{21}$$

and

$$\frac{1}{2}\partial_t \langle \overline{\mathbf{A}}\cdot\overline{\mathbf{B}}\rangle = \langle \overline{\mathcal{E}}\cdot\overline{\mathbf{B}}\rangle - \nu_M \langle \overline{\mathbf{B}}\cdot\nabla\times\overline{\mathbf{B}}\rangle - \frac{1}{2}\nabla\cdot(c\overline{\Phi}\overline{\mathbf{B}} + c\overline{\mathbf{E}}\times\overline{\mathbf{A}}). \tag{22}$$

where $\overline{\mathcal{E}} \equiv \overline{\mathbf{v}\times\mathbf{b}} = -\overline{\mathbf{E}} + \eta\overline{\mathbf{J}} = -\mathbf{E} + \mathbf{e} + \eta(\mathbf{J} - \mathbf{j})$ is the turbulent electromotive force. The simplest expression for $\overline{\mathcal{E}}$ that connects 20th century dynamo theory to 21st century makes use of the 'tau' or 'minimal tau' closure approach for incompressible MHD (Blackman and Field 2002; Brandenburg and Subramanian 2005). This means replacing triple correlations by a damping term on the grounds that the EMF $\overline{\mathcal{E}}$ should decay in the absence of $\overline{\mathbf{B}}$. The result is

$$\partial_t\overline{\mathcal{E}} = \overline{\partial_t\mathbf{v}\times\mathbf{b}} + \overline{\mathbf{v}\times\partial_t\mathbf{b}} = \frac{\alpha}{\tau}\overline{\mathbf{B}} - \frac{\beta}{\tau}\nabla\times\overline{\mathbf{B}} - \overline{\mathcal{E}}/\tau, \tag{23}$$

where τ is a damping time and

$$\alpha \equiv \frac{\tau}{3}\left(\frac{\langle\mathbf{b}\cdot\nabla\times\mathbf{b}\rangle}{4\pi\rho} - \langle\mathbf{v}\cdot\nabla\times\mathbf{v}\rangle\right) \quad \text{and} \quad \beta \equiv \frac{\tau}{3}\langle v^2\rangle, \tag{24}$$

and we assume $\langle\mathbf{v}\cdot\nabla\times\mathbf{v}\rangle \simeq \overline{\mathbf{v}\cdot\nabla\times\mathbf{v}}$ and $\langle v^2\rangle \simeq \overline{v^2}$.

Keeping the time evolution of $\overline{\mathcal{E}}$ as a separate equation to couple into the theory and solve allows for oscillations and phase delays between extrema of field strength and extrema of $\overline{\mathcal{E}}$ (Blackman and Field 2002) and are of observational relevance (as they have been used to explain the phase shift between spiral arms and dominant large scale mean field magnetic polarization in the galaxy; Chamandy et al. 2013). But simulations of magnetic field evolution in the simplest forced isotropic helical turbulence reveal that a good match to the large scale magnetic field evolution can be achieved even when the left side of Eq. (23) is ignored and τ is taken as the eddy turnover time associated with the forcing scale. We adopt that approximation here and so Eq. (23) then gives

$$\overline{\mathcal{E}} = \alpha \overline{\mathbf{B}} - \beta \nabla \times \overline{\mathbf{B}}. \tag{25}$$

Equations (21) and (22) then become

$$\frac{1}{2}\partial_t \langle \overline{\mathbf{a} \cdot \mathbf{b}} \rangle = -\alpha \langle \overline{B}^2 \rangle + \beta \langle \overline{\mathbf{B}} \cdot \nabla \times \overline{\mathbf{B}} \rangle - \nu_M \langle \mathbf{b} \cdot \nabla \times \mathbf{b} \rangle \tag{26}$$

and

$$\frac{1}{2}\partial_t \langle \overline{\mathbf{A} \cdot \mathbf{B}} \rangle = \alpha \langle \overline{B}^2 \rangle - \beta \langle \overline{\mathbf{B}} \cdot \nabla \times \overline{\mathbf{B}} \rangle - \nu_M \langle \overline{\mathbf{B}} \cdot \nabla \times \overline{\mathbf{B}} \rangle. \tag{27}$$

The energy associated with the small scale magnetic field does not enter $\overline{\mathcal{E}}$ (Gruzinov and Diamond 1996) so it does not enter Eqs. (26) and (27). It arises as a higher order hyper diffusion correction (Subramanian 2003) which we ignore. However, the energy density in the large scale field $\propto \overline{B}^2$ *does* enter (26) and (27). In general we need a separate equation for the energy associated with the energy of the mean field. Fortunately, for the simplest α^2 dynamo discussed in Sect. 4.2 below, the non-helical large scale field does not grow even when the additional equation is added. And for the decay problem of Sect. 4.3, the non helical part of the magnetic energy decays very rapidly. As such, it is acceptable to assume that the large scale field is fully helical for present purposes.

The essential implications of the coupled Eqs. (26) and (27) for a closed or periodic system are revealed in standard approaches where the large scale overbarred mean magnetic quantities are now indicated with subscript "1", small scale quantities by subscript "2". The kinetic forcing scale is indicated by subscript f. In the usual two-scale model for the α^2 dynamo, the kinetic forcing wavenumber k_f is assumed to be same as that for the small scale magnetic fluctuations k_2. Relaxing this provides some versatility, but we take $k_f = k_2$ here for simplicity. We assume that the wave number k_1 associated with the spatial variation scale of large scale quantities satisfies $k_1 \ll k_2$, where k_2 is the wave number associated with small scale quantities. Applying these approximations to a closed or periodic system, we then use $\langle \overline{\mathbf{B}} \cdot \nabla \times \overline{\mathbf{B}} \rangle = k_1^2 \langle \overline{\mathbf{A}} \cdot \overline{\mathbf{B}} \rangle \equiv k_1^2 H_1$, $\langle \overline{B}^2 \rangle = k_1 |H_1|$, along with $\langle \mathbf{b} \cdot \nabla \times \mathbf{b} \rangle = k_2^2 \langle \mathbf{a} \cdot \mathbf{b} \rangle = k_2^2 H_2$. We will assume that $H_1 \geq 0$ for the example cases studied below.

Non-dimensionalizing by scaling lengths in units of k_2^{-1}, and time in units of τ, where we assume $\tau = (k_2 v_2)^{-1}$, we have

$$h_1 \equiv \frac{k_2 H_1}{4\pi \rho v_2 2}, \qquad h_2 \equiv \frac{k_2 H_2}{4\pi \rho v_2^2}, \qquad h_v \equiv \frac{H_V}{k_2 v_2^2}, \qquad R_M \equiv \frac{v_2}{\nu_M k_2}.$$

With these non-dimensional quantities, Eqs. (26) and (27) become

$$\partial_\tau h_1 = \frac{2}{3}(h_2 - h_v)\frac{k_1}{k_2}h_1 - \frac{2}{3}\left(\frac{k_1}{k_2}\right)^2 h_1 - \frac{2}{R_M}\left(\frac{k_1}{k_2}\right)^2 h_1, \tag{28}$$

$$\partial_\tau h_2 = -\frac{2}{3}(h_2 - h_v)\frac{k_1}{k_2}h_1 + \frac{2}{3}\left(\frac{k_1}{k_2}\right)^2 h_1 - \frac{2}{R_M}h_2. \tag{29}$$

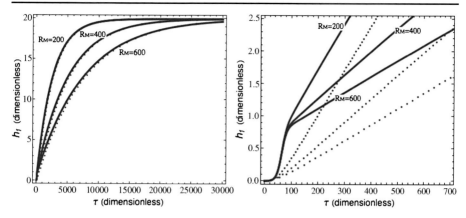

Fig. 9 Solutions of Eqs. (28) and (29) for the α^2 dynamo problem discussed in the text (figs. based on Field and Blackman 2002 and Blackman and Field 2002). The calculation maintains $h_v = -1$ and the initial conditions are $h_2(0) = 0$ and $h_1(0) = 0.001$ and for three different magnetic Reynolds numbers as shown (order of R_M in the *solid curves* are the same for the right curves as the left). Time (*x*-axis) is in units of eddy turnover times at the forcing scale $k = 5$. *Left panel* shows the solution over a long time period highlighting the analytical result that all curves eventually converge toward the same final value of h_1, but the higher R_M cases (those which are slower to dissipate the offending small scale magnetic helicity) take longer to get there. The *dotted curves* show the quasi-empirical fit formula used by Brandenburg (2001) to fit simulations at late times. The dynamical theory of Eqs. (28) and (29) do very well to match this empirical fit formula which in turn matched simulations. *Right panel* shows the solution only through $\tau = 700$. This highlights that at early times, before h_2 has grown significantly, the growth of h_1 is independent of R_M. The *dotted lines* are the *artificial* extension of the empirical fit formula of Brandenburg (2001) beyond its region of validity. The R_M dependence of the dynamical solution only arises at late times, and the empirical fit formula is only applicable in the R_M dependent regime

Equations (28) and (29) comprise a powerful set for capturing basic helicity dynamics for closed or periodic volume low lowest order in turbulent anisotropy. They help to conceptually unify a range of physical process depending on the initial conditions. Examples are described in the next subsections.

4.2 Large Scale Field Growth: The α^2 Dynamo Example

Consider a closed or periodic system with an initially weak seed large scale magnetic helicity at wave number k_1, with initially $0 < h_1(\tau = 0) \ll 1$, and with $h_2(\tau = 0) = 0$. Suppose the system is steadily forced isotropically with kinetic helicity at wave number $k_2 \gg k_1$ such that $h_v = -1$ in Eqs. (28) and (29). Solutions to this problem are shown for different time ranges as the solid lines in Fig. 9(left and right panels) for $k_2/k_1 = 5$ and three different magnetic Reynolds numbers (as in Field and Blackman 2002 rather than Blackman and Field 2002 since $\partial_t \overline{\mathcal{E}}$ is ignored).

    The basic interpretation of the curves is this: At early times, the R_M dependent terms in Eqs. (28) and (29) are small and h_1 grows exponentially with a growth rate $\gamma = \frac{2k_1}{3k_2}(h_v - \frac{k_1}{k_2})$ from Eq. (28). The first three terms on the right of Eq. (29) have the same magnitude but opposite sign as those on the right of Eq. (28), so h_2 grows with opposite sign as h_1 with the same growth rate during this kinematic, R_M independent growth phase. This phase lasts until the compensating growth of h_2 becomes large enough to significantly offset the driving from h_v and reduce γ to a level for which the R_M terms become influential. The R_M term of Eq. (28) is k_2^2/k_1^2 times that Eq. (29), so h_2 is quenched by its R_M term earlier, allowing h_1 to continue growing, albeit now at an R_M dependent rate. In the $R_M \gg 1$ limit, the growth rate

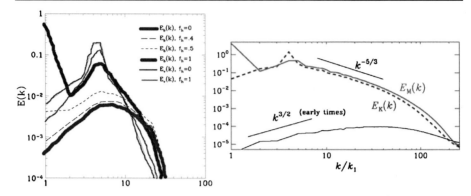

Fig. 10 Example steady-state saturated spectra from direct numerical simulations of helically forced MHD turbulence. The *left panel* is adapted from Maron and Blackman (2002) for 64^3 simulation and magnetic Prandtl number 3 with forcing wavenumber $k = 5$. The *thick red* and *blue lines* are the magnetic and kinetic energy spectra for fractional kinetic helicity $f_h = 1$. The *thin red* and *blue lines* are the magnetic and kinetic energy spectra for $f_h = 0$. Large scale field growth (at $k = 1$) is dramatic in the $f_h = 1$ case and negligible for $f_h = 0$. The *right panel* is a 512^3 simulation for $f_h = 1$ for unit magnetic Prandtl number and forcing wavenumber $k = 4$ from Brandenburg et al. (2012). In the *right panel*, *blue* indicates kinetic energy and *red* indicates magnetic energy. The *thick* and *thin red lines* of the *left panel* thus correspond respectively to the *red* and *blue lines* of the *right panel* in that these are all for the case of $f_h = 1$. The *right panel* also shows large scale $k = 1$ field for helical forcing. The essential features of the growth of the large scale field to saturation in such simulations are captured by Eqs. (28) and (29) the solutions of which are shown in Fig. 9

past the initial R_M independent regime is extremely small and generally not astrophysically relevant.

At the end of the kinematic phase, it can be shown analytically that the energy in the helical field grows to $B_1^2 \sim k_1 H_1 = \frac{k_1}{k_2}(1 - \frac{k_1}{k_2})v_2^2$. At this point the three different R_M curves in Fig. 9(right panel) diverge with the lowest R_M curve being the faster to reach the asymptotic steady state. By setting the left sides of Eqs. (28) and (29) to zero, the asymptotic saturation value can be shown to be $B_1^2 = k_1 H_1 = \frac{k_2}{k_1}(1 - \frac{k_1}{k_2})v_2^2$, or $h_1 = 20$ for $k_2 = 5k_1$ as seen in Fig. 9(left panel). The final value of h_1 is independent of R_M even though the time to get there is longer for larger R_M.

In Fig. 9(left panel), the dotted lines represent an R_M dependent empirical fit formula to the late time data in numerical simulations of Brandenburg (2001) of the α^2 dynamo. The agreement between the solutions to Eqs. (28) and (29) and the empirical fit formula to the simulations is quite good at late times. The dotted lines Fig. 9(right panel) represent the *artificial extension* of this empirical fit formula to early times where it does not apply. The asymptotic regime has an R_M dependence whereas the early time regime does not. More recent simulations (Brandenburg 2009) have confirmed that the large scale growth rate varies by only 16 % in the kinematic regime when R_M varies by a factor of 100. The success of these two simple equations in capturing saturation features of dynamo simulations highlights the importance of coupling magnetic helicity into the dynamics of field growth.

Examples of the large scale magnetic and kinetic energy spectra for the α^2 dynamo from direct numerical simulations in a periodic box starting with an initial weak seed field and forced with helical forcing at $k = 5$ are shown in Fig. 10. The left panel shows a 64^3 simulation and the saturated end state of magnetic and kinetic energies for different forcing fractions of kinetic helicity $f_h = |\langle \mathbf{v} \cdot \nabla \times \mathbf{v} \rangle|/k_f \langle v^2 \rangle$ and magnetic Prandtl number $= 3$ from Maron and Blackman (2002). In the left panel, the thick red and blue lines are the magnetic and kinetic energy spectra for the case of $f_h = 1$. The thin red and blue lines are the magnetic and kinetic energy spectra for the case of $f_h = 0$. The right panel is a 512^3 simulation from Brandenburg et al. (2012) for $f_h = 1$ for unit magnetic Prandtl number. In the right panel, blue indicates kinetic energy spectrum and red indicates magnetic energy spectrum. The thick and thin red lines of the left panel thus correspond respectively to the red and blue lines of the right panel in that these are all for the case of $f_h = 1$. Both panels show the dramatic emergence of the large scale $k = 1$ field for helical forcing. The thick blue curve in the left panel shows the absence of the large scale $k = 1$ magnetic field for $f_h = 0$.

4.3 Helical Field Decay

A second use of Eqs. (28) and (29) is to study how helical fields decay. As alluded to earlier, when a magnetic activity cycle period with field reversals is evident, the need for an in situ LSD is unambiguous. But when no cycle period is detected, the question of whether otherwise inferred large scale fields are LSD produced or merely a fossil field often arises. The question of how efficiently magnetic fields decay in the presence of turbulence is important because if a fossil field would have to survive this diffusion to avoid the need for an in situ dynamo.

Most work on the diffusion of large scale fields has not distinguished between the diffusion of helical vs. non-helical large scale fields. Yousef et al. (2003) and Kemel et al. (2011) found that fully helical large scale fields decay more slowly than non-helical large scale fields in numerical simulations. Blackman and Subramanian (2013, hereafter BS13) analyzed (28) and (29) for the field decay problem and identified a critical helical large scale magnetic energy above which decay is slow and below which decay is fast when the small scale helicity h_2 is initially zero. Bhat et al. (2014) further developed this theory and tested the results with numerical simulations and emphasized a distinction between two problems: the case studied by BS13 and the case in which the initial field decays slowly and then it transitions to fast decay.

The basic result in BS13 is captured by the solution to Eqs. (28) and (29) for initial conditions for which $h_1 > 0$, $h_2 = 0$, $h_v = 0$. This corresponds to a case in which there is no kinetic helicity, just a driving turbulent kinetic energy that causes turbulent diffusion of h_1 through the penultimate term on the right of Eq. (28). The question studied is how does h_1 evolve for these circumstances as a function of its initial value? The solutions are shown in Fig. 11 adapted form Bhat et al. (2014). From top to bottom the curves correspond to the initial values of large scale helical field energy $k_1 h_{1,0}/k_2$ in units of equipartition with the turbulent kinetic energy. For the chosen ratio $k_2/k_1 = 5$, the value $k_1 h_{1,0}/k_2 = 0.04 = (k_1/k_2)^2$ and marks a threshold initial value below which $k_1 h_1/k_2$ decays rapidly via turbulent diffusion and above which it decays restively slowly (actually, at twice the resistive diffusion rate). Since $k_1 h_1/k_2$ is the dimensionless magnetic energy associated with h_1, the energy in the helical field need only be at least k_1^2/k_2^2 times that of the turbulent kinetic energy to avoid fast decay. The implication is that helical field energy above a modest value decays resistively slowly even in the presence of turbulent diffusion. This contrasts the behavior of the non-helical part of the large scale magnetic, which always decays at the turbulent diffusion rate independent of the presence or absence of a helical component (BS13).

Why should the helical field resist turbulent diffusion and what determines the critical value? The answer is as follows (BS13; Bhat et al. 2014): Slow decay of h_1 occurs when the last term on the right of (28) is no smaller than the sum of the first two terms on the right ($h_v = 0$ for the present case). For large R_M, each of those first two terms is separately much larger than the last term so their combination would have to nearly cancel to meet the aforementioned condition. These same terms also appear in the equation for h_2 with

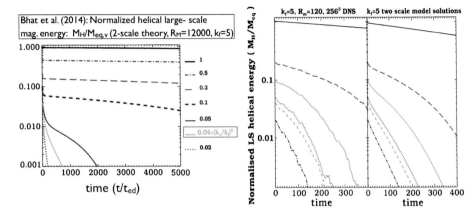

Fig. 11 Figure is from Bhat et al. (2014) and addresses the resilience of large scale helical fields to decay. The *left panel* shows solutions of Eqs. (28) and (29) for $M = k_1 h_1/k_2$, non-dimentionalized to the turbulent equipartition value and for the initial values shown, subjected to steady turbulent forcing with $h_v = 0$. Slow decay occurs when initially $h_1 > k_1/k_2$ and fast decay occurs when the initial value is below this value. The critical value of M is boxed in yellow. In the *left panel*, $k_2/k_1 = 5$ and $R_M = 12000$. The *right panel* shows a comparison between simulations and theory for the same problem at lower R_M. The agreement looks good. A subtlety however, is that the R_M accessible in the simulations is too low to identify the transition value $h_1 = k_1/k_2$ (see text of Sect. 4.3)

opposite sign. Initially, $h_1 > 0$ and $h_2 = 0$ and if we seek the initial value of h_1 for slow decay, we note that slow decay can only occur after a very rapid evolutionary phase (with negligible dependence on R_M) where a swift buildup of h_2 leads to an approximate balance between these aforementioned two terms. The needed amount of h_2 to abate decay of h_1 can be estimated by balancing first two terms on the right of either (28) or (29) for our case of $h_v = 0$. This gives $h_2 \sim (k_1/k_2)$. But since the only source of h_2 is h_1, this value of h_2 is also the minimum value of the initial $h_{1,0}$ needed for slow decay of h_1. That is, $h_{1,0} > h_c \equiv k_1/k_2$ for slow decay.

A qualitative comparison between the numerical simulations and the analytic solutions for the case just described is shown in Fig. 9(right panel). The behavior in the simulations looks similar to that of the analytic model. The one caveat is that the simulations do not achieve the resolution needed to ensure that the last term of Eq. (27) is smaller than the penultimate term for h_1 near the small critical value of k_1/k_2. However, as mentioned above, in addition to the case just described where the threshold initial value of h_1 for slow decay is sought, Bhat et al. (2014) also studied the case for which the field is initially above the threshold for slow decay and later transitions to fast decay. In that case they show that the value of h_1 at the transition is independent of k_1/k_2, unlike the case discussed above. In this second case, h_2 has already saturated by the time h_1 makes the transition to fast decay and so the estimate of the threshold of h_2 (and thus h_1) for the previous case does not apply. This result and the distinction between the two cases are both contained within the analytic framework of Eqs. (28) and (29) as discussed in Bhat et al. (2014) where theory and simulation are shown to agree. Confidence in the overall theory and physical interpretation of both cases is bolstered by this correspondence.

Taken at face value, the survival of helical fields to turbulent diffusion may provide rejuvenated credence to pre-galactic mechanisms of large scale field production that produce sufficiently strong helical fields (Field and Carroll 2000; Copi et al. 2008; Díaz-Gil et al. 2008; Semikoz et al. 2012; Tevzadze et al. 2012; Kahniashvili et al. 2013) because such helical fields could then avoid decay by supernova driven turbulent diffusion over a galactic lifetime in the absence of boundary terms. Although most energy in large scale galactic magnetic fields resides in non-helical toroidal fields, as long as the turbulent decay time for the non-helical field exceeds the linear shear time, we can expect a predominance of non-helical field in a steady state, even without an in situ dynamo to regenerate the poloidal fields: The helical field provides a minimum value below which the toroidal field cannot drop. The toroidal field enhancement over the poloidal field would be that which can be linearly shear amplified in a non-helical field diffusion time. Similar considerations regarding the survival of helical fields could apply for the large scale fields of stars and accretion disks.

Another implication of slow diffusion of helical fields is that the observation of a helical large scale field in jets from Faraday rotation (Asada et al. 2008; Gabuzda et al. 2008, 2012) does not guarantee magnetic energy domination (Lyutikov et al. 2005) on the observed scales (BS13).

The calculations just discussed do not include buoyancy or other boundary loss terms that could extract large scale helicity at a rate that may still need to be re-supplied from within the rotator. If such terms are important, then both helical and non-helical large scale fields would deplete, and an in situ dynamo would be needed for replenishment. But this shifts the focus from turbulent diffusion to that of boundary loss terms in assessing the necessity of in situ dynamos. More on flux and boundary terms will be discussed in the next section.

4.4 Dynamical Magnetic Relaxation

The resilience of helical fields to turbulent diffusion of the previous section is actually the result of the current helicity part of the α effect in the language of dynamo theory rather than an intrinsic change of the diffusion coefficient β. In this respect, the resilience of large scale helical field to decay is very rooted in the basic principles of Sect. 2. Namely, magnetic helicity has the lowest energy when on the largest scale. Diffusing it to small scales while conserving magnetic helicity is fighting against this relaxation.

In fact we can also use Eqs. (28) and (29) to study yet a different problem. Starting with $h_1 \ll 1$ and $h_2 \gg h_1$, and $h_v = 0$ we can solve for the evolution of h_1. Indeed, Blackman and Field (2004). Kemel et al. (2011), and Park and Blackman (2012) carried out such calculations using versions of Eqs. (28) and (29) where the system is initially driven with h_2 and found that it does indeed capture the dynamical relaxation of magnetic helicity to large scales: h_1 grows exponentially as the helicity is transferred from h_2. In this case, the large scale field grows with the same sign as h_2, and the combination of the turbulent diffusion and a modestly growing h_v emerge as the back reactors, contrasting the α^2 dynamo case of Sect. 4.2 above.

Although the dynamics of magnetic helicity and dynamical relaxation have often not been explicitly discussed in the context of fossil field origin models of stars (Braithwaite and Spruit 2004), the same basic principles are prevalent.

5 Helicity Fluxes

The previous section did not include the role of flux or boundary terms in the evolution equations. While the role of helicity fluxes in sustaining magnetic relaxation dynamos in laboratory fusion plasmas has been long-studied as essential (Strauss 1985, 1986; Bhattacharjee and Hameiri 1986; Bellan 2000) the awareness of its importance for astrophysical contexts has emerged more recently (Blackman and Field 2000a, 2000b; Vishniac and Cho 2001; Blackman 2003; Shukurov et al. 2006; Käpylä et al. 2008; Ebrahimi and Bhattacharjee 2014).

Laboratory plasma helical dynamos in a reversed field pinch (RFP, Ji and Prager 2002) for example, typically involve a magnetically dominated initial state with a dominant mean magnetic toroidal magnetic field. When an external toroidal electric field is applied along this torioidal field, a current is driven along the magnetic field which injects magnetic helicity of one sign on small scales. This generates a poloidal field. For sufficiently strong applied electric fields, the system is driven far enough from its relaxed state that helical tearing or kink mode instabilities occur. The consequent fluctuations produce a turbulent EMF $\overline{\mathcal{E}}$ that drives the system back toward the relaxed state. As discussed in Sect. 2, the relaxed state is the state in which the magnetic helicity is at the largest scale possible, subject to boundary conditions. When the helicity injection is externally sustained, a dynamical equilibrium with oscillations can incur as the system evolves toward and away from the relaxed state. The time-averaged $\overline{\mathcal{E}}$ is maintained by a spatial (radial) flux of small scale magnetic helicity within the plasma. The injection of helicity is balanced by the dynamo relaxation, so the dynamo sustains the large scale field configuration against decay.

The simplest circumstance revealing the importance of flux terms is a steady state for which the left hand side of Eqs. (21) and (22) vanish. Then, if divergences do not vanish, the magnetic field aligned EMF would be sustained by helicity fluxes, whose divergences are equal and opposite for the large and small scale contributions, that is

$$0 = 2\langle \overline{\mathcal{E}} \cdot \overline{\mathbf{B}} \rangle - c\nabla \cdot \langle \overline{\Phi}\overline{\mathbf{B}} + \overline{\mathbf{E}} \times \overline{\mathbf{A}} \rangle \tag{30}$$

and

$$0 = -2\langle \overline{\mathcal{E}} \cdot \overline{\mathbf{B}} \rangle - c\nabla \cdot \langle \phi\mathbf{b} + \mathbf{e} \times \mathbf{a} \rangle. \tag{31}$$

Combining these two equations reveals that the divergences of large and small scale helicity through the system are equal in magnitude but oppositely signed.

The specific observational interpretation of flux divergence terms depends on the averaging procedure. If the averaging is taken over the entire interior of a rotator such as a star or disk, then such non-zero fluxes in a steady-state would imply equal magnitude but oppositely signed rates of large and small scale magnetic helicity flow through the boundary into the corona (Blackman and Field 2000b). In a steady state, each hemisphere would receive both signs of magnetic helicity but the respective signs on large and small scales would be reversed in the two hemispheres.

Complementarily, Blackman (2003, Fig. 9 therein) showed how an imposed preferential small scale helicity flux could reduce quenching in the α^2 dynamo and the result is shown in Fig. 12 to a larger value of R_M. A simple term of the form $-\lambda h_2$ was added to Eq. (29) to make these plots and solutions for different values of λ are shown. The left panel shows the late time saturation value is increased with increasing λ and the right panel shows the tendency that for large enough λ the kinematic regime (the regime independent of R_M) can be extended. Del Sordo et al. (2013) have studied numerically the relative role of advective and diffusive fluxes. Their generalized α^2 dynamos can become oscillatory with even a weak advective wind, due to the spatial dependence of the imposed kinetic helicity. They do indeed find that for $R_M > 1000$ the advective flux dominates the diffusive flux and helps alleviate the resistive quenching. However predicting analytically the exact value of the critical R_M where this occurs requires further work.

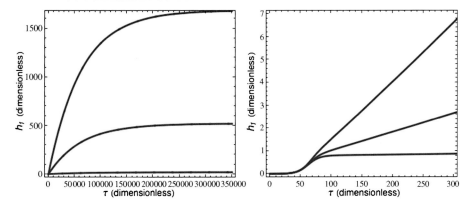

Fig. 12 Figure updated from Blackman (2003): This shows the conceptual role of a simple advective type helicity flux in increasing the h_1 saturation value of the α^2 dynamo, and the trend toward extending the growth of h_1 before resistive quenching incurs. The *three curves* in each panel correspond solutions of Eqs. (28) and (29) modified by the addition of a loss term $-\lambda h_2$ to Eq. (29). All curves correspond to $R_M = 5000$. From top to bottom in each panel the curves have $\lambda = 1/30, 1/100$ and 0 respectively. *Left panel* is the late time regime: The *bottom curve* saturates at the same value as those in Fig. 9(left panel), as those have no loss term (although Fig. 9(left panel) has faster growth because the R_M values are smaller). The *right panel* is the early time regime, and shows that for $\lambda = 1/30$ the resistive turnover in the h_1 curve is nearly avoided. This holds true even more dramatically for all larger values of λ (not shown)

5.1 Helicity Fluxes in Galactic and Stellar Contexts

The α^2 dynamo has no shear but helicity flux in the presence of shear is particularly important because generalizations of Eqs. (28) and (29) to the $\alpha - \Omega$ LSD otherwise lead to large scale field decay. Figure 12 can be contrasted with Fig. 13 in this regard. Figure 13 shows the results of Shukurov et al. (2006) for a model of the Galactic dynamo. The equations shown represent the generalization of Eqs. (28) and (29) to the $\alpha - \Omega$ dynamo with only vertical z derivatives retained and with an advective flux term that ejects small scale magnetic helicity from the system. The figure shows that the large scale field decays when this flux is too small, illustrating its important role in sustaining the EMF. In the solar context, a similar circumstance arises: Käpylä et al. (2008) found from simulations that an LSD is produced by convection + shear when surfaces of constant shear were aligned toward open boundaries allowing a helicity flux whereas Tobias et al. (2008) found no LSD when shear was aligned toward periodic boundaries disallowing helicity fluxes.

The role and potential observability of such global helicity fluxes are exemplified schematically in Fig. 14, originally in the context of the sun (Blackman and Brandenburg 2003). The figure illustrates the same principles as the comparison between Figs. 7 and 8 but with shear and buoyancy. In the northern hemisphere, the field exhibits a right handed writhe and a left handed small scale twist as the structure buoyantly emerges into the corona. The emergent separation of scales of magnetic helicity was verified by simulation of a buoyant, writhed tube. Such structure might in fact be evident in the TRACE image of coronal loop shown in Fig. 15 (from Gibson et al. 2002). The figure shows left hand twist and right handed writhe. A key point is that such ejections could be an essential part of sustaining the fast solar cycle, not just an independent consequences of magnetic field generation.

This segregation of helicity signs is consistent with other evidence that the northern hemisphere exhibits primarily small scale left handed twist and larger scale right handed writhe, with these reversed in southern hemisphere (Rust and Kumar 1996; Pevtsov et al. 2008;

Shukurov et al. (2006):

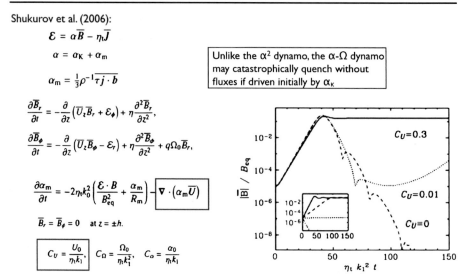

$$\mathcal{E} = \alpha\overline{B} - \eta_t\overline{J}$$

$$\alpha = \alpha_K + \alpha_m$$

$$\alpha_m = \tfrac{1}{3}\rho^{-1}\overline{\tau j \cdot b}$$

Unlike the α^2 dynamo, the α-Ω dynamo may catastrophically quench without fluxes if driven initially by α_K

$$\frac{\partial\overline{B}_r}{\partial t} = -\frac{\partial}{\partial z}\left(\overline{U}_z\overline{B}_r + \mathcal{E}_\phi\right) + \eta\frac{\partial^2\overline{B}_r}{\partial z^2},$$

$$\frac{\partial\overline{B}_\phi}{\partial t} = -\frac{\partial}{\partial z}\left(\overline{U}_z\overline{B}_\phi - \mathcal{E}_r\right) + \eta\frac{\partial^2\overline{B}_\phi}{\partial z^2} + q\Omega_0\overline{B}_r,$$

$$\frac{\partial\alpha_m}{\partial t} = -2\eta_t k_0^2\left(\frac{\mathcal{E}\cdot B}{B_{eq}^2} + \frac{\alpha_m}{R_m}\right) - \nabla\cdot\left(\alpha_m\overline{U}\right)$$

$$\overline{B}_r = \overline{B}_\phi = 0 \quad \text{at } z = \pm h.$$

$$C_U = \frac{U_0}{\eta_t k_1}, \quad C_\Omega = \frac{\Omega_0}{\eta_t k_1^2}, \quad C_\alpha = \frac{\alpha_0}{\eta_t k_1}$$

Fig. 13 From Shukurov et al. (2006), this provides a simple model illustrating the importance of helicity fluxes in sustaining the $\alpha - \Omega$ dynamo when magnetic helicity dynamics and fluxes are coupled into the theory. They considered an advective contribution to the helicity flux, as shown boxed in *red*, and quantified its contribution by C_U. The *plots on the right* show that only above a threshold value of C_U does the large scale mean field (plotted in units of the field strength corresponding to equipartition with the turbulent kinetic energy) sustain. Without these fluxes, the field decays. This contrasts the situation of the α^2 dynamo in Fig. 12 in which flux terms increase the saturation value but do not determine the difference between growth and decay

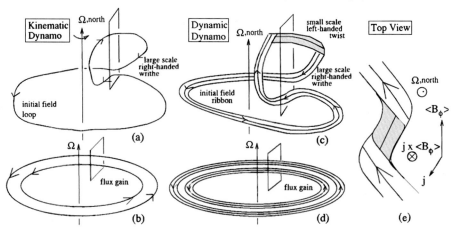

Fig. 14 From Blackman and Brandenburg (2003), a comparison of the classic picture of $\alpha - \Omega$ dynamo without magnetic helicity conservation (panel (**a**)) to that with magnetic helicity conservation (panel (**c**)). Panels (**b**) and (**d**) show the gain in toroidal field after ejection of the large poloidal loop. For panel (**d**), this ejection alleviates the twist that would otherwise build up. This figure is a conceptual generalization of the concepts addressed by the comparison of Figs. 7 and 8 to include shear and buoyancy and to motivate why e.g. coronal mass ejections of the sun, or galaxies may represent the irreversible loss of small scale magnetic helicity that allows LSD action to sustain. See also Fig. 13

Fig. 15 TRACE 195 Å image of a solar sigmoid of the Northern hemisphere from Gibson et al. (2002). It is tempting to interpret the large scale right-handed writhe and small scale left handed striations along the sigmoid being consistent with what a magnetic helicity conserving dynamo would predict when the fluxes eject both small and large scale helicities from the interior. The ejection of the small scale helicity in sigmoids or coronal mass ejections may be fundamental to the operation of the solar dynamo, not just a consequence of large scale field generation

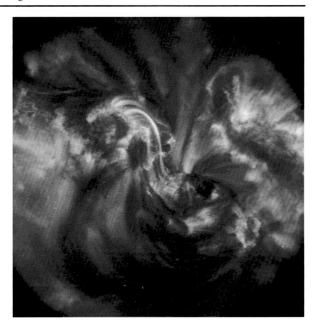

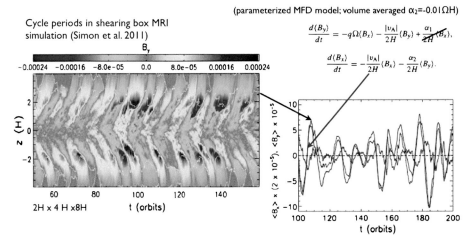

Fig. 16 Example of evidence for large scale dynamo action in MRI simulations from Simon et al. (2011), compared with an empirically tuned $\alpha - \Omega$ dynamo model. The *left panel* shows the net torioidal field vs. time, volume averaged over all x and y and over $|z| < 0.5H$. Outflow vertical boundaries were used and periodic boundaries in the other two dimensions. The cycle period of ~ 10 orbits is evident. The *black line* in the *right panel* shows the mean toroidal field as function of time and the *blue line* corresponds to the model semi-empirical fit equation of the $\alpha - \Omega$ dynamo. The sign of the required dynamo α coefficient is opposite to that expected from kinetic helicity

Zhang et al. 2010; Hao and Zhang 2011). These features seem to be invariant with respect to solar cycle, as would be predicted by helical dynamos, even when the field itself reverses sign. Note also that $H\alpha$ filaments seem to exhibit "dextral" (right handed) twist in North and "sinstral" (left handed) in south (Martin and McAllister 1997) BUT: right handed $H\alpha$ filaments may be supported by left handed fields and vice versa (Rust 1999).

There have been efforts to measure the rate of injection of magnetic helicity (Chae et al. 2004; Schuck 2005; Lim et al. 2007), particularly, its gauge invariant cousin: the "relative magnetic helicity" (Berger and Field 1984; the difference between actual magnetic helicity and that of a potential field) by tracking footprint motions. It can be shown that the footpoint motions provide a direct measure of this input rate (Démoulin and Berger 2003). There have also been efforts to relate the current helicity to the injection rate of magnetic helicity (Zhang et al. 2012). Ultimately, measuring the detailed spectra of current helicity and relative magnetic helicity injection into the corona and solar wind (Brandenburg et al. 2011) is a highly desirable enterprise for the future. Commonly, observational work has focused on the component of twist with the current along the line of sight, but measuring the full current and magnetic helicities require all three components of the field.

5.2 Helicity Fluxes in Accretion Disks and Shearing Boxes

Essentially all MRI unstable simulations with large enough vertical domains, whether stratified, unstratified, local or global, show the generation of large scale toroidal fields of the same flux for cycle periods of ~ 10 orbits (Brandenburg et al. 1995; Lesur and Ogilvie 2008; Davis et al. 2010; Simon et al. 2011; Guan and Gammie 2011; Sorathia et al. 2012; Suzuki and Inutsuka 2013; Ebrahimi and Bhattacharjee 2014). The patterns indicate a large scale dynamo operating contemporaneously with the small scale dynamo. The key unifying property of all of these cases is mean field aligned EMF, that is $\langle \overline{\mathcal{E}} \cdot \overline{\mathbf{B}} \rangle$. The explicit form of $\overline{\mathcal{E}}$ and the terms that contribute to it can depend on the boundary and stratification conditions and on the particular procedure for large scale averaging, but there is considerable similarity among the dynamos operating in these simulations. Sorting out whether and which flux terms are important for specific averaging and initial conditions is an ongoing task. To see the issues at hand, I highlight approaches to LSD modeling in shearing boxes without using helicity fluxes and then compare to those that do.

The left panel of Fig. 16, adapted from Simon et al. (2011) shows the toroidal magnetic field from a shearing box MRI simulation. The simulation used vertical stratification and periodic boundaries in the x and y (radial and azimuthal) and outflow boundaries in z (vertical) directions. The toroidal field was calculated by averaging over x and y and over $z \leq 5H$ where H is a density scale height. There was an initial net toroidal field in the box, but the left panel shows that the toroidal field reverses every 10 orbits in the saturated state. The right panel shows the use of model equations from an $\alpha - \Omega$ dynamo that Simon et al. (2011) adopted from Guan and Gammie (2011) tuned to match the simulation. The equations have only a shear term and a loss term from buoyancy in the torioidal field equation. The radial field equation has a buoyancy loss term and the $\alpha = \alpha_2$ dynamo term. There are no helicity dynamics in this empirical set of equations; those dynamics would at most y be hidden in the empirically determined α_2. Nevertheless, this quasi-empirical set of equations does well to model the field seen in the simulations.

In this example of Simon et al. (2011), like those of the first analyses of cycle periods in shearing box simulations (Brandenburg et al. 1995; Brandenburg and Donner 1997), the sign of the $\alpha = \alpha_2$ coefficient is found to be inconsistent with the standard 20th century textbook kinetic helicity. Additional features such as density stratification and rotation lead to higher order turbulent anisotropy and inhomogeneity that change the dominant sign of α appropriately (Rüdiger and Kichatinov 1993), and is consistent with the role of magnetic buoyancy (Brandenburg 1998). In this context, Gressel (2010) further looked at the behavior

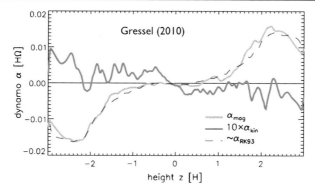

Fig. 17 Example of $\alpha - \Omega$ dynamo model for large scale fields compared to vertically stratified shearing box, MRI simulations from Gressel (2010). The vertical dependences of the α dynamo coefficients are shown. This coefficient is the proportionality between the EMF and the mean field (averaged in radius and azimuth). The values α_{kin} and α_{mag} are proportional to the kinetic and magnetic helcities respectively. The value α_{RK93} comes from Rüdiger and Kichatinov (1993) and is derived for stratified rotating turbulence. The figure shows that α_{RK93} or α_{mag} are much better fits to the dynamo in the shearing box than the traditional α_{kin} of 20th century textbooks. It would seem that α_{RK93} is therefore capturing the α_{mag} contribution and may result from magnetic buoyancy

of the dynamo α coefficient in stratified MRI shearing box (with outflow vertical boundaries) simulations and found that the sign of the dynamo α coefficient was consistent with the generalizations of Rüdiger and Kichatinov (1993), and also consistent with the sign of the current helicity correction term to α in Eq. (24), as if the current helicity contribution may be sourced by magnetic buoyancy. Gressel's results are shown in Fig. 17 and his averaging was taken over radius and azimuth.

The aforementioned approaches to MRI LSDs do not involve helicity fluxes, but calculations of EMF sustaining fluxes in this context have been emerging (e.g. Vishniac and Cho 2001; Vishniac 2009; Käpylä and Korpi 2011; Ebrahimi and Bhattacharjee 2014) (see also Vishniac and Shapovalov 2014 for an isotropically forced case with shear). In particular, Ebrahimi and Bhattacharjee (2014) studied an MRI unstable cylinder with conducting boundaries. They wrote down mean quantities averaged over azimuthal and vertical directions, leaving the radial direction unaveraged. They looked explicitly at the terms in the helicity conservation equation (21) and found that the electromotive force is well matched by the local flux terms measured from direct numerical simulations. Their results are shown in Fig. 18.

To calculate the specific form of these fluxes by brute force in a mean field theory involves expanding the fluctuating quantities in terms of mean field quantities via the dynamical equations for those fluctuating quantities and a closure (e.g. Brandenburg and Subramanian 2005). Such efforts are ongoing. One such flux that emerges from such a procedure that of Vishniac and Cho (2001). Ebrahimi and Bhattacharjee (2014) also plot this latter flux as seen in the let panel of Fig. 18. They find that it is much smaller than the total flux term directly calculated from the simulations, and thus is subdominant. Recall that Shukurov et al. (2006) in Fig. 13 considered an advective flux term that is another candidate flux term that emerges from mean field theory. Sur et al. (2007) also assessed the role of the Vishniac-Cho flux semi-analyticially in the galactic context and found it can be helpful if the mean magnetic field is above a threshold value to begin with. Vishniac and Shapovalov (2014) considered an isotropically forced periodic box with linear shear, (without the coriolis force) and find that the Vishniac-Cho flux is dominant. Their averaging procedure involves averaging over

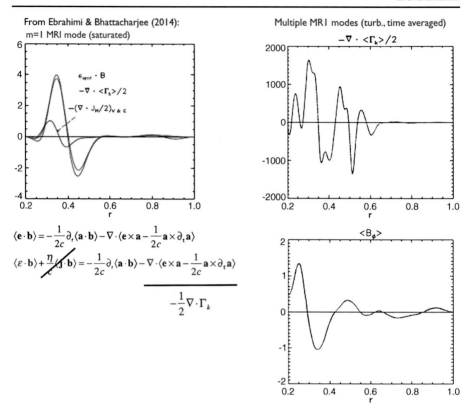

Fig. 18 Figures adapted from Ebrahimi and Bhattacharjee (2014), representing $R_M \gg 1$, and unit magnetic Prandtl number, shear flow driven simulations of an MRI unstable cylinder with perfectly conducting boundaries. This shows evidence for the generation of large scale field when averages are taken over height and azimuth, leaving the radial variable unaveraged. No initial net torioidal field was present in the box but one emerges as a result of LSD action. *Left panel*: Strong evidence for the importance of the radial flux of small scale magnetic helicity in sustaining the EMF needed for dynamo action for a single saturated unstable MRI mode is shown. The *black curve* is a measure of the field aligned EMF and the *red curve* is the helicity flux in shown in the equation. The *blue curve* is the Vishniac and Cho (2001) flux which is too small to match the total EMF sustaining flux. *Right pair of figures* shows the radial correspondence between the net toroidal field and helicity flux divergence additionally time-averaged for a turbulent state in which multiple MRI unstable modes interact. The correspondence provides further evidence for the importance of local helicity fluxes in sustaining the LSD

the entire box and filtering by wavenumber to distinguish contributions from mean and fluctuating components. This procedure is different from that of Ebrahimi and Bhattacharjee (2014) discussed above. See also Hubbard and Brandenburg (2011) however, who suggest that the choice of gauge influences whether the Vishniac-Cho flux is important in numerical simulations.

In general, the different circumstances and averaging procedures of mean field quantities between simulations of e.g. Simon et al. (2011), Ebrahimi and Bhattacharjee (2014) and those of Vishniac and Shapovalov (2014) highlight the need for clarity in tailoring the specific mean field theory to capture the dominant contributions for different combinations of forcing, boundary conditions, and averaging procedures. There is also an opportunity to combine mean field theories for local transport (e.g. Ogilvie 200; Pessah et al. 2006) with those of mean field dynamos and large scale transport to fully model angular momentum transport in accretion disks.

6 Gauge Issues

A subtle aspect of the helicity density equations (21) and (22) is the issue of gauge non-invariance. In the absence of boundary flux terms or time dependent terms, the magnetic helicity is gauge invariant relative to an arbitrary choice of initial value. This is a straight-forward consequence of Eq. (7). Since $\mathbf{E} \cdot \mathbf{B}$ is gauge invariant under any circumstance, the sum of the time dependent term on the left minus the divergence term on the right is always gauge invariant (even though the flux itself need not itself be gauge invariant). If the flux term vanishes, then the time derivative term is gauge invariant. If the time derivative term vanishes, then the leftover flux term is gauge invariant. This was used by Blackman and Field (2000b) to estimate an energy associated with the ejected magnetic helicity into coronae by a steady state dynamo. Mitra et al. (2010) show numerically that indeed the diffusive magnetic helicity fluxes that arise naturally across the mid plane in a system forced with oppositely signed kinetic helicities are gauge invariant. In the steady-state the divergence is invariant at every point so one can obtain the spatial dependence of the flux. These principles were also verified in Hubbard and Brandenburg (2010) and apply even to oscillatory dynamos by first identifying a gauge for which the helicity is steady (Del Sordo et al. 2013), which eliminates the dependent terms and the divergence term emerges as gauge invariant.

Additional subtleties of gauge invariance in the different context of shearing boxes are discussed in Candelaresi et al. (2011) and Hubbard and Brandenburg (2011).

Gauge non-invariance is closely related to the fact that for open boundaries, the amount of external field linkage is not in general specified. That is, the amount of field linkage inside the boundary can be the same for different amounts of exterior linkage. Fixing the gauge for the vector potential removes this ambiguity in that gauge. In actuality, one can choose a gauge, work in this gauge to study helicity dynamics, and then calculate physical quantities that are gauge invariant. The gauge non-invariance does not change the role of magnetic helicity as an intermediate conceptual tool.

However to interpret physically the magnetic helicity, gauge invariant versions can be helpful. Subramanian and Brandenburg (2006) developed a generalized local helicity density whose evolution reduces to the above Eqs. (21) and (22) in the absence of flux terms. In the presence of flux terms, their equation has the same form but with a different helicity density that is gauge invariant. For a turbulent system, with large scale separation between fluctuating and mean quantities, their gauge invariant helicity density associated with small scale quantities is similar to what is obtained for the usual magnetic helicity density in the Coulomb gauge because the gauge variant boundary terms become small in their averaging procedure. In the context of the Galaxy, Shukurov et al. (2006) solved the mean field induction equation for $\overline{\mathbf{B}}$ using this gauge invariant helicity density.

The gauge invariant relative magnetic helicity (Berger and Field 1984; Finn and Antonsen 1985; Bellan 2000) referred to earlier, was developed for more direct interpretations of observations originally in the solar context. This quantity involves separating global space into two parts, a region of physical interest and the exterior to this region. The relative magnetic helicity is specifically the difference between the magnetic helicity of the system integrated over global space minus that associated with an integral over global space where the field in region of physical interest is replaced by a vacuum, potential field, namely $H_R \equiv \int \mathbf{A} \cdot \mathbf{B} dV - \int \mathbf{A}_p \cdot \mathbf{B}_p dV$, where $\mathbf{B}_p = \nabla \times \mathbf{A}_p$ and $\mathbf{B}_p$ is the potential field. In this way the external linkage is removed, and what remains is gauge invariant. The time evolution equation for H_R is then (Berger and Field 1984)

$$\partial_t H_R = -2c \int \mathbf{E} \cdot \mathbf{B} dV - \int \nabla \cdot (\mathbf{E} \times \mathbf{A}_p) dV. \qquad (32)$$

All terms in this generalized relative magnetic helicity conservation equation are gauge invariant. Not only is the divergence term gauge invariant but the flux itself is gauge invariant. Démoulin and Berger (2003) have shown how the rate of injection of relative magnetic helicity into the solar corona depends on measurable quantities at the footprints of the anchoring fields. Sorting out the relation between the relative magnetic helicity, the current helicity, and the gauge variant magnetic helicity in theoretical calculations warrants further attention.

7 Summary and Conclusions

Tracking magnetic helicity in MHD systems is an important, unifying tool to understand the processes by which large scale magnetic fields form and evolve in both astrophysical and laboratory plasmas. The purpose of this review has been to provide one path through the forest as a conceptual primer to the literature. Three key principles underlie the role of magnetic helicity in all contexts: (1) magnetic helicity is a measure of twist, linkage or writhe; (2) magnetic helicity is better conserved than magnetic energy under most circumstances for a closed system with or without velocity flows; (3) the energy in a helical magnetic field is minimized when the field relaxes to the largest scale available consistent with the boundary conditions.

The importance of magnetic helicity for large scale field generation is evident from early studies of helical MHD turbulence (Pouquet et al. 1976), but incorporating its role into a 21st century dynamical mean field theory has emerged only in the pas decade or so. The 20th century textbook dynamos, unlike 21st century theory, do not conserve magnetic helicity and as such were unable to predict or reveal how LSDs saturate. A key aspect of 21st century theory is that the growth of large scale fields, facilitated by an EMF, involves growth of large scale magnetic helicity and small scale magnetic helicity of the opposite sign. In the absence of helicity fluxes, the small scale build up suppresses further growth of large scale field. In the case of sheared rotators, unless helicity fluxes can remove the offending small scale magnetic helicity, the large scale field not only saturates but may even decay. Alternatively expressed, it seems that astrophysical dynamos, like laboratory plasma dynamos may involve an EMF that is commonly aided or sustained by the divergence of small scale helicity fluxes. In a quasi-steady state for the sun, a crude minimalist prediction is that both signs of helical magnetic fields should appear in the northern hemisphere with small scale left handed structures and right handed large scale structures with the reversed combination in the southern hemisphere. More efforts to measure the spectral distribution of helical fields in the solar corona and wind would be valuable. Large scale dynamo models for the galaxy and for accretion disks that incorporate magnetic helicity dynamics have also been emerging.

Accretion disks pose an interestingly rich opportunity of study for helicity dynamics and large scale dynamos because traditionally large scale dynamo theory has been studied independent of theories of angular momentum transport. The ubiquity of large scale dynamos seen in simulations and the ubiquity of observed astrophysical coronae and jets indicates that a significant contribution to angular momentum transport comes from large scale fields.

This needs to be incorporated into a combined mean field accretion disk theory that captures local and large scale angular momentum transport, and large scale field growth.

Coronae of stars, disks, and laboratory plasmas are all magnetically energy dominated. In magnetically dominated environments the principles of magnetic helicity evolution have long been helpful to understand the evolution of magnetic structures subject to their foot-point motions. The helicity injection by foot-point motions is analogous to injection of small scale helicity in laboratory devices, where the system responds by relaxing the helicity to large scales In astrophysical coronae, it is likely that some contribution from both signs rather than a single sign are injected, so the relaxation process must take this into consideration globally, even if local structures are injected with primarily one sign. It was in fact in the context of laboratory plasma magnetic relaxation where the importance of helicity fluxes was first identified.

Finally, as reviewed herein, the basic properties of magnetic helicity also underlie its role in making large scale helical fields resilient to turbulent diffusion. Recent work on this topic for closed systems may strengthen the potential efficacy of fossil field origin of large scale fields in some astrophysical contexts. A helical field (in the absence of global helicity fluxes) is much more resilient to turbulent diffusion than non-helical large scale fields. The effect is best understood not as reduction of the turbulent diffusion coefficient, but rather as a competition between the unfettered turbulent diffusion and additional competing tendency for helicity to relax back toward the largest scales. The driver for this inverse transfer is the very small scale magnetic helicity that is sourced by initial diffusion from the large scale helicity in the first place.

Acknowledgements I acknowledge NSF grant AST-1109285, and thank the organizers of the ISSI Workshop on "Multi-Scale Structure Formation and Dynamics of Cosmic Plasmas", and the organizers of the Lyman Spitzer 100th birthday conference for engaging meetings in Bern and Princeton respectively. I also acknowledge particular discussions with P. Bhat, A. Bhattacharjee, A. Brandenburg, F. Ebrahimi, G. Field, A. Hubbard, F. Nauman, J. Stone and K. Subramanian.

References

K. Asada, M. Inoue, M. Nakamura, S. Kameno, H. Nagai, Astrophys. J. **682**, 798 (2008)
S.A. Balbus, Annu. Rev. Astron. Astrophys. **41**, 555 (2003)
S.A. Balbus, J.F. Hawley, Astrophys. J. **376**, 214 (1991)
S.A. Balbus, J.F. Hawley, Rev. Mod. Phys. **70**, 1 (1998)
R. Beck, Space Sci. Rev. **166**, 215 (2012)
P.M. Bellan, *Spheromaks* (Imperial College Press, London, 2000)
M.A. Berger, G.B. Field, J. Fluid Mech. **147**, 133 (1984)
P. Bhat, K. Subramanian, Mon. Not. R. Astron. Soc. **429**, 2469 (2013)
P. Bhat, E.G. Blackman, K. Subramanian, Mon. Not. R. Astron. Soc. **438**, 2954 (2014)
A. Bhattacharjee, E. Hameiri, Phys. Rev. Lett. **57**, 206 (1986)
E.G. Blackman, Recent developments in magnetic dynamo theory, in *Lecture Notes in Physics*, vol. 614, 2003, p. 432
E.G. Blackman, Plasma Phys. Control. Fusion **46**, 423 (2004)
E.G. Blackman, G.B. Field, Astrophys. J. Lett. **534**, 984 (2000a)
E.G. Blackman, G.B. Field, Mon. Not. R. Astron. Soc. **318**, 724 (2000b)
E.G. Blackman, G.B. Field, Phys. Rev. Lett. **89**, 265007 (2002)
E.G. Blackman, G.B. Field, Phys. Plasmas **11**, 3264 (2004)
E.G. Blackman, A. Brandenburg, Astrophys. J. **579**, 359 (2002)
E.G. Blackman, A. Brandenburg, Astrophys. J. Lett. **584**, L99 (2003)
E.G. Blackman, A. Hubbard, Mon. Not. R. Astron. Soc. (2014, drafted for submission)
E.G. Blackman, M.E. Pessah, Astrophys. J. **704**, L113 (2009)
E.G. Blackman, K. Subramanian, Mon. Not. R. Astron. Soc. **429**, 1398, BS13 (2013)

E.G. Blackman, A. Frank, C. Welch, Astrophys. J. **546**, 288 (2001)
R.D. Blandford, D.G. Payne, Mon. Not. R. Astron. Soc. **199**, 883 (1982)
J. Braithwaite, H.C. Spruit, Nature **431**, 819 (2004)
A. Brandenburg, in *Theory of Black Hole Accretion Disks*, ed. by M.A. Abramowicz, G. Bjornsson, J.E. Pringle (Cambridge University Press, Cambridge, 1998), p. 61
A. Brandenburg, Astrophys. J. **550**, 824 (2001)
A. Brandenburg, Astrophys. J. **625**, 539 (2005)
A. Brandenburg, Astrophys. J. **697**, 1206 (2009)
A. Brandenburg, K.J. Donner, Mon. Not. R. Astron. Soc. **288**, L29 (1997)
A. Brandenburg, A. Lazarian, Space Sci. Rev. **178**, 163 (2013)
A. Brandenburg, K. Subramanian, Phys. Rep. **417**, 1 (2005)
A. Brandenburg, A. Nordlund, R.F. Stein, U. Torkelsson, Astrophys. J. **446**, 741 (1995)
A. Brandenburg, D. Sokoloff, K. Subramanian, Space Sci. Rev. **169**, 123 (2012)
A. Brandenburg, K. Subramanian, A. Balogh, M.L. Goldstein, Astrophys. J. **734**, 9 (2011)
A. Brandenburg, D. Sokoloff, K. Subramanian, Space Sci. Rev. **169**, 123 (2012)
V. Bujarrabal, A. Castro-Carrizo, J. Alcolea, C. Sánchez Contreras, Astron Astrophys. **377**, 868 (2001)
S. Candelaresi, A. Hubbard, A. Brandenburg, D. Mitra, Phys. Plasmas **18**, 012903 (2011)
J. Chae, Y.-J. Moon, Y.-D. Park, Sol. Phys. **223**, 39 (2004)
L. Chamandy, K. Subramanian, A. Shukurov, Mon. Not. R. Astron. Soc. **428**, 3569 (2013)
P. Charbonneau, SASS **39** (2013)
C.J. Copi, F. Ferrer, T. Vachaspati, A. Achúcarro, Phys. Rev. Lett. **101**, 171302 (2008)
A. Díaz-Gil, J. García-Bellido, M. García Pérez, A. González-Arroyo, Phys. Rev. Lett. **100**, 241301 (2008)
S.W. Davis, J.M. Stone, M.E. Pessah, Astrophys. J. **713**, 52 (2010)
P. Démoulin, M.A. Berger, Sol. Phys. **215**, 203 (2003)
F. Del Sordo, G. Guerrero, A. Brandenburg, Mon. Not. R. Astron. Soc. **429**, 1686 (2013)
M. Dikpati, P.A. Gilman, Space Sci. Rev. **144**, 67 (2009)
F. Ebrahimi, A. Bhattacharjee, Phys. Rev. Lett. **112**, 125003 (2014)
G. Field, AIPC **144**, 324 (1986)
G.B. Field, E.G. Blackman, Astrophys. J. **572**, 685 (2002)
G.B. Field, S.M. Carroll, Phys. Rev. D, Part. Fields **62**, 103008 (2000)
G.B. Field, R.D. Rogers, Astrophys. J. **403**, 94 (1993)
J.M. Finn, T.M. Antonsen, Comments Plasma Phys. Control. Fusion **9**, 111123 (1985)
D.C. Gabuzda, D.M. Christodoulou, I. Contopoulos, D. Kazanas, Int. J. Mod. Phys. Conf. Ser. **355**, 012019 (2012)
D.C. Gabuzda, V.M. Vitrishchak, M. Mahmud, S.P. O'Sullivan, Mon. Not. R. Astron. Soc. **384**, 1003 (2008)
S.E. Gibson et al., Astrophys. J. **574**, 1021 (2002)
G.A. Glatzmaier, Annu. Rev. Earth Planet. Sci. **30**, 237 (2002)
O. Gressel, Mon. Not. R. Astron. Soc. **405**, 41 (2010)
A.V. Gruzinov, P.H. Diamond, Phys. Plasmas **3**, 1853 (1996)
X. Guan, C.F. Gammie, Astrophys. J. **728**, 130 (2011)
J. Hao, M. Zhang, Astrophys. J. **733**, L27 (2011)
T. Heinemann, J.C. McWilliams, A.A. Schekochihin, Phys. Rev. Lett. **107**, 255004 (2011)
A. Hubbard, A. Brandenburg, Geophys. Astrophys. Fluid Dyn. **104**, 577 (2010)
A. Hubbard, A. Brandenburg, Astrophys. J. **727**, 11 (2011)
H. Ji, S.C. Prager, Magnetohydrodynamics **38**, 191 (2002)
T. Kahniashvili, A.G. Tevzadze, A. Brandenburg, A. Neronov, Phys. Rev. D **87**, 083007 (2013)
A.P. Kazantsev, J. Exp. Theor. Phys. **26**, 1031 (1968)
P.J. Käpylä, M.J. Korpi, A. Brandenburg, Astron. Astrophys. **491**, 353 (2008)
P.J. Käpylä, M.J. Korpi, Mon. Not. R. Astron. Soc. **413**, 901 (2011)
K. Kemel, A. Brandenburg, H. Ji, Phys. Rev. E, Stat. Nonlinear Soft Matter Phys. **84**, 056407 (2011)
N.I. Kleeorin, A.A. Ruzmaikin, Magnetohydrodynamics **18**, 116 (1982)
F. Krause, K.H. Rädler, *Mean Field Magnetohydrodynamics and Dynamo Theory* (Pergamon, Elmsford, 1980)
A. Konigl, Astrophys. J. **342**, 208 (1989)
R.M. Kulsrud, E.G. Zweibel, Rep. Prog. Phys. **71**, 046901 (2008)
R.M. Kulsrud, R. Cen, J.P. Ostriker, D. Ryu, Astrophys. J. **480**, 481 (1997)
G. Lesur, G.I. Ogilvie, Astron. Astrophys. **488**, 451 (2008)
E.-K. Lim, H. Jeong, J. Chae, Y.-J. Moon, Astrophys. J. **656**, 1167 (2007)
R.V.E. Lovelace, D.M. Rothstein, G.S. Bisnovatyi-Kogan, Astrophys. J. **701**, 885 (2009)
D. Lynden-Bell, Mon. Not. R. Astron. Soc. **369**, 1167 (2006)
M. Lyutikov, V.I. Pariev, D.C. Gabuzda, Mon. Not. R. Astron. Soc. **360**, 869 (2005)

J. Maron, E.G. Blackman, Astrophys. J. **566**, L41 (2002)
S.F. Martin, A.H. McAllister, Geophys. Monogr. **99**, 127 (1997)
D. Mitra, A. Brandenburg, Mon. Not. R. Astron. Soc. **420**, 2170 (2012)
D. Mitra, S. Candelaresi, P. Chatterjee, R. Tavakol, A. Brandenburg, Astron. Nachr. **331**, 130 (2010)
H.K. Moffatt, *Magnetic Field Generation in Electrically Conducting Fluids* (Cambridge University Press, Cambridge, 1978)
S. Ortolani, D.D. Schnack, *Magnetohydrodynamics of Plasma Relaxation* (World Scientific, Singapore, 1993)
K. Park, E.G. Blackman, Mon. Not. R. Astron. Soc. **423**, 2120 (2012)
E.N. Parker, Clarendon/Oxford University Press, Oxford/New York, 1979, 858 pp.
R.F. Penna, R. Narayan, A. Sądowski, Mon. Not. R. Astron. Soc. **436**, 3741 (2013)
M.E. Pessah, C.-K. Chan, D. Psaltis, Phys. Rev. Lett. **97**, 221103 (2006)
A.A. Pevtsov, R.C. Canfield, T. Sakurai, M. Hagino, Astrophys. J. **677**, 719 (2008)
A. Pouquet, U. Frisch, J. Leorat, J. Fluid Mech. **77**, 321 (1976)
H. Ji, S.C. Prager, J.S. Sarff, Phys. Rev. Lett. **74**, 2945 (1995)
R.E. Pudritz, M.J. Hardcastle, D.C. Gabuzda, Space Sci. Rev. **169**, 27 (2012)
P.H. Roberts, E.M. King, Rev. Plasma Phys. **76**, 096801 (2013)
G. Rüdiger, L.L. Kichatinov, Astron. Astrophys. **269**, 581 (1993)
D.M. Rust, Geophys. Monogr. **111**, 221 (1999)
D.M. Rust, A. Kumar, Astrophys. J. **464**, L199 (1996)
A.A. Ruzmaikin, D.D. Sokolov, A.M. Shukurov (eds.), *Magnetic Fields of Galaxies* (Kluwer Academic, Dordrecht, 1988)
A. Shukurov, Lect. Notes Phys. **664**, 113 (2005)
A. Shukurov, D. Sokoloff, K. Subramanian, A. Brandenburg, Astron. Astrophys. **448**, L33 (2006)
A.A. Schekochihin, S.C. Cowley, G.W. Hammett, J.L. Maron, J.C. McWilliams, New J. Phys. **4**, 84 (2002)
C.J. Schrijver, C. Zwaan, *Solar and Stellar Magnetic Activity* (Cambridge University Press, Cambridge, 2000)
P.W. Schuck, Astrophys. J. **632**, L53 (2005)
V.B. Semikoz, D.D. Sokoloff, J.W.F. Valle, J. Cosmol. Astropart. Phys. **6**, 8 (2012)
J.B. Simon, J.F. Hawley, K. Beckwith, Astrophys. J. **730**, 94 (2011)
K.A. Sorathia, C.S. Reynolds, J.M. Stone, K. Beckwith, Astrophys. J. **749**, 189 (2012)
S. Sridhar, N.K. Singh, ArXiv e-prints (2013)
H.R. Strauss, Phys. Fluids **28**, 2786 (1985)
H.R. Strauss, Phys. Fluids **29**, 3008 (1986)
K. Subramanian, Phys. Rev. Lett. **90**, 245003 (2003)
K. Subramanian, Astron. Nachr. **331**, 110 (2010)
K. Subramanian, A. Brandenburg, Astrophys. J. Lett. **648**, L71 (2006)
S. Sur, A. Shukurov, K. Subramanian, Mon. Not. R. Astron. Soc. **377**, 874 (2007)
T.K. Suzuki, S.-i. Inutsuka, arXiv:1309.6916 (2013)
J.B. Taylor, Phys. Rev. Lett. **33**, 1139 (1974)
J.B. Taylor, Rev. Mod. Phys. **58**, 741 (1986)
A.G. Tevzadze, L. Kisslinger, A. Brandenburg, T. Kahniashvili, Astrophys. J. **759**, 54 (2012)
S.M. Tobias, F. Cattaneo, N.H. Brummell, Astrophys. J. **685**, 596 (2008)
C.L. Van Eck et al., Astrophys. J. **728**, 97 (2011)
E.T. Vishniac, Astrophys. J. **696**, 1021 (2009)
E.T. Vishniac, A. Brandenburg, Astrophys. J. **475**, 263 (1997)
E.T. Vishniac, J. Cho, Astrophys. J. **550**, 752 (2001)
E.T. Vishniac, D. Shapovalov, Astrophys. J. **780**, 144 (2014)
Y.-M. Wang, N.R. Sheeley Jr., Astrophys. J. **599**, 1404 (2003)
L.M. Widrow, D. Ryu, D.R.G. Schleicher, K. Subramanian, C.G. Tsagas, R.A. Treumann, Space Sci. Rev. **166**, 37 (2012)
L. Woltjer, Proc. Natl. Acad. Sci. USA **44**, 489 (1958a)
L. Woltjer, Proc. Natl. Acad. Sci. USA **44**, 833 (1958b)
T.A. Yousef, A. Brandenburg, G. Rüdiger, Astron. Astrophys. **411**, 321 (2003)
T.A. Yousef, T. Heinemann, A.A. Schekochihin, N. Kleeorin, I. Rogachevskii, A.B. Iskakov, S.C. Cowley, J.C. McWilliams, Ph R L **100**, 184501 (2008)
H. Zhang, S. Yang, Y. Gao, J. Su, D.D. Sokoloff, K. Kuzanyan, Astrophys. J. **719**, 1955 (2010)
H. Zhang, D. Moss, N. Kleeorin, K. Kuzanyan, I. Rogachevskii, D. Sokoloff, Y. Gao, H. Xu, Astrophys. J. **751**, 47 (2012)

DOI 10.1007/978-1-4939-3547-5_4
Reprinted from *Space Science Reviews* Journal, DOI 10.1007/s11214-014-0045-7

Large-Scale Structure Formation: From the First Non-linear Objects to Massive Galaxy Clusters

S. Planelles · D.R.G. Schleicher · A.M. Bykov

Received: 14 January 2014 / Accepted: 5 April 2014 / Published online: 26 April 2014
© Springer Science+Business Media Dordrecht 2014

Abstract The large-scale structure of the Universe formed from initially small perturbations in the cosmic density field, leading to galaxy clusters with up to 10^{15} $M_\odot$ at the present day. Here, we review the formation of structures in the Universe, considering the first primordial galaxies and the most massive galaxy clusters as extreme cases of structure formation where fundamental processes such as gravity, turbulence, cooling and feedback are particularly relevant. The first non-linear objects in the Universe formed in dark matter halos with 10^5–10^8 $M_\odot$ at redshifts 10–30, leading to the first stars and massive black holes. At later stages, larger scales became non-linear, leading to the formation of galaxy clusters, the most massive objects in the Universe. We describe here their formation via gravitational processes, including the self-similar scaling relations, as well as the observed deviations from such self-similarity and the related non-gravitational physics (cooling, stellar feedback, AGN). While on intermediate cluster scales the self-similar model is in good agreement with the observations, deviations from such self-similarity are apparent in the core regions, where numerical simulations do not reproduce the current observational results. The latter indicates that the interaction of different feedback processes may not be correctly accounted for

S. Planelles (✉)
Department of Astronomy, University of Trieste, via Tiepolo 11, 34143 Trieste, Italy
e-mail: susana.planelles@oats.inaf.it

S. Planelles
INAF—National Institute for Astrophysics, Trieste, Italy

D.R.G. Schleicher
Institut für Astrophysik, Georg-August-Universität Göttingen, Friedrich-Hund-Platz 1,
37077 Göttingen, Germany
e-mail: dschleic@astro.physik.uni-goettingen.de

A.M. Bykov
A.F. Ioffe Institute for Physics and Technology, 194021 St. Petersburg, Russia
e-mail: byk@astro.ioffe.ru

A.M. Bykov
St. Petersburg State Politechnical University, St. Petersburg, Russia

 Springer

in current simulations. Both in the most massive clusters of galaxies as well as during the formation of the first objects in the Universe, turbulent structures and shock waves appear to be common, suggesting them to be ubiquitous in the non-linear regime.

Keywords Cosmology: theory · Early Universe · Numerical simulations · Galaxies: clusters · Hydrodynamics · X-ray: galaxies

1 Introduction

The current hierarchical paradigm of structure formation is set within the spatially flat *Λ-Cold Dark Matter* model (*ΛCDM*; Blumenthal et al. 1984) with cosmological constant, also known as the *concordance* model. Tight constraints on the parameters of the underlying cosmological model have now been placed thanks to the combination of different observational probes (see, e.g. Voit 2005; Allen et al. 2011; Hamilton 2013, for recent reviews). In the resulting scenario (see Ade et al. 2013a, and references therein) the Universe, whose age is estimated to be ~13.8 Gyr, is composed of dark energy ($\Omega_\Lambda \approx 0.7$), dark matter ($\Omega_{DM} \approx 0.25$) and baryonic matter ($\Omega_b \approx 0.05$), with a Hubble constant given by $H_0 \approx 67$ km/s/Mpc. In addition, the primordial matter power spectrum seems to be characterized by a power-law index $n \approx 0.96$ with an amplitude $\sigma_8 \approx 0.83$.

Within this paradigm, the formation of the first structures in the Universe is driven by the gravitational collapse of small inflation-induced matter density perturbations existing in the primordial matter density field. Predictions from N-body simulations (e.g. Klypin and Shandarin 1983) have confirmed that the growth of these perturbations gives rise to the formation of a complex network of cosmic structures interconnected along walls and filaments concerning a wide range of scales.

The first structures in the Universe are expected to form at redshifts of 10–30 in dark matter (DM) halos of 10^5–10^8 $M_\odot$ (Tegmark et al. 1997; Barkana and Loeb 2001; Glover 2005; Bromm et al. 2009). A crucial condition for these DM halos to form stars or galaxies is the ability of their gas to cool in a Hubble time. To address this question, Tegmark et al. (1997) have modeled the cooling in DM halos of different virial temperatures, showing that a virial temperature of at least 1000 K is required so that efficient cooling via molecular hydrogen can occur. Such a temperature corresponds to a mass scale of

$$M_{H_2} \sim 10^{6.5} \left(\frac{10}{1+z} \right)^{3/2} M_\odot. \tag{1}$$

Halos of this or slightly higher masses are typically referred to as the so-called minihalos, which are generally assumed to harbor the first primordial stars in the Universe. Their formation has been explored through detailed numerical simulations starting from cosmological initial conditions, following the formation of the first minihalos and their gravitational collapse, including gas chemistry and cooling, down to AU-scales or below (Abel et al. 2002; Bromm and Larson 2004; Yoshida et al. 2008). The first such simulations typically followed only the formation of the first peak during the gravitational collapse, hinting at the formation of rather massive isolated stars of ~100–300 $M_\odot$ due to the rather high accretion rates of ~10^{-3} $M_\odot$ yr^{-1}. Subsequent studies have explored the formation of self-gravitating disks and their fragmentation at later stages (Stacy et al. 2010; Clark et al. 2011; Greif et al. 2011, 2012; Latif et al. 2013c), indicating the formation of star clusters and binaries rather than isolated stars. The resulting initial mass function (IMF) of these stars is expected to be top-heavy, with characteristic masses in the range of 10–100 $M_\odot$. The studies involving sink particles further suggest that low-mass protostars can be ejected from the center of the halo via 3-body interactions, thus implying the potential presence of primordial stars with less than a solar mass that could survive until the present day. Radiative feedback seems to imply an upper mass limit of 50–100 $M_\odot$ (Hosokawa et al. 2011; Susa 2013).

In DM halos with virial temperatures above 10^4 K, an additional cooling channel is present via atomic hydrogen. In such DM halos, also referred to as atomic cooling halos, cooling is always possible via atomic hydrogen lines, helium lines or recombination cooling, while the minihalos may not be able to cool if their molecular hydrogen content is destroyed by photodissociating backgrounds (Machacek et al. 2001; Johnson et al. 2007, 2008; Schleicher et al. 2010b; Latif et al. 2011). Such halos are also more robust with respect to the first supernova explosions (Wise and Abel 2008; Greif et al. 2010), and may thus give rise to a self-regulated mode of star formation. In the presence of a strong radiative background, for instance from a nearby galaxy, they may remain metal-free and collapse close to isothermally at $\sim$8000 K (Omukai 2001; Spaans and Silk 2006; Schleicher et al. 2010b; Shang et al. 2010; Latif et al. 2011, 2013b; Prieto et al. 2013). While the initial studies followed on the collapse of the first peak (Wise et al. 2008; Regan and Haehnelt 2009a; Shang et al. 2010), Regan and Haehnelt (2009b) aimed at following the longer-term evolution confirming the formation of a self-gravitating disk. Latif et al. (2013a) recently pursued the first high-resolution investigation on the fragmentation of such halos on AU scales, finding that fragmentation may occur, but does not inhibit the growth of the resulting central objects. For the high accretion rates of $\sim$1 $M_\odot$ yr^{-1} measured in their simulations, radiative feedback is expected to be negligible (Hosokawa et al. 2012) and the formation of very massive objects with up to 10^5 $M_\odot$ seems feasible (Schleicher et al. 2013). Such supermassive stars are expected to collapse via the post-Newtonian instability and form the progenitors of supermassive black holes (SMBHs; Shapiro and Teukolsky 1986).

Depending on previous metal enrichment and the ambient radiation field, atomic cooling halos may also gather the proper conditions for the formation of the first galaxies. The formation and evolution of these galaxies, directly connected to the formation of the first stars and their associated radiative or supernova feedback, represent a crucial and complicated aspect of the whole cosmic history. The main focus of this review will concern the formation of the first stars and SMBHs, while the reader is referred to the reviews by Bromm et al. (2009) and Bromm and Yoshida (2011) for the formation and properties of the first galaxies.

In the hierarchical paradigm of structure formation, the first objects are the building blocks of subsequent structure formation, leading to larger galaxies and galaxy clusters through accretion and mergers (e.g. Somerville et al. 2012). As a consequence of this connection, regardless of the wide range of involved scales, a number of physical processes, such as the generation of turbulence during collapse and the relevance of cooling and feedback processes, seem to be common in the formation of the different cosmic structures. Roughly speaking, the cosmic hierarchy is delimited, in terms of mass and formation time, by the first galaxies in the early Universe and the most massive galaxy clusters at the present day, whereas the bulk of galaxies generally lie in-between these extreme cases. However, a full understanding of galaxy evolution represents a complex and fundamental topic in cosmology that is being currently investigated by a considerable number of authors (see Silk and Mamon 2012, for a recent review on the current status of galaxy formation). Given the complexity of this topic and the limited space available for this review, we avoid any description of galaxy evolution. Instead, we are mostly interested in the role that the physics of plasma plays on the formation of cosmic structures. We will focus both on the formation of the first objects, i.e. the first stars and massive black holes, as well as on the large galaxy clusters at the present day. These extreme scenarios will allow us to illustrate the importance of cooling, turbulence and feedback during structure formation independently of the considered scales.

Galaxy clusters are the largest nonlinear objects in the Universe today and thus a central part of the large-scale structure (LSS). Clusters of galaxies, whose total masses vary from 10^{13} up to 10^{15} $M_\odot$, are characterized by very deep gravitational potential wells containing a large number of galaxies ($\sim 10^2$–10^3) over a region of a few Mpc (see, e.g. Sarazin 1988, for an early review on galaxy clusters). Although most of the mass in clusters is in the form of DM, a very hot and diffuse plasma, the intra-cluster medium (ICM), resides within the space between galaxies in clusters. The ICM, where the thermal plasma coexists with magnetic fields and relativistic particles, holds the major part of the baryonic matter in clusters. This cluster environment affects the evolution of the hosted galaxies by means of a number of dynamical processes such as harassment, ram-pressure stripping or galaxy mergers (e.g. see Mo et al. 2010, for a textbook on galaxy formation and evolution). The intra-cluster plasma, with typical temperatures of $T \sim 10^7$–10^8 K, strongly emits X-ray radiation, causing clusters of galaxies to have high X-ray luminosities, $L_X \sim 10^{43}$–10^{45} erg/s. In addition, the ICM is quite tenuous, with electron number densities of $n_e \sim 10^{-4}$–10^{-2} cm^{-3} and, although it is formed mainly of hydrogen and helium, it also holds a mean abundance of heavier elements of about $\sim 1/3$ of the solar abundance.

Given their typical extensions and their deep gravitational potential wells, clusters of galaxies are fundamental for our comprehension of the Universe, marking the transition between cosmological and galactic scales. Whereas on cosmological scales the growth of perturbations is mainly driven by the effects of gravity on the DM component, on galactic scales gravity operates in connection with a number of gas dynamical and astrophysical phenomena. Given such an scenario, galaxy clusters and, in particular, the hot intra-cluster plasma represent a fascinating and complex environment harboring a wide range of astrophysical and dynamical processes related to both the gravitational collapse and the baryonic physics: gravitational shock waves, gas radiative cooling, star formation (SF), gas accretion onto SMBHs hosted by massive cluster galaxies, feedback from supernovae (SNe) or active galactic nuclei (AGN), shock acceleration, magnetohydrodynamical (MHD) processes, gas turbulence, ram-pressure stripping of galaxies, thermal conduction processes, energetics associated to the populations of cosmic ray (CR) electrons and protons, etc.

All these processes are manifested by a number of cluster observables such as the thermal X-ray emission, the Sunyaev-Zel'dovich effect (SZ; Sunyaev and Zeldovich 1972), the spectra of galaxies, or the radio synchrotron and gamma-ray emissions associated to the population of non-thermal particles. As a consequence, galaxy clusters reside in an incomparable position within astrophysics and cosmology: while the number and distribution of clusters can be used to place constraints on the current model of cosmic structure formation, a thorough understanding of the complicated processes determining the properties of the hot intra-cluster plasma seems to be crucial to fully understand galaxy cluster observations.

In this review, we describe the formation of the large-scale structure of the Universe in the framework of the ΛCDM model. A particular focus is both on the formation of the first objects, i.e. the first stars and massive black holes, as well as on the large galaxy clusters at the present day. In both applications, we emphasize the role of gravitational as well as non-gravitational plasma physics such as turbulence, cooling, magnetic fields or feedback processes. The overall structure of this review is as follows: in Sect. 2 we start by reviewing the basic concepts of cosmic structure formation, from the early linear evolution of small density perturbations out to the complex collapse of real overdensities; in Sect. 3 we overview the relevance for cosmology of a proper calibration of the halo mass function; in

Sect. 4 we describe the formation of the first halos in the early Universe; a brief description of the self-similar model of the intra-cluster plasma is done in Sect. 5, whereas in Sect. 6, the role played by non-gravitational heating and cooling processes in altering the predictions of such a model is discussed; finally, we summarize the results presented in Sect. 7.

Given the limited space available for this review, we refer the reader to recent reviews about early structure formation in the Universe (e.g. Bromm and Yoshida 2011) and cosmology with clusters of galaxies (e.g. Allen et al. 2011; Kravtsov and Borgani 2012) for a more extensive discussion of these topics.

2 Theory of Structure Formation

In this Section we outline the main theory of cosmic structure formation through the process of gravitational instability of small initial density perturbations. We refer the reader to previous reviews or cosmology textbooks for a more detailed analysis (e.g. Peebles 1993, Coles and Lucchin 2002; see as well Borgani 2008).

2.1 Linear Evolution of Density Perturbations

The gravitational instability of a uniform and non-evolving medium versus small perturbations was first addressed by Jeans (1902). Applying this theory to an expanding Universe in the linear regime, while density perturbations are small, provides a general picture of cosmic structure formation.

Let us consider an initial density perturbation field characterized by its dimensionless density contrast:

$$\delta(\mathbf{x}) = \frac{\rho(\mathbf{x}) - \bar{\rho}}{\bar{\rho}}, \tag{2}$$

where $\rho(\mathbf{x})$ is the matter density field at the position $\mathbf{x}$, and $\bar{\rho} = \langle \rho \rangle$ is the mean mass density of the background universe. The primordial properties of this field are determined during the inflationary epoch. In general, inflationary models predict a homogeneous and isotropic Gaussian random fluctuation field (e.g. Guth and Pi 1982), which appears to be confirmed by observed fluctuations in the Cosmic Microwave Background (CMB; e.g. Ade et al. 2013b).

To resolve the evolution of the initial density perturbations in an expanding Universe, the perturbed Friedmann's equations need to be solved. However, during the linear evolution the problem can be simplified. Consider that a self-gravitating and pressureless fluid dominates the matter content of an expanding Universe. In principle, these assumptions are valid if the perturbation is unstable, that is, if its scale is larger than the characteristic Jeans scale,[1] and if we deal with the evolution of DM perturbations. If the fluid is also assumed to be non-relativistic, the Newtonian treatment can be applied. In this case, the evolution of density perturbations is described by the continuity, the Euler, and the Poisson equations:

$$\frac{\partial \delta}{\partial t} + \nabla \cdot \left[(1 + \delta)\mathbf{u} \right] = 0 \tag{3}$$

---

[1] The Jeans length, the characteristic length scale for the self-gravity of the gas, is defined as $\lambda_J = \sqrt{\frac{15 k_B T}{4 \pi G \mu \rho_{gas}}}$, with k_B the Boltzmann constant, T the gas temperature, G the Newton's constant, μ the mean molecular weight and ρ_{gas} the mass density of the gas.

$$\frac{\partial \mathbf{u}}{\partial t} + 2H(t)\mathbf{u} + (\mathbf{u} \cdot \nabla)\mathbf{u} = -\frac{\nabla \phi}{a^2} \tag{4}$$

$$\nabla^2 \phi = 4\pi G \bar{\rho} a^2 \delta, \tag{5}$$

where spatial derivatives are with respect to the comoving coordinate $\mathbf{x}$, $a(t)$ is the cosmic expansion factor such that $\mathbf{r} = a(t)\mathbf{x}$ is the proper coordinate, $\mathbf{v} = \dot{\mathbf{r}} = \dot{a}\mathbf{x} + \mathbf{u}$ is the total velocity of a fluid element (with $\dot{a}\mathbf{x}$ and $\mathbf{u} = a(t)\dot{\mathbf{x}}$ giving the Hubble flow and the peculiar velocities, respectively), $\phi(\mathbf{x})$ is the gravitational potential and $H(t) = \dot{a}/a = E(t)H_0$ is the time-dependent Hubble parameter. In the case of a ΛCDM cosmology, when relativistic species are neglected, $E(z)$ is given by

$$E(z) \equiv \frac{H(t)}{H_0} = \left[(1+z)^3 \Omega_m + (1+z)^2(1 - \Omega_m - \Omega_\Lambda) + \Omega_\Lambda\right]^{1/2}. \tag{6}$$

When small density fluctuations ($\delta \ll 1$) are considered, all the non-linear terms with respect to δ and $\mathbf{u}$ can be ignored and, therefore, the above equations can be written as

$$\frac{\partial^2 \delta}{\partial t^2} + 2H(t)\frac{\partial \delta}{\partial t} = 4\pi G \bar{\rho} \delta. \tag{7}$$

This relation represents one of the most fundamental equations within the linear theory of gravitational collapse: it delineates the Jeans instability of a fluid with no pressure under the counter-effect of the cosmic expansion (represented by the $2H(t)\partial\delta/\partial t$ term). Since Eq. 7 is a second order differential equation in time t, its solution can be written as

$$\delta(\mathbf{x}, t) = \delta_+(\mathbf{x}, t_i)D_+(t) + \delta_-(\mathbf{x}, t_i)D_-(t), \tag{8}$$

where $D_+(t)$ and $D_-(t)$ are, respectively, the growing and decaying modes of $\delta(\mathbf{x}, t)$, and $\delta_+(\mathbf{x}, t_i)$ and $\delta_-(\mathbf{x}, t_i)$ the corresponding spatial distribution of the primordial matter field. Given that the density growing modes only depend on time, the density fluctuations will evolve at the same pace throughout the cosmic volume. However, these density growing modes depend on the particular underlaying cosmology in such a way that, in different Friedmann–Lemaitre–Robertson–Walker (FLRW) universes structures will grow in a different manner.

As an example, in the case of a flat matter-dominated Einstein–de-Sitter universe (EdS, $\Omega_m = 1$, $\Omega_\Lambda = 0$), given that $H(t) = 2/(3t)$, $D_+(t) = (t/t_i)^{2/3} \propto a(t)$ and $D_-(t) = (t/t_i)^{-1}$. Therefore, in this particular case, cosmic expansion and gravitational instability proceed at the same rate. Contrarily, it can be shown that, in the case of a cosmological model with $\Omega_m < 1$, such as a flat one with $\Omega_m = 0.3$, there is an epoch, when the cosmological constant begins to be significant, at which the characteristic time-scale of expansion turns out to be shorter than in the EdS case. As a consequence, after that epoch, cosmic expansion proceeds faster than gravitational collapse, generating a minor evolution in the number of collapsed objects between $z \sim 0.6$ and $z = 0$ (e.g. Borgani and Guzzo 2001). These results indicate that the observational determination of the level of evolution of collapsed regions (such as galaxy clusters) provides important constraints on cosmological parameters.

We can define the two-point correlation function of $\delta(\mathbf{x})$ as $\xi(r) = \langle\delta(\mathbf{x}_1)\delta(\mathbf{x}_2)\rangle$, which depends only on the distance between the considered points, $r = |\mathbf{x}_1 - \mathbf{x}_2|$. $\xi(r)$ describes whether the density field is more ($\xi(r) > 0$) or less ($\xi(r) < 0$) concentrated than the mean. In addition, a convenient description of $\delta(\mathbf{x})$ is given by its Fourier representation $\delta(\mathbf{k}) =$

$(2\pi)^{-3/2} \int d\mathbf{x} \delta(\mathbf{x}) e^{i\mathbf{k}\cdot\mathbf{x}}$. If we also express $\xi(r)$ in Fourier space, it is easily demonstrable that its Fourier transform corresponds to the power spectrum of density fluctuations:

$$P(k) = \langle |\delta(\mathbf{k})|^2 \rangle = \frac{1}{2\pi^2} \int dr\, r^2 \xi(r) \frac{\sin kr}{kr}, \tag{9}$$

which does not depend on the orientation of the wave-vector $\mathbf{k}$ but on its modulus.

$P(k)$ is a fundamental quantity that provides a full statistical description of a uniform and isotropic Gaussian field. Inflationary models predict a perturbation power spectrum of the form $P(k) = Ak^n$, where A is the normalization and n the spectral index. More precisely, inflation provides a nearly Gaussian density perturbation field characterized by a scale-invariant spectrum with $n \simeq 1$ (e.g. Guth and Pi 1982), which appears to be confirmed by measured CMB anisotropies (e.g. Ade et al. 2013b).

A common practice in the analysis of cosmological density fields is that of using filtering functions to define spatial scales. To analyze the collapse of primordial fluctuations on scales $R \propto (M/\bar{\rho})^{1/3}$, giving rise to objects of mass M, it is useful to define a window function, $W_R(r)$, which filters out the modes on smaller scales, and the corresponding smoothed density field, $\delta_R(\mathbf{x}) = \delta_M(\mathbf{x}) = \int \delta(\mathbf{y}) W_R(|\mathbf{x} - \mathbf{y}|) d\mathbf{y}$. If the Fourier transform of the window function[2] is $W_R(k)$, the variance of the perturbation field at the scale R is given by

$$\sigma_R^2 = \sigma_M^2 = \langle \delta_R^2 \rangle = \frac{1}{2\pi^2} \int dk\, k^2 P(k) W_R^2(k). \tag{10}$$

In principle, whereas the functional form of $P(k)$ depends on Ω_m, Ω_b, and H_0 (e.g. Eisenstein and Hu 1999), its normalization, which is related to σ_R^2, needs to be determined by observations of the cosmic LSS or of the CMB anisotropies. The most widely used parameter for this normalization is σ_8, which is the variance estimated within comoving spheres of radius $R = 8\,h^{-1}$ Mpc, roughly matching the typical scale of massive clusters.[3] As we discuss in Sect. 3, an estimate of σ_8 is given by the halo mass function.

This linear approximation applies after recombination, while $\delta \ll 1$, to describe the evolution of density fluctuations on an initial mass scale $M \gtrsim M_J(z_{rec}) \sim 10^5\ M_\odot$. However, this linear theory can not be used to study the growth of structures in the strongly non-linear regime, where typical fluctuations reach amplitudes of about unity and overdensities with $\delta \gg 1$ are plausible (as an example, a cluster of galaxies corresponds to $\delta \gtrsim 100$). In this case, non-linear models or numerical simulations are required to solve the evolution.

2.2 Non-linear Evolution of Spherical Perturbations: The Spherical Top-Hat Collapse

The only situation in which the non-linear evolution can be precisely calculated is the one addressed by the simple spherically symmetric collapse model (e.g. Gunn and Gott 1972; Bertschinger 1985). This model resolves the evolution of a spherical density perturbation of radius R into a virialized halo. Initially, the spherical perturbation is assumed to have constant overdensity and, since it is expanding with the background universe, null velocity

---

[2]The functional form of the window function, which depends on the particular choice of filter, provides the connection between mass and smoothing scale. Two common filter functions are $W_R(k) = 3[\sin(kR) - kR\cos(kR)]/(kR)^3$ and $W_R(k) = \exp(-(kR)^2/2)$ corresponding to the top-hat and the Gaussian windows, respectively. For each of these filters, the correspondent relation between mass and smoothing scale is given by $M = (4\pi/3)R^3\bar{\rho}$ and $M = (2\pi R^2)^{3/2}\bar{\rho}$.

[3]Early redshift surveys showed that $\sigma \sim 1$ for spheres of $R = 8\,h^{-1}$ Mpc (e.g. Davis and Peebles 1983).

at its border. The symmetry of this configuration allows us to treat the spherical perturbation as an isolated FLRW universe, meaning that we can describe the growth of the overdensity using the same Friedmann's equations as for cosmology.

For simplicity we consider that the background universe is described by a close EdS model, in which the perturbation radius $R(t)$ behaves in the same way as the expansion factor. Within such a model, after a short period of time the growing mode will dominate the evolution of the perturbation. At the initial time t_i, by imposing the condition of null velocities at the edge of the spherical region, the linear growing mode is given by $D_+(t_i) = (3/5)\delta(t_i)$ and, thus, the corresponding density parameter is $\Omega_p(t_i) = \Omega(t_i)(1 + \delta_i)$, where the suffix p stands for the perturbation itself, and the other quantities for the unperturbed background universe.

The perturbation will grow until reaching its maximum expansion at a given turn-around time, t_{ta}. If at this moment the spherical perturbation detaches from the expansion of the background and, instead, initiates to collapse, the structure will be formed. By solving the Friedmann's equations it can be shown that, if $\Omega_p(t_i) > 1$, the perturbation will recollapse. At t_{ta}, the corresponding perturbation overdensity is given by[4] $\delta_+(t_{ta}) \simeq 4.6$.

After t_{ta}, the perturbation decouples from the underlying cosmic expansion and recollapses, reaching an equilibrium state at the time t_{vir} at which the virial condition between the kinetic K and the potential U energy of the perturbation is satisfied, that is, $U = -2K$. Assuming energy conservation during the evolution into this equilibrium state, the virial condition can be used to derived that $R_{ta} = 2R_{vir}$ and that the density at t_{vir} is $\rho_p(t_{vir}) = 8\rho_p(t_{ta})$. Hence, the non-linear overdensity of the perturbation at the virialization is given by

$$\Delta_{vir} = \frac{\rho_p(t_{vir})}{\rho(t_{vir})} = 18\pi^2 \simeq 178. \tag{11}$$

Despite the simplicity of the spherical collapse model, this result is quite encouraging since N-body simulations find that a density contrast of $\sim$100–200 is quite successful in defining DM halos. Indeed, in these simulations a common definition of halo mass is given by M_{200}, defined as the mass enclosed by a sphere with an overdensity equal to 200.

On the contrary, the linear-theory extrapolation predicts a smaller value for the required overdensity at the time of collapse:

$$\delta_c = \delta_+(t_{vir}) = \delta_+(t_{ta})\left(\frac{t_{vir}}{t_{ta}}\right)^{2/3} \simeq 1.69. \tag{12}$$

As we will see later, this is a key discriminant that determines the halo mass function.

The above equations, valid for an EdS cosmology, can be extended to any other cosmological model. In this sense, the overdensity at virialization can be defined as Δ_c or as Δ_{vir} depending on whether it is referred to the critical, $\rho_c(z)$, or to the mean background matter density, $\rho_m(z)$. These overdensities relate to each other as $\Delta_{vir} = \Delta_c/\Omega_m(z)$ (see Bryan and Norman 1998 for an estimation of Δ_{vir} in open and flat ΛCDM universes).

2.3 Non-linear Evolution of Real Overdensities

Given the simplicity of the spherical collapse model, it is not adequate to properly describe the non-spherical collapse of actual overdense regions. To this end, cosmological simulations (see, e.g. Dolag et al. 2008; Borgani and Kravtsov 2011, for details on the numerical techniques) are essential to deepen into the main properties of the real gravitational collapse.

---

[4]On the contrary, the linear-theory extrapolation to t_{ta} yields a smaller value: $\delta_+(t_{ta}) \simeq 1.07$.

As an example, the left panel of Fig. 1 displays the evolution of the DM density field from $z \simeq 4$ until $z = 0$ as obtained from a cosmological hydrodynamical simulation. At early epochs, collapsed objects with low masses populate the proto-cluster region. As the evolution proceeds, these objects merge into larger structures at later times. As can be inferred from this figure, the actual collapse of overdense regions shows a number of complexities in comparison to the top-hat collapse model: severe departures from spherical symmetry and constant density edges, filamentary matter accretion, and the existence of smaller overdensities within larger overdense already collapsing regions. Given that different overdense regions have different spatial extensions and their time evolution proceeds differently, the actual collapse of a cluster-scale overdensity is a process prolonged in time (e.g. Diemand et al. 2007). Besides, the non-linear nature of this picture gives rise to merging events and interactions between overdensities at different scales, leading to an important matter redistribution within the considered regions.

Despite the complexities associated with the process of non-linear collapse, as the evolution proceeds, the resulting collapsing regions tend to reach an equilibrium state. This state is described differently depending on the collisional or collisionless nature of the considered component. Indeed, the equilibrium state of the collisional gas component can be approximated by the condition of hydrostatic equilibrium (HE from now on), $\nabla \phi(\mathbf{x}) = -\nabla p(\mathbf{x})/\rho_{gas}(\mathbf{x})$, under which pressure gradients and gravitational forces compensate each other. On the contrary, the equilibrium configuration for the collisionless dark matter component is provided by the Jeans equation (Binney and Tremaine 2008). With the additional assumptions of spherical symmetry and an ideal gas equation of state for the ICM gas, the resulting equations (see, e.g. Kravtsov and Borgani 2012, for a complete description) are commonly used to derive cluster masses (e.g. Ettori et al. 2013). However, a number of processes, such as continuous matter accretion or merging events, can keep clusters away from equilibrium, introducing systematic uncertainties when applying the above assumptions.

Given the complexities inherent to the non-linear process of halo formation, the definition of a DM halo is not trivial and, as a consequence, there is no single definition that is agreed upon in the literature. This incongruity has resulted in a number of different halo finding algorithms based on different halo boundaries and mass definitions (see Knebe et al. 2011, 2013; Onions et al. 2012, for recent comparisons of different halo finders). In this sense, two of the most widely used halo definitions are those based on the Friends-of-Friends (FoF; Davis et al. 1985) and the Spherical Overdensity (SO; Lacey and Cole 1994) algorithms.[5]

3 The Halo Mass Function

The halo mass function (HMF) is the number density of collapsed objects, at redshift z, with mass between M and $M + dM$ in a given comoving volume. While from an observational point of view the HMF is difficult to determine with high precision (e.g. Rozo et al. 2010), it can approximately be analyzed through analytic models (e.g. Press and Schechter 1974), and it is relatively simple to study by means of N-body cosmological simulations (e.g. Cohn and White 2008; Tinker et al. 2008; Crocce et al. 2010; Courtin et al. 2011; Bhattacharya et al. 2011; Angulo et al. 2012; Watson et al. 2013; Murray et al. 2013, for recent studies).

---

[5]While the FoF algorithm identifies DM halos with groups of DM particles separated by a distance shorter than a given linking length parameter, the SO algorithm is based on the mean overdensity criterion.

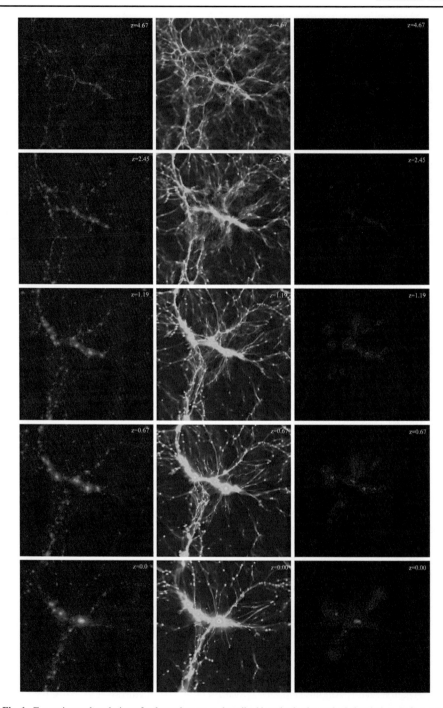

Fig. 1 Formation and evolution of galaxy clusters as described by a hydrodynamical simulation. *Left, central* and *right columns* show, respectively, the evolution of the dark matter, gas and stellar densities from $z \simeq 4$ (*top panels*) until $z = 0$ (*bottom panels*). At $z = 0$, the biggest cluster formed has a virial mass of $\sim 10^{15}$ M$_\odot$ and a radius of ~ 3 Mpc. The simulation was performed with the Eulerian-AMR cosmological code MASCLET (Quilis 2004). *Each panel* is 64 comoving Mpc length per edge and 5 comoving Mpc depth

In this Section, after introducing the HMF as originally derived by Press and Schechter (1974), we overview how cosmological simulations are currently used to provide more precise calibrations of this important prediction.

3.1 The Press-Schechter Approach

Press and Schechter (1974, PS from now on), based on the spherical collapse model, performed the first analytical attempt to derive the HMF. The main idea of this formalism is that, any collapsed object with mass $\geq M$ by redshift z stems from regions where $\delta_M \geq \delta_c$, being δ_M the linearly extrapolated density field (smoothed on a mass scale M), and δ_c the critical overdensity for collapse. Motivated by the spherical collapse model (see Eq. 12), $\delta_c \simeq 1.69$, being z-independent only in an EdS universe.[6] Assuming a Gaussian distribution for the initial density fluctuations, the probability of a given point to be within a region of scale R satisfying the above conditions is:

$$F(M, z) = \frac{1}{\sqrt{2\pi}\sigma_M(z)} \int_{\delta_c}^{\infty} \exp\left(-\frac{\delta_M^2}{2\sigma_M(z)^2}\right) d\delta_M, \tag{13}$$

where $\sigma_M(z)$ is the corresponding rms density fluctuation.

From the above equation, the HMF is estimated as $\partial F(M, z)/\partial M$ (the fraction of independent regions evolving into objects with mass between M and $M + dM$) divided by $M/\bar{\rho}$. An inherent implication of the PS approach is that, only overdense regions participate in the spherical collapse and, consequently, only half of the total mass content of the Universe is considered. Thus, including a missing factor of 2, the PS mass function is given by:

$$\frac{dn(M, z)}{dM} = \frac{2}{(M/\bar{\rho})} \frac{\partial F(M, z)}{\partial M} = \sqrt{\frac{2}{\pi}} \frac{\bar{\rho}}{M^2} \frac{\delta_c}{\sigma_M(z)} \left|\frac{d\log\sigma_M(z)}{d\log M}\right| \exp\left(-\frac{\delta_c^2}{2\sigma_M(z)^2}\right)$$
$$= \frac{\bar{\rho}}{M} \psi(\nu), \tag{14}$$

which only depends on the peak height $\nu \equiv \delta_c(z)/\sigma_M(z)$.

It has been shown that the functional form of $\psi(\nu)$ provided by the PS approach diverts significantly from the one derived from cosmological simulations (e.g. Sheth and Tormen 1999; Jenkins et al. 2001; Tinker et al. 2010). In order to alleviate these discrepancies, a number of changes in the original model have been introduced (Bond et al. 1991; Lacey and Cole 1993). In this regard, the HMF has been analyzed accounting for the ellipsoidal collapse (Audit et al. 1997; Sheth and Tormen 1999; Sheth et al. 2001) or, within the excursion set theory, for non-Gaussian primordial conditions (Maggiore and Riotto 2010). However, given the relatively simple assumptions on which these analytical prescriptions rely, their accuracy to properly describe the HMF is limited. In spite of these limitations, current analytical models are reaching higher accuracies (e.g. Paranjape et al. 2013; Paranjape 2014), providing an important way to analyse the HMF.

3.2 Using Cosmological Simulations to Calibrate the Halo Mass Function

Cosmological simulations represent a powerful means to accurately calibrate the HMF (see Murray et al. 2013, for a recent comparison of different HMF available in the literature).

Given the exponential dependence of the HMF on mass and redshift, a precise calibration is extremely useful to place constraints on cosmological parameters (e.g. Allen et al. 2011;

---

[6]In the general case, however, this overdensity depends weakly on redshift and cosmology (for example, $\delta_c \simeq 1.675$ in a ΛCDM model at $z = 0$).

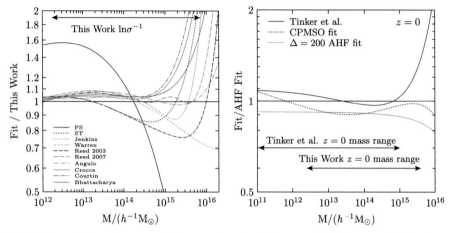

Fig. 2 *Left panel*: Comparison, at $z = 0$, between the universal fit proposed by Watson et al. (2013) for the FoF HMF and several FoF fits available in the literature. *Right panel*: Comparison, at $z = 0$, between the mass function obtained by Watson et al. (2013) with two different SO halo finders (AHF and CPMSO) and the fit by Tinker et al. (2008). Figures from Watson et al. (2013). In *both figures*, the label 'This Work' refers to the work by Watson et al. (2013)

Weinberg et al. 2013, for recent reviews). As an example, whereas at low redshift the mass function of massive objects can be used to set limits on the combination of σ_8 and Ω_m, the redshift evolution of the HMF helps in breaking this degeneracy.

On the other hand, however, this exponential dependence makes the HMF very sensitive to the particular mass definition or to the small variations of $\delta_c(z)$ with redshift and cosmology. In this regard, many analysis of halo abundance by means of DM-only simulations consider that $\delta_c(z)$ is constant. As a consequence, they generally obtain a HMF that can be written as an almost 'universal' function of ν, i.e., independent of redshift or cosmology (e.g. Sheth and Tormen 1999; Jenkins et al. 2001; Evrard et al. 2002; White 2002; Warren et al. 2006; Tinker et al. 2008; Crocce et al. 2010; Bhattacharya et al. 2011; Courtin et al. 2011). However, small deviations in $\delta_c(z)$ can induce important changes in the HMF and, therefore, a precise calibration of the HMF requires to account for the correct dependencies of $\delta_c(z)$ (e.g. Courtin et al. 2011).

In the last years, in order to reach an accurate description of the shape of the HMF, large N-body simulations have been used to evaluate the uncertainties induced by different redshifts, cosmological models and halo definitions. As an example, Watson et al. (2013), with a set of large cosmological DM-only simulations, examined the redshift evolution (out to $z = 30$) of the HMF and its dependence on the FoF and SO halo mass definitions (see Knebe et al. 2011, for a comparison of the HMF provided by different halo finders). In this work, they showed that the SO HMF clearly evolves with redshift and obtained a z-parameterized fit suitable for the whole redshift interval to within $\sim$20 %. The right panel of Fig. 2 shows the comparison of several SO mass functions that they obtained at $z = 0$. On the contrary, a weaker z-evolution was found for the FoF HMF. In this case, as it is shown in the left panel of Fig. 2, they obtained a 'universal' fit function[7] that agrees to within $\sim$10 %

---

[7]The fit obtained for the HMF based on the FoF halos takes the form: $f(\sigma) = A[(\frac{\beta}{\sigma})^\alpha + 1]e^{-\gamma/\sigma^2}$, with $A = 0.282$, $\alpha = 2.163$, $\beta = 1.406$, and $\gamma = 1.210$. This fit holds for $-0.55 \leq \ln\sigma^{-1} < 1.31$, corresponding to masses within $[1.8 \times 10^{12}, 7.0 \times 10^{15}]\, h^{-1}\, M_\odot$ at $z = 0$.

with a number of fits available in the literature at $z = 0$, a degree of deviation in accordance with the values reported in previous works (e.g. Reed et al. 2007; Lukić et al. 2007; Tinker et al. 2008; Courtin et al. 2011).

More recently, hydrodynamical simulations have demonstrated that baryonic cooling and heating processes can also affect the HMF (e.g. Rudd et al. 2008; Stanek et al. 2009; Cui et al. 2012; Cusworth et al. 2014; Cui et al. 2014). In this regard, Cui et al. (2014) recently found that, in comparison to DM-only simulations, hydrodynamical simulations accounting for cooling, star formation and SN feedback produce an increase of the HMF, while simulations including as well AGN feedback tend to reduce it by an overdensity-dependent amount. This reduction is a consequence of the changes that AGN feedback induces in gas density profiles and, therefore, in halo masses. However, given our limited understanding of the physics of baryons, and in view of the large galaxy cluster surveys coming in the near future, the importance of this effect needs to be further and accurately investigated.

4 Structure Formation in the Early Universe

The first objects in the Universe were expected to form in halos with 10^5–10^8 $M_\odot$ at redshifts 10–30. As discussed in the introduction, one may distinguish between the so-called minihalos, with virial temperatures of a few ~ 1000 K and cooling via molecular hydrogen, and the so-called atomic cooling halos with virial temperatures of 10^4 K. Depending on previous metal enrichment and the ambient radiation field, the latter may form the first galaxies or directly collapse into supermassive black holes. The main focus of this review will be on the latter scenario, and the reader is referred to the reviews by Bromm et al. (2009) and Bromm and Yoshida (2011) concerning the formation and properties of the first galaxies.

4.1 Primordial Star Formation in the First Minihalos

A central ingredient for star formation in the first minihalos is their ability to cool via molecular hydrogen. The latter is possible since H_2 may form even in primordial gas via the so-called H^- channel

$$H + e \rightarrow H^- + \gamma, \tag{15}$$

$$H + H^- \rightarrow H_2 + e \tag{16}$$

and the H_2^+ channel

$$H^+ + H \rightarrow H_2^+ + \gamma, \tag{17}$$

$$H_2^+ + H \rightarrow H_2 + H^+. \tag{18}$$

In both channels, electrons need to act as catalysts in order to promote H_2 formation, which is possible due to the electron freeze-out during cosmic recombination, leading to a ionization fraction of $\sim 10^{-4}$ in the primordial Universe (Peebles 1968). As shown by Saslaw and Zipoy (1967), the latter may promote molecular hydrogen formation in molecular clouds, allowing them to cool and collapse within a Hubble time (Tegmark et al. 1997). The fact that the chemistry is out of equilibrium is thus crucial for the formation of the first structures, and detailed non-equilibrium calculations for the homogeneous Universe have been performed (e.g. Galli and Palla 1998; Stancil et al. 1998; Puy and Signore 2007; Schleicher et al. 2008;

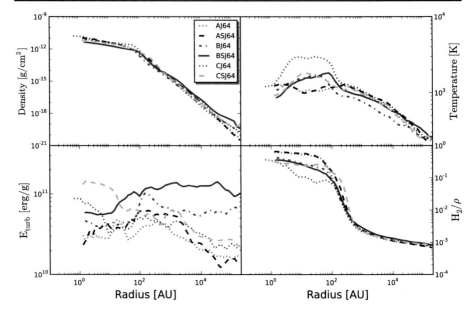

Fig. 3 Gravitational collapse in the first minihalos. Shown are radial profiles of the baryon density (*upper left*), gas temperature (*upper right*), specific turbulent energy density (*lower left*) and the H_2 abundance (*lower right*). *The figure* shows results for three different halos A–C with and without the turbulence subgrid-scale model. Figure by Latif et al. (2013c).

Coppola et al. 2012) to determine the chemical initial conditions for subsequent structure formation.

The first cosmological simulations following the formation and subsequent collapse of the first minihalos, including detailed models for the primordial chemistry and cooling, have been performed by Abel et al. (2002) and Bromm et al. (2002). These calculations were exploring the initial collapse phase where no fragmentation occurred. Yoshida et al. (2006) were the first to incorporate a more detailed treatment of the microphysics at high densities in such simulations. In particular, at number densities of $\sim 10^8$ cm^{-3}, three-body H_2 formation rates become relevant, which turn the hydrogen gas from a predominantly atomic into a fully molecular state. The dominant three-body reactions are given as

$$H + H + H \rightarrow H_2 + H, \tag{19}$$

$$H_2 + H + H \rightarrow 2H_2. \tag{20}$$

In this review, we illustrate the collapse dynamics based on the recent simulation by Latif et al. (2013c). They employed the cosmological hydrodynamics code Enzo (O'Shea et al. 2004; Bryan et al. 2014) including a detailed network for primordial chemistry (Abel et al. 1997) with updated rates and cooling functions, as well as a subgrid-scale model for hydrodynamical turbulence (Schmidt et al. 2006). Their simulation box had a total size of 300 h^{-1} kpc with a root grid resolution of 128^3, and two additional nested grids of the same resolution centered on the most massive halo. The simulation further employed 27 adaptive refinement levels, ensuring a minimum resolution of 64 cells per Jeans length. While Truelove et al. (1997) argued for a minimum resolution of 4 cells per Jeans length to avoid artificial fragmentation, more recent studies indicate a minimum resolution of 32 to 64 cells

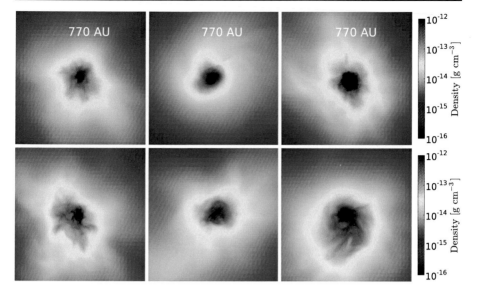

Fig. 4 Projections of the gas density in the central 770 AU for three different halos. *The top panel* shows purely hydrodynamical simulations, while the simulations in *the bottom panel* include the subgrid-scale turbulence model by Schmidt et al. (2006). Figure by Latif et al. (2013c)

to capture the main properties of turbulence (Federrath et al. 2011; Turk et al. 2012; Latif et al. 2013b).

In Fig. 3, we show the central region of this halo when the highest refinement level is reached. The gas density follows approximately an isothermal profile with $\rho \propto r^{-2}$, and flattens on scales comparable to the Jeans length at the density peak. The temperature increases towards smaller radii due to gravitational compression and increasing optical depth, and reaches a temperature of ~ 1000 K at densities of $\sim 10^9$ cm^{-3} when three-body H_2 formation becomes relevant. The specific turbulent energy appears almost independent of scale, implying that turbulence is continuously re-generated during gravitational collapse via shocks and shear flows. The H_2 abundance is of the order 10^{-3} on larger spatial scales, and increases rather steeply at ~ 100 AU when the gas becomes fully molecular as a result of the three-body reactions. The corresponding density projections in the central 770 AU are given in Fig. 4 for three different halos in simulations with and without the turbulence subgrid-scale model. At this stage of the simulation, the central regions are approximately spherical with turbulent fluctuations in the density field. As also reported by Turk et al. (2012), there is no disk during this stage of the evolution.

Following the evolution beyond the first peak is not straightforward, as the Jeans length always needs to be resolved with at least 4 cells to avoid artificial fragmentation (Truelove et al. 1997), thus requiring more and more refinement levels as collapse proceeds. The latter can be avoided by replacing the high-density gas with sink particles, or by turning on an adiabatic equation of state at high densities. Latif et al. (2013c) followed the latter approach to pursue the subsequent evolution for five free-fall times. As shown in Fig. 5, disk structures can be clearly recognized at the end of the simulation both in the simulations with and without the turbulence subgrid-scale model. In one of the simulations, fragmentation has already occurred, while most of the runs show a central massive object with approximately $10 \, M_\odot$. Previous simulations employing a lower resolution per Jeans length indeed followed

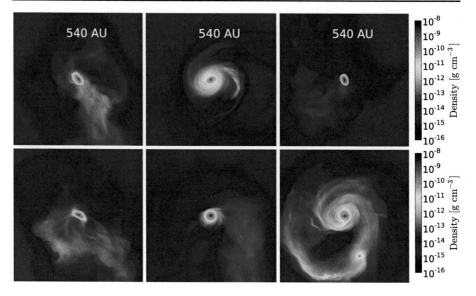

Fig. 5 Projections of the gas density in the central 540 AU for three different halos approximately 40 years (5 free-fall times) after reaching the highest refinement level. *The top panel* shows purely hydrodynamical simulations, while the simulations in *the bottom panel* include the subgrid-scale turbulence model by Schmidt et al. (2006). Figure by Latif et al. (2013c)

the evolution for longer times, indicating that fragmentation is expected to occur in the majority of these systems (Stacy et al. 2010; Clark et al. 2011; Greif et al. 2011).

As for the properties of these disks at the final stage of the simulation, we note that high accretion rates of 10^{-3}–10^{-2} $M_\odot$ yr^{-1} are found in the central 100 AU, with considerable spatial fluctuations as expected in a real system. On the other hand, the rotational velocities range from a few up to 10 km/s, while the radial infall velocities are of order 1 km/s.

While most of the evolution depends only on the initial conditions and the subsequent dynamics, the three-body H_2 formation rate has been a major uncertainty until recently, as the rates derived by different groups (e.g. Palla et al. 1983; Abel et al. 2002; Flower and Harris 2007) showed differences by about three orders of magnitude, implying significant uncertainties for the densities at which the transition to the fully molecular state is expected to occur. A detailed description of these uncertainties was provided by Glover et al. (2008), and their implications in 3D simulations have been explored by Turk et al. (2011). Since recently, a new accurate determination of this rate is however available from quantum-molecular calculations by Forrey (2013) which considerably reduces the uncertainties discussed here.

The impact of this rate compared to the previous rates employed in the literature has been explored by Bovino et al. (2014) in three cosmological halos using the chemistry package KROME[8] (Grassi et al. 2014). A representative example is given here in the top panel of Fig. 6. For the halo shown here, the resulting H_2 fraction lies in-between the simulations based on Abel et al. (2002) and Palla et al. (1983). We however note that in some cases, the result can be closer to one of these two. The same is also true for the abundance of atomic hydrogen. The abundances of the electrons and H^- are lower compared to the other

---

[8]http://kromepackage.org/.

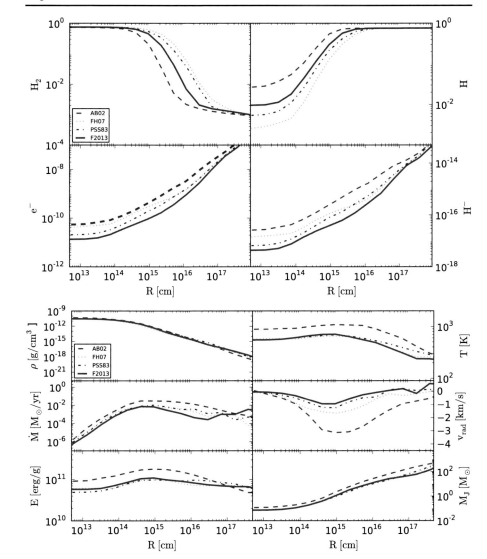

Fig. 6 A comparison of different three-body H$_2$ formation rates based on Palla et al. (1983), Abel et al. (2002), Flower and Harris (2007) and Forrey (2013). Figures by Bovino et al. (2014). *Top panel*: Shown is the H$_2$ abundance (*top left*), the electron abundance (*bottom left*), the atomic hydrogen abundance (*top right*) and the abundance of H$^-$ (*bottom right*). *Bottom panel*: Shown is the gas density (*top left*), accretion rate (*middle left*), specific energy (*bottom left*), gas temperature (*top right*), radial velocity (*middle right*) and the Jeans mass (*bottom right*)

simulations, which is likely a result of the different dynamical evolution as a result of the new rates. We note that this behavior varies strongly from halo to halo without a clear trend.

The impact of the chemical evolution on the dynamics is illustrated in the bottom panel of Fig. 6. We note that central quantities like the gas density have a very small dependence on the rates considered here. The gas temperature shows only minor differences for most of the rates we explored, but the simulation based on Abel et al. (2002) appears to have a clearly enhanced temperature compared to the other cases. The radial velocity depends rather sensitively on the adopted three-body rate and is again larger for Abel et al. (2002). The same is true for the resulting accretion rates. The Jeans mass, on the other hand, is very similar in the cases considered here, with a small enhancement for the rate of Abel et al. (2002).[9] We however note that relevant fluctuations in these quantities are still expected due to locally varying initial conditions.

We further note that the impact of additional physical processes can be expected to be relevant during primordial star formation. Magnetic fields are expected to form rapidly via the small-scale dynamo during the initial collapse (Schleicher et al. 2010a; Sur et al. 2010, 2012; Schober et al. 2012; Turk et al. 2012), and subsequent ordering may occur as a result of large-scale dynamos in the accretion disk (Pudritz and Silk 1989; Tan and Blackman 2004; Silk and Langer 2006). It thus needs to be determined whether jet formation can occur under realistic conditions (Machida et al. 2006). Recent studies further indicate that radiative feedback can potentially set an upper limit on the stellar masses of order 50–100 $M_\odot$ (Hosokawa et al. 2011; Susa 2013), but more realistic 3D investigations are required for a final conclusion.

4.2 Black Hole Formation in Massive Primordial Halos

The potential pathways to the formation of SMBHs were already sketched by Rees (1984), and a detailed discussion would be largely beyond the scope of this review (see Volonteri and Bellovary 2012, for a more general discussion). Here, we consider the formation of massive black holes as one potential outcome of the gravitational collapse in massive primordial halos in the presence of strong photodissociating backgrounds that destroy the molecular hydrogen leading to an almost isothermal collapse regulated by atomic hydrogen.

Such scenarios were put forward by Koushiappas et al. (2004), Begelman et al. (2006), Spaans and Silk (2006), Lodato and Natarajan (2007) and Schleicher et al. (2010b) (see as well Bromm and Loeb 2003a, for an earlier work). The first numerical simulations following the gravitational collapse of these halos were reported by Wise et al. (2008), finding that gravitational instabilities may transport angular momentum and allow the formation of massive central objects. The thresholds required to fully dissociate the H_2 were carefully investigated by Shang et al. (2010), while Regan and Haehnelt (2009b) pursued the first numerical simulation exploring the evolution beyond the formation of the first peak and confirming the formation of a disk at late times. While these simulations typically employed only a moderate resolution per Jeans length, Latif et al. (2013b) demonstrated the appearance of extended turbulent structures once a high resolution per Jeans length is employed. Such turbulence can further aid the amplification of magnetic fields via the small-scale dynamo (Schleicher et al. 2010a; Latif et al. 2013d).

A central question is indeed how often the right ambient conditions exist in order to allow for such a close-to-isothermal collapse as required here. While a first investigation by Dijkstra et al. (2008) indicated that their abundance is likely sufficient to explain the observed population of SMBHs, more recent results (Agarwal et al. 2012; Johnson et al. 2013b) suggest that appropriate conditions typically occur for at least one halo in a box of $1\ \mathrm{Mpc}^{-3}$. It is thus highly relevant to explore if such conditions lead to the formation of massive central objects.

The latter question was explored by Latif et al. (2013a) with the cosmological hydrodynamics code ENZO (O'Shea et al. 2004; Bryan et al. 2014). Their simulation setup started

---

[9]Based on the Forrey (2013) calculation, the chemical uncertainties are thus strongly reduced.

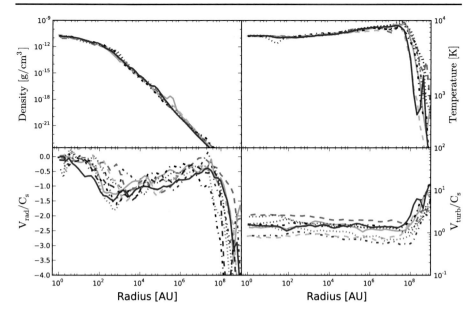

Fig. 7 The initial collapse in primordial massive halos. Shown are radial profiles of the gas density (*top left*), the radial velocity relative to the sound speed (*bottom left*), the gas temperature (*top right*) and the turbulent velocity normalized relative to the sound speed (*bottom right*). Figure by Latif et al. (2013a)

from cosmological initial conditions at $z = 100$ in a box of $1 \, h^{-1}$ Mpc, with a root grid of 128^3, two initial nested grids of the same resolution centered on the most massive halo of $\sim 10^7 \, M_\odot$, as well as 27 additional refinement levels. The refinement level adopted here ensured a minimum resolution of at least 64 cells per Jeans length. The simulations included primordial chemistry in the presence of a strong photodissociating background, parametrized via $J = J_{21} \cdot 10^{-21} \, \mathrm{erg \, cm^{-3} \, s^{-1} \, Hz^{-1} \, sr^{-1}}$, with $J_{21} = 1000$. The simulations further included the subgrid-scale turbulence model by Schmidt et al. (2006). In order to follow the evolution beyond the first peak, the equation of state was adiabatic for densities higher than $10^{14} \, \mathrm{cm^{-3}}$. A total of nine such simulations has been pursued employing a different seed for the initial conditions.

The initial state of the simulation when reaching the highest refinement level is shown in Fig. 7. The gas density follows the expected isothermal profile with $\rho \propto r^{-2}$, which flattens on the highest refinement level on scales comparable to the Jeans scale. The gas temperature is shock heated to $\sim 10^4$ K when falling onto the halo and then remains approximately constant during the collapse, as the atomic hydrogen cooling acts as a thermostat. The radial velocity normalized in terms of the sound speed increases during the initial infall, and subsequently becomes approximately constant. The infall appears enhanced on scales of 100 AU, potentially due to the central mass, and decreases on smaller scales where the gas is not self-gravitating. Also the turbulent velocity appears approximately constant throughout the collapse, implying that turbulence is continuously regenerated via accretion shocks and shear flows.

In order to study fragmentation and the accretion onto the central object, we have followed the simulations for four free-fall times beyond the formation of the first peak. The resulting density distribution is shown in Fig. 8 for nine different halos. In all cases, the central object has already reached a mass of $\sim 1000 \, M_\odot$ within the central 30 AU, with still

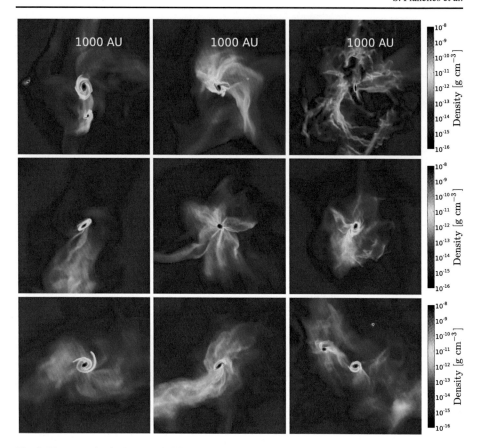

Fig. 8 Fragmentation in the central 1000 AU for 9 different halos. Only 3 out of 9 halos fragment (one is not seen in this projection). Shown are density projections 4 free-fall times after the initial collapse. Figure from Latif et al. (2013a)

ongoing and high accretion rates of $\sim$1 $M_\odot$ yr^{-1}. In these simulations, we find fragmentation in three out of nine halos (one is not visible in the projection given here). It is however possible that some of the clumps will subsequently merge, and also additional clumps may still form. The accretion occurs via self-gravitating disk structures surrounding the most massive objects. A more detailed investigation by Latif et al. (2013a) further revealed that the presence of such self-gravitating disks occurred only in simulations employing the turbulence subgrid-scale model, which provided additional support for the stability of these disks, while a more filamentary accretion mode occurs in purely hydrodynamical runs.

The hydrodynamical evolution can be further assessed via Fig. 9. The figure shows the accretion rates of more than $\sim$1 $M_\odot$ yr^{-1}. The enclosed mass scales as the radius on larger scales, corresponding to an isothermal profile, and varies more rapidly as r^3 on small scales where the density is approximately constant. The rotational velocity in the disk is of the order several 10 km/s, while the radial velocity corresponds to $\sim$10 km/s. Similar results are also found by Regan et al. (2014).

For such accretion rates, stellar evolution calculations indicate that the resulting protostar behaves as a red giant, implying a highly extended envelope and a rather cool temperature of $\sim$5000 K (Hosokawa et al. 2012). Considerations of the typical contraction timescales

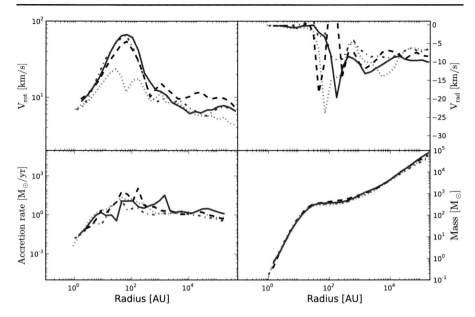

Fig. 9 Properties of the central region, 4 free-fall times after the initial collapse. Shown are radial profiles of the rotational velocity (*top left*), the accretion rate (*bottom left*), the radial velocity (*top right*) and the enclosed mass (*bottom right*). Figure from Latif et al. (2013a)

in the star indicate that such states can be maintained as long as the accretion rate is high, implying that radiative feedback is weak and leading to typical masses of $\sim 10^5$ $M_\odot$ for the resulting protostars (Schleicher et al. 2013). Even assuming that the protostars contract faster and reach the main sequence, the resulting feedback is however likely not sufficient to overcome the accretion rates (Omukai and Inutsuka 2002; Johnson et al. 2013a). It therefore appears that the formation of very massive objects is feasible. These may collapse via general-relativistic instabilities to become SMBHs (Shapiro and Teukolsky 1986).

The black holes forming in these early stages may later become the supermassive black holes observed at $z \sim 6$–7 (Fan et al. 2001, 2003; Mortlock et al. 2011), the supermassive black holes observed in galactic centers (e.g. Magorrian et al. 1998; Häring and Rix 2004). The feedback of such black holes may have a significant impact both on the evolution of galaxies (Somerville et al. 2008; Silk 2013) as well as on the evolution in galaxy clusters. The latter will be explained in further detail in Sect. 6 of this manuscript.

5 Galaxy Clusters: Self-Similar Model

Galaxy clusters, the largest and most massive objects in our Universe, form from the smaller units into a sequence of mergers. In this Section we briefly introduce the main predictions of the self-similar model (Kaiser 1986) which, based on relatively simple assumptions, is able of estimating the main ICM properties and important correlations between them.

The self-similar model relies on several basic assumptions. On the one hand, an EdS background cosmology and a power-law shape for the power spectrum of fluctuations are assumed. These conditions imply that there is not a characteristic scale of collapse. On the other hand, since gravity is supposed to be the unique driver of halo collapse and gas heating, there are not additional characteristic scales introduced in the process of cluster formation. This model also assumes clusters to have spherical symmetry and to be in HE (e.g. Borgani et al. 2008; see also Kravtsov and Borgani 2012 for a recent review and extensions of this model).

Under these assumptions, the mass inside a spherical region of radius R_{Δ_c} enclosing a mean density equal to $\Delta_c \rho_c(z)$ at redshift z is given by $M_{\Delta_c} = (4\pi/3)\Delta_c \rho_c(z) R_{\Delta_c}^3$, where $\rho_c(z) = \rho_{c0} E^2(z)$ is the critical cosmic density at z, being $E(z)$ given in Eq. 6 and ρ_{c0} the critical density at $z = 0$. Therefore, the cluster radius scales as $R_{\Delta_c} \propto M_{\Delta_c}^{1/3} E^{-2/3}(z)$. In addition, if the condition of HE is valid and the gas in clusters is distributed in a similar way to the DM, then it is satisfied that $k_B T \propto M_{\Delta_c}/R_{\Delta_c}$, which can be used to include in the above relation the dependence on the ICM temperature as

$$M_{\Delta_c} \propto T^{3/2} E^{-1}(z). \tag{21}$$

This equation can now be used to derive additional scaling relations between different X-ray observables. For instance, the X-ray luminosity goes like the product of the emissivity and the cluster volume, that is, $L_X \propto \rho_{gas}^2 \Lambda(T, Z) R_{\Delta_c}^3$, where ρ_{gas} is the gas density and $\Lambda(T, Z)$, which depends on the gas temperature and metallicity, is the cooling function associated to a particular emission process. The X-ray emission of the ICM plasma is mainly contributed by thermal Bremsstrahlung and line emission. At high temperatures ($T \gtrsim 2$ keV), where thermal Bremsstrahlung dominates, the cooling function goes like $\Lambda(T) \propto T^{1/2}$ (e.g. Sarazin 1986; Peterson and Fabian 2006) and, therefore, L_X scales with temperature as

$$L_X \propto M_{\Delta_c} \rho_c T^{1/2} \propto T^2 E(z). \tag{22}$$

For $T \lesssim 2$ keV, however, the dependence on temperature is more intricate because line emission becomes more important than the free-free radiation and, as a consequence, the above relation has to be adjusted by accounting for the metal contribution.

An additional key quantity describing the ICM thermodynamics is the entropy (Voit 2005) which is usually defined as $K = k_B T n_e^{-2/3}$, with n_e being the electron number density. Therefore, for pure gravitational heating the entropy scales as

$$K_{\Delta_c} \propto T E(z)^{-4/3}. \tag{23}$$

It is important to stress that the shape of the above relations and their z-dependence are a natural consequence of both the particular assumptions of the self-similar model and the redshift dependence of $\Delta_c \rho_c(z)$ associated to the assumed SO mass definition (Kravtsov and Borgani 2012). Therefore, given that the standard cosmological model only introduces minor departures from self-similarity, observational deviations from these predictions can be used to determine the effects of additional physical processes other than gravity.

In general, hydrodynamical simulations including only the effects of gravity (e.g. Navarro et al. 1995; Eke et al. 1998; Nagai et al. 2007b) are able to reproduce the shape of the above X-ray scaling relations. However, as we discuss below, a number of observations show some important deviations from these predictions, indicating an additional contribution from non-gravitational processes.

5.1 Observational Deviations from Self-Similarity

A number of X-ray observations point against the simple self-similar scenario, especially at the scale of small clusters and groups. Indeed, galaxy clusters observations have confirmed that, despite the simplicity and the important predictions provided by the adiabatic model just described, there are still some important issues that deserve a further investigation:

The Cooling Flow Problem   At high temperatures, when the free-free radiation dominates the ICM X-ray emission, the characteristic time scale of the gas to cool down can be written as $t_{cool} \simeq 6.9 \times 10^{10}$ yr $(n_e/10^{-3}$ cm$^{-3})^{-1}$ $(T/10^8$ K$)^{1/2}$ (e.g. Sarazin 2008). According to this expression, as the gas temperature decreases, gas cooling becomes faster. In addition, given its dependence on the gas density, whereas t_{cool} in outer cluster regions is usually longer than the age of the Universe, it can reach much shorter values $(t_{cool} \sim 10^8$–10^9 yr) in denser, central regions of cooling core clusters.

Some of the main features of these cooling core clusters are the following (see Sarazin 1986, for a review and references therein): (i) they have X-ray surface brightness with very high central values; (ii) their gas temperature, which is very low at the center, increases with radius; (iii) within the cooling core regions, t_{cool} is lower than the Hubble time; and (iv) they show an increasing iron abundance towards the interior. These general features are detected in a considerable number of clusters which are the so-called cool core clusters (CC). In general, observations report that these properties are most often found in dynamically relaxed clusters (e.g. Fabian et al. 1994; Chen et al. 2007).

These observations led to the classical *cooling flow model* (e.g. Sarazin 1986, 1988). This model is based on the assumption that there is not a heating source to compensate the gas radiative cooling. Therefore, the gas cools and falls inward subsonically, eventually reaching very low central temperatures. When these temperatures go below the characteristic X-ray emitting values, strong emission lines are produced.

However, the well-known cooling flow problem stems from the observation of several facts: (i) the expected cooling rates predict emission lines stronger than observed; (ii) the ratio between the central and mean cluster temperatures only remains at a factor $\sim 1/3$ (e.g. Peterson et al. 2003); (iii) the mass deposition rates, the amounts of cool gas and the star formation rates are generally smaller than estimated from the expected cooling rates (e.g. Edge and Frayer 2003). These discrepancies, which imply the existence of a central heating source, gave rise to the well-known *cooling flow problem*: which mechanism (or mechanisms) prevents the intra-cluster gas from cooling down to low temperatures?

Self-Similar Scaling Relations?   Contrary to what is expected from pure gravitational models, early X-ray observations of galaxy clusters demonstrated that observational scaling laws do not scale self-similarly (see Giodini et al. 2013, for a recent review).

The first indication of self-similarity breaking was the $L_X - T$ relation (e.g. Markevitch 1998; Allen et al. 2001; Ettori et al. 2004; Pratt et al. 2009; Maughan et al. 2012), for which observations report a steeper slope (≈ 3) than the expected self-similar value of 2 (e.g. Markevitch 1998; Osmond and Ponman 2004), a departure that becomes even larger for systems with $k_B T \lesssim 3.5$ keV (Ponman et al. 1996; Balogh et al. 1999; Maughan et al. 2012). In general, the observed $L_X - T$ relation shows a relatively large scatter (e.g. Pratt et al. 2009), which is mainly due to both the strong emission associated to CC clusters and to unrelaxed systems for which the HE condition is not a good approximation (e.g. Maughan et al. 2012). A common practice to reduce this scatter consists in excluding either cluster cores (e.g. Markevitch 1998; Pratt et al. 2009), or CC systems (e.g. Arnaud and Evrard 1999). As an example, the top panel of Fig. 10 shows a comparison of the luminosity-temperature relation as obtained from different X-ray observational samples (Arnaud and Evrard 1999; Markevitch 1998; Helsdon and Ponman 2000) and from the radiative simulations with SN feedback by Borgani et al. (2004). As we can see, numerical results are consistent with observations for $k_B T \gtrsim 2$ keV. However, the agreement is not so encouraging at $k_B T \lesssim 2$ keV,

Fig. 10 *Top panel*: Comparison of the $L_X - T$ relation as obtained from different X-ray observational samples and from the radiative simulations by Borgani et al. (2004) (*green dots*). These simulations assume null metallicity and include cooling, star formation and SN feedback. Figure from Borgani et al. (2004). *Middle panel*: Relation between gas entropy and temperature for a sample of galaxy clusters and groups. *The solid line* represents the observational relation, whereas *the dotted line* stands for the self-similar prediction. Figure from Ponman et al. (2003). *Bottom panel*: Observations of the baryonic, gas and stellar mass fractions as a function of total mass. The power-law fits obtained for the baryonic and stellar mass fractions are given by *the black* and *blue solid lines*, respectively. For comparison, the corresponding best-fits derived by Lin et al. (2003) and Giodini et al. (2009) are also included. *The lower panel* shows the ratio between observational and WMAP-7 baryon mass fraction. Figure from Laganá et al. (2011)

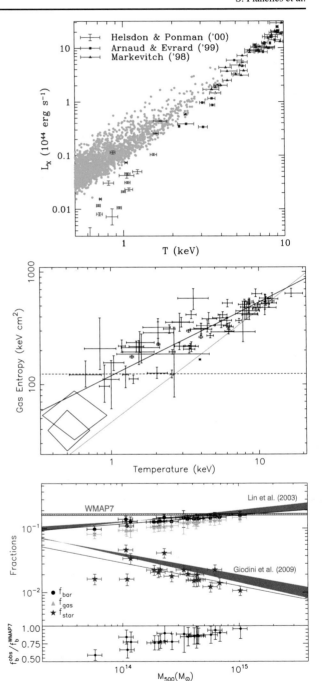

where simulations may include additional or more efficient feedback processes to further reduce the X-ray emission.

Consistently with the $L_X - T$ scaling, the relation between X-ray luminosity and mass is also steeper than expected from self-similarity (e.g. Reiprich and Böhringer 2002; Chen et al. 2007). As shown in the bottom panel of Fig. 10, a possible reason for this may be related with the observations of an increasing trend of the gas mass fraction with total mass (e.g. Balogh et al. 2001; Lin et al. 2003; Sanderson et al. 2003; Vikhlinin et al. 2006; Pratt et al. 2009; Dai et al. 2010; Laganá et al. 2011). The lower gas content observed in low mass systems would generate a lower X-ray emission and, therefore, a steepening of the scaling laws.

As can be inferred from the middle panel of Fig. 10, observations of central regions of small galaxy clusters and groups report higher entropy values than expected from pure gravitational predictions, thereby generating a flattening of the observed $K - T$ relation (Ponman et al. 2003). This increase of the gas entropy in low mass systems prevents the gas from falling into the center, reducing the central amounts of gas and, as a consequence, leading to the steepening of the observed $L_X - T$ relation (e.g. Evrard and Henry 1991; Tozzi and Norman 2001; Borgani et al. 2001; Voit et al. 2002).

As for the total mass-temperature relation, observations generally report a self-similar behavior for massive galaxy clusters (e.g. Arnaud et al. 2005; Vikhlinin et al. 2009), whereas a slightly steeper slope is obtained for smaller objects (e.g. Arnaud et al. 2005, Sun et al. 2009; see also Kettula et al. 2013 for a recent weak lensing calibration of this relation). In addition, given its small intrinsic scatter, which is mainly due to the existence of substructures (e.g. O'Hara et al. 2006; Yang et al. 2009), the $M_{tot} - T$ relation turns out to be extremely useful for cosmological applications with galaxy clusters.

Temperature and Entropy Radial Profiles   Independent X-ray observations have confirmed that the ICM temperature distribution is not isothermal. Indeed, as it is shown in the left panel of Fig. 11, the cluster temperature radial profiles are characterized by negative gradients at $r \gtrsim 0.1 R_{180}$ (e.g. Piffaretti et al. 2005; Pratt et al. 2007), being nearly self-similar out to the most external regions (e.g. Markevitch 1998; De Grandi and Molendi 2002; Vikhlinin et al. 2005). However, the shape of these profiles in inner cluster regions partially depends on the dynamical state of the considered systems: dynamically relaxed clusters generally show decreasing temperature profiles towards the core, while unrelaxed systems show, instead, different patterns. In principle, while gas radiative cooling might be able to generate low temperatures in the central regions of CC clusters, some source of energy feedback may avoid an excess of cooling, thus reducing the resulting star formation rates in these systems.

As for the entropy radial profiles, pure gravitational models predict that, in outer cluster regions, entropy scales with radius as $K \propto r^{1.1}$ (Tozzi and Norman 2001). However, recent observations (e.g. Simionescu et al. 2011; Walker et al. 2012; Eckert et al. 2013) have reported that these profiles flatten in central cluster regions and their dependence on radius at larger radii is gentler than predicted (see right panel of Fig. 11). The particular radius at which the profiles become flat depends on a number of factors, being shorter for CC than for non cool-core (NCC) clusters (e.g. Sanderson et al. 2009; Pratt et al. 2010). Moreover, while high-mass clusters show a higher mean core entropy (e.g. Cavagnolo et al. 2009) and nearly self-similar profiles in outer regions, smaller systems are characterized by shallower profiles. This distinction between low and high mass systems, which is a clear indication of the higher efficiency of non-gravitational processes on smaller objects (e.g. Cavagnolo et al. 2009; Pratt et al. 2010; Maughan et al. 2012), highlights that entropy profiles provide an important fingerprint of the different physical processes breaking cluster self-similarity (e.g. Voit et al. 2002, 2003).

The existing disparity between the self-similar predictions and the observations clearly indicates that, in addition to the effects of gravity, additional processes related to the physics of baryons should be also taken into account. Therefore, additional non-gravitational processes operating within galaxy clusters and groups, such as radiative cooling, star formation

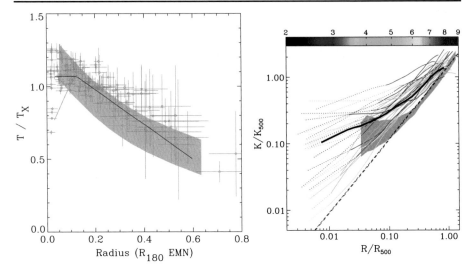

Fig. 11 *Left panel*: Comparison of the dimensionless temperature profiles from *XMM-Newton* observations of 15 nearby clusters by Pratt et al. (2007) (*dots* with error bars) with the average profiles from *ASCA* (grey band, Markevitch 1998), and the observations of cooling core clusters from *BeppoSAX* (green line, De Grandi and Molendi 2002) and *Chandra* (red line, Vikhlinin et al. 2005). Figure from Pratt et al. (2007). *Right panel*: Comparison of the scaled entropy profiles of the REXCESS sample (colored lines according to the temperature of the clusters, Pratt et al. 2010) with the theoretical results from the non-radiative simulations by Voit (2005). *The thick black line* represents the median of all the observed profiles, whereas *the dashed line* stands for the best power law fit to the median profile of the simulated clusters. Figure from Pratt et al. (2010)

and its inferred energy feedback, or AGN feedback, should interact among them in order to solve the aforementioned problems. As we discuss below, efforts to obtain a completely consistent and satisfactory model able to reproduce the observations are still ongoing (e.g. Borgani et al. 2004; Voit 2005; McCarthy et al. 2008).

6 The Physics of the Intra-Cluster Plasma

Gravitational processes lead the evolution of initial density fluctuations into DM halos. However, as we review in this Section, the thermodynamics of the ICM is determined by the combined action of gravity, acting on both DM and baryonic components, and a number of non-gravitational processes such as radiative cooling, star formation and AGN feedback.

6.1 Structure Formation Shock Waves

As explained in Sect. 2.3, within the hierarchical ΛCDM paradigm the formation of DM halos is driven by the gravitational collapse of initial matter density perturbations. Figure 1, which displays the formation and evolution of galaxy clusters as obtained from hydrodynamical cosmological simulations, indicates that in addition to DM (left panels), the evolution of the hot diffuse gas (middle panels) and the stellar (right panels) components are significantly connected to each other. At $z = 0$, galaxy clusters are identified as high-overdense DM regions at the intersection of large matter filaments. Whereas the gas distribution follows closely the DM distribution, the pattern of the stellar component is much clumpier.

As a consequence of the hierarchical process of structure growth, clusters of galaxies often undergo major merger episodes, which are one of the most energetic phenomena in the Universe (Sarazin 1988). These mergers, associated to the collisions of proto-cluster structures of similar masses at velocities of several thousands of kilometers per second, generate shocks and compression waves in the ICM. On large scales, these hydrodynamical shocks are very efficient in releasing an important amount of the energy associated with the collision ($\sim 10^{61}$–10^{65} ergs) as thermal energy in the final system, heating and compressing the hot intra-cluster gas and, therefore, increasing its entropy (e.g. Quilis et al. 1998; Miniati et al. 2000). On smaller scales within already collapsed structures, additional weaker shocks are developed as a consequence of subhalo mergers, accretion processes, or random gas flows. These internal shocks are relevant in the thermalization of the intra-cluster gas (McCarthy et al. 2007), contributing significantly to the virialization of halos.

In addition, shocks also generate ICM turbulence and mixing (Norman and Bryan 1999; Ricker and Sarazin 2001; Nagai et al. 2003; Dolag et al. 2005), redistribution and amplification of magnetic fields (Roettiger et al. 1999), and are likely sources of non-thermal emission in galaxy clusters (e.g. Bykov et al. 2000). Indeed, cosmological simulations have shown that an important non-thermal pressure support in galaxy clusters is provided by ICM turbulence (e.g. Dolag et al. 2005; Vazza et al. 2011), relativistic CR particles (e.g. Pfrommer et al. 2007; Vazza et al. 2012), and magnetic fields (Dolag et al. 1999). In general, none of these processes alone can regulate cooling, but they may be more efficient when AGN feedback (Voit 2011) or plasma instabilities (e.g. Sharma et al. 2012) are also considered.

Therefore, it is clear that cosmological shocks leave their imprint on relevant ICM plasma properties (see Bykov et al. 2008, for a review). First, the rise of the gas entropy and thermal pressure support induced by shocks represents a natural alternative way of breaking the self-similar scaling (e.g. Bykov et al. 2008, for a multi-fluid accretion shock model, derived $K \propto T^{0.8}$). Second, shocks surrounding mildly overdense, non-linear, non-collapsed structures, such as sheets and filaments, heat the intergalactic medium and determine the evolution of the warm-hot intergalactic medium (WHIM; e.g. Cen and Ostriker 1999; Davé et al. 2001). Finally, the diffusive shock acceleration process (DSA; e.g. Drury and Falle 1986; Blandford and Eichler 1987) can convert a part of the thermal population of particles into non-thermal CRs, resulting therefore in both thermal and non-thermal energy constituents (Kang et al. 2002; Kang and Jones 2005; Kang et al. 2007). Whereas the CR electrons may be responsible of the observed radio halos (e.g. Ferrari et al. 2008; Giovannini et al. 2009) and radio relics (e.g. Ensslin et al. 1998) in clusters (e.g. Pfrommer et al. 2008), the CR protons may generate γ-ray emission.

From an observational point of view, detecting large-scale shocks is non-trivial because they take place in the periphery of galaxy clusters where the X-ray emission, due to the relatively low gas density, is weak. In spite of this, large-scale shocks have been identified in the form of radio relics in more than 30 clusters (e.g. Bonafede et al. 2009; van Weeren et al. 2009). Some merger-induced internal shocks have been also detected with Mach numbers[10] $\mathcal{M} \sim 1.5$–3 (e.g. Markevitch and Vikhlinin 2007). From the theoretical point of view, several approaches to study shocks have been pursued using semi-analytical models (e.g. Press and Schechter 1974; Sheth and Tormen 1999) as well as numerical simulations based on both grid-based, single-grid (e.g. Quilis et al. 1998; Miniati et al. 2000; Ryu et al. 2003; Vazza et al. 2009) and AMR schemes (e.g. Skillman et al. 2008; Planelles and Quilis 2013),

---

[10]The Mach number, which characterizes the strength of a shock, is given by $\mathcal{M} = v_s/c_s$, where v_s is the shock speed and c_s is the sound speed ahead of the shock.

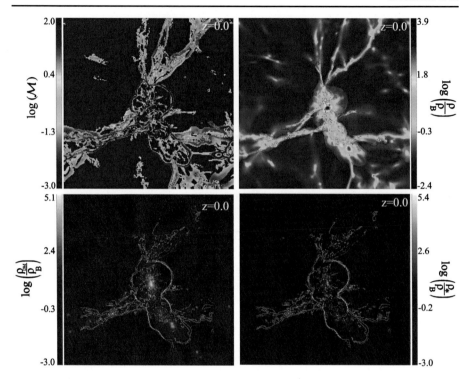

Fig. 12 Distribution of shock Mach numbers (*upper left panel*) and its comparison with the DM (ρ_{DM}/ρ_B, *lower left*), gas (ρ/ρ_B, *upper right*) and stellar (ρ_*/ρ_B, *lower right*) overdensities at $z = 0$. In *all the panels*, the contours of the shock waves with high Mach numbers are overplotted. *Each panel* represents a projection of 0.2 Mpc thickness and 64 Mpc side length. Figure from Planelles and Quilis (2013)

and particle-based codes (e.g. Pfrommer et al. 2006). As an example of the results obtained from these numerical analysis, Fig. 12 shows the distribution of the shock waves detected in a cosmological AMR simulation compared with the distributions of DM, gas and stellar overdensities at $z = 0$. This figure clearly highlights that cosmological shock waves, which occupy the whole simulated volume in a complex way, accurately trace the cosmic web. External shocks, with quasi-spherical shapes and at relatively large distances from the center of the structure where they appear, are characterized by high Mach numbers ($\mathcal{M} \gtrsim 20$). On the contrary, within collapsed objects, weaker shocks with $\mathcal{M} < 2$–3 are present, contributing significantly to the dissipation of kinetic energy (e.g. Skillman et al. 2008).

In spite of all the studies and of the long-standing debate about the use of different numerical techniques (e.g. Agertz et al. 2007), the identification of cosmological shocks still represents an issue.

6.2 Non-gravitational Processes

'Gravitational feedback', mainly in the form of shock and compression waves, contributes to most of the intra-cluster gas heating. However, as discussed in Sect. 5.1, the discrepancies with the self-similar model observed in small clusters and groups and in inner cluster regions, indicate the existence of additional non-gravitationally induced cooling and heating processes. In the following, we explain some of the main non-gravitational processes that need to be incorporated in hydrodynamical simulations to reproduce the self-similarity breaking.

6.2.1 Gas Radiative Cooling

Radiative cooling, which plays a major role in the ICM emissivity, can significantly contribute to break self-similarity to the observed level. To understand the role of cooling, it is useful to express the cooling time in terms of the gas entropy and temperature, which in the Bremsstrahlung regime can be approximated by (Sarazin 2008)

$$t_{cool} \approx 17 \left(\frac{K}{130 \,\text{keV}\,\text{cm}^2} \right)^{3/2} \left(\frac{k_B T}{2 \,\text{keV}} \right)^{-1} \text{Gyrs.} \tag{24}$$

Therefore, for a galaxy cluster with $k_B T \sim 2.5$ keV, the cooling time is lower than the Hubble time for an entropy of $K \lesssim 130$ keV cm^2. This means that, the lower entropy gas within galaxy clusters will cool before, being evacuated earlier from the hot gas in cluster core regions. This low-entropy gas will be superseded by higher entropy gas coming from outer regions, thus increasing the mean gas entropy (Voit and Bryan 2001).

Hydrodynamical simulations including radiative cooling support this prediction (e.g. Pearce et al. 2000; Muanwong et al. 2001; Davé et al. 2002; Valdarnini 2002; Kay et al. 2004; Nagai et al. 2007a). As an example, Fig. 13 shows the comparison between the mean ICM profiles of relaxed clusters derived from cosmological simulations by Nagai et al. (2007a) and the observations by Vikhlinin et al. (2006). In particular, there are two sets of simulations, one accounting only for non-radiative physics (*NR*) and another taking into account radiative cooling, star formation and metal enrichment (*CSF*). From the analysis of this figure we infer that, whereas the *NR* simulations produce ICM profiles in conflict with observations, the *CSF* runs produce a better match, at least outside cluster core regions. Specifically, at $r \gtrsim 0.5 r_{500}$, the *CSF* runs produce both an increase in the gas entropy in accordance with the observed values, and nearly self-similar pressure profiles. Another effect of cooling is the reduction of the gas density in inner regions (see top-left panel of Fig. 13).

However radiative cooling has also some undesirable effects. As it is shown in the top-right panel of Fig. 13, an unexpected effect of cooling is that of increasing the ICM temperature towards the cluster center together with the steepening of the temperature profiles. This effect results from the dearth of central pressure support induced by cooling, which makes outer gas to fall sub-sonically to the center while being heated adiabatically. In addition, given the runaway nature of radiative cooling, it suffers from overcooling, thus transforming quite large amounts of gas into stars. In fact, radiative simulations including different forms of stellar feedback obtain that around 35–40 % of the cluster baryon content is converted into stars (e.g. Nagai and Kravtsov 2004), a value which is significantly larger than observed ($\sim$10–15 %; e.g. Balogh et al. 2001; Lin et al. 2003; Gonzalez et al. 2007). In principle, these two shortcomings of cooling represent two sides of the same problem (Borgani and Kravtsov 2011). A proper source of gas heating (or most likely, a combination of several) able of counterbalancing cooling, keeping the gas pressurized in inner regions, and controlling the star formation, may provide the solution to this complicated issue. However, as it will be discussed in the following sections, unearthing such a mechanism is non-trivial and represents nowadays an important challenge in the numerical description of clusters.

6.2.2 Star Formation and Its Associated Feedback

SN explosions contribute to both the heating of the surrounding medium and the distribution of metals from star-forming regions into the hotter intra-cluster plasma. Given that

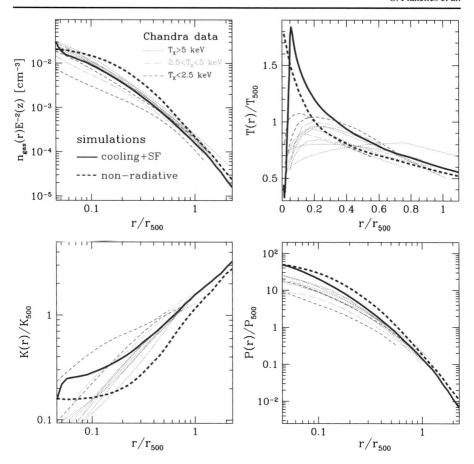

Fig. 13 From *top* to *bottom* clockwise: Mean radial profiles of gas density, temperature, pressure, and entropy of relaxed clusters at $z = 0$ identified in cosmological simulations. There are two sets of simulations, one non-radiative (*red thick dashed lines*) and another accounting for *cooling + SF* (*blue thick solid lines*). The observed *Chandra* cluster sample by Vikhlinin et al. (2006) is used for comparison (*thin lines* in different colors indicating the temperature of the systems). Figure from Nagai et al. (2007a)

SN-driven winds are a by-product of the star formation process, SN feedback was suggested to produce a realistic and self-regulated cosmic star formation rate (e.g. Springel and Hernquist 2003). We note that even beyond feedback models, appropriate subgrid-scale models for hydrodynamical turbulence may be required for the description of the hydrodynamical state (e.g. Schmidt et al. 2006; Iapichino and Niemeyer 2008; Maier et al. 2009). Moreover, given the relatively low resolutions reached by present-day cosmological simulations, the physics of the interstellar medium can not be properly described (Borgani and Kravtsov 2011), a problem that persists even in relatively well-resolved simulations of individual galaxies (e.g. Tasker and Tan 2009, Dobbs et al. 2010, Wang et al. 2010, Bournaud et al. 2010; see also Mayer et al. 2008 for a review). Given the current limitations, stellar feedback is usually incorporated in cluster simulations via phenomenological prescriptions of star formation and SN heating (e.g. Braun and Schmidt 2012) at moderately low resolutions.

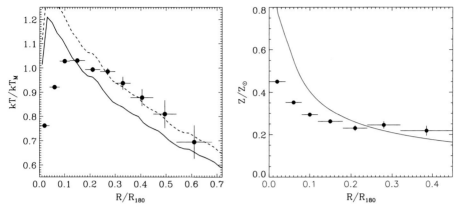

Fig. 14 *Left panel*: Comparison between the mean temperature profile of a sample of about 50 local ($z \lesssim 0.3$) and hot ($k_B T_X > 3$ keV) *XMM-Newton* clusters (dots with error bars; Leccardi and Molendi 2008b) and the mean profile obtained from cosmological simulations including radiative cooling, star formation and SN feedback (solid line; Borgani et al. 2004). *The dashed line* stands for the mean simulated profile rescaled by 10 %. Figure from Leccardi and Molendi (2008b). *Right panel*: Comparison of the mean metallicity profile for the same sample of *XMM-Newton* clusters (*dots* with error bars) with the one derived by Fabjan et al. (2008) from simulations performed with the SPH code GADGET-2 (Springel 2005) assuming the chemical enrichment model by Tornatore et al. (2007). Figure from Leccardi and Molendi (2008a)

In general, cosmological simulations show that SN feedback can help in partially offsetting radiative cooling, flattening the temperature profiles, and reducing the cluster stellar mass fractions (see Borgani and Kravtsov 2011, and references therein). However, in spite of these promising results, stellar feedback alone is not efficient enough to produce the observed thermal structure of CC clusters. As an example, the left panel of Fig. 14 shows the comparison between the temperature profiles of a sample of relaxed clusters as derived from observations and from simulations including SN feedback. We see that, whereas in outer cluster regions, $r \gtrsim 0.2 R_{180}$, simulations recover quite well the observed profiles, within inner regions the agreement is not so satisfactory. The low efficiency of SN feedback in compensating the cooling properly produces additional undesirable results (e.g. Kravtsov and Borgani 2012): in general, the levels of core entropy, although reduced, remain significantly larger than reported by observations; the BCGs have stellar masses larger than observed; and, the excessive star formation is also translated into an excessive metal production in cluster central regions (see right panel of Fig. 14).

6.2.3 AGN Feedback

Currently, the most favored mechanism to explain the ICM self-similarity breaking and the cooling flow problem is the AGN heating resulting from gas accretion onto a central SMBH. Indeed, many cluster observations confirm the effects of AGN heating on the ICM plasma (e.g. McNamara and Nulsen 2007; Chandran et al. 2009). First, the existence of SMBHs at the nuclei of galaxies (Magorrian et al. 1998) and the observed correlations between the BH masses and the halo and bulge properties of the host galaxies (Ferrarese and Merritt 2000) point to an scenario in which galaxy formation and BH growth must proceed together. Second, as already reported by many observations (e.g. Burns 1990; Ball et al. 1993; Sanderson et al. 2006), almost every dynamically relaxed CC cluster has an active central radio emitting source, which has been associated by X-ray observations with the presence of cavities or bubbles in the X-ray emitting gas around the central galaxy (e.g. Bîrzan et al. 2004).

Third, there is a clear connection between the ICM X-ray luminosity within the core of clusters and the mechanical (e.g. Bîrzan et al. 2004) and radio luminosities (e.g. Eilek 2004) of the central AGN. Another important point in favor of this heating mechanism is that AGN feedback is a self-regulated process, compensating in a natural way radiative cooling (e.g. Rosner and Tucker 1989). This is due to the fact that the efficiency of AGN feedback is proportional to the rate at which the central SMBH accretes intra-cluster gas while radiative cooling takes place. Therefore, if the feedback efficiency is too large, the ICM is naturally over-heated and the gas accretion is reduced. On the contrary, if the gas accretion rate is too low, the intra-cluster plasma cools faster, the accretion rate onto the central BH increases and, correspondingly, the associated AGN feedback efficiency. In addition, AGN heating is supposed to be strong enough to reduce the star formation in the BCGs.

However, despite the strong reasons in favor of this AGN feedback cycle, implementing such a self-regulated mechanism in simulations represents a challenging task (e.g. Borgani and Kravtsov 2011, and references therein). In this sense, the first attempts to build competent models of AGN heating consisted in theoretical studies accounting for the effects of AGN feedback out of cosmological context (e.g. Churazov et al. 2001; Quilis et al. 2001). In the last years, however, different implementations and refinements of AGN feedback models have been included in cosmological simulations (e.g. Springel et al. 2005; Sijacki et al. 2007; Puchwein et al. 2008, 2010; McCarthy et al. 2010; Fabjan et al. 2010; Short et al. 2010; Battaglia et al. 2013; Martizzi et al. 2012; Ragone-Figueroa et al. 2013). Given the limited spatial and temporal resolutions achievable by current simulations, phenomenological models are needed to include this form of energy feedback. In these models, the rates of AGN energy injection are usually computed by adopting the Bondi gas accretion onto the central SMBHs (Bondi 1952).[11] In addition to thermal AGN feedback, some observations report that BHs in the center of galaxies generate relativistic jets that shock and heat the neighboring ICM. In the light of these observations, the effects of kinetic AGN feedback in the form of AGN-driven winds have been also analyzed by several authors (e.g. Omma et al. 2004; Dubois et al. 2011; Gaspari et al. 2011; Barai et al. 2014).

Simulations including different prescriptions of these phenomenological models have indeed reported some promising achievements (e.g. Sijacki et al. 2007; Puchwein et al. 2008; Fabjan et al. 2010; Martizzi et al. 2012; Planelles et al. 2013a,b; Le Brun et al. 2013). As it is shown in the right panel of Fig. 15, AGN feedback seems to be very efficient in attenuating the star formation in high-mass galaxy clusters, producing therefore stellar mass fractions in better agreement with observational data. In addition, as shown in the left panel of Fig. 15, AGN heating can also reduce the amount of hot gas in small clusters and groups, thus reproducing better the observed $L_X - T$ relation and partially resolving the disagreement that otherwise existed for small systems (see top panel of Fig. 10). Besides, AGN feedback has been also shown to be quite effective in dispersing heavy elements throughout the intra-cluster plasma, producing a better consistency with the observed ICM metallicity profiles (e.g. Fabjan et al. 2010; McCarthy et al. 2010; Planelles et al. 2013b).

Despite the above successes, a number of discrepancies between observations and simulations still exist. As an example, Fig. 16 shows the mean temperature and entropy radial profiles for the sample of relaxed and unrelaxed massive clusters identified in the AGN simulations by Planelles et al. (2013b). The lack of diversity between the simulated profiles of

---

[11]It is important to point out that the Bondi approach is the simplest model of gas accretion. A number of studies (e.g. Hobbs et al. 2012; Gaspari et al. 2013) have already highlighted the main drawbacks of this approach and the necessity of adopting alternative and more realistic schemes.

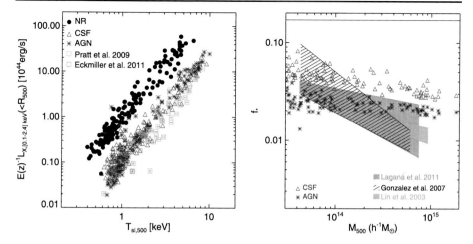

Fig. 15 *Left panel*: $L_X - T$ relation for the sample of groups and clusters identified in the simulations by Planelles et al. (2013b). Results are shown for a set of non-radiative simulations (*NR*), and for two sets of radiative simulations, one including cooling, star formation, SN feedback and metal enrichment (*CSF*), and another one accounting as well for the effects of AGN feedback (*AGN*). Figure from Planelles et al. (2013b). *Right panel*: Stellar mass fraction as a function of cluster mass as obtained in the simulations by Planelles et al. (2013a). Results from radiative simulations with (*AGN*) and without (*CSF*) AGN feedback are shown. *The horizontal continuous line* stands for the assumed baryon mass fraction in the simulations. Figure from Planelles et al. (2013a). In *both panels*, data from different observational samples is used for comparison

relaxed and unrelaxed systems is at odds with the observed profiles of CC and NCC clusters. This indicates that, even including AGN feedback and accounting for metal-dependent cooling rates, simulations are still not able to produce the correct cooling/heating interplay in cluster cores. The entropy values in inner regions are also higher than reported by observations. In addition, although the stellar masses of the BCGs obtained in these simulations are reduced, they are still much larger than observed (Ragone-Figueroa et al. 2013; see as well Kravtsov et al. 2014 for a recent observational analysis of the stellar mass-halo mass relation). Moreover, in a recent work, Gaspari et al. (2014) investigated the isolated effect of kinetic and thermal AGN feedback on the $L_X - T$ relation of galaxy clusters and groups. They showed that, even with different parameterizations of these commonly used AGN models, it is not possible to break self-similarity to the desired level without actually breaking as well the cool-core structure of the considered systems.

These results suggest that, in order to describe the observational properties of the intra-cluster plasma in inner core regions and beyond, a proper scheme of AGN feedback may be complemented by additional physical processes. In this sense, a number of mechanisms, such as the effects of CRs in AGN-induced bubbles (e.g. Sijacki et al. 2008), the heating induced by galaxy motions (e.g. Kim et al. 2005), or thermal conduction (e.g. Zakamska and Narayan 2003), have been suggested. However, further investigation is required to find the correct interaction between these and additional plasma physical processes.

6.3 Additional Plasma Physical Processes

Intergalactic gas heating at the cluster formation stage occurs by means of conversion of the energy of a cold gravitationally accelerated baryonic matter into the energy of hot thermal gas at cosmological shocks. The process of relaxation of the kinetic energy of the cold plasma flow to the quasi-equilibrium thermal distribution in rarefied cosmic plasmas

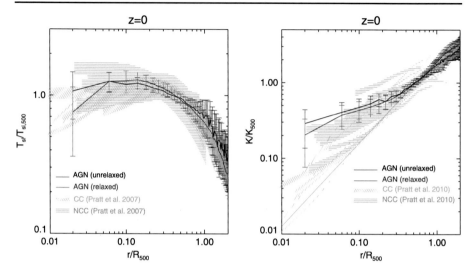

Fig. 16 Mean temperature (*left panel*) and entropy (*right panel*) radial profiles for the sample of relaxed/unrelaxed galaxy clusters identified in a set of simulations including AGN feedback (adapted from Planelles et al. 2013b). *Error bars* indicate $\pm 1 - \sigma$ dispersion around the mean profile. In *both panels*, radial profiles of CC and NCC clusters from the REXCESS sample (Pratt et al. 2007, 2010), which are represented by *colored shadowy areas*, are used for comparison. *The dotted black line* in *the right panel* shows the self-similar expectation for the entropy ($K \propto r^{1.1}$)

is non-trivial. This is because the Coulomb collision rate is slow in the rarefied intergalactic medium and the relaxation processes are collisionless which means they are due to the collective plasma wave-particle interactions. Therefore, the standard textbook single fluid hydrodynamic and MHD approaches are not *a priori* valid for the description of the collisionless plasma flows.

6.3.1 Gas Heating, Magnetic Field Amplification and Particle Acceleration in Collisionless Cosmological Shocks

The Coulomb mean free path of a proton of velocity v_7 (measured in 100 km s^{-1}) in the WHIM of overdensity δ can be estimated as $\lambda_p \approx 3.5 \times 10^{21} \cdot v_7^4 \cdot \delta^{-1} \cdot (1 + z)^{-3} \cdot (\Omega_b h^2/0.02)^{-1}$ cm, where Ω_b is the baryon density parameter and we assume the Coulomb logarithm to be about 40. The protons are magnetized in the flow (i.e. their gyro-frequencies are higher than the mean frequencies of the Coulomb collisions) if the magnetic field magnitude exceeds about 10^{-18} G.

The microscopic plasma scale, called the ion inertial length, is defined as $l_i = c/\omega_{pi} \approx 2.3 \times 10^7 n^{-0.5}$ cm, where ω_{pi} is the ion plasma frequency and n is the ionized ambient gas number density measured in cm^{-3}. This scale determines the widths of the transition region of the supercritical collisionless shock waves. In the strong enough collisionless shocks (typically of a Mach number above a few) resistivity cannot provide energy dissipation fast enough to create a standard shock transition on a microscopic scale (e.g. Kennel et al. 1985). Ion instabilities are important in such shocks, the so-called *supercritical* shocks. At the microscopic plasma scale the front of a supercritical shock wave is a transition region occupied by magnetic field fluctuations of an amplitude $\delta B/B \sim 1$ and characteristic frequencies of about the ion gyro-frequency. Generation of the fluctuations is due to instabilities in the interpenetrating multi-flow ion movements. The viscous transition region

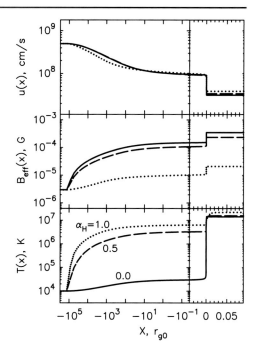

Fig. 17 Profile of a strong shock in WHIM simulated by Vladimirov et al. (2008) in the non-linear Monte Carlo model with different values of the fraction α_H of CR driven magnetic turbulence dissipated in the precursor. *The solid, dashed* and *dotted lines* correspond, respectively, to $\alpha_H = 0, 0.5$ and 1.0. The plotted quantities are the bulk flow speed $u(x)$, the effective amplified magnetic field $B_{eff}(x)$ and the thermal gas temperature $T(x)$. The shock is located at $x = 0$, and note the change from the logarithmic to the linear scale at $x = -0.05\, r_{g0}$

width is typically a few hundreds ion inertial lengths for a parallel shock, while it is about ten times shorter for a transverse shock. The ion inertial length in the WHIM can be estimated as $l_i \approx 5.1 \times 10^{10} \cdot \delta^{-1/2} \cdot (1 + z)^{-3/2} \cdot (\Omega_0\, h^2/0.02)^{-1/2}$ cm, providing the width of the collisionless shock transition region is smaller by many orders of magnitude than the Coulomb mean free path (that is in the kiloparsec range). The question whether collisionless shocks can form in plasmas with magnetic pressure much smaller than the plasma pressure is of fundamental importance. Two-dimensional particle-in-cell (PIC) simulations of the structure of non-relativistic collisionless shocks in unmagnetized electron-ion plasmas performed by Kato and Takabe (2008, 2010) revealed that the energy density of the magnetic field generated by the Weibel-type instability within the shock transition region reaches typically 1–2 % of the upstream bulk kinetic energy density. The width of the shock transition region was found in their simulation to be about 100 ion inertial lengths l_i, independent of the shock velocity. A tiny fraction (much less than a percent) of the incoming protons can be reflected from the collisionless shock transition region and these particles are subject of Fermi acceleration if the shock upstream flow carries magnetic fluctuations. In the case of strong shocks with high Alfvén and sonic Mach numbers the accelerated particles can get a substantial part (tens of percent) of the shock ram pressure. The pressure of non-thermal accelerated particles may mediate the shock flow as it was apparently observed by Voyager 2 in the heliosphere termination shock (see, e.g. Florinski et al. 2009). Moreover, evidences of strong non-adiabatic amplification of fluctuating magnetic fields by anisotropic distributions of accelerated particles observed in strong shocks of young supernova remnants (see, e.g. Helder et al. 2012).

Hybrid plasma simulations with kinetic treatment of ions and fluid electron description (see, e.g. Winske et al. 1990; Treumann 2009; Burgess and Scholer 2013) allow us to study domains of some thousands gyroradii of incoming proton around the non-relativistic shocks. Recent two-dimensional hybrid simulations by Gargaté and Spitkovsky (2012), who modeled quasi-parallel shock with the Alfvén Mach number $\mathcal{M}_a = 6$ revealed energetic power-law ion distribution of index about -2 in the shock downstream. The energetic non-thermal particle population contained about 15 % of the incoming upstream flow. Limited dynamical ranges of PIC and hybrid simulations do not allow us yet to study the formation of extended non-thermal tails of relativistic particles accelerated by non-relativistic shocks. On the other hand, simulations of DSA based on kinetic and Monte-Carlo modeling indicated the formation of the extended tails of the non-thermal particles. The energetic particles accelerated by a strong shock have hard spectral indexes and therefore, the CR pressure is dominated by the high-energy end of particle distribution. These particles may penetrate into far upstream and modify the shock flow by the CR-pressure gradient (see, e.g. Blandford and Eichler 1987; Jones and Ellison 1991; Malkov and Drury 2001; Amato and Blasi 2006; Vladimirov et al. 2008). Energetic particles of the highest energy escape into the upstream region providing energy outflow and allowing the shock compression ratio to be higher while the post-shock gas temperature and entropy appear significantly reduced comparing to that in the standard single fluid shock. It is important to note that the DSA is a very complicated highly non-equilibrium non-linear process with a strong coupling between the thermal and non-thermal components. Fast growing instabilities of the anisotropic distributions of energetic particles result in efficient production of strong magnetic turbulence in the shock upstream (see e.g. McKenzie and Voelk 1982; Bell 2004; Bykov et al. 2012; Schure et al. 2012).

In Fig. 17 we illustrate the effects mentioned above with simulated velocity, magnetic field, and gas temperature profiles of a strong shock of velocity 5000 km s^{-1} with the far upstream gas temperature of 10^4 K and magnetic field of about μG. The simulation was made with the non-linear Monte-Carlo model by Vladimirov et al. (2008, 2009) which accounts for efficient CR acceleration, strong non-adiabatic magnetic field amplification due to CR-driven instabilities in the shock upstream with magnetic turbulence dissipation, and the escape of highest energy particles to the shock upstream. The models of non-linear DSA predict hard spectra of accelerated relativistic particles, which often show concave spectral shapes instead of the power-laws. Strong amplification of the fluctuating magnetic fields in the upstream flow by CR-driven instabilities is expected in the models of DSA. Important physical effects to be learned from the strong collisionless shock modeling are: (i) potentially sizeable (above ten percent) energy leakage from the system with the ultra-relativistic particles accelerated at the shock and escaping through the shock upstream back into inter-galactic medium, and (ii) a possibility of strong super-adiabatic magnetic field amplification by CR-driven instabilities (see for a review Bykov et al. 2013).

These effects may strongly affect the thermal properties of shocks with high sonic and Alfvenic Mach numbers as it is expected to be the case in the external accreting shocks at cluster outskirts. In Fig. 18 we illustrate the possible effect of the non-thermal components on the ion temperature (left panel) and the entropy (right panel) in the downstream of the multi-fluid shock simulated with the non-linear Monte-Carlo model described in Vladimirov et al. (2009). Both post-shock ion temperature and gas entropy K_{mf} in the multi-fluid collisionless shock can be strongly reduced compared to that in the standard single fluid Rankine-Hugoniot adiabat. This is because of a substantial increase of the gas compression ratio in a strong collisionless shock converting a sizable part of the shock ram pressure into relativistic particles. Some fraction of the accelerated particles escape the system thus allowing the gas compression ratio to be much larger than 4. As it was discussed in Sect. 6.1 most of the kinetic energy dissipation occurs at the cluster inner shocks with the modest Mach numbers where the effects discussed above are likely much less prominent. Indeed, the ratio of the thermal gas heating to CR acceleration rates in weak shocks of sonic Mach number $\mathcal{M}_s < 2$ is proportional to $(\mathcal{M}_s - 1)\beta_p$, providing inefficient CR acceleration by weak shocks in the

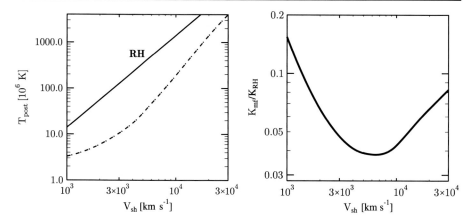

Fig. 18 *Left panel*: The post-shock proton temperature as a function of shock velocity simulated by a non-linear Monte Carlo model with an account of the efficient particle acceleration and magnetic field amplification. The far upstream gas temperature was about 2×10^4 K corresponding to the photo-ionized intergalactic gas accreting by a cluster. The post-shock temperature (shown with *dash-dotted line*) was simulated for a model with turbulent cascade of CR-driven magnetic fluctuations (Vladimirov et al. 2009). *The solid curve* (labeled as RH) is the standard Rankine-Hugoniot single fluid post-shock temperature (it is presented for a comparison). *Right panel*: The ratio of the post-shock gas entropy (labeled as K_{mf} multi-fluid) to the standard single fluid (Rankine-Hugoniot K_{RH}) post-shock gas entropy as a function of shock velocity simulated for cosmic ray modified collisionless shock

case of the large ratio of the thermal and magnetic pressures $\beta_p \gg 1$ expected in the inner regions of the cluster. A recent search of γ-ray emission from stacked Fermi-LAT count maps of some dozens of clusters of galaxies (Ackermann et al. 2013; Dutson et al. 2013; Huber et al. 2013) established stringent upper limits on the average CR to thermal pressure ratio to be below of a few percent within the radius r_{200} depending on the assumed index of the power-law CR distribution and γ-ray background models. Detection of γ-ray emission from extended regions around the external accretion shocks in the vicinity of filaments and clusters is a challenging task given its low surface brightness because of the low gas density.

7 Concluding Remarks and Open Problems

In this review, we have discussed recent results on structure formation focusing our attention on the first objects in the Universe and the most massive clusters of galaxies at the present day. These extreme scenarios allow us to clearly illustrate the relevance of the physics of plasma on the formation of cosmic structures along a wide range of spatial and temporal scales. In the hierarchical paradigm of structure formation, the first objects are the building blocks of subsequent structure formation, leading to larger galaxies and galaxy clusters through accretion and merger events (e.g. Somerville et al. 2012). Despite the disparity of the involved scales, a number of physical processes, such as radiative cooling, turbulence and feedback, appear to be common, suggesting them to be ubiquitous in the non-linear regime of cosmic structure formation.

In the early Universe, the first objects are expected to form in halos with 10^5-10^8 M$_\odot$ at redshift 10–30 (e.g. Bromm et al. 2009; Bromm and Yoshida 2011). Here we distinguish the so-called minihalos with virial temperatures above 1000 K from the atomic cooling halos with virial temperatures above 10^4 K. Minihalos are the expected formation sites for the first

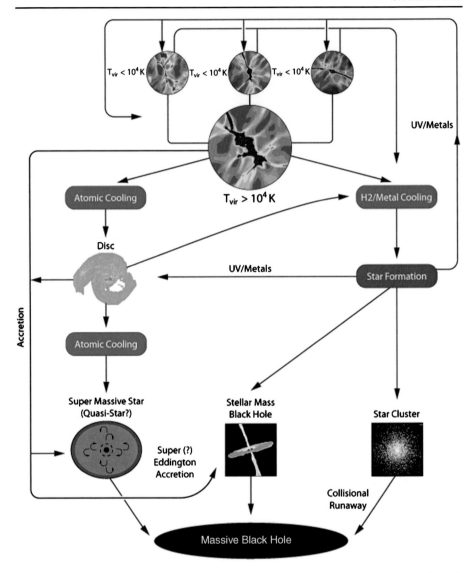

Fig. 19 Flowchart summarizing possible paths for the formation of the first SMBHs in high-redshift atomic cooling halos. Figure from Regan and Haehnelt (2009a)

primordial stars, with typical masses in the range of 10–100 $M_\odot$ (Abel et al. 2002; Bromm et al. 2002; Yoshida et al. 2008; Clark et al. 2011; Greif et al. 2011; Hosokawa et al. 2011; Turk et al. 2012; Latif et al. 2013c; Susa 2013). Their formation is governed by the chemistry and cooling of molecular hydrogen, as well as additional processes such as turbulence (e.g. Turk et al. 2012; Latif et al. 2013c), radiative feedback (e.g. Hosokawa et al. 2011; Susa 2013) and magnetic fields (e.g. Tan and Blackman 2004; Machida et al. 2006; Sur et al. 2010, 2012; Schober et al. 2012; Turk et al. 2012).

The atomic cooling halos show a more complex evolution depending on their local conditions, in particular regarding their metallicity and dust content. In this review, we restricted ourselves to the formation of massive black holes in primordial halos (see Fig. 19 for an illustrative summary), while a more general discussion is given by Bromm and Yoshida (2011). In the presence of strong photodissociating backgrounds, H_2 formation is suppressed (Omukai 2001; Machacek et al. 2001; Johnson et al. 2008; Schleicher et al. 2010b; Shang et al. 2010; Latif et al. 2011), leading to a close-to-isothermal collapse regulated via atomic hydrogen lines. Recent numerical simulations confirm that massive central objects can indeed form, due to the high accretion rates of more than 1 $M_\odot \, yr^{-1}$ (Latif et al. 2013a). In the presence of such accretion rates, feedback can be expected to be weak (Hosokawa et al. 2012; Schleicher et al. 2013) and does not impede the accretion. Indeed, even trace amounts of dust, corresponding to 10^{-5}–10^{-3} times the dust-to-gas ratio in the solar neighborhood, may already trigger strong cooling and fragmentation at high densities (Schneider et al. 2004; Omukai et al. 2005), but also stimulate the formation of molecular hydrogen at low to moderate densities (Cazaux and Spaans 2009; Latif et al. 2012). The extremely metal poor star SDSS J1029151+172927 (Caffau et al. 2011) shows chemical abundances at which metal line coolant is inefficient, and where only trace amounts of dust grains were able to trigger cooling and fragmentation (Klessen et al. 2012; Schneider et al. 2012).[12] For metallicities above 10^{-2} solar, on the other hand, metal line cooling can be expected to be significant (Bromm and Loeb 2003b; Omukai et al. 2005). The fragmentation of such metal-enriched atomic cooling halos is in fact poorly understood (see e.g. Safranek-Shrader et al. 2010, 2014, for recent simulations) and needs to be investigated in further detail. Overall, constraints on the high-redshift black hole population can be obtained from the infrared background (Yue et al. 2013, 2014).

In the hierarchical paradigm of structure formation, where the first objects are the building blocks of subsequent structure development, clusters of galaxies, with masses of up to 10^{15} $M_\odot$ at $z = 0$, occupy the most massive extreme of the cosmic hierarchy. The formation and evolution of galaxy clusters is a complex and non-linear event resulting from the intricate interaction of a number of physical processes acting on a wide range of scales (see Kravtsov and Borgani 2012, for a recent review and references therein). As an example, Fig. 20 shows a simplified summary of some of the main processes operating in galaxy clusters. On large scales, the hierarchical process of structure formation induces the development of strong cosmological shocks, surrounding galaxy clusters and filaments, that contribute to heat and compress the hot intra-cluster plasma. Within galaxy clusters, weaker internal shocks, mainly originated by subhalo mergers or accretion phenomena, change the energetic balance of the gas and allow the halos to virialize. These shocks can also generate ICM turbulence and mixing, amplify magnetic fields, and accelerate thermal distributions of particles giving rise to a non-thermal population of CRs. In dense regions within galaxies and galaxy clusters, the intra-cluster gas can cool radiatively, leading to both star formation and gas accretion onto SMBHs residing at the center of massive cluster galaxies. These processes can then provide a significant energy contribution to the ICM in the form of SNe explosions or AGN feedback. As shown in Fig. 20, all these processes, which are highly interconnected between them, are manifested by means of different observational channels.

In the last years, the new generations of supercomputing and programming facilities have been crucial to deepen in our understanding of the complicated physical processes taking place within the intra-cluster plasma and shaping the observational properties of

---

[12]An alternative formation scenario for the star SDSS J1029151+172927 has been recently proposed by MacDonald et al. (2013), who suggest that it may have been a subgiant formed with significantly higher metallicity in the vicinity of a SN-Ia.

Cluster observables: Physical processes in clusters:

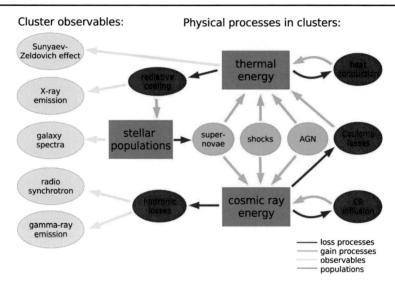

Fig. 20 Flowchart summarizing the connections between the main physical processes taking place in galaxy clusters together with the different observational channels through which they can be detected. Figure from Pfrommer et al. (2008)

galaxy clusters. In order to explain the observations, cosmological hydrodynamical simulations have tried to implement the most relevant physical processes self-consistently with the cosmic evolution. In particular, in addition to gravitationally-induced phenomena inherent to structure formation, the standard non-gravitational processes commonly included in these simulations are radiative cooling, star formation and SN feedback. In the last years, the inclusion of the effects of thermal and/or kinetic AGN feedback is also becoming a common practice, despite the fact that the particularities of the heating mechanism are still uncertain. In spite of the relatively simplicity employed in modeling these complex processes, simulations have been able to significantly reproduce most of the observational cluster properties, at least for massive systems at relatively outer cluster regions ($0.1 R_{500} \lesssim r \lesssim R_{500}$), where clusters are assumed to be nearly self-similar. However, inner cluster regions and smaller systems show a number of significant issues that still need to be solved. In these inner regions, simulations still show an excess of gas cooling, which produces an excess in both the star formation and the metal production. In addition, simulations are still not able to solve the cooling flow problem or to reproduce the diversity of the observed temperature and entropy radial profiles of relaxed and unrelaxed systems. On the other hand, cluster outskirts ($r \gtrsim R_{500}$) are also affected by strong deviations from hydrostatic equilibrium caused, primarily, by sources of non-thermal pressure support such as CRs or magnetic fields, which generally are not modeled in simulations. These results indicate that, in addition to the processes already included, a number of additional physical processes, mainly related with the complex physics of plasma, such as turbulence, viscosity or thermal conduction, must be also properly taken into account. Therefore, although AGN feedback seems to be the most favored energy source to regulate cooling in clusters, a subtle interplay with a number of supplementary physical phenomena may be needed to explain the observational properties of galaxy clusters and groups, from the core regions out to the outskirts.

In the near future, a significant numerical effort will be aimed at performing larger and better-resolved cosmological simulations with a more accurate modeling of the physics of

⌐Ⓓ Springer

galaxy evolution. In addition to these technical improvements, forthcoming instruments, like the *JWST* (Gardner et al. 2006) and the new generation of large ground-based telescopes, are expected to detect light from the first galaxies, contributing to interpret early structure formation. Besides, a number of large observational surveys in different wavebands, such as *eROSITA* (Pillepich et al. 2012), *Euclid* (Laureijs et al. 2012), *WFXT* (Pareschi and Campana 2011) or the *LSST* (Abell et al. 2009), will provide a significantly large number of clusters. These numerical and observational efforts, together with a more accurate treatment of the physics of plasma, will definitely shed some more light on the nature of the physical processes governing the formation of structures in the Universe, from the first non-linear objects to the present-day massive galaxy clusters.

Acknowledgements We would like to thank the ISSI staff for their hospitality and for providing an inspiring atmosphere at the International Space Science Institute Workshop in Bern in 2013. We also would like to thank the anonymous referee for his/her constructive comments. S. Planelles acknowledges support by the PRIN-INAF09 project "Towards an Italian Network for Computational Cosmology" and by the PRIN-MIUR09 "Tracing the growth of structures in the Universe". D.R.G. Schleicher thanks for funding from the German Science Foundation (DFG) in the DFG priority program SPP 1573 "Physics of the Interstellar Medium" under grant SCHL 1964/1-1, and via the collaborative research center (CRC) 963/1 "Astrophysical flow instabilities and turbulence" (project A12). A.M. Bykov was supported in part by RAS Presidium and OFN 15 and 17 programs. We further thank for stimulating discussions with Stefano Borgani, Stefano Bovino, Muhammad Latif, Wolfram Schmidt, Jens Niemeyer and Barbara Sartoris.

References

T. Abel, P. Anninos, Y. Zhang, M.L. Norman, New Astron. **2**, 181 (1997)

T. Abel, G.L. Bryan, M.L. Norman, Science **295**, 93 (2002)

P.A. Abell, J. Allison et al. (LSST Science Collaboration), arXiv e-print (arXiv:0912.0201) (2009)

M. Ackermann, M. Ajello et al. (The Fermi-LAT Collaboration), arXiv e-prints (arXiv:1308.5654) (2013)

P.A.R. Ade, N. Aghanim et al. (Planck Collaboration), arXiv e-prints (arXiv:1303.5076) (2013a)

P.A.R. Ade, N. Aghanim et al. (Planck Collaboration), arXiv e-prints (arXiv:1303.5083) (2013b)

B. Agarwal, S. Khochfar, J.L. Johnson et al., Mon. Not. R. Astron. Soc. **425**, 2854 (2012)

O. Agertz, B. Moore, J. Stadel et al., Mon. Not. R. Astron. Soc. **380**, 963 (2007)

S.W. Allen, R.W. Schmidt, A.C. Fabian, Mon. Not. R. Astron. Soc. **328**, L37 (2001)

S.W. Allen, A.E. Evrard, A.B. Mantz, Annu. Rev. Astron. Astrophys. **49**, 409 (2011)

E. Amato, P. Blasi, Mon. Not. R. Astron. Soc. **371**, 1251 (2006)

R.E. Angulo, V. Springel, S.D.M. White et al., Mon. Not. R. Astron. Soc. **426**, 2046 (2012)

M. Arnaud, A.E. Evrard, Mon. Not. R. Astron. Soc. **305**, 631 (1999)

M. Arnaud, E. Pointecouteau, G.W. Pratt, Astron. Astrophys. **441**, 893 (2005)

E. Audit, R. Teyssier, J.-M. Alimi, Astron. Astrophys. **325**, 439 (1997)

R. Ball, J.O. Burns, C. Loken, Astron. J. **105**, 53 (1993)

M.L. Balogh, A. Babul, D.R. Patton, Mon. Not. R. Astron. Soc. **307**, 463 (1999)

M.L. Balogh, F.R. Pearce, R.G. Bower, S.T. Kay, Mon. Not. R. Astron. Soc. **326**, 1228 (2001)

P. Barai, M. Viel, G. Murante, S. Borgani, Mon. Not. R. Astron. Soc. **437**, 1456 (2014)

R. Barkana, A. Loeb, Phys. Rep. **349**, 125 (2001)

N. Battaglia, J.R. Bond, C. Pfrommer, J.L. Sievers, Astrophys. J. **777**, 123 (2013)

M.C. Begelman, M. Volonteri, M.J. Rees, Mon. Not. R. Astron. Soc. **370**, 289 (2006)

A.R. Bell, Mon. Not. R. Astron. Soc. **353**, 550 (2004)

E. Bertschinger, Astrophys. J. Suppl. Ser. **58**, 39 (1985)

S. Bhattacharya, K. Heitmann, M. White et al., Astrophys. J. **732**, 122 (2011)

J. Binney, S. Tremaine, *Galactic Dynamics*, 2nd edn. (Princeton University Press, Princeton, 2008)

L. Bîrzan, D.A. Rafferty, B.R. McNamara, M.W. Wise, P.E.J. Nulsen, Astrophys. J. **607**, 800 (2004)

R. Blandford, D. Eichler, Phys. Rep. **154**, 1 (1987)

G.R. Blumenthal, S.M. Faber, J.R. Primack, M.J. Rees, Nature **311**, 517 (1984)

A. Bonafede, G. Giovannini, L. Feretti, F. Govoni, M. Murgia, Astron. Astrophys. **494**, 429 (2009)

J.R. Bond, S. Cole, G. Efstathiou, N. Kaiser, Astrophys. J. **379**, 440 (1991)

H. Bondi, Mon. Not. R. Astron. Soc. **112**, 195 (1952)

S. Borgani, A pan-chromatic view of clusters of galaxies and the large-scale structure, in *Lecture Notes in Physics*, vol. 740, ed. by M. Plionis, O. López-Cruz, D. Hughes (Springer, Berlin, 2008), p. 287

S. Borgani, L. Guzzo, Nature **409**, 39 (2001)

S. Borgani, A. Kravtsov, Adv. Sci. Lett. **4**, 204 (2011)

S. Borgani, F. Governato, J. Wadsley et al., Astrophys. J. Lett. **559**, L71 (2001)

S. Borgani, G. Murante, V. Springel et al., Mon. Not. R. Astron. Soc. **348**, 1078 (2004)

S. Borgani, A. Diaferio, K. Dolag, S. Schindler, Space Sci. Rev. **134**, 269 (2008)

F. Bournaud, B.G. Elmegreen, R. Teyssier, D.L. Block, I. Puerari, Mon. Not. R. Astron. Soc. **409**, 1088 (2010)

S. Bovino, D.R.G. Schleicher, T. Grassi, Astron. Astrophys. **561**, A13 (2014)

H. Braun, W. Schmidt, Mon. Not. R. Astron. Soc. **421**, 1838 (2012)

V. Bromm, R.B. Larson, Annu. Rev. Astron. Astrophys. **42**, 79 (2004)

V. Bromm, A. Loeb, Astrophys. J. **596**, 34 (2003a)

V. Bromm, A. Loeb, Nature **425**, 812 (2003b)

V. Bromm, N. Yoshida, Annu. Rev. Astron. Astrophys. **49**, 373 (2011)

V. Bromm, P.S. Coppi, R.B. Larson, Astrophys. J. **564**, 23 (2002)

V. Bromm, N. Yoshida, L. Hernquist, C.F. McKee, Nature **459**, 49 (2009)

G.L. Bryan, M.L. Norman, Astrophys. J. **495**, 80 (1998)

G.L. Bryan, M.L. Norman, B.W. O'Shea et al., Astrophys. J. Suppl. Ser. **211**, 19 (2014)

D. Burgess, M. Scholer, Space Sci. Rev. **178**, 513 (2013)

J.O. Burns, Astron. J. **99**, 14 (1990)

A.M. Bykov, H. Bloemen, Y.A. Uvarov, Astron. Astrophys. **362**, 886 (2000)

A.M. Bykov, K. Dolag, F. Durret, Space Sci. Rev. **134**, 119 (2008)

A.M. Bykov, D.C. Ellison, M. Renaud, Space Sci. Rev. **166**, 71 (2012)

A.M. Bykov, A. Brandenburg, M.A. Malkov, S.M. Osipov, Space Sci. Rev. **178**, 201 (2013)

E. Caffau, P. Bonifacio, P. François et al., Nature **477**, 67 (2011)

K.W. Cavagnolo, M. Donahue, G.M. Voit, M. Sun, Astrophys. J. Suppl. Ser. **182**, 12 (2009)

S. Cazaux, M. Spaans, Astron. Astrophys. **496**, 365 (2009)

R. Cen, J.P. Ostriker, Astrophys. J. **514**, 1 (1999)

B.D.G. Chandran, P. Sharma, I.J. Parrish, astro2010: the astronomy and astrophysics decadal survey. Astronomy **2010**, 41 (2009)

Y. Chen, T.H. Reiprich, H. Böhringer, Y. Ikebe, Y.-Y. Zhang, Astron. Astrophys. **466**, 805 (2007)

E. Churazov, M. Brüggen, C.R. Kaiser, H. Böhringer, W. Forman, Astrophys. J. **554**, 261 (2001)

P.C. Clark, S.C.O. Glover, R.J. Smith et al., Science **331**, 1040 (2011)

J.D. Cohn, M. White, Mon. Not. R. Astron. Soc. **385**, 2025 (2008)

P. Coles, F. Lucchin, *Cosmology: The Origin and Evolution of Cosmic Structure*, 2nd edn. (2002)

C.M. Coppola, R. D'Introno, D. Galli, J. Tennyson, S. Longo, Astrophys. J. Suppl. Ser. **199**, 16 (2012)

J. Courtin, Y. Rasera, J.-M. Alimi et al., Mon. Not. R. Astron. Soc. **410**, 1911 (2011)

M. Crocce, P. Fosalba, F.J. Castander, E. Gaztañaga, Mon. Not. R. Astron. Soc. **403**, 1353 (2010)

W. Cui, S. Borgani, K. Dolag, G. Murante, L. Tornatore, Mon. Not. R. Astron. Soc. **423**, 2279 (2012)

W. Cui, S. Borgani, G. Murante, arXiv e-prints (arXiv:1402.1493) (2014)

S.J. Cusworth, S.T. Kay, R.A. Battye, P.A. Thomas, Mon. Not. R. Astron. Soc. **439**, 2485 (2014)

X. Dai, J.N. Bregman, C.S. Kochanek, E. Rasia, Astrophys. J. **719**, 119 (2010)

R. Davé, R. Cen, J.P. Ostriker et al., Astrophys. J. **552**, 473 (2001)

R. Davé, N. Katz, D.H. Weinberg, Astrophys. J. **579**, 23 (2002)

M. Davis, P.J.E. Peebles, Astrophys. J. **267**, 465 (1983)

M. Davis, G. Efstathiou, C.S. Frenk, S.D.M. White, Astrophys. J. **292**, 371 (1985)

S. De Grandi, S. Molendi, Astrophys. J. **567**, 163 (2002)

J. Diemand, M. Kuhlen, P. Madau, Astrophys. J. **667**, 859 (2007)

M. Dijkstra, Z. Haiman, A. Mesinger, J.S.B. Wyithe, Mon. Not. R. Astron. Soc. **391**, 1961 (2008)

C.L. Dobbs, C. Theis, J.E. Pringle, M.R. Bate, Mon. Not. R. Astron. Soc. **403**, 625 (2010)

K. Dolag, M. Bartelmann, H. Lesch, Astron. Astrophys. **348**, 351 (1999)

K. Dolag, F. Vazza, G. Brunetti, G. Tormen, Mon. Not. R. Astron. Soc. **364**, 753 (2005)

K. Dolag, S. Borgani, S. Schindler, A. Diaferio, A.M. Bykov, Space Sci. Rev. **134**, 229 (2008)

L.O. Drury, S.A.E.G. Falle, Mon. Not. R. Astron. Soc. **223**, 353 (1986)

Y. Dubois, J. Devriendt, R. Teyssier, A. Slyz, Mon. Not. R. Astron. Soc. **417**, 1853 (2011)

K.L. Dutson, R.J. White, A.C. Edge, J.A. Hinton, M.T. Hogan, Mon. Not. R. Astron. Soc. **429**, 2069 (2013)

D. Eckert, S. Molendi, F. Vazza, S. Ettori, S. Paltani, Astron. Astrophys. **551**, A22 (2013)

A.C. Edge, D.T. Frayer, Astrophys. J. Lett. **594**, L13 (2003)

J.A. Eilek, in *The Riddle of Cooling Flows in Galaxies and Clusters of Galaxies*, ed. by T. Reiprich, J. Kempner, N. Soker (2004), p. 165

D.J. Eisenstein, W. Hu, Astrophys. J. **511**, 5 (1999)
V.R. Eke, J.F. Navarro, C.S. Frenk, Astrophys. J. **503**, 569 (1998)
T.A. Ensslin, P.L. Biermann, U. Klein, S. Kohle, Astron. Astrophys. **332**, 395 (1998)
S. Ettori, P. Tozzi, S. Borgani, P. Rosati, Astron. Astrophys. **417**, 13 (2004)
S. Ettori, A. Donnarumma, E. Pointecouteau et al., Space Sci. Rev. **177**, 119 (2013)
A.E. Evrard, J.P. Henry, Astrophys. J. **383**, 95 (1991)
A.E. Evrard, T.J. MacFarland, H.M.P. Couchman et al., Astrophys. J. **573**, 7 (2002)
A.C. Fabian, C.S. Crawford, A.C. Edge, R.F. Mushotzky, Mon. Not. R. Astron. Soc. **267**, 779 (1994)
D. Fabjan, L. Tornatore, S. Borgani, K. Dolag, Mon. Not. R. Astron. Soc. **386**, 1265 (2008)
D. Fabjan, S. Borgani, L. Tornatore et al., Mon. Not. R. Astron. Soc. **401**, 1670 (2010)
X. Fan, V.K. Narayanan, R.H. Lupton et al., Astron. J. **122**, 2833 (2001)
X. Fan, M.A. Strauss, D.P. Schneider et al., Astron. J. **125**, 1649 (2003)
C. Federrath, S. Sur, D.R.G. Schleicher, R. Banerjee, R.S. Klessen, Astrophys. J. **731**, 62 (2011)
L. Ferrarese, D. Merritt, Astrophys. J. Lett. **539**, L9 (2000)
C. Ferrari, F. Govoni, S. Schindler, A.M. Bykov, Y. Rephaeli, Space Sci. Rev. **134**, 93 (2008)
V. Florinski, R.B. Decker, J.A. le Roux, G.P. Zank, Geophys. Res. Lett. **36**, 12101 (2009)
D.R. Flower, G.J. Harris, Mon. Not. R. Astron. Soc. **377**, 705 (2007)
R.C. Forrey, Astrophys. J. Lett. **773**, L25 (2013)
D. Galli, F. Palla, Astron. Astrophys. **335**, 403 (1998)
J.P. Gardner, J.C. Mather, M. Clampin et al., Space Sci. Rev. **123**, 485 (2006)
L. Gargaté, A. Spitkovsky, Astrophys. J. **744**, 67 (2012)
M. Gaspari, C. Melioli, F. Brighenti, A. D'Ercole, Mon. Not. R. Astron. Soc. **411**, 349 (2011)
M. Gaspari et al., arXiv e-print (arXiv:1301.3130) (2013)
M. Gaspari, F. Brighenti, P. Temi, S. Ettori, Astrophys. J. Lett. **783**, L10 (2014)
S. Giodini, D. Pierini, A. Finoguenov et al., Astrophys. J. **703**, 982 (2009)
S. Giodini, L. Lovisari, E. Pointecouteau et al., Space Sci. Rev. **177**, 247 (2013)
G. Giovannini, A. Bonafede, L. Feretti et al., Astron. Astrophys. **507**, 1257 (2009)
S. Glover, Space Sci. Rev. **117**, 445 (2005)
S.C.O. Glover, P.C. Clark, T.H. Greif et al., in *IAU Symposium*, ed. by L.K. Hunt, S.C. Madden, R. Schneider. IAU Symposium, vol. 255 (2008), pp. 3–17
A.H. Gonzalez, D. Zaritsky, A.I. Zabludoff, Astrophys. J. **666**, 147 (2007)
T. Grassi, S. Bovino, D.R.G. Schleicher et al., Mon. Not. R. Astron. Soc. **439**, 2386 (2014)
T.H. Greif, S.C.O. Glover, V. Bromm, R.S. Klessen, Astrophys. J. **716**, 510 (2010)
T.H. Greif, V. Springel, S.D.M. White et al., Astrophys. J. **737**, 75 (2011)
T.H. Greif, V. Bromm, P.C. Clark et al., Mon. Not. R. Astron. Soc. **424**, 399 (2012)
J.E. Gunn, J.R. Gott III, Astrophys. J. **176**, 1 (1972)
A.H. Guth, S.-Y. Pi, Phys. Rev. Lett. **49**, 1110 (1982)
J.-C. Hamilton, arXiv e-prints (arXiv:1304.4446) (2013)
N. Häring, H.-W. Rix, Astrophys. J. Lett. **604**, L89 (2004)
E.A. Helder, J. Vink, A.M. Bykov et al., Space Sci. Rev. **173**, 369 (2012)
S.F. Helsdon, T.J. Ponman, Mon. Not. R. Astron. Soc. **315**, 356 (2000)
A. Hobbs, C. Power, S. Nayakshin, A.R. King, Mon. Not. R. Astron. Soc. **421**, 3443 (2012)
T. Hosokawa, K. Omukai, N. Yoshida, H.W. Yorke, Science **334**, 1250 (2011)
T. Hosokawa, K. Omukai, H.W. Yorke, Astrophys. J. **756**, 93 (2012)
B. Huber, C. Tchernin, D. Eckert et al., Astron. Astrophys. **560**, A64 (2013)
L. Iapichino, J.C. Niemeyer, Mon. Not. R. Astron. Soc. **388**, 1089 (2008)
J.H. Jeans, Philos. Trans. R. Soc. Lond. Ser. A **199**, 1 (1902)
A. Jenkins, C.S. Frenk, S.D.M. White et al., Mon. Not. R. Astron. Soc. **321**, 372 (2001)
J.L. Johnson, T.H. Greif, V. Bromm, Astrophys. J. **665**, 85 (2007)
J.L. Johnson, T.H. Greif, V. Bromm, Mon. Not. R. Astron. Soc. **388**, 26 (2008)
J.L. Johnson, V.C. Dalla, S. Khochfar, Mon. Not. R. Astron. Soc. **428**, 1857 (2013a)
J.L. Johnson, D.J. Whalen, H. Li, D.E. Holz, Astrophys. J. **771**, 116 (2013b)
F.C. Jones, D.C. Ellison, Space Sci. Rev. **58**, 259 (1991)
N. Kaiser, Mon. Not. R. Astron. Soc. **222**, 323 (1986)
H. Kang, T.W. Jones, Astrophys. J. **620**, 44 (2005)
H. Kang, T.W. Jones, U.D.J. Gieseler, Astrophys. J. **579**, 337 (2002)
H. Kang, D. Ryu, R. Cen, J.P. Ostriker, Astrophys. J. **669**, 729 (2007)
T.N. Kato, H. Takabe, Astrophys. J. Lett. **681**, L93 (2008)
T.N. Kato, H. Takabe, Astrophys. J. **721**, 828 (2010)
S.T. Kay, P.A. Thomas, A. Jenkins, F.R. Pearce, Mon. Not. R. Astron. Soc. **355**, 1091 (2004)

C.F. Kennel, J.P. Edmiston, T. Hada, *A Quarter Century of Collisionless Shock Research*. American Geophysical Union Geophysical Monograph Series, vol. 34 (1985), p. 1
K. Kettula, A. Finoguenov, R. Massey et al., Astrophys. J. **778**, 74 (2013)
W.-T. Kim, A.A. El-Zant, M. Kamionkowski, Astrophys. J. **632**, 157 (2005)
R.S. Klessen, S.C.O. Glover, P.C. Clark, Mon. Not. R. Astron. Soc. **421**, 3217 (2012)
A.A. Klypin, S.F. Shandarin, Mon. Not. R. Astron. Soc. **204**, 891 (1983)
A. Knebe, S.R. Knollmann, S.I. Muldrew et al., Mon. Not. R. Astron. Soc. **415**, 2293 (2011)
A. Knebe, F.R. Pearce, H. Lux et al., Mon. Not. R. Astron. Soc. **435**, 1618 (2013)
S.M. Koushiappas, J.S. Bullock, A. Dekel, Mon. Not. R. Astron. Soc. **354**, 292 (2004)
A.V. Kravtsov, S. Borgani, Annu. Rev. Astron. Astrophys. **50**, 353 (2012)
A. Kravtsov, A. Vikhlinin, A. Meshscheryakov, arXiv e-prints (arXiv:1401.7329) (2014)
C. Lacey, S. Cole, Mon. Not. R. Astron. Soc. **262**, 627 (1993)
C. Lacey, S. Cole, Mon. Not. R. Astron. Soc. **271**, 676 (1994)
T.F. Laganá, Y.-Y. Zhang, T.H. Reiprich, P. Schneider, Astrophys. J. **743**, 13 (2011)
M.A. Latif, D.R.G. Schleicher, M. Spaans, S. Zaroubi, Astron. Astrophys. **532**, A66 (2011)
M.A. Latif, D.R.G. Schleicher, M. Spaans, Astron. Astrophys. **540**, A101 (2012)
M.A. Latif, D.R.G. Schleicher, W. Schmidt, J. Niemeyer, Mon. Not. R. Astron. Soc. **433**, 1607 (2013a)
M.A. Latif, D.R.G. Schleicher, W. Schmidt, J. Niemeyer, Mon. Not. R. Astron. Soc. **430**, 588 (2013b)
M.A. Latif, D.R.G. Schleicher, W. Schmidt, J. Niemeyer, Astrophys. J. Lett. **772**, L3 (2013c)
M.A. Latif, D.R.G. Schleicher, W. Schmidt, J. Niemeyer, Mon. Not. R. Astron. Soc. **432**, 668 (2013d)
R. Laureijs, P. Gondoin, L. Duvet et al., *Society of Photo-Optical Instrumentation Engineers (SPIE) Conference Series*, vol. 8442 (2012)
A.M.C. Le Brun, I.G. McCarthy, J. Schaye, T.J. Ponman, arXiv e-prints (arXiv:1312.5462) (2013)
A. Leccardi, S. Molendi, Astron. Astrophys. **487**, 461 (2008a)
A. Leccardi, S. Molendi, Astron. Astrophys. **486**, 359 (2008b)
Y.-T. Lin, J.J. Mohr, S.A. Stanford, Astrophys. J. **591**, 749 (2003)
G. Lodato, P. Natarajan, Mon. Not. R. Astron. Soc. **377**, L64 (2007)
Z. Lukić, K. Heitmann, S. Habib, S. Bashinsky, P.M. Ricker, Astrophys. J. **671**, 1160 (2007)
J. MacDonald, T.M. Lawlor, N. Anilmis, N.F. Rufo, Mon. Not. R. Astron. Soc. **431**, 1425 (2013)
M.E. Machacek, G.L. Bryan, T. Abel, Astrophys. J. **548**, 509 (2001)
M.N. Machida, K. Omukai, T. Matsumoto, S.-i. Inutsuka, Astrophys. J. Lett. **647**, L1 (2006)
M. Maggiore, A. Riotto, Astrophys. J. **711**, 907 (2010)
J. Magorrian, S. Tremaine, D. Richstone et al., Astron. J. **115**, 2285 (1998)
A. Maier, L. Iapichino, W. Schmidt, J.C. Niemeyer, Astrophys. J. **707**, 40 (2009)
M.A. Malkov, L. Drury, Rep. Prog. Phys. **64**, 429 (2001)
M. Markevitch, Astrophys. J. **504**, 27 (1998)
M. Markevitch, A. Vikhlinin, Phys. Rep. **443**, 1 (2007)
D. Martizzi, R. Teyssier, B. Moore, Mon. Not. R. Astron. Soc. **420**, 2859 (2012)
B.J. Maughan, P.A. Giles, S.W. Randall, C. Jones, W.R. Forman, Mon. Not. R. Astron. Soc. **421**, 1583 (2012)
L. Mayer, F. Governato, T. Kaufmann, Adv. Sci. Lett. **1**, 7 (2008)
I.G. McCarthy, R.G. Bower, M.L. Balogh et al., Mon. Not. R. Astron. Soc. **376**, 497 (2007)
I.G. McCarthy, A. Babul, R.G. Bower, M.L. Balogh, Mon. Not. R. Astron. Soc. **386**, 1309 (2008)
I.G. McCarthy, J. Schaye, T.J. Ponman et al., Mon. Not. R. Astron. Soc. **406**, 822 (2010)
J.F. McKenzie, H.J. Voelk, Astron. Astrophys. **116**, 191 (1982)
B.R. McNamara, P.E.J. Nulsen, Annu. Rev. Astron. Astrophys. **45**, 117 (2007)
F. Miniati, D. Ryu, H. Kang et al., Astrophys. J. **542**, 608 (2000)
H. Mo, F.C. van den Bosch, S. White, *Galaxy Formation and Evolution* (2010)
D.J. Mortlock, S.J. Warren, B.P. Venemans et al., Nature **474**, 616 (2011)
O. Muanwong, P.A. Thomas, S.T. Kay, F.R. Pearce, H.M.P. Couchman, Astrophys. J. Lett. **552**, L27 (2001)
S.G. Murray, C. Power, A.S.G. Robotham, Mon. Not. R. Astron. Soc. **434**, L61 (2013)
D. Nagai, A.V. Kravtsov, in *Outskirts of Galaxy Clusters: Intense Life in the Suburbs*, ed. by A. Diaferio. IAU Colloq., vol. 195 (2004), pp. 296–298
D. Nagai, A.V. Kravtsov, A. Kosowsky, Astrophys. J. **587**, 524 (2003)
D. Nagai, A.V. Kravtsov, A. Vikhlinin, Astrophys. J. **668**, 1 (2007a)
D. Nagai, A. Vikhlinin, A.V. Kravtsov, Astrophys. J. **655**, 98 (2007b)
J.F. Navarro, C.S. Frenk, S.D.M. White, Mon. Not. R. Astron. Soc. **275**, 720 (1995)
M.L. Norman, G.L. Bryan, in *The Radio Galaxy Messier 87*, ed. by H.-J. Röser, K. Meisenheimer. Lecture Notes in Physics, vol. 530 (Springer, Berlin, 1999), p. 106
T.B. O'Hara, J.J. Mohr, J.J. Bialek, A.E. Evrard, Astrophys. J. **639**, 64 (2006)
H. Omma, J. Binney, G. Bryan, A. Slyz, Mon. Not. R. Astron. Soc. **348**, 1105 (2004)
K. Omukai, Astrophys. J. **546**, 635 (2001)

K. Omukai, S.-i. Inutsuka, Mon. Not. R. Astron. Soc. **332**, 59 (2002)
K. Omukai, T. Tsuribe, R. Schneider, A. Ferrara, Astrophys. J. **626**, 627 (2005)
J. Onions, A. Knebe, F.R. Pearce et al., Mon. Not. R. Astron. Soc. **423**, 1200 (2012)
B.W. O'Shea, G. Bryan, J. Bordner et al., arXiv Astrophysics e-prints (arXiv:astro-ph/0403044) (2004)
J.P.F. Osmond, T.J. Ponman, Mon. Not. R. Astron. Soc. **350**, 1511 (2004)
F. Palla, E.E. Salpeter, S.W. Stahler, Astrophys. J. **271**, 632 (1983)
A. Paranjape et al., arXiv e-prints (arXiv:1210.1483) (2013)
A. Paranjape, arXiv e-prints (arXiv:1403.3402) (2014)
G. Pareschi, S. Campana, Mem. Soc. Astron. Ital. Suppl. **17**, 16 (2011)
F.R. Pearce, P.A. Thomas, H.M.P. Couchman, A.C. Edge, Mon. Not. R. Astron. Soc. **317**, 1029 (2000)
P.J.E. Peebles, Astrophys. J. **153**, 1 (1968)
P.J.E. Peebles, *Principles of Physical Cosmology* (1993)
J.R. Peterson, A.C. Fabian, Phys. Rep. **427**, 1 (2006)
J.R. Peterson, S.M. Kahn, F.B.S. Paerels et al., Astrophys. J. **590**, 207 (2003)
C. Pfrommer, V. Springel, T.A. Enßlin, M. Jubelgas, Mon. Not. R. Astron. Soc. **367**, 113 (2006)
C. Pfrommer, T.A. Enßlin, V. Springel, M. Jubelgas, K. Dolag, Mon. Not. R. Astron. Soc. **378**, 385 (2007)
C. Pfrommer, T.A. Enßlin, V. Springel, Mon. Not. R. Astron. Soc. **385**, 1211 (2008)
R. Piffaretti, P. Jetzer, J.S. Kaastra, T. Tamura, Astron. Astrophys. **433**, 101 (2005)
A. Pillepich, C. Porciani, T.H. Reiprich, Mon. Not. R. Astron. Soc. **422**, 44 (2012)
S. Planelles, V. Quilis, Mon. Not. R. Astron. Soc. **428**, 1643 (2013)
S. Planelles, S. Borgani, K. Dolag et al., Mon. Not. R. Astron. Soc. **431**, 1487 (2013a)
S. Planelles, S. Borgani, D. Fabjan et al., Mon. Not. R. Astron. Soc. (2013b). doi:10.1093/mnras/stt2141. arXiv:1311.0818
T.J. Ponman, P.D.J. Bourner, H. Ebeling, H. Böhringer, Mon. Not. R. Astron. Soc. **283**, 690 (1996)
T.J. Ponman, A.J.R. Sanderson, A. Finoguenov, Mon. Not. R. Astron. Soc. **343**, 331 (2003)
G.W. Pratt, H. Böhringer, J.H. Croston et al., Astron. Astrophys. **461**, 71 (2007)
G.W. Pratt, J.H. Croston, M. Arnaud, H. Böhringer, Astron. Astrophys. **498**, 361 (2009)
G.W. Pratt, M. Arnaud, R. Piffaretti et al., Astron. Astrophys. **511**, A85 (2010)
W.H. Press, P. Schechter, Astrophys. J. **187**, 425 (1974)
J. Prieto, R. Jimenez, Z. Haiman, Mon. Not. R. Astron. Soc. **436**, 2301 (2013)
E. Puchwein, D. Sijacki, V. Springel, Astrophys. J. Lett. **687**, L53 (2008)
E. Puchwein, V. Springel, D. Sijacki, K. Dolag, Mon. Not. R. Astron. Soc. **406**, 936 (2010)
R.E. Pudritz, J. Silk, Astrophys. J. **342**, 650 (1989)
D. Puy, M. Signore, New Astron. Rev. **51**, 411 (2007)
V. Quilis, Mon. Not. R. Astron. Soc. **352**, 1426 (2004)
V. Quilis, J.M.A. Ibanez, D. Saez, Astrophys. J. **502**, 518 (1998)
V. Quilis, R.G. Bower, M.L. Balogh, Mon. Not. R. Astron. Soc. **328**, 1091 (2001)
C. Ragone-Figueroa, G.L. Granato, G. Murante, S. Borgani, W. Cui, Mon. Not. R. Astron. Soc. **436**, 1750 (2013)
D.S. Reed, R. Bower, C.S. Frenk, A. Jenkins, T. Theuns, Mon. Not. R. Astron. Soc. **374**, 2 (2007)
M.J. Rees, Annu. Rev. Astron. Astrophys. **22**, 471 (1984)
J.A. Regan, M.G. Haehnelt, Mon. Not. R. Astron. Soc. **396**, 343 (2009a)
J.A. Regan, M.G. Haehnelt, Mon. Not. R. Astron. Soc. **393**, 858 (2009b)
J.A. Regan, P.H. Johansson, M.G. Haehnelt, Mon. Not. R. Astron. Soc. **439**, 1160 (2014)
T.H. Reiprich, H. Böhringer, Astrophys. J. **567**, 716 (2002)
P.M. Ricker, C.L. Sarazin, Astrophys. J. **561**, 621 (2001)
K. Roettiger, J.M. Stone, J.O. Burns, Astrophys. J. **518**, 594 (1999)
R. Rosner, W.H. Tucker, Astrophys. J. **338**, 761 (1989)
E. Rozo, R.H. Wechsler, E.S. Rykoff et al., Astrophys. J. **708**, 645 (2010)
D.H. Rudd, A.R. Zentner, A.V. Kravtsov, Astrophys. J. **672**, 19 (2008)
D. Ryu, H. Kang, E. Hallman, T.W. Jones, Astrophys. J. **593**, 599 (2003)
C. Safranek-Shrader, V. Bromm, M. Milosavljević, Astrophys. J. **723**, 1568 (2010)
C. Safranek-Shrader et al., arXiv e-print (arXiv:1307.1982) (2014)
A.J.R. Sanderson, T.J. Ponman, A. Finoguenov, E.J. Lloyd-Davies, M. Markevitch, Mon. Not. R. Astron. Soc. **340**, 989 (2003)
A.J.R. Sanderson, T.J. Ponman, E. O'Sullivan, Mon. Not. R. Astron. Soc. **372**, 1496 (2006)
A.J.R. Sanderson, E. O'Sullivan, T.J. Ponman, Mon. Not. R. Astron. Soc. **395**, 764 (2009)
C.L. Sarazin, Rev. Mod. Phys. **58**, 1 (1986)
C.L. Sarazin, Sky Telesc. **76**, 639 (1988)
C.L. Sarazin, A pan-chromatic view of clusters of galaxies and the large-scale structure, in *Lecture Notes in Physics*, vol. 740, ed. by M. Plionis, O. López-Cruz, D. in Hughes (Springer, Berlin, 2008), pp. 1–4020

W.C. Saslaw, D. Zipoy, Nature **216**, 976 (1967)
D.R.G. Schleicher, D. Galli, F. Palla et al., Astron. Astrophys. **490**, 521 (2008)
D.R.G. Schleicher, R. Banerjee, S. Sur et al., Astron. Astrophys. **522**, A115 (2010a)
D.R.G. Schleicher, M. Spaans, S.C.O. Glover, Astrophys. J. Lett. **712**, L69 (2010b)
D.R.G. Schleicher, F. Palla, A. Ferrara, D. Galli, M. Latif, Astron. Astrophys. **558**, A59 (2013)
W. Schmidt, J.C. Niemeyer, W. Hillebrandt, Astron. Astrophys. **450**, 265 (2006)
R. Schneider, A. Ferrara, R. Salvaterra, Mon. Not. R. Astron. Soc. **351**, 1379 (2004)
R. Schneider, K. Omukai, M. Limongi et al., Mon. Not. R. Astron. Soc. **423**, L60 (2012)
J. Schober, D. Schleicher, C. Federrath et al., Astrophys. J. **754**, 99 (2012)
K.M. Schure, A.R. Bell, L. O'C Drury, A.M. Bykov, Space Sci. Rev. **173**, 491 (2012)
C. Shang, G.L. Bryan, Z. Haiman, Mon. Not. R. Astron. Soc. **402**, 1249 (2010)
S.L. Shapiro, S.A. Teukolsky, *Black Holes, White Dwarfs and Neutron Stars: The Physics of Compact Objects* (1986)
P. Sharma, M. McCourt, E. Quataert, I.J. Parrish, Mon. Not. R. Astron. Soc. **420**, 3174 (2012)
R.K. Sheth, G. Tormen, Mon. Not. R. Astron. Soc. **308**, 119 (1999)
R.K. Sheth, H.J. Mo, G. Tormen, Mon. Not. R. Astron. Soc. **323**, 1 (2001)
C.J. Short, P.A. Thomas, O.E. Young et al., Mon. Not. R. Astron. Soc. **408**, 2213 (2010)
D. Sijacki, V. Springel, T. Di Matteo, L. Hernquist, Mon. Not. R. Astron. Soc. **380**, 877 (2007)
D. Sijacki, C. Pfrommer, V. Springel, T.A. Enßlin, Mon. Not. R. Astron. Soc. **387**, 1403 (2008)
J. Silk, Astrophys. J. **772**, 112 (2013)
J. Silk, M. Langer, Mon. Not. R. Astron. Soc. **371**, 444 (2006)
J. Silk, G.A. Mamon, Res. Astron. Astrophys. **12**, 917 (2012)
A. Simionescu, S.W. Allen, A. Mantz et al., Science **331**, 1576 (2011)
S.W. Skillman, B.W. O'Shea, E.J. Hallman, J.O. Burns, M.L. Norman, Astrophys. J. **689**, 1063 (2008)
R.S. Somerville, P.F. Hopkins, T.J. Cox, B.E. Robertson, L. Hernquist, Mon. Not. R. Astron. Soc. **391**, 481 (2008)
R.S. Somerville, R.C. Gilmore, J.R. Primack, A. Domínguez, Mon. Not. R. Astron. Soc. **423**, 1992 (2012)
M. Spaans, J. Silk, Astrophys. J. **652**, 902 (2006)
V. Springel, Mon. Not. R. Astron. Soc. **364**, 1105 (2005)
V. Springel, L. Hernquist, Mon. Not. R. Astron. Soc. **339**, 289 (2003)
V. Springel, T. Di Matteo, L. Hernquist, Mon. Not. R. Astron. Soc. **361**, 776 (2005)
A. Stacy, T.H. Greif, V. Bromm, Mon. Not. R. Astron. Soc. **403**, 45 (2010)
P.C. Stancil, S. Lepp, A. Dalgarno, Astrophys. J. **509**, 1 (1998)
R. Stanek, D. Rudd, A.E. Evrard, Mon. Not. R. Astron. Soc. **394**, L11 (2009)
M. Sun, G.M. Voit, M. Donahue et al., Astrophys. J. **693**, 1142 (2009)
R.A. Sunyaev, Y.B. Zeldovich, Comments Astrophys. Space Phys. **4**, 173 (1972)
S. Sur, D.R.G. Schleicher, R. Banerjee, C. Federrath, R.S. Klessen, Astrophys. J. Lett. **721**, L134 (2010)
S. Sur, C. Federrath, D.R.G. Schleicher, R. Banerjee, R.S. Klessen, Mon. Not. R. Astron. Soc. **423**, 3148 (2012)
H. Susa, Astrophys. J. **773**, 185 (2013)
J.C. Tan, E.G. Blackman, Astrophys. J. **603**, 401 (2004)
E.J. Tasker, J.C. Tan, Astrophys. J. **700**, 358 (2009)
M. Tegmark, J. Silk, M.J. Rees et al., Astrophys. J. **474**, 1 (1997)
J. Tinker, A.V. Kravtsov, A. Klypin et al., Astrophys. J. **688**, 709 (2008)
J.L. Tinker, B.E. Robertson, A.V. Kravtsov et al., Astrophys. J. **724**, 878 (2010)
L. Tornatore, S. Borgani, K. Dolag, F. Matteucci, Mon. Not. R. Astron. Soc. **382**, 1050 (2007)
P. Tozzi, C. Norman, Astrophys. J. **546**, 63 (2001)
R.A. Treumann, Astron. Astrophys. Rev. **17**, 409 (2009)
J.K. Truelove, R.I. Klein, C.F. McKee et al., Astrophys. J. Lett. **489**, L179 (1997)
M.J. Turk, P. Clark, S.C.O. Glover et al., Astrophys. J. **726**, 55 (2011)
M.J. Turk, J.S. Oishi, T. Abel, G.L. Bryan, Astrophys. J. **745**, 154 (2012)
R. Valdarnini, Astrophys. J. **567**, 741 (2002)
R.J. van Weeren, H.J.A. Röttgering, J. Bagchi et al., Astron. Astrophys. **506**, 1083 (2009)
F. Vazza, G. Brunetti, A. Kritsuk et al., Astron. Astrophys. **504**, 33 (2009)
F. Vazza, G. Brunetti, C. Gheller, R. Brunino, M. Brüggen, Astron. Astrophys. **529**, A17 (2011)
F. Vazza, M. Brüggen, C. Gheller, G. Brunetti, Mon. Not. R. Astron. Soc. **421**, 3375 (2012)
A. Vikhlinin, M. Markevitch, S.S. Murray et al., Astrophys. J. **628**, 655 (2005)
A. Vikhlinin, A. Kravtsov, W. Forman et al., Astrophys. J. **640**, 691 (2006)
A. Vikhlinin, R.A. Burenin, H. Ebeling et al., Astrophys. J. **692**, 1033 (2009)
A.E. Vladimirov, A.M. Bykov, D.C. Ellison, Astrophys. J. **688**, 1084 (2008)
A.E. Vladimirov, A.M. Bykov, D.C. Ellison, Astrophys. J. Lett. **703**, L29 (2009)

G.M. Voit, Rev. Mod. Phys. **77**, 207 (2005)

G.M. Voit, Astrophys. J. **740**, 28 (2011)

G.M. Voit, G.L. Bryan, Nature **414**, 425 (2001)

G.M. Voit, G.L. Bryan, M.L. Balogh, R.G. Bower, Astrophys. J. **576**, 601 (2002)

G.M. Voit, M.L. Balogh, R.G. Bower, C.G. Lacey, G.L. Bryan, Astrophys. J. **593**, 272 (2003)

M. Volonteri, J. Bellovary, Rep. Prog. Phys. **75**, 124901 (2012)

S.A. Walker, A.C. Fabian, J.S. Sanders, M.R. George, Mon. Not. R. Astron. Soc. **424**, 1826 (2012)

H.-H. Wang, R.S. Klessen, C.P. Dullemond, F.C. van den Bosch, B. Fuchs, Mon. Not. R. Astron. Soc. **407**, 705 (2010)

M.S. Warren, K. Abazajian, D.E. Holz, L. Teodoro, Astrophys. J. **646**, 881 (2006)

W.A. Watson, I.T. Iliev, A. D'Aloisio et al., Mon. Not. R. Astron. Soc. **433**, 1230 (2013)

D.H. Weinberg, M.J. Mortonson, D.J. Eisenstein et al., Phys. Rep. **530**, 87 (2013)

M. White, Astrophys. J. Suppl. Ser. **143**, 241 (2002)

D. Winske, V.A. Thomas, N. Omidi, K.B. Quest, J. Geophys. Res. **95**, 18821 (1990)

J.H. Wise, T. Abel, Astrophys. J. **685**, 40 (2008)

J.H. Wise, M.J. Turk, T. Abel, Astrophys. J. **682**, 745 (2008)

H.-Y.K. Yang, P.M. Ricker, P.M. Sutter, Astrophys. J. **699**, 315 (2009)

N. Yoshida, K. Omukai, L. Hernquist, T. Abel, Astrophys. J. **652**, 6 (2006)

N. Yoshida, K. Omukai, L. Hernquist, Science **321**, 669 (2008)

B. Yue et al., arXiv e-print (arXiv:1305.5177) (2013)

B. Yue et al., arXiv e-print (arXiv:1402.5675) (2014)

N.L. Zakamska, R. Narayan, Astrophys. J. **582**, 162 (2003)

DOI 10.1007/978-1-4939-3547-5_5
Reprinted from *Space Science Reviews* Journal, DOI 10.1007/s11214-014-0129-4

Structures and Components in Galaxy Clusters: Observations and Models

A.M. Bykov[1,2,3] · E.M. Churazov[4,5] · C. Ferrari[6] ·
W.R. Forman[7] · J.S. Kaastra[8] · U. Klein[9] ·
M. Markevitch[10] · J. de Plaa[8]

Received: 23 November 2014 / Accepted: 22 December 2014 / Published online: 4 February 2015
© Springer Science+Business Media Dordrecht 2015

Abstract Clusters of galaxies are the largest gravitationally bounded structures in the Universe dominated by dark matter. We review the observational appearance and physical models of plasma structures in clusters of galaxies. Bubbles of relativistic plasma which are inflated by supermassive black holes of AGNs, cooling and heating of the gas, large scale plasma shocks, cold fronts, non-thermal halos and relics are observed in clusters. These con-

✉ A.M. Bykov
byk@astro.ioffe.ru

E.M. Churazov
churazov@mpa-garching.mpg.de

C. Ferrari
chiara.ferrari@oca.eu

W.R. Forman
forman@cfa.harvard.edu

J.S. Kaastra
j.kaastra@sron.nl

U. Klein
uklein@astro.uni-bonn.de

M. Markevitch
maxim.markevitch@nasa.gov

J. de Plaa
J.de.Plaa@sron.nl

[1] A.F. Ioffe Institute for Physics and Technology, 194021, St. Petersburg, Russia

[2] St. Petersburg State Politecnical University, St. Petersburg, Russia

[3] International Space Science Institute, Bern, Switzerland

[4] Max Planck Institute for Astrophysics, Karl-Schwarzschild-Str. 1, 85741 Garching, Germany

[5] Space Research Institute (IKI), Profsoyuznaya 84/32, Moscow 117997, Russia

[6] Laboratoire Lagrange, UMR7293, Université de Nice Sophia-Antipolis, CNRS, Observatoire de la Côte d'Azur, Boulevard de l'Observatoire, 06300 Nice, France

stituents are reflecting both the formation history and the dynamical properties of clusters of galaxies. We discuss X-ray spectroscopy as a tool to study the metal enrichment in clusters and fine spectroscopy of Fe X-ray lines as a powerful diagnostics of both the turbulent plasma motions and the energetics of the non-thermal electron populations. The knowledge of the complex dynamical and feedback processes is necessary to understand the energy and matter balance as well as to constrain the role of the non-thermal components of clusters.

Keywords Clusters of galaxies · Radiation mechanisms: non-thermal · Radio continuum · X-rays: galaxies: clusters

1 Introduction

Clusters of galaxies grow by gravitational collapse to the most massive objects of the Universe. While the total mass is dominated by dark matter (~ 80 %), there is a significant baryonic contribution of 20 % and the processes where baryons are involved prominently determine the evolutionary physics and the observational appearance of clusters.

In this paper we focus on the baryonic component of clusters and in particular on the hot gas that, with its thermal and non-thermal constituents, comprises the majority of these baryons. Cluster galaxies are embedded in this hot intracluster medium (ICM), but represent a small fraction of both the volume and the baryonic mass.

For most clusters, the hot gas reaches temperatures of 10^7–10^8 K. At the low temperature end, below temperatures of about 2×10^7 K, it is more common to speak about groups of galaxies rather than clusters of galaxies, but the transition is of course smooth.

From an X-ray perspective, clusters are found in two variants: those with a cool core and those lacking such a core. In the cool core clusters the density of the gas in the center reaches values of $\sim 10^{-2}$ cm^{-3} (that is, for instance, only 1 % of the typical density of the interstellar medium of a galaxy). The density decreases rapidly towards the outskirts down to levels of order 10^{-4} cm^{-3} or less.

The density in the cool cores is high enough to cause significant cooling over cosmological time scales through thermal bremsstrahlung emission observed in X-rays. Without any heating mechanism, this gas would cool down further and do form stars. However, it appears that active nuclei in the core of the dominant cluster galaxies can emit so much power that the associated heating compensates the cooling of the central gas. This leads to a physically interesting but complex feedback loop between the central supermassive black hole and the cluster gas. These processes and the associated plasma structures will be discussed in this review.

Clusters are by no means static entities, they still grow. Large scale violent processes occur, like cluster mergers or, more frequently, the capture of groups or individual galaxies by massive systems. Due to the supersonic velocities involved, shocks are produced at various locations within colliding clusters. This leads to local heating, particle acceleration and modification of the magnetic fields, and we will also discuss these processes.

[7] Harvard Smithsonian Center for Astrophysics, 60 Garden Street, Cambridge, MA 02138, USA

[8] SRON Netherlands Institute for Space Research, Sorbonnelaan 2, 3584 CA Utrecht, Netherlands

[9] Argelander-Institut für Astronomie, University of Bonn, Bonn, Germany

[10] Astrophysics Science Division, NASA/Goddard Space Flight Center, Greenbelt, MD 20771, USA

Another part of the bulk motions may cascade downwards in scale in the form of turbulence, that represents a contribution to the total thermal pressure of the order of 5–15 % and is stronger in the outer parts of clusters (see for review Kravtsov and Borgani 2012; Dolag et al. 2008). In these external regions, infalling galaxies and groups will give rise to density inhomogeneities which, due to the relatively low density in cluster outskirts, take a long time to disappear.

Mixing of merging components also leads to interesting processes at the interface between cold and hot gas. Further, when galaxies are being captured by a cluster, they may lose their chemically enriched gas to the intracluster medium by ram pressure or interaction with other galaxies. The ICM itself is a rich archive of the past chemical history of the cluster: it contains information on the distribution of stars and the relative frequencies of specific subclasses.

Clusters, that are the largest gravitationally bound systems in the Universe, form a rich and living laboratory to study all kinds of processes that shape their appearance. Our review is devoted to discussion of plasma structures of different scales and origin in clusters of galaxies. More general discussion and many important ideas about the evolution and cosmological importance of clusters of galaxies can be found in recent reviews (see e.g. Böhringer and Werner 2010; Allen et al. 2011; Kravtsov and Borgani 2012; Planelles et al. 2014, and references therein).

2 AGNs in Galaxy Clusters. Feedback Processes

The gas in the cores of galaxy clusters has a radiative cooling time of the order of 10^9 years or less, opening the possibility of forming an extremely massive central galaxy. This occurs in numerical simulations with radiative cooling, but is generally not observed in nature, with a few observed exceptions, such as, e.g., the Phoenix cluster (McDonald et al. 2012). Instead, observations suggest that the mechanical energy released by a central AGN regulates the thermal state of the gas, preventing it from catastrophic cooling. Bubbles of relativistic plasma are inflated by a supermassive black hole and rise buoyantly through the gaseous atmosphere, leading to a number of spectacular phenomena such as expanding shocks, X-ray dim and radio bright cavities, old and "dead" cavities and filaments in the wakes of the rising bubbles formed by entrained low entropy gas.

Simple arguments based on the energy content of bubbles and their life-time show that the amount of mechanical energy supplied by AGNs matches approximately the gas cooling losses in objects vastly different in size and luminosity. This hints at some form of self-regulation of the AGN power. How the mechanical energy, provided by the AGN, is dissipated depends on the ICM microphysics (e.g. magnetic fields, viscosity, conduction etc.). However it is plausible that close to 100 % of the mechanical energy is eventually dissipated in the cluster core, regardless of the particular physical process involved.

AGN feedback is plausibly a key process for the formation of massive ellipticals at $z \sim$ 2–3, as suggested by the correlation of galaxy bulge properties and the mass of the SMBH (Ferrarese and Merritt 2000; Gebhardt et al. 2000). Galaxy clusters offer us a zoomed view on this process at $z \sim 0$. Three pre-requisites are needed for this scenario to work (i) a hot gaseous atmosphere in the galaxy is present, (ii) the black hole is sufficiently massive and (iii) a large fraction of AGN energy is in mechanical form. The latter depends critically on the physics of accretion, in particular, on the transition of the SMBH energy output from the radiation-dominated mode to the mechanically-dominated mode when the accretion rate drops below a fraction of the Eddington value (e.g., Churazov et al. 2005). Given that the coupling constant of these two forms of energy output with the ICM can differ by a factor of 10^4–10^5, this change in the accretion mode may explain the switch of a SMBH (and its parent galaxy) from the QSO-type behavior and an intense star formation epoch to the radiatively inefficient AGN and essentially passive evolution of the parent galaxy.

Below we briefly outline only most general features of the AGN Feedback model in galaxy clusters and do not discuss results from numerical simulations. Extended reviews on the AGN Feedback in clusters can be found in, e.g., McNamara and Nulsen (2007), Fabian (2012).

2.1 Cluster Cores Without AGN Feedback

The radiative cooling time of the gas in the central parts of rich galaxy clusters ($t_{\text{cool}} = \frac{\gamma}{\gamma-1} \frac{nkT}{n^2 \Lambda(T)} \sim$ few 10^8 – 10^9 yr) is shorter than the Hubble time (e.g., Lea 1976; Cowie and Binney 1977; Fabian and Nulsen 1977). Here n and T are the density and temperature of the ICM, respectively, γ is the adiabatic index and $\Lambda(T)$ is the radiative cooling function. Since $t_{\text{cool}} \propto 1/n$, the cooling time is short in the center, but rapidly increases with radius. The radius r_{cool} where $t_{\text{cool}} \sim t_{\text{Hubble}}$ is usually referred to as the cooling radius. Without an external source of energy, the gas inside r_{cool} must cool and flow towards the center, forming a so-called "cooling flow" (see Fabian 1994, for a review of the scenario without AGN feedback). Observations (e.g., Peterson and Fabian 2006; David et al. 2001), however, suggest that the net rate of gas cooling to low temperatures is a small fraction (~ 10 % or below) of the straightforward estimate:

$$\dot{M}_{\text{cool}} = \frac{L_{\text{cool}}}{\frac{\gamma}{\gamma-1} kT} \mu m_{\text{p}} \sim 10^2 - 10^3 \ M_\odot \ \text{yr}^{-1}, \tag{1}$$

where L_{cool} is the total cooling rate within r_{cool}, μ is the mean particle atomic weight. This implies that a source of heat is needed to offset ICM cooling losses. In the late 90's and early 2000's it became clear that a supermassive black hole sitting at the center of the dominant cluster galaxy could operate as such source.

The above "cooling flow" problem applies not only to rich clusters and groups, but also to individual hot gas rich galaxies (e.g., Thomas et al. 1986). The fact that these galaxies do not show significant star formation argues for feedback from these low mass systems (some with SMBH masses as large as those in central galaxies in rich clusters) up to the most massive and X-ray luminous clusters.

2.2 Evidence for AGN Mechanical Feedback and Its Energetics

Massive ellipticals at the cores of rich clusters host black holes with masses larger than $10^9 \ M_\odot$. In terms of energetics, such black holes, accreting at the Eddington rate, could release up to 10^{47} erg s^{-1}—much more than needed to reheat the gas. However, we do not find extremely bright AGNs in nearby clusters, and the coupling of radiation to the fully ionized ICM (via Compton scattering) is weak. Based on radio observations, Pedlar et al. (1990) argued that mechanical power of jets in NGC1275 (dominant galaxy in the Perseus cluster) is likely much higher than their radiative power and it might be comparable to the gas cooling losses. For less massive systems Tabor and Binney (1993) and Binney and Tabor (1995) suggested that the mechanical power of jets may have a strong impact on the thermal state of the gas in the central region. But it was only the combination of X-ray and radio data that convincingly demonstrated a crucial role of AGN feedback in cluster cores.

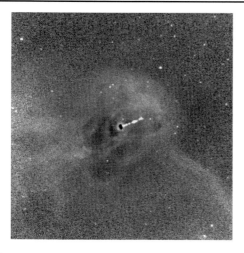
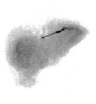

Fig. 1 X-ray (Forman et al. 2007) and 6 cm radio (Hines et al. 1989) images of the M87 core (3′ × 3′). The X-ray image (*left*) was flat-fielded to emphasize various non-axisymmetric features. Images are shown to scale, centered at the supermassive black hole with a kpc scale jet going to the NW. X-ray cavities, matching the 6 cm image of the inner radio lobes are clearly seen in the *left image*

By now, Chandra and XMM-Newton found signs of AGN/ICM interactions in a large fraction of relaxed clusters (e.g., McNamara et al. 2000; Dunn and Fabian 2006; Rafferty et al. 2006; Hlavacek-Larrondo et al. 2012; Bîrzan et al. 2012). But clear evidence of this process in the Perseus cluster and in M87/Virgo had previously been seen in ROSAT images (Boehringer et al. 1993; Bohringer et al. 1995). Based on these sketchy X-ray and radio images, the basic features of the mechanical feedback model and its energetics, inspired by analogy with powerful atmospheric explosions, were outlined (Churazov et al. 2000) just before Chandra and XMM-Newton launch.

This impact of the SMBH mechanical power on the ICM can be best illustrated for M87—the X-ray brightest elliptical galaxy in the nearby Virgo cluster. While M87/Virgo is not a very luminous system (L_{cool} is of the order 10^{43} erg s^{-1}, e.g. Peres et al. 1998, Owen et al. M87), its proximity (distance ~ 16 Mpc) offers an exquisitely detailed view on the processes in the very core of the galaxy (e.g. Forman et al. 2005, 2007; Million et al. 2010; Werner et al. 2010).

Shown in Fig. 1 are the X-ray and 6 cm radio images of M87 (central 3′ × 3′). Synchrotron emission for a jet coming from the SMBH is clearly visible. Except for the jet, the X-ray image is dominated by the thermal emission of the optically thin plasma at a temperature 1–2 keV (unambiguously proven by the observed spectra). The radio image on the right shows synchrotron emission from the jet and radio lobes filled with relativistic plasma. The radio-bright lobes at 6 cm nicely correspond to the X-ray dim regions that are defined by X-ray bright shells, suggesting that thermal plasma is displaced by the relativistic plasma within the central 1′ from the core. A minimum energy, required to (slowly) inflate a bubble of a given volume at a constant pressure is the enthalpy:

$$E_{bub} = \frac{\gamma}{\gamma - 1} PV, \qquad (2)$$

where γ is the adiabatic index of the fluid inside the bubble ($\gamma = 4/3$ and $5/3$ for mono-atomic relativistic and non-relativistic fluids respectively), P is the ICM pressure and V is the volume. This gives a lower limit on the amount of mechanical energy produced by the SMBH. Since we do not see evidence for a very strong shock surrounding the lobes (see, however, below), the true amount of energy should not be far from this lower limit.

To evaluate the mechanical power of the AGN, we need to estimate the life-time of the bubbles t_{bub}. The simplest recipe (Churazov et al. 2000) comes from the analysis of the buoyancy driven evolution of the bubble. The importance of buoyancy for the radio lobes, filled with relativistic plasma, was pointed out by Gull and Northover (1973). Comparing the expansion velocity (which depends on the AGN power and the size of the bubble) and the buoyant rise time (which depends on the gravity and the size of the bubble) one gets an upper limit on the life time of the bubble. For a steady AGN power L_{AGN} deposited into a small volume, the initial expansion of the bubble is supersonic, but it slows as the bubble grows. Soon after the expansion velocity becomes subsonic, the bubble is deformed by the Rayleigh-Taylor instability and rises under the action of buoyancy. The terminal velocity of the rising bubble v_{rise} is set by the balance of the ram pressure (assuming low viscosity of the ICM)and the buoyancy force:

$$g\frac{4}{3}\pi r^3 \rho_{gas} \approx C\pi r^2 \rho_{gas} v^2, \tag{3}$$

where g is the gravitational acceleration, r is the bubble radius, ρ_{gas} is the ICM density and C is a dimensionless constant of order unity. Thus $v_{rise} \sim \sqrt{gr}$. At the same, time AGN activity drives the expansion of the bubble with the velocity v_{exp} set by the AGN power L_{AGN}, e.g., from Eq. (2) $v_{exp} \sim \frac{\gamma-1}{\gamma} \frac{L_{AGN}}{4\pi P r^2}$, provided that the expansion velocity is subsonic. The condition $v_{exp} \gtrsim v_{rise}$ sets the lower limit on the AGN power needed to ignore the role of buoyancy. For M87 the size of the inner lobes suggests the jet mechanical power $L_{AGN} \sim \frac{\gamma}{\gamma-1} \frac{4\pi P r^2}{\sqrt{gr}} \sim$ few 10^{43} erg s^{-1}. Of course such estimates (and various modifications e.g., McNamara and Nulsen 2007) are only accurate to within a factor of a few. Nevertheless, it is clear that mechanical input of the SMBH is sufficient to offset gas cooling losses, i.e. $L_{AGN} \sim 10^{43}$ erg s$^{-1} \sim L_{cool}$.

One can expect that the initial phase of the bubble expansion is supersonic and it will drive a shock into the ICM. As the expansion slows, the shock weakens and moves ahead of the expanding boundary. These shocks offer yet another way to measure the AGN power. In M87 we see at least two generations of shocks (Forman et al. 2005, 2007) at 0.6′ and 2.7′ from the center (Fig. 2). The hard emission (3.5–7.5 keV) shows a ring of emission with an outer radius ranging from 2.5′ to 2.85′ (11.6–13.3 kpc). This ring of hard emission provides an unambiguous signature of a weak shock. The gas temperature in the ring rises from ~ 2.0 keV to ~ 2.4 keV implying a Mach number of $\mathcal{M} \sim 1.2$ (shock velocity $v = 880$ km s^{-1} for a 2 keV thermal gas). At the shock, the density jump is 1.33 which yields a Mach number of 1.22, consistent with that derived from the temperature jump. The total energy needed to drive this shock can be readily estimated from 1D hydro simulations $E \sim 5 \times 10^{57}$ erg. The age of the outburst that gave rise to the shock is ~ 12 Myr. If we divide the energy of the outburst by the age of the shock we get a mean energy release $\sim 10^{43}$ erg s^{-1}—broadly in agreement with the ICM cooling losses.

2.3 Dissipation of Mechanical Energy

Once the bubble is detached from the central source, its evolution is governed by buoyancy. During the rise the bubble may transform into a toroidal structure as the "ear-like" structure in M87 (see Figs. 3 and 4) which resembles a mushroom formed by a powerful atmospheric explosion. As in case of the atmospheric explosion, the bubble is able to entrain

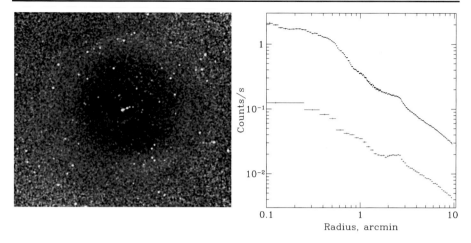

Fig. 2 *Left*: Chandra image of M87 ($\sim 7' \times 7'$) in the 3.5–7.5 keV energy band divided by a spherically symmetric model. This energy band shows pressure variations in the gas (Forman et al. 2007). A nearly perfect ring at $\sim 2.75'$ (12.8 kpc) is clearly seen. This is a characteristic signature of a shock, driven by an outburst from the central SMBH. *Right*: the surface brightness profiles (Forman et al. 2007) in the 1.2–2.5 keV (*upper curve*) and 3.5–7.5 keV (*lower curve*) bands show a prominent feature at 2–3′, along with a fainter feature at 0.6′

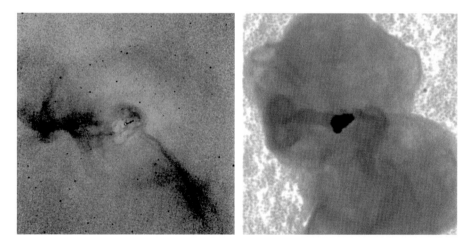

Fig. 3 X-ray (Forman et al. 2007) and 90 cm radio (Owen et al. M87) images (*right*) of the core of M87 ($8' \times 8'$). Filaments of cool gas are entrained by the buoyantly rising bubbles

large amounts of ambient gas from the core of cluster and transport it to large distance from the cluster center (e.g., Churazov et al. 2001; Fabian et al. 2003; Werner et al. 2010).

Note that adiabatic expansion of rising bubbles leads to a rapid decrease of the radio emission, since both the magnetic field strength, and the density and energy of relativistic particles are decreasing at the same time. This decrease is especially strong if the aging break in the distribution of electrons is brought by adiabatic expansion into the observable frequency range. Thus, unless there is continuous reacceleration of electrons, the radio bright bubbles should evolve into a radio dim objects. Since the pressure support inside the bubble could still come from magnetic fields and low energy electrons and protons (Lorentz factor

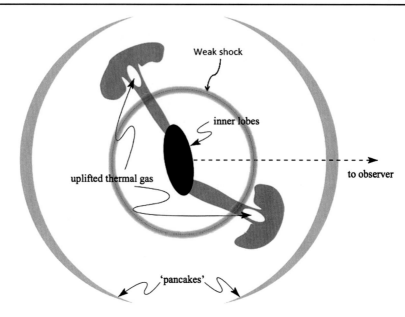

Fig. 4 Schematic picture of major signs of AGN/ICM interaction (adapted and modified from Churazov et al. 2001), inspired by analogy with mushroom clouds produced by powerful atmospheric explosions. The *black region in the center* denotes the inner radio lobes, driven by the SMBH mechanical power. The circular structure is a weak shock wave produced by these inner lobes. *Gray "mushrooms"* correspond to the buoyant bubbles already transformed into tori, and the *gray lens-shaped structures* are the pancakes formed by the older bubbles (cf. Fig. 3)

of 1000 or lower), the bubble is still seen as an X-ray cavity, but is very dim in radio. Such ghost bubbles are believed to be widespread in the cluster cores and one can expect many of them to be detected with a new generation of a low frequency instruments (Enßlin and Heinz 2002).

The bubble rise velocity v_{rise} is smaller than the ICM sound speed, and it is much smaller than the sound speed of the relativistic fluid inside the bubble. Adiabatic expansion of the rising bubble implies that its enthalpy is decreasing according to the ambient gas pressure $H = \frac{\gamma}{\gamma-1} P V \propto P^{\frac{\gamma-1}{\gamma}}$. This means that after crossing a few pressure scale heights, much of the energy originally stored as the enthalpy of the relativistic bubble is transferred to the gas. Subsonic motion with respect to the ambient gas guarantees that only a fraction of this energy is "lost" as sound waves, which may leave the cluster core. Thus we can conclude that a fraction of energy of order unity is transferred to the ICM. This leads to the conjecture that essentially all of the mechanical energy is dissipated in the cluster core, which acts as a calorimeter of AGN activity (Churazov et al. 2002).

Details of the dissipation process depend sensitively on the properties of the ICM. For instance, this energy could drive turbulent motions in the wake of the rising bubble and excite gravity waves (e.g., Churazov et al. 2001; Omma et al. 2004). These motions will eventually dissipate into heat. Alternatively (if the ICM viscosity is high), the energy can be dissipated directly in the flow around the bubble. However in either case, the energy does not escape from the cluster core.

An interesting recent development came from the analysis of X-ray surface brightness fluctuations in the Perseus and M87/Virgo clusters (Zhuravleva et al. 2014). If the observed fluctuations are interpreted as weak perturbations of a nearly hydrostatic cluster atmosphere,

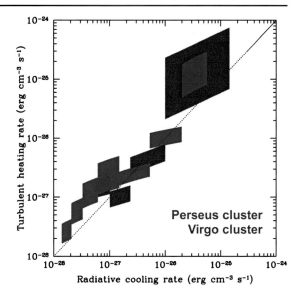

Fig. 5 Turbulent heating versus gas cooling rates in the Perseus and Virgo cores. *Shaded regions* show the heating and cooling rates estimated at different distances from the cluster center. The size of each region reflects estimated statistical and stochastic uncertainties. (Adapted from Zhuravleva et al. 2014)

then one can link their amplitude to the characteristic gas velocities at different spatial scales. Since in the canonical Kolmogorov turbulence the energy flow is constant across the inertial range, it is sufficient to know the velocity V at one scale l (within the inertial range) to estimate the dissipation rate as $Q_{turb} \sim \rho \frac{V^3}{l}$. Applying this approach to Perseus and M87/Virgo leads to a tantalizing conclusion that the turbulent dissipation approximately matches the gas cooling rate in these clusters (Fig. 5). While a number of assumptions enter these calculations, the result is encouraging. Future ASTRO-H measurements of the gas velocities in these clusters should be able to verify these findings.

A substantial fraction of energy released by the SMBH can go into quasi-spherical sound waves propagating through the ICM. Unlike strong shocks the dissipation of the energy carried by sound waves depends on the ICM microphysics, but it is plausible that their energy will be dissipated before the wave leaves the core of the cluster (Fabian et al. 2006). The attractiveness of this model is that the energy can be evenly distributed over large regions. Whether sound waves provide the dominant source of heat to the ICM depends critically on what fraction of AGN energy goes to sound waves. In M87 the fraction of energy, which went into the weak shock is $\lesssim 25$ % (Forman et al. 2007).

2.4 Self-regulation

Similar signs of SMBH-ICM interaction are observed for objects having vastly different sizes and luminosities. Two examples are shown in Fig. 6—these are $3' \times 3'$ X-ray images of the elliptical galaxy NGC5813 (Randall et al. 2011) and the Perseus cluster (Fabian et al. 2000). In each case, X-ray cavities of approximately the same angular size are clearly seen. The distances to NGC5813 and the Perseus cluster are 32 and 70 Mpc, respectively. Therefore the volumes of the cavities differ by a factor of 10. A more extreme example is the MS0735.6+7421 cluster (McNamara et al. 2005) at redshift $z = 0.22$, which has cavities with volume $\sim 10^4$ times larger than in NGC5813.

A systematic comparison of the AGN mechanical power and the gas cooling losses has been done for several dozens of objects (e.g., Rafferty et al. 2006; Hlavacek-Larrondo et al.

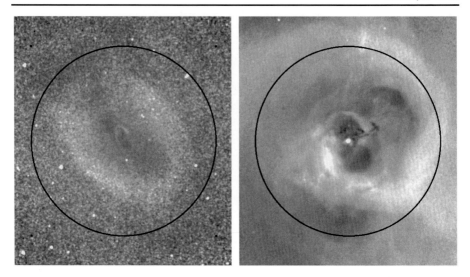

Fig. 6 Central $3' \times 3'$ region of the Chandra 0.6–2 keV images of NGC5813 (left) and the Perseus cluster (*right*). Cavities inflated by AGNs are clearly visible in both images. For NGC5813 two (or even three) generations of cavities are easily identifiable. The volume of the cavities in Perseus is a factor of ~ 10 larger than in NGC5813

Fig. 7 Comparison of the estimated AGN mechanical power and the ICM cooling losses for a sample of clusters. Adopted from Hlavacek-Larrondo et al. (2012). The correlation is evident, albeit with substantial scatter, suggesting that mechanical output is regulated to match the cooling losses

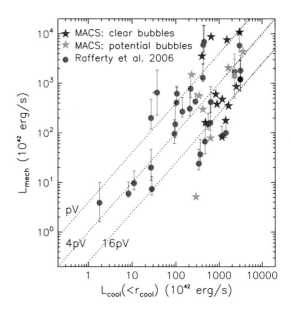

2012). These studies suggest an approximate balance between AGN energy input and cooling losses (see Fig. 7), implying that some mechanism is regulating the AGN power to maintain this balance. A natural way to establish such regulation is to link the accretion rate onto SMBHs with the thermodynamic parameters of the gas. Two scenarios are outlined below.

In the first scenario, known as "hot accretion", the classic Bondi formula (Bondi 1952) regulates the accretion rate of the hot gas onto the black hole and provides a link between the gas parameters and the mechanical power of an AGN (e.g., Churazov et al. 2002; Böhringer et al. 2002; Di Matteo et al. 2003).

The rate of spherically symmetric adiabatic gas accretion without angular momentum onto a point mass can be written as:

$$\dot{M} = 4\pi\lambda(GM)^2 c_s^{-3}\rho \propto s^{-3/2}, \tag{4}$$

where λ is a numerical coefficient, which depends on the gas adiabatic index γ (for $\gamma = 5/3$ the maximal valid value of λ allowing steady spherically symmetric solution is $\lambda_c = 0.25$), G is the gravitational constant, M is the mass of the black hole, $c_s = \sqrt{\gamma \frac{kT}{\mu m_p}}$ is the gas sound speed, ρ is the mass density of the gas and $s = \frac{T}{n^{2/3}}$ is the entropy index of the gas.

Thus the Bondi accretion rate is proportional to $s^{-3/2}$ which increases when the gas entropy decreases. Assuming that the heating is directly proportional to the SMBH accretion rate (Heating $= \epsilon c^2 \dot{M}$), there must be a value of gas entropy such, that heating balances cooling:

$$s \approx 3.5 \left(\frac{M}{10^9\,M_\odot}\right)^{4/3}\left(\frac{\epsilon}{0.1}\right)^{2/3}\left(\frac{L_X}{10^{43}\,\mathrm{erg\,s^{-1}}}\right)^{-2/3}\mathrm{keV\,cm^2} \tag{5}$$

A lower/higher entropy than this value implies too much/little heating and therefore an overall increase/decrease of the accretion rate. Since in stable hydrostatic equilibrium the low entropy gas falls to the bottom of the potential well (the location of the SMBH), the energy input is controlled by the minimum value of the gas entropy in the whole central region. This provides a natural mechanism for self-regulation of the cooling and heating. Interestingly, in many nearby systems this simple prescription leads to an order of magnitude balance between cooling and heating (e.g., Churazov et al. 2002; Böhringer et al. 2002; Allen et al. 2006).

Another possibility is that some (sufficiently small) amount of gas is first able to cool from the hot phase down to low temperatures (e.g. down to 10^4 K or below). Cold gas blobs then move in the potential well of the central galaxy, collide, lose angular momentum and feed the black hole (e.g., Pizzolato and Soker 2005; Gaspari et al. 2012). This model is known as a "cold accretion" scenario, as opposed to Bondi-type accretion straight from the "hot" phase. While these scenarios differ strongly in the physical process involved, they both advocate a negative feedback loop, when the SMBH affects the thermal state of the gas, which in turn affects the accretion onto the black hole.

2.5 Link to Evolution of Galaxies

A correlation of galaxy bulge properties and the mass of the SMBH (e.g., Ferrarese and Merritt 2000; Gebhardt et al. 2000) implies that the black hole and its parent galaxy affect each other. The clear evidence of mechanical feedback in nearby galaxy clusters suggests that the same mechanism may be relevant for the formation of massive ellipticals and for the growth of their SMBHs at $z \sim 2$–3. Three pre-requisites are needed for this scenario to work (i) a hot gaseous atmosphere in the galaxy is present, (ii) the black hole is sufficiently massive and (iii) large fraction of AGN energy is in mechanical form and the coupling of mechanical energy to the ICM is strong.

One can parametrize the magnitude of the feedback/gas heating H with a simple expression:

$$H(M_{BH}, \dot{M}) = \left[\alpha_M\epsilon_M(\dot{m}) + \alpha_R\epsilon_R(\dot{m})\right]0.1\dot{M}c^2, \tag{6}$$

where $\epsilon_M(\dot{m})$ and $\epsilon_R(\dot{m})$ characterize the efficiency of the transformation of accreted rest mass per unit time $\dot{M}c^2$ into mechanical energy and radiation respectively, while α_M and α_R are AGN-ICM coupling constants—the fraction of the released energy which is eventually transferred to the gas. The first pair of coefficients ($\epsilon_M(\dot{m}), \epsilon_R(\dot{m})$) should come from accretion physics, while the second pair (α_M, α_R) depends on the properties of the ICM and on the details of the AGN-ICM interaction. The value of α_R is typically very low $\alpha_R \lesssim 10^{-4}$ (Sazonov et al. 2004, 2005), while α_M can be close to unity (see above). This difference between α_R and α_M is the most important element of the mechanical feedback scenario.

Let us assume that the system (SMBH + gaseous atmosphere) evolves to the state where the heating by the black hole is equal to the gas cooling losses (if such a state does exist). Thus the mass accretion rate is the solution of the equation:

$$H(M_{\mathrm{BH}}, \dot{M}) = L_{\mathrm{cool}}, \tag{7}$$

provided $\dot{m} = \frac{\dot{M}}{\dot{M}_{\mathrm{Edd}}} \leq 1$. In the opposite case, we set $\dot{M} = \dot{M}_{\mathrm{Edd}}$. In other words, if the black hole cannot offset ISM cooling losses even at the Eddington rate then it will keep accreting gas at this rate. If there is a solution of Eq. (7) at $\dot{m} \leq 1$ then at this rate an equilibrium between heating and cooling is possible. We will see below that several distinct solutions of Eq. (7) are possible (some of them are unstable).

Let us now assume that the mechanical output $\epsilon_M(\dot{m})$ is large at low accretion rate, but decreases at high accretion rates (so-called "radio mode"). The radiative output $\epsilon_R(\dot{m})$ is on the contrary large at high accretion rates and decreases at low accretion rates. The mechanical and radiative outputs corresponding to this scenario are shown in Fig. 8 with the thin blue and red curves respectively. The above assumptions are motivated by X-ray and radio observations of several X-ray binaries (e.g., Gallo et al. 2003) and AGNs (e.g., Owen et al. M87) and can also be supported by theoretical arguments for radiatively inefficient accretion flows (e.g., Ichimaru 1977; Rees et al. 1982; Narayan and Yi 1994). For instance, the nucleus of M87 can be regarded as the prototypical example of an AGN in the low accretion rate mode, when kinetic power of radio emitting outflows exceeds its radiative power. At the same time, there are black holes in binary systems in the high accretion regime, which are very bright in X-rays, but show no evidence for a powerful outflow.

We now set $\alpha_M \sim 0.7$ and $\alpha_R \sim 10^{-4}$. The ICM heating by mechanical and radiative power (obtained by multiplying the curves by 0.7 and 10^{-4}) is shown by the thick blue and red curves respectively. Finally the total ICM heating rate (the sum of two curves) is plotted as a thick black curve. Once the black hole is sufficiently massive, there are two possible solutions with the heating balancing ICM cooling losses (points marked as A and B in Fig. 8). The point B is certainly unstable since an increase of the mass accretion rate causes a decrease of the heating rate. The system will evolve from state B into one with lower or higher accretion rate. If the system goes into a high accretion rate mode then the feedback power drops and enhanced cooling boosts the accretion rate towards the Eddington value. Finally, the black hole mass reaches the level where AGN heating exceeds ICM cooling at any (sufficiently high) accretion rate. The gas entropy increases, the mass accretion rate drops and the system then switches to the stable state at low accretion rate, low radiative efficiency and high mechanical efficiency. This is the so called "radio-mode" which presumably describes M87 now. In the simple scenario outlined above the black hole at the center of the cooling core first looks like a QSO and then jumps to a state of a low luminosity AGN as shown with green line in Fig. 8.

The time evolution of the black hole mass, accretion rate and radiative luminosity is sketched in the right panel of Fig. 8. The black hole initially accretes at the Eddington rate

Fig. 8 *Top*: illustration of gas heating and cooling in elliptical galaxies (Churazov et al. 2005). The *thick solid line* shows, as a function of the SMBH accretion rate, the heating rate due to outflow, which is complemented/dominated by radiative heating near the Eddington limit. *Horizontal dashed lines* show the gas cooling rate. The *upper cooling line* represents a young galaxy in which a large amount of gas is present and/or the black hole is small. Feedback from the black hole is not able to compensate for gas cooling losses and the black hole is in the QSO stage with a near-critical accretion rate, high radiative efficiency and weak feedback. As the black hole grows it moves down in this plot. The *black solid dot marks* the termination of this stage, when the black hole is first able to offset gas cooling, despite the low gas heating efficiency. The *lower cooling line* illustrates present day ellipticals: a stable solution exists at low accretion rates when mechanical feedback from the black hole compensates gas cooling losses. The radiative efficiency of accretion is very low and the black hole growth rate is very slow. *Bottom*: possible time evolution, corresponding to the *figure above*: the black hole accretes at the Eddington rate until its mass is large enough so that even weak feedback does not allow a stable solution. At later times the AGN switches into the low accretion rate mode, and the radiative efficiency drops dramatically

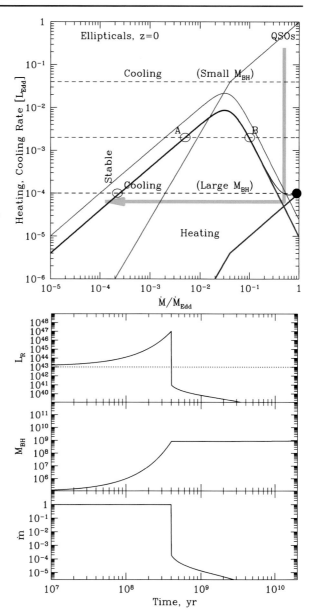

and grows exponentially with limited impact on the ICM. This fast growth and QSO-type behavior of the black hole continues until its mass is large enough so that even with pure radiative feedback, heating can offset cooling losses. After the system moves to the low accretion state the luminosity and the black hole growth rate drops by a large factor of order the ratio of the coupling constants $\alpha_M/\alpha_R \sim 10^4$ (or more if an ADAF-type scenario is adopted).

The above consideration is of course an overly simplified picture. Various prescriptions of the "radio-mode" feedback have been tested (e.g., Croton et al. 2006; Bower et al. 2006) in the semianalytic models coupled with the numerical simulations of structure formation of the Universe. In general, the key element of mechanical AGN feedback—the ability to provide good coupling of the AGN and the ICM—seems to be able to resolve the difficult issue of over-cooling and excess star formation in the most massive halos.

2.6 Conclusions About AGN Feedback

AGN Feedback in galaxy clusters is a rapidly developing area in astrophysics. It depends on a combination of various physical processes, ranging from physics of accretion to cosmological evolution of the most massive systems in our Universe. Much of the physics involved is still poorly understood. At the present epoch, AGN feedback prevents gas cooling in massive elliptical galaxies and clusters, but its role at higher redshifts is only beginning to emerge today (e.g. Hlavacek-Larrondo et al. 2012; McDonald et al. 2013). All this leaves much opportunity for future observational and theoretical studies.

3 Galactic Feedback and Metal Enrichment in Clusters

During the formation history of clusters of galaxies, metals have been continuously produced by the stars in the member galaxies. In the earliest epoch of star formation, about 500 Myr after the 'Big bang', Population III stars formed from the primordial gas. The nature of this stellar population is still uncertain, but probably it consisted of intermediate and high-mass stars (Vangioni et al. 2011). The first metals produced by this population were mostly ejected into the surrounding medium, leading to an initial metal abundance of about 10^{-4} times Solar (Matteucci and Calura 2005). This relatively low metal abundance was enough to allow the gas to cool more efficiently through spectral line emission, resulting in an epoch of increased star formation, which peaked around a redshift of $z \sim 2$–3.

This peak of the Universal star formation rate roughly coincides with the build-up of a hot Intra-Cluster Medium in the massive galaxy clusters. Dilute metal-poor gas accreted from the cluster surroundings is mixed with the gas expelled by galactic winds driven by supernova explosions. A combination of compression, accretion shocks, supernova heating and AGN feedback boosted the ICM temperature to more than 1 million Kelvin. In such a hot environment, star formation in the cluster was effectively quenched (Gabor et al. 2010). While the build-up of the ICM continued, the star formation rate dropped to very low levels, which explains the old stellar populations observed in local galaxy clusters today.

Due to their deep gravitational potential wells, clusters have retained the metals ejected into their ICM. Metal abundances observed in the ICM of local clusters thus provide an integrated record of their enrichment history. The low star-formation rates in cluster galaxies since $z \sim 2$–3 form an interesting contrast to the enrichment history of our own galaxy, where the star-formation rate showed a smaller decline. Clusters of galaxies therefore provide a unique insight in the chemical enrichment due to stellar populations before $z \sim 2$.

3.1 Sources of Metals

Most of the elements heavier than beryllium are produced in supernovae. Some elements, like nitrogen and sodium, are ejected into the medium by Asymptotic Giant Branch (AGB) stars. There are two distinct types of supernova explosions, type Ia and core-collapse, that each have a separate role in metal enrichment. Core-collapse supernovae yield the bulk of

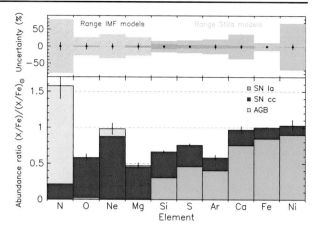

Fig. 9 Expected abundances measured in a 120 ks XMM-Newton observation of Sérsic 159-03 (*bottom panel*), which is a typical bright local cluster. The statistical error bars were obtained from de Plaa et al. (2006). The estimates for the SNIa, SNcc, and AGB contributions are based on a sample of 22 clusters de Plaa et al. (2007) and two elliptical galaxies (Grange et al. 2011). The *top panels* show the typical range in SNIa and IMF models with respect to the statistical error bars in the observation

the elements in the mass range from oxygen to silicon, while type Ia supernovae produce mostly elements from silicon to nickel (see Fig. 9).

Supernova type Ia models, however, do not agree well on the exact yields for each element. Uncertainties on the yields of, for example, calcium and nickel are about 40–60 %. The underlying cause is the very interesting progenitor problem of type Ia supernovae. Recent optical observations of type Ia supernovae show variations in their spectral properties. So far, it has been challenging to link an observed type Ia supernova to a progenitor type (see e.g. Howell 2011). Two main progenitor scenarios are being considered. The first is the 'classical' type Ia, where a white dwarf accretes matter from a companion star in the Red Giant phase. If the accretion rate is right, the mass and temperature of the white dwarf can grow to a point where carbon fusion ignites. This is close to the Chandrasekhar limit of 1.4 solar masses. The carbon ignites explosively and unbinds the entire white dwarf. The second progenitor channel is a scenario where two white dwarfs merge. They spiral toward each other through the emission of gravitational radiation and the less massive star is accreted onto the more massive one, until carbon is ignited and the star explodes. It is clear that these progenitor channels allow a range of possible explosion scenarios and metal yields, which makes accurate predictions challenging.

The yields of core-collapse supernovae are currently better established. The main uncertainty in the core-collapse yield of an entire stellar population is the Initial-Mass Function (IMF). To obtain the total yield for a population, the yields calculated for individual masses need to be integrated over the IMF (Tsujimoto et al. 1995). Therefore, the observed abundances of metals in the oxygen to silicon mass range can constrain the IMF of the cluster's stellar population above ~ 8 solar masses.

Below 8 $M_\odot$, intermediate-mass stars in their AGB phase are a source of nitrogen and sodium. Yields for these sources for several mass bins are available (Karakas 2010) and also need to be integrated over the IMF, like core-collapse yields, to obtain the yield for the whole stellar population. Measuring the abundances of AGB products therefore puts constraints the low-mass end of the IMF.

3.2 Abundance Measurements in X-Rays

The soft X-ray band between 0.1 and 10 keV is very suitable for abundance studies because it in principle contains spectral lines from all elements between carbon and zinc. In clusters of galaxies, the hot plasma is in (or is very close to) collisional ionization equilibrium,

Fig. 10 Abundance ratios fitted with supernova yields from the WDD2 SNIa model (Iwamoto et al. 1999) and a SNcc model with an initial metallicity $Z = 0.02$ and a Salpeter IMF. The calcium abundance appears to be underestimated. It can not explain the Ar/Ca ratio measured in this XMM-Newton sample of 22 clusters. (Adapted from de Plaa et al. 2007)

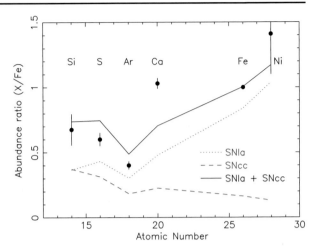

which makes it relatively easy to predict the emitted X-ray spectrum. The advance of X-ray spectroscopy has enabled the study of abundances in clusters and yielded interesting results (see e.g. Werner et al. 2008; de Plaa 2013, for more extended reviews on this topic).

The first attempt to link cluster abundances to supernova yields was done with the ASCA satellite (Mushotzky et al. 1996). With this instrument it was already possible to measure the abundances of O, Ne, Mg, Si, S, Ar, Ca, Fe and Ni. From these, and other ASCA studies the picture emerged of an ICM that was enriched early in its formation history with core-collapse supernova products and later by type Ia supernovae.

With the launch of XMM-Newton, a telescope with a substantial effective area and spectral resolution (through the Reflection Grating Spectrometer, RGS) became available, which enabled deep abundance studies in larger cluster samples. In de de Plaa et al. (2007), for example, 22 clusters were analyzed and abundances were measured in their core regions. If the average abundances of the sample are compared to supernova type Ia models, the calcium abundance appeared to be systematically higher than expected by the models (see Fig. 10). Possible explanations for this high calcium abundance include an unexpected difference in the type Ia explosion mechanism or an increased importance of progenitor systems where helium is accreted on the white dwarf, because explosive He fusion is expected to yield more calcium. Although a systematic error in the determination of the Ca abundance cannot yet be fully excluded (detailed analysis did not show any problem), this measurement can constrain supernova type Ia models. Note that the fit shown in Fig. 10 includes Ca. If Ca is excluded a better fit for S and Ar is obtained.

If a combination of a type Ia model and core-collapse model fits, their ratio is an indication of the relative contribution of type Ia supernovae to the enrichment. Bulbul et al. (2012) developed an extention to the APEC model that is able to fit the type Ia to core-collapse ratio directly to the spectra. This ratio depends strongly on the used models (de Grandi and Molendi 2009). But despite of this uncertainty, multiple groups (e.g. Bulbul et al. 2012; Sato et al. 2007) report a type Ia contribution of ~ 30–40 %, which is consistent with optical data.

In cool clusters and groups, the RGS spectrometer aboard XMM-Newton is able to measure carbon and nitrogen abundances. These elements are not produced in large quatities in supernovae, but appear to originate from metal-poor massive stars or AGB stars. The exact origin is still subject of debate (Romano et al. 2010). RGS observations by Werner et al. (2006a, 2006b) and Grange et al. (2011) have shown nitrogen to be very abundant around el-

Fig. 11 Measured cluster abundances versus redshift. The *top panel* shows the abundances up to $0.6r_{500}$ and the lower panel shows the abundances when the inner region up to $0.15r_{500}$ is ignored. Adapted from Baldi et al. (2012)

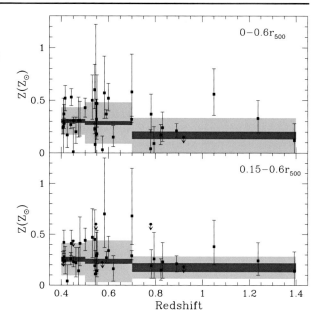

liptical galaxies. The high abundance of nitrogen cannot be explained by supernovae alone. A population of intermediate-mass AGB stars is therefore likely responsible for the high nitrogen content of the ICM.

With the Japanese Suzaku satellite, launched into low-earth orbit in 2005, abundance studies are performed in the outskirts of clusters (see Reiprich et al. 2013, for a recent and complete review of Suzaku interesting results). In these low surface-brightness areas, the low Suzaku background level is favorable above XMM-Newton. Recently, in an elaborate study of the outskirts of the Perseus cluster with Suzaku, Werner et al. (2013) found that the iron abundance distribution is surprisingly smooth in the outskirts. This points toward a scenario of an early enrichment of the hot X-ray gas with iron, well before $z \sim 2$. This result confirms earlier indications that chemical enrichment occured very early in the development of the universe and probably also before the formation of clusters.

One of these earlier indications was found in a study of a sample of high redshift clusters between $z = 0.3$–1.3 by Baldi et al. (2012). Although the scatter in the measured metal abundance as a function of redshift is large, a significant trend in the metal abundance up to $z = 1.3$ was not found (Fig. 11), indicating that the chemical enrichment mechanisms have not added a lot to the enrichment since $z = 1.3$. Deeper observations of high redshift clusters would be necessary to build a large enough sample to confirm the lack of a metallicity trend with redshift.

4 Magnetic Feedback Processes

Galaxies play a key role in the enrichment of the ICM or IGM, not only as far as heavy elements are concerned, but also regarding the magnetization. According to the standard bottom-up scenario of lowest galactic masses, primeval galaxies must have injected much of their ISM into the IGM during the initial bursts of star formation, thereby "polluting" large volumes of intergalactic space because of their high number density.

Kronberg et al. (1999) were the first to raise the question whether low-mass galaxies could have made a significant contribution to the magnetization of the IGM (apart from more massive starburst galaxies and AGN). Owing to their large number (observed and predicted in a CDM cosmology) and their injection of relativistic particles, they could have played a cardinal role in the context of this cosmologically important scenario. If true, it is to be expected that dwarf galaxies are "wrapped" in large envelopes of previously highly relativistic particles and magnetic fields, which are pushed out of them during epochs of vigorous star formation. Bertone et al. (2006) have discussed this more quantitatively and made predictions for the strengths of magnetic seed fields to be then amplified by large-scale dynamos over cosmic time. In particular, they also predict the existence of magnetic voids.

Donnert et al. (2009) and Dubois and Teyssier (2010) performed numerical models of supernova-driven winds in dwarf galaxies. Their simulations can provide an understanding of the origin of intergalactic magnetic fields at the level of 10^{-4} μG. Beck et al. (2013) presented a first numerical model of supernova-driven seeding of magnetic fields by proto-galaxies.

The existence of winds in low-mass galaxies is inferred from the observed kinematics of the gas (measured with slit spectroscopy), but can arguably be also inferred from measurements of the temperature of the hot (X-ray-emitting) gas. For instance, Martin (1998) found the outflow velocities in NGC1569 to exceed the escape speed, and della Ceca et al. (1996) derived the temperature of its hot, X-ray-emitting gas to exceed the virial temperature. The transport of a relativistic plasma out of this galaxy is strongly suggested by radio continuum observations of dwarf galaxies. Kepley et al. (2010) studied the radio halo in NGC1569, which extends out to about 2 kpc at 1.4 GHz. The dwarf irregular NGC4449 also possesses a low-frequency radio halo (Klein et al. 1996).

Of course, in magnetizing the ICM/IGM, low-mass galaxies have been competing with AGN (Rees 1987). Judging from the radio luminosities of the "culprits", it is clear that nevertheless low-mass galaxies may have contributed significantly. While a typical starburst dwarf galaxy emits a monochromatic radio luminosity of $P_{1.4\ GHz} \approx 10^{20.5}$ W Hz^{-1}, radio galaxies in the FRI/II transition regime have luminosities of $P_{1.4\ GHz} \approx 10^{24.7}$ W Hz^{-1}. Hence, the radio power produced by AGN is about 10^4 times larger than that of dwarf galaxies. On the other hand, ΛCDM cosmology with bottom-up structure formation implies that dwarf galaxies must have been formed in huge numbers.

The role of massive black holes in the build-up of strong magnetic fields in galaxies was addressed by Chakrabarti et al. (1994). They pointed out that galactic winds or collimated jets were able to disperse such magnetic fields over large volumes of the host galaxies and beyond. Xu et al. (2010) presented magneto-hydrodynamical simulations in which they studied the evolution of magnetic fields ejected by an AGN shortly after the formation of a galaxy cluster. They showed that, if the magnetic fields are ejected before any major mergers occurring in the forming cluster, they can be spread throughout the cluster, with further subsequent amplification by turbulence in the ICM. It should be noted at this point that central so-called "mini-halos", which are bright extended radio sources located in the centers of cooling-flow clusters (e.g. Perseus A, Hydra A, Virgo A), cannot possibly magnetize large cluster volumes, as they are pressure-confined (e.g. de Gasperin et al. 2012).

5 Cold Fronts in Galaxy Clusters

Among the first results from high-resolution cluster images obtained with Chandra was the discovery of sharp edges in the X-ray surface brightness in merging clusters A2142 and

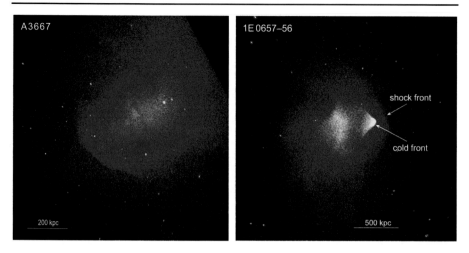

Fig. 12 Prominent cold fronts in the Chandra X-ray images of A3667 (MV07) and the Bullet cluster (Markevitch 2006). In the Bullet cluster, the surface of the cool "bullet" is a cold front; its supersonic motion generates a prominent bow shock

A3667 (Markevitch et al. 2000) (M00); (Vikhlinin et al. 2001a) (V01). Figure 12 shows deep Chandra images of A3667 and the Bullet cluster, both of which exhibit such brightness edges. The characteristic shape of the brightness profiles across these edges corresponds to a projected abrupt jump of the gas density at the boundary of a roughly spherical body (M00). The Bullet cluster shows two prominent edges, one at the nose of the dense cool "bullet" and another ahead of it. Their physical nature is revealed by radial profiles of the gas density, pressure and specific entropy, shown in Fig. 13. As one can guess already from the image, the outer edge is a bow shock—the dense side of the edge is hotter, with the pressure jump satisfying the Rankine-Hugoniot jump conditions (Markevitch 2006).

At the same time, the boundary of the bullet has a different physical nature—it separates two gas phases with very different specific entropies that are in approximate pressure equilibrium at the boundary. The denser side of this edge has a lower temperature, opposite to what's expected for a shock front, which is how the two phenomena can be distinguished observationally. The edge in A3667 and those in A2142 have the same sign of the temperature jump as that at the bullet boundary. These features in clusters they have been named "cold fronts" (V01). The term "contact discontinuity" is sometimes used, but it implies continuous pressure and velocity between the gas phases, whereas these structures in clusters may have discontinuous tangential velocity, as we will see below. Cold fronts turned out to be much more ubiquitous than shocks and have been observed with Chandra and XMM-Newton in many clusters and even galaxies (e.g., Machacek et al. 2005; see Markevitch and Vikhlinin 2007 (MV07), for a detailed review). If the dense gas cloud is moving with respect to the ambient gas, there will be a ram pressure component contributing to the pressure balance near the front (M00), which makes it possible to estimate the velocity of the cloud (V01).

Cluster cold fronts have two main physical causes, both related to mergers. The obvious one, originally proposed by M00 for the two fronts in A2142, involves an infalling subcluster ram-pressure stripped of its outer gas layers, which leaves a sharp boundary between the dense remnant of the subcluster's core and the less dense, hotter gas of the main cluster flowing around it. The bullet in the merging Bullet cluster and the NGC1404 galaxy falling

Fig. 13 Schematic radial profiles of the gas density, pressure and specific entropy in a sector crossing the bullet and the bow shock in the Bullet cluster and centered on the bullet's center of curvature (see Fig. 12). The shock front and the bullet boundary have sharp density jumps of similar amplitudes; however, the shock at $r \approx 50''$ exhibits a large pressure jump and a slight entropy increase, while the bullet boundary at $r \approx 12''$ is in near pressure equilibrium but separates gases with very different entropies—it is a cold front (reproduced from MV07)

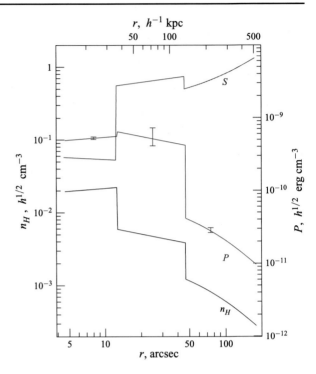

into the Fornax cluster (Machacek et al. 2005) appear to be such "stripping" fronts.[1] However, cold fronts have also been observed in the cores of the majority of *relaxed* clusters that show no signs of recent merging (Mazzotta et al. 2001; Markevitch et al. 2001, M01; Markevitch et al. 2003; Ghizzardi et al. 2010). These fronts are typically more subtle in terms of the density jump than those in mergers, and occur close to the center ($r \lesssim 100$ kpc), with their arcs usually curved around the central gas density peak. An example is seen in the Ophiuchus cluster (Fig. 14a). A detailed study of such a front in A1795 by M01 has shown that the gas on two sides of the front has different centripetal acceleration. This led those authors to propose that the dense gas of the cool core is "sloshing" around the center of the cluster gravitational potential, perhaps as a result of a disturbance of the potential by past mergers. While M01 envisioned radial "sloshing", Keshet et al. (2010) offered a more plausible scenario with a tangential flow of the cool gas beneath the fronts being responsible for centripetal acceleration.

Ascasibar and Markevitch (2006) (A06) have reproduced this phenomenon in high-resolution hydrodynamic simulations of idealized binary mergers. To explain the absence of the gas disturbance on large scales, the small infalling subcluster should have no gas (only the collisionless dark matter—perhaps having lost its gas during previous stages of infall), thus disturbing only the gravitational potential of the main cluster without generating shocks in the gas. Such a disturbance results in a displacement between the gas peak and the collisionless dark matter peak, which sets off sloshing, which continues for several

---

[1] It appears that the original M00 scenario for A2142, which involved two surviving gas cores, is not correct (Tittley and Henriksen 2005; MV07). Instead, the two original fronts, the front discovered at a large radius (Rossetti et al. 2013), and yet another front seen close to the center in a deep Chandra observation—all concentric—are "sloshing" fronts discussed below. The "stripping" scenario does work in some other clusters.

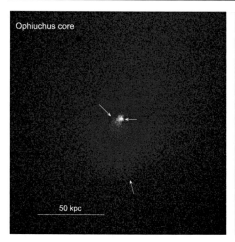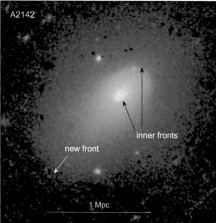

Fig. 14 *Left*: a series of "sloshing" cold fronts at different radii in the cool core of the Ophiuchus cluster, shown by arrows in this archival Chandra image. (The subtle outermost front is seen better in a more coarsely binned image.) *Black cross marks* the center of the cD galaxy. *Right*: a cold front discovered in A2142 at a distance of 1 Mpc from the center (marked "new front") by Rossetti et al. (2013) in this XMM-Newton dataset. The original cold fronts in the central region (M00) are shown by *black arrows*

billion years, generating concentric cold fronts in a spiral pattern (if the merger had any angular momentum), as is often observed in cool cores. All that is required is a radial gradient of the specific entropy (which is always present in relaxed clusters) and an initial gas displacement. The idea that this class of cold fronts is the result of oscillations of the dark matter peak caused by a merger has been first proposed by Tittley and Henriksen (2005); the initial displacement may also be caused by the passage of a mild shock (Churazov et al. 2003; Fujita et al. 2004). Sloshing fronts are easily detected in cool cores, often delineating their boundary, but they are not confined to cool cores—in A06, a sufficiently strong initial disturbance caused detectable fronts to propagate to large radii. Indeed, cold fronts of this nature (that is, not associated with any infalling subclusters) have recently been found at $r \sim 0.7$–1 Mpc in Perseus (Simionescu et al. 2012) and in A2142 (Rossetti et al. 2013), see Fig. 14b. Note that this does not mean that the cluster gas oscillates from the center all the way out to those large distances. While sloshing does begin as a physical displacement of the gas density peak from the potential peak, the fronts propagate outwards as waves—see Fig. 8 in A06 and Nulsen and Roediger (2013), who describe sloshing and the resulting cold fronts as superposition of g-mode oscillations in the stratified cluster atmosphere.

Core sloshing has a number of important effects on clusters. One is seen in the X-ray image of the Ophiuchus cluster (Fig. 14a; a matching snapshot from the numeric simulations can be seen in Fig. 7c of A06). The gas density peak (which contains the lowest-entropy gas of the cool core) is completely displaced from the center of the cD galaxy (shown by a cross). This would temporarily starve the central AGN of its fuel and may stop its activity. ZuHone et al. (2010) showed that sloshing can also facilitate heat transport from the hot reservoir outside the core into the cool core via mixing, which can compensate for most of radiative cooling in the core, provided that gas mixing is not suppressed. Keshet et al. (2010) and ZuHone et al. (2011) showed that tangential velocity shear in a sloshing core should amplify and reorder the initially tangled magnetic field. This effect has observable consequences for the shapes of cold fronts, which offers an independent tool to study those magnetic fields, as discussed below. There is also a physical connection between gas sloshing and cold fronts

Fig. 15 X-Ray surface brightness profile across the cold front in A3667 (V01, MV07). *Red line* shows a best-fit model of a density jump that is smoothed with $\sigma = 11$ kpc, which is the m.f.p. for Coulomb diffusion from the dense side to the less dense side of the front. If diffusion were present, the front would have been smeared by several times this width; such diffusion is clearly excluded by the data

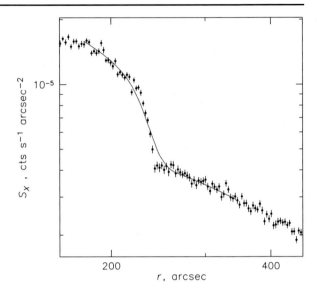

on one side and the radio-emitting ultrarelativistic electrons in the cluster cores on the other (Giacintucci et al. 2014; ZuHone et al. 2013a).

5.1 Physics of Cold Fronts

Cold fronts can be used to study the microphysics of the intracluster plasma. One property of the cold fronts observed with Chandra that is immediately striking is their sharpness. Figure 15 shows an X-ray brightness profile across the prominent front in A3667 (Fig. 12a). Vikhlinin et al. (2001a) pointed out that the front is sharper than the mean free path for Coulomb collisions. Indeed, red line in Fig. 15 shows a best-fit model that includes broadening of the front with a width equal to the Coulomb m.f.p. for diffusion from the cool inner side to the hot outer side of the front. Such broadening is clearly inconsistent with the data, which means the diffusion across the front is suppressed. Ettori and Fabian (2000) pointed out that the existence of the observed temperature jumps at cold fronts implies that thermal conduction is also strongly suppressed. Vikhlinin et al. (2001b) proposed (within the "subcluster stripping" scheme of M00) that the motion of the infalling subcluster through a tangled magnetic field frozen into the ambient gas of the main cluster would naturally form an insulating layer of the field oriented strictly along the front surface, as a result of the field "draping" around an obstacle, as originally proposed by Alfvén to explain the comet tails. This effect is shown in a simulation by Asai et al. (2005) in Fig. 16. As pointed out by Lyutikov (2006), such draping can amplify the field in the narrow layer immediately outside the front to values approaching equipartition with thermal pressure (compared to magnetic pressures of order 1 % of thermal pressure in the rest of the cluster). Such a layer would completely suppress diffusion and thermal conduction across the front, while the magnetic tension of such a layer may stabilize the front against Kelvin-Helmholtz instability (Vikhlinin et al. 2001b; MV07), which we will see below in a simulation.

If a cluster with a cool core initially has a tangled magnetic field, when the core is disturbed and starts sloshing, the field is rapidly stretched by the tangential gas velocity shear (ZuHone et al. 2011), forming layers of amplified field oriented along the surface of the cold front form. These layers are located *under* the front surfaces, unlike for the draping effect

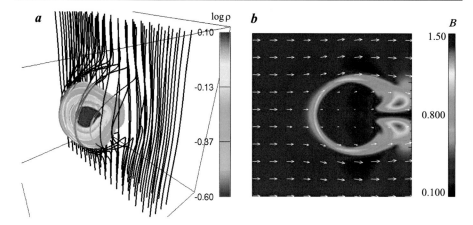

Fig. 16 Magnetic field draping around an infalling subcluster (panel **a**). This results in a layer of amplified and ordered field immediately outside the cold front (panel **b**). The field is initially uniform, but the effect for a tangled field is similar. (Reproduced from Asai et al. 2005)

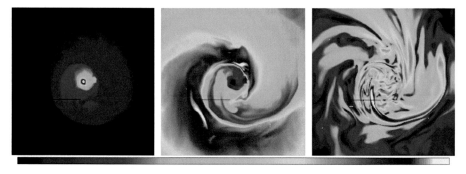

Fig. 17 The magnetic field around a "sloshing" cold front. *Left panel* shows X-ray surface brightness, *middle panel* shows gas temperature, and *right panel* shows the magnetic field strength. Unlike for the "stripping" front (Fig. 16), the field is strongly amplified *inside* the front. (Reproduced from ZuHone et al. 2011)

in a "stripping" front. A snapshot from the simulations of this process is shown in Fig. 17. Such layers can suppress KH instability of the front surface (Fig. 18) similarly to a draping layer. However, there is an important difference for the effective thermal conduction across the front.

If a subcluster falls into a cluster from a large distance, it will be thermally insulated from the ambient gas by a magnetic draping layer for as long as the subcluster survives as a coherent structure, because the initially disjoint magnetic field structures of the cluster and subcluster cannot connect at any stage of the infall (as long as there is no magnetic reconnection)—even when the gases are geometrically separated only by a thin cold front. Since heat is conducted only along the field lines, it is not surprising to see a cool infalling subcluster survive the immersion into the hot gas of the bigger cluster (e.g., the group falling into A2142 discovered by Eckert et al. 2014). However, for a sloshing cold front, thermal conductivity across the front may not be completely suppressed by the magnetic layer, as shown by simulations of ZuHone et al. (2013b). The reason is that prior to the onset of sloshing, the regions inside and outside the cluster core are connected by the field lines,

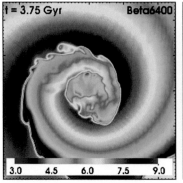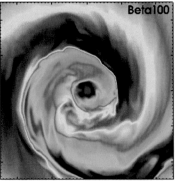

Fig. 18 The stabilizing effect of the magnetic field on cold fronts. Panel size is 500 kpc; color shows gas temperature (the scale is in keV). *Left panel* shows a simulation with a weak field (initial plasma $\beta = 6400$), *right panel* shows a stronger, more realistic field ($\beta = 100$). The realistic field suppresses K-H instabilities, leaving relatively undisturbed cold fronts, similar to those observed. (Reproduced from ZuHone et al. 2011)

and while sloshing stretches most of them tangentially, it does not sever the connection completely. Those authors suggested that the existence of the temperature jumps across the sloshing cold fronts may therefore be used to constrain heat conduction *along* the field lines.

Another interesting (and completely unknown) plasma property that may be constrained by the observations of cold fronts is viscosity (MV07). Figure 19 (from ZuHone et al. 2015) shows a simulated sloshing core with the plasma viscosity modeled in different ways. The viscosity, either isotropic or anisotropic (it is likely to be the latter in the presence of magnetic fields), acts to suppress the KH instabilities, and the effect should be observable. Of course, the effect of viscosity is superimposed on the stabilizing effect of the magnetic layers discussed above, so the observations will most likely be able to constrain some combination of the two. Based on simulations without the magnetic field, Roediger et al. (2013) concluded that the effective isotropic viscosity should be significantly suppressed to explain the disturbed appearance of a cold front in Virgo.

This area of research is currently under rapid development, with high-quality observations of cold fronts being obtained and tailored numeric simulations being constructed, so interesting constraints on the microphysical properties of the cluster plasma can be expected soon.

6 Shock Fronts and Non-thermal Components

6.1 Shock Fronts

Shock waves in clusters of galaxies are the main agents to convert the kinetic energy of supersonic and superalfvénic plasma flows produced by gas accretion, merging substructures and AGN outflows into both the thermal and the non-thermal components. The shocks are essential to heat the gas and to produce the observed thermal and non-thermal radiation. The shock structures in plasma and their ability to create the non-thermal components— energetic charged particles and electromagnetic fields—depends on whether the shock is collisional or collisionless. The mean free path of a proton due to the Coulomb collisions is $\lambda_p \approx 7 \times 10^{21} v_8^4 n_{-3}^{-1}$ (here v_8 is proton velocity in thousands $\mathrm{km\,s^{-1}}$ and n_{-3} is the ambient gas number density in $10^{-3}\,\mathrm{cm^{-3}}$). Note that an ion of charge Z and atomic weight A have a

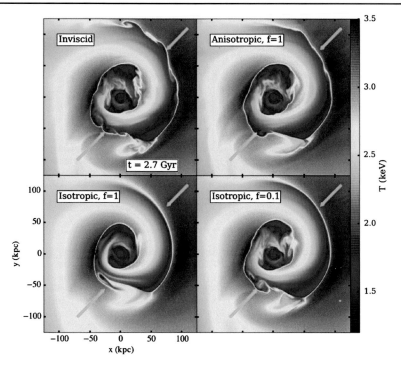

Fig. 19 Simulation of a sloshing core with different physical models for plasma viscosity: inviscid (with a weak magnetic field present), isotropic Spitzer viscosity, anisotropic Braginskii viscosity along the field lines, and isotropic viscosity suppressed by factor $f = 0.1$. A highly suppressed viscosity is required to produce visible K-H eddies at cold fronts. (Reproduced from ZuHone et al. 2015)

gyroradius $r_{gi} = 3.3 \times 10^9 (AT_{keV})^{1/2} (ZB_{\mu G})^{-1}$ cm in a magnetic field $B_{\mu G}$ measured in μG. The Coulomb mean free path is much larger than the proton gyroradius (as well as the ion inertial length which we shall introduce later) at all particle energies of interest in the cluster and thus the plasma shocks in clusters are expected to be collisionless as it is the case in the hot interstellar and interplanetary plasmas.

A specific feature of the collisionless shocks is the mechanism of the flow momentum and the energy dissipation by means of the excitation and damping of collective electromagnetic fluctuations providing numerous degrees of freedom with a very broad range of the relaxation times and spatial scales. Moreover, the collisionless shock may accelerate particles to ultra-relativistic energies resulting in a broad particle energy spectra formation. The multi-scale nature of the shock with a strong coupling between the scales makes the problem of laboratory studies and theoretical modeling of the structure of the collisionless shock to be very complicated. However, some basic features of the collisionless shock physics were established from direct interplanetary plasma observations, imaging and spectroscopy of shocks in supernova remnants and computer simulations (see e.g. the review by Treumann 2009).

The shock flow dissipation mechanisms depend on the flow velocity, magnetization and shock obliquity. Shocks of low enough Mach numbers $\mathcal{M}_s < \mathcal{M}_{crit}$ are able to dissipate the kinetic energy of the flow by anomalous Joule dissipation. In the lack of Coulomb collisions the dissipation is usually associated with the anomalous resistivity provided by wave-particle interactions. Such shocks are called subcritical and $\mathcal{M}_{crit}$ is dubbed the first critical Mach

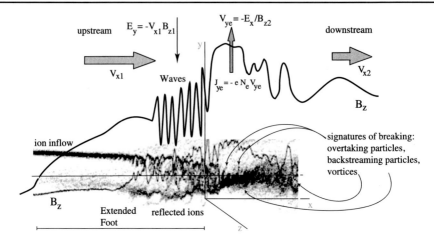

Fig. 20 Schematic structure of a perpendicular supercritical shock. The profiles of the magnetic field and plasma density at the shock transition region dominated by the ion reflection dissipation mechanism are shown. Charge separation over an ion gyro-radius in the shock ramp magnetic field produce the electric field along the shock normal which may reflect the incoming ions back upstream. The electric field along the shock front produced by conductive plasma motion in the magnetic field may accelerate particles by the electric drift acceleration mechanism. The magnetic field of the current carried by the accelerated back-streaming ions causes the magnetic foot in front of the shock ramp. Reproduced from Treumann (2009)

number (see e.g. Kennel et al. 1985). The structure of the shock transition in this case may be smooth and laminar, but this mainly occurs for low β shocks. The magnetization parameter $\beta = 8\pi n T/B^2 = \mathcal{M}_a^2/\mathcal{M}_s^2 \approx 40 n_{-3} T_{\text{keV}} B_{\mu\text{G}}^{-2}$, characterizes the ratio of the thermal and magnetic pressures. The first critical Mach number is maximal for a quasi-transverse shock in the plasma with low magnetization $\beta > 1$ and it is below 2.76. At large β typical for the hot intracluster plasmas some transverse shocks as well as very weak shocks can be subcritical, while most of the shocks are supercritical. The width of the viscous transition in the collisionless subcritical shock wave can be estimated as $l_e/\sqrt{\mathcal{M}_s - 1}$, where $l_e = c/\omega_{\text{pe}} \approx 1.7 \times 10^7 n_{-3}^{-0.5}$ cm. Here ω_{pe} is the electron plasma frequency.

If the shock Mach number exceeds $\mathcal{M}_{\text{crit}}$ the anomalous resistivity is unable to provide the required dissipation rate to satisfy the Rankine-Hugoniot conservation laws at the shock. Then the shock became supercritical and its structure is turbulent. The dissipation mechanism of a supercritical shock is usually dominated by ion reflection. Some fraction of the incoming ions are reflected by a force which is a combination of electrostatic and magnetic fields and the reflected ion behavior depends on angle between the local upstream magnetic field and the local shock normal. In Fig. 20, taken from Treumann (2009), we illustrate the basic processes in the collisionless perpendicular shocks.

The width of the supercritical shock transition in case of a quasi-parallel shock may reach a few hundreds ion inertial lengths which is defined as $l_i = c/\omega_{\text{pi}} \approx 7.2 \times 10^8 n_{-3}^{-0.5}$ cm. Here ω_{pi} is the ion plasma frequency and n is the ionized ambient gas number density measured in cm^{-3}. The shock transition structure is unsteady in the case of magnetized shocks being a subject of shock front reformations which however are expected to be suppressed in the hot intercluster plasmas. The widths of viscous transition in the collisionless shocks is orders of magnitude less the Coulomb mean free path and it is below the spatial resolution of optical and X-ray observations even for a few kpc distance galactic supernova shocks.

Non-thermal energetic electrons are responsible for most of the observed radio emission from clusters, but the exact nature of particle acceleration process and the origin of the observed $\sim$ µG strength magnetic fields in clusters are still under debate. From the energetic ground shock waves are the most natural accelerating agent in hot weakly magnetized plasmas. As we have learned from multi-wavelength observations of strong shocks in young supernova remnants in order for diffusive shock acceleration (DSA) to be fast enough to reach 100 TeV regime particle energies observed in the sources, significant non-adiabatic magnetic field amplification is required (see e.g. Bell 2004; Amato and Blasi 2009; Bykov et al. 2012; Schure et al. 2012). The cosmic ray driven instabilities may provide a source of free energy for strong magnetic field amplification transferring a few percent of ram pressure of a strong shock (with $\mathcal{M}_s \gg 1$) into fluctuating magnetic fields (see Fig. 11 in Bykov et al. 2014). This is indeed enough to provide µG-magnetic fields behind the strong large scale accretion shock. The observed µG magnetic field strength is well above the field amplitude produced by the adiabatic compression of the intergalactic field at an accretion shock. Therefore it is possibly produced by CR driven instabilities at the strong accretion shock. The field amplification mechanism driven by instabilities of the anisotropic CR distribution produced by DSA is efficient if CRs get a substantial fraction of the shock ram pressure, which is expected in strong shocks with hard CR spectra. In the case of strong amplification of the fluctuating magnetic field in the shock precursor, the direction of the local magnetic field just before the viscous plasma shock transition will vary with time providing periods of both quasi-perpendicular and quasi-parallel configurations. This may differ from the case of weak shocks with steep spectra of accelerated particles which may not contain enough free energy in the high energy end of particle spectra to provide strong long-wavelength turbulence in the shock upstream.

Because of much higher pre-shock densities, the internal shocks with modest Mach numbers $\mathcal{M}_s < 4$ in clusters, which are propagating in a very hot intracluster plasma, dissipate more energy than the strong accretion shocks. Internal shocks with $2 < \mathcal{M}_s < 4$ were shown in simulations by Ryu et al. (2003) to produce about a half of the total kinetic energy dissipation, while the internal shocks as a whole are responsible in this model for about 95 % of gas thermalization. Microscopic simulations of electron acceleration in quasi-perpendicular shocks of $\mathcal{M}_s < 5$ were performed by Guo et al. (2014a). This particle-in-cell plasma modeling demonstrated that the repeated cycles of shock drift acceleration (SDA) may form power-law electron energy spectra. In a particular run with a quasi-perpendicular pre-shock magnetic field and $\mathcal{M}_s = 3$ they found that about 15 % of the electrons were accelerated and a power-law electron energy spectrum with a slope of $q \approx 2.4$ was formed. The energy density carried by the accelerated energetic electrons in this simulation was about 10 % of the bulk kinetic energy density of the incoming ions. This electron acceleration efficiency is much higher than that estimated from supernova remnant observations. The transverse box size in this simulation was about one ion gyro-radius. It is important to confirm the interesting result with larger simulation boxes since the shock structure is determined by ions and the model have to fulfill the requirements of the theory of charged particle motion in an electromagnetic field with one ignorable spatial coordinate by Jones et al. (1998).

The very long (Mpc scale) highly polarized radio structure observed in the merging galaxy cluster CIZA J2242.8+5301 by van Weeren et al. (2010) can be understood in this way, as the associated weak shock may not disturb the initial inter-cluster magnetic field in the shock upstream. The observed polarization favors a transverse shock configuration which may provide efficient shock drift acceleration of relativistic electrons (Guo et al. 2014a,b). This would require a very uniform magnetic field in the shock upstream. The radio structures and the X-ray emission of CIZA J2242.8+5301 cluster illustrating the location and the extension of the large scale shocks of moderate strengths are shown in Fig. 21 from Akamatsu et al. (2014).

Fig. 21 Smoothed 0.5–2.0 keV band X-ray image (*red*) and WSRT 1.4 GHz image of CIZA J2242.8+5301 (cyan). The *thin yellow lines* depict the approximate locations of the shock fronts confirmed by Suzaku. The figure is courtesy Akamatsu et al. (2014)

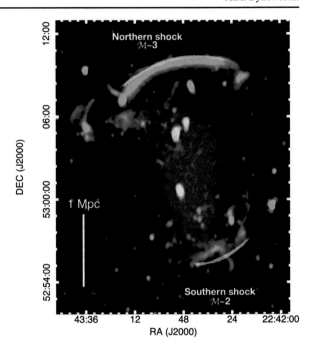

6.2 Intracluster Cosmic Rays

Both electrons and ions are likely accelerated at the shock fronts produced by mergers or supersonic outflows, but their subsequent evolution differ markedly. Since the ultra relativistic electrons have a radiative lifetime much shorter than the age of the cluster, they rapidly radiate most of their energy away and then may comprise a long lived population at Lorentz factors of around 100 where both Coulomb and radiation losses are longer than 10^9 years (Petrosian et al. 2008). On the other hand, ions (protons mostly) lose only a small fraction of their energy during the lifetime of the cluster, and their diffusion time out of the cluster is even larger, so that they are stored in clusters for a cosmological time scale (Völk et al. 1996; Berezinsky et al. 1997).

The long lived non-thermal particle populations may be the subject of a consequent re-acceleration by multiple shocks of different strengths as it is likely the case in galactic superbubbles and starburst regions (cf. Bykov 2001, 2014). Relativistic electrons can be re-accelerated in-situ by a long-wavelength MHD turbulence (Bykov and Toptygin 1993) which may be generated in the ICM during cluster mergers (Brunetti and Lazarian 2011; Brunetti and Jones 2014). The energy contained in both populations depends on the energy spectra of particles. Electrons of energies at about 50 MeV are difficult to constrain from the observational point of view. Gamma-ray observations are used to constrain the energy density in relativistic particles (see e.g. Ackermann et al. 2014). Recently, Prokhorov and Churazov (2014) analyzed *Fermi-LAT* photons above 10 GeV collected from the stacked 55 clusters selected from a sample of the X-ray brightest clusters. They obtained an upper limit of the pressure of relativistic protons to be 1.5 % relative to the gas thermal energy density, provided that the spectral index q of relativistic proton power-law distribution is 2.1, while for $q = 2.4$ the limit is already about 6 %. These estimations assume that relativistic and thermal components are mixed. Similar results were reported by Ackermann et al. (2014) who analyzed another set of clusters at the photon energies starting from 500 MeV.

The observations reported above have placed stringent limits on the pressure contained in the high-energy particle populations with hard spectra, while the constraints on the non-thermal components with steep enough spectra to be dominated by the sub-GeV particles remain to be established.

The most direct evidence of a non-thermal component within the intracluster medium comes from radio observations, that have revealed the presence of two main classes of Mpc-scale diffuse synchrotron sources, generally called "radio halos" and "radio relics" (see e.g. Ferrari et al. 2008, 2012). The origin of cosmic rays in radio halos is still elusive, even though their existence has been known for decades. The morphological, spectral and polarization properties of the radio halos were attributed by Brunetti et al. (2001) to radiation of relativistic electrons accelerated by ICM turbulence. Alternatively, the radiating leptons could be of the secondary origin i.e. a product of hadronic collisions of (long-lived) non-thermal energetic protons with thermal nuclei (Dennison 1980; Blasi and Colafrancesco 1999). In this case, one should also expect γ-ray emission, which has not been seen so far. MHD simulations carried out by Donnert et al. (2010) indicated some problems in explaining the observed steepening of the synchrotron spectra of halos. However, hadronic models with energy-dependent CR proton transport coefficients may be able to reproduce the steep spectra (Enßlin et al. 2011).

Radio relics are considered to be related to DSA, even if some of current observational results suggest the need to review or refine electron acceleration models within this class of sources (van Weeren et al. 2012; Ogrean et al. 2014). Radio observations of most kinds of relic sources in clusters of galaxies show rather steep synchrotron spectra ($S_\nu \propto \nu^{-\alpha}$), with $\alpha \gtrsim 1$ (see e.g. Miley 1980; Ferrari et al. 2008, 2012). As expected in DSA models, the spectral index α can change within the source, with a steepening from the front towards the back of the shock (see e.g. the spectacular radio relic in the cluster CIZA J2242.8, whose spectral index is ranging from 0.7 to about 1.7; Stroe et al. 2013). The corresponding indices of the relativistic lepton power-law spectra are $q \gtrsim 2.5$. The soft particle spectra are expected to be produced by relatively weak shocks in both the diffusive shock acceleration model (see e.g. Brüggen et al. 2012) and in the shock drift electron acceleration model by Guo et al. (2014a). The simulated statistical distribution of the merging shock strengths (see e.g. Ryu et al. 2003) peaks at rather weak shocks, and therefore the constrains on the pressure in relativistic leptons of Lorentz factors of about 100 and the non-thermal sub- and semi-relativistic protons of soft energy spectra still need further substantiation.

Apart from radio and gamma-ray observations discussed above fine X-ray spectroscopy has a potential to study non-thermal components. To illustrate the effects of such a supra-thermal electron distribution on data, Kaastra et al. (2009) simulated two X-ray micro-calorimeter spectra extracted from a circular region with a radius of $1'$ centered on the core of a bright cluster with a 0.3–10 keV luminosity of 6.3×10^{37} W within the extraction region, at an assumed redshift of $z = 0.055$. In the simulation of the spectrum of a deep 100 ks X-ray micro-calorimeter observation, they assumed a post-shock downstream electron distribution for a Mach number of $\mathcal{M} = 2.2$ and pre-shock temperature of $kT = 8.62$ keV (i.e. 10^8 K) calculated with a kinetic model of electron heating/acceleration by a collisionless shock developed by Bykov and Uvarov (1999). In Fig. 22 taken from Kaastra et al. (2009) we present the 6.9–7.0 keV part of the spectrum (rest-frame energies) which shows the Fe XXVI Lyα lines and the Fe XXV j-satellite line simulated for a 100 ks observation with the X-ray micro-calorimeter aboard the *ASTRO-H* mission and also for that proposed for the future *Athena* mission. Enhanced equivalent widths of satellite lines are shown to be a good indicators of non-thermal electrons. The satellite line in the simulated spectrum is clearly stronger than that predicted by the thermal model with a Maxwellian electron distribution. The energy density in the supra-thermal electrons modeled by Guo et al. (2014a)

Fig. 22 X-ray micro-calorimeter spectra from a 1′ region at the center of a bright cluster with a mixture of the thermal and supra-thermal electrons as simulated by Kaastra et al. (2009). Simulations made for 100 ks observations with Athena (*top panel*) and ASTRO-H (*bottom panel*). The energy distribution of electrons accelerated at $\mathcal{M} = 2.2$ Mach number shock was calculated with the kinetic electron acceleration model by Bykov and Uvarov (1999). *Crosses* are simulated data points. *Solid line*: best-fit model to a pure Maxwellian plasma, with temperature 16.99 keV. Note the excess emission of satellite lines in the data, in particular the Fe XXVI-line

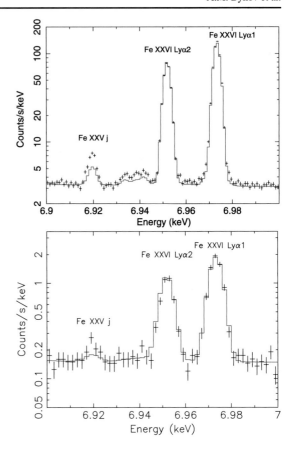

is larger than that was used in the simulations by Kaastra et al. (2009) and therefore it can be observationally tested in this way. This illustrates a good potential of high-resolution spectra obtained by future satellites with a large effective area to observationally reveal non-Maxwellian tails in the electron distributions.

6.3 Intracluster Magnetic Fields

The study of the linearly polarized emission from radio images is of cardinal importance in order to constrain the properties of intra-cluster magnetic fields.

Vacca et al. (2010) studied the power spectrum of the magnetic field associated with the giant radio halo in the galaxy cluster A665. They performed sensitive observations of this cluster with the Very Large Array at 1.4 GHz and compared these with simulations of random three-dimensional turbulent magnetic fields, trying at the same time to reproduce the observed radio continuum emission from the halo. They constrained the strength and structure of the intracluster magnetic field by assuming that it follows a Kolmogorov power-law spectrum and that it is in local equipartition with relativistic particles. A central magnetic field strength of about 1.3 μG was inferred for A665. The azimuthally averaged brightness profile of its radio emission suggests the energy density of the magnetic field to follow the thermal gas density, leading to an averaged magnetic field strength over the central 1 Mpc^3 of about 0.75 μG. An outer scale of the magnetic field power spectrum of ~ 450 kpc was estimated from the observed brightness fluctuations of the radio halo.

Bonafede et al. (2011) used the Northern VLA Sky Survey to analyze the fractional polarization of radio sources out to 10 core radii from the centers of 39 massive galaxy clusters with the aim to find out how different magnetic field strengths affect the observed polarized emission along different sight lines through the clusters. They found the fractional polarization to increase towards the cluster peripheries, and their findings are in accord with a magnetic-field strength of a few µG in the centers. A statistical test gives no hint at any differences in depolarization for clusters with and without radio halos, but indicates significant differences of the depolarization of sources seen through clusters with and without cool cores.

Govoni et al. (2013) investigated the potential of new-generation radio telescopes (e.g. the Square Kilometer Array, SKA and its pathfinders) in detecting the polarized emission of radio halos in galaxy clusters. To this end, they used magneto-hydrodynamical simulations conducted by Xu et al. (2011, 2012), Beck et al. (2013) to predict the expected polarized emission of radio halos at 1.4 GHz. The synthetic maps of radio polarization were compared with the detection limits on polarized emission from radio halos set by current and upcoming radio interferometers. They show that both the angular resolution and sensitivity expected in future sky surveys at 1.4 GHz using the SKA precursors and pathfinders (like APERTIF, ASKAP, Meerkat) are rather promising for the detection of the polarized emission of the most powerful ($L_{1.4\,\mathrm{GHz}} > 10^{25}$ W Hz) radio halos. Furthermore, the upgraded JVLA has the potential to detect polarized emission from strong radio halos already now, though with relatively low angular resolution. However, the detection of polarization signal in faint radio halos ($L_{1.4\,\mathrm{GHz}} < 10^{25}$ W Hz) has to await the fully deployed SKA.

7 Turbulence in Clusters

Turbulence and density fluctuations affect the basic assumptions behind hydrostatic equilibrium and hence the estimates of important cosmological parameters such as cluster masses. For this reason, but even more because of the interesting astrophysical processes involved, we discuss in more detail the various observational signatures of turbulence and density inhomogeneities.

We consider here constraints from measurements in the radio band and pressure maps, surface brightness distributions, resonant scattering and line broadening in the X-ray range.

7.1 X-Ray View: Pressure and Surface Brightness Mapping

Large scale plasma motions and turbulence in clusters primarily concerns fluctuations in the velocity field. However, they also lead to density and pressure fluctuations that can be measured with X-ray telescopes with sufficient spatial resolution and only modest spectral resolution (to measure temperatures).

Schuecker et al. (2004) described this method in a seminal paper on the Coma cluster. From a mosaic of XMM-Newton observations of Coma, spatial scales between 20 kpc and 2.8 Mpc could be sampled. Their maps show—superimposed on the normal radial density and pressure gradients—fluctuations up to 2 % in temperature and 5 % in intensity. These fluctuations are correlated and agree with the expected adiabatic fluctuations of the pressure. The spatial power spectrum of these pressure fluctuations is in agreement with a Kolmogorov-like spectrum. They found that between 10–25 % of the total intracluster pressure is in the form of turbulence.

More recently, Churazov et al. (2012) extended this work by also including higher spatial resolution observations of Coma with Chandra. Contrary to Schuecker et al. (2004), they consider density fluctuations. They find that the relative density fluctuations in Coma have amplitudes of 5 and 10 % at scales of about 30 and 300 kpc, respectively. They also consider several other explanations for the observed density fluctuations. For instance, at large scales, gravitational perturbations due to the large cD galaxies and entropy variations due to infalling cold gas may produce apparent density fluctuations.

7.1.1 Surface Brightness Profiles

In the cluster outskirts, models predict that the X-ray emission should become more clumpy due to the shallow gravitational potential and the ongoing strong cluster evolution associated, for instance, with infall of galaxies or groups towards the growing mass concentrations. One may define a cluster clumping factor C as

$$C^2 = \langle n^2 \rangle / \langle n \rangle^2, \tag{8}$$

with n the gas density and the brackets denoting averaging over volume.

Simionescu et al. (2011) used Suzaku observations of the outskirts of the Perseus cluster to determine the cluster properties around the virial radius and beyond. They find an apparent baryon fraction higher than the cosmic value at these large radii, which they reconcile by assuming a very clumpy medium at these distances.

Interestingly, Eckert et al. (2013), by combining pressure measurements from Planck with density profiles obtained from ROSAT, found a much lower clumping factor C of about 1.2 at R_{200}, compared to 3–4 by Simionescu et al. (2011). These differences may be associated to the precise way of averaging used in determining C.

7.1.2 Direct Measurements of Line Broadening

The most direct way to measure turbulence in clusters is to measure the spectral line broadening caused by the turbulence. Motion of the line-emitting ions will produce Doppler shifts. However, also thermal motion of the ions produces Doppler shifts. The width σ_T for thermal motion of the ions (for a Gaussian velocity profile $\sim \exp(-\Delta v^2 / 2\sigma_T^2)$) is given by

$$\sigma_T / c = 0.00103 T_{\text{ion}} / \sqrt{M}_{\text{ion}}, \tag{9}$$

where the ion temperature T_{ion} is expressed in keV and the ion mass M_{ion} in atomic units. Thus, for hydrogen and an ion temperature of 1 keV the thermal broadening is larger than the turbulent broadening if the turbulent velocity dispersion in the line of sight σ_v is less than 300 km s^{-1}. For the more relevant case of iron (atomic mass 56.25) the thermal velocity dispersion is always less than 200 km s^{-1} (Fig. 23), and hence turbulence is relatively easily detected provided that the detector has sufficient energy resolution to measure the line broadening.

At this moment the only instrument capable of measuring line broadening in clusters is the RGS reflection grating spectrometer on board of XMM-Newton. However, because this is a slit-less dispersive instrument, the spatial extent of a cluster gives an additional line broadening effect, and this must be modeled properly. Only some cool core clusters have sufficiently small size to allow for such a study. In addition, there is some uncertainty in the calibration of the instrumental line spread function. With these caveats in mind, Sanders

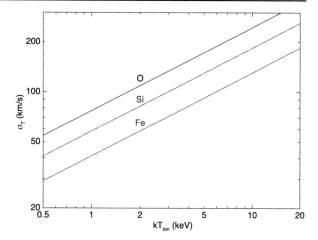

Fig. 23 Thermal velocity dispersion in the *line* of sight corresponding to the thermal motion of emitting ions, for oxygen silicon and iron ions, as a function of gas temperature

et al. (2011) studied a sample of 62 clusters, groups and individual ellipticals. They found upper limits for the turbulent velocity of 200 km s^{-1} for five of their studied cases, and one case (Klemola 44) with a measured broadening of 1500 km s^{-1}. Optical observations show this latter cluster to be a very disturbed system. For the sample as a whole, 15 sources appear to have less than 20 % of their thermal energy in the form of turbulence.

Clearly, better measurements are needed. ASTRO-H (expected launch 2015) will contain a calorimeter with about 6 eV spectral resolution (300 km s^{-1} at the iron lines), that does not suffer from the spatial blurring inherent to the RGS. Turbulent broadening of spectral lines is expected to be measurable down to 50–100 km s^{-1}, ultimately limited by the calibration of the instrument. Measuring cluster turbulence is one of the key scientific questions that this satellite will address.

While ASTRO-H has sufficient spectral resolution to measure the turbulence, its spatial resolution of the order of an arcmin will limit the amount of mapping that is possible. In the longer future, the proposed ESA mission Athena (launch 2028) will have a significantly higher spatial resolution, in addition to a spectral resolution of a few eV and a much higher effective area. This will allow to study turbulence also for a significant number of clusters at much higher redshifts. In addition, the spectral resolution will be sufficient to study line profiles in detail (e.g. Inogamov and Sunyaev 2003).

7.1.3 Resonance Scattering

Resonance scattering offers a unique tool to study turbulence in clusters. The hot plasma in clusters of galaxies is in general optically thin, which means that all radiation that it produces can freely escape. However, a few of the strongest resonance lines have an optical depth of the order of unity. Photons emitted in such lines can be scattered several times before they finally escape. For clusters of galaxies, the method was pioneered by Gilfanov et al. (1987), see Churazov et al. (2010) for a review. Essentially, line photons from resonance lines emitted from the center of the cluster towards the observer will be scattered out of the line of sight, but are not lost. This leads to an apparent dimming of the line towards the cluster center, but a brightening towards the outer parts. Because of the higher surface brightness near the cluster center, this dimming of resonance lines towards the center is the easiest to observe.

Fig. 24 Simulated Astro-H spectra from the core (0.5–1.5 arcmin annulus) of the Perseus Cluster for 100 ks observation. Points show the simulated spectra with and without resonant scattering for different levels of turbulence, parametrized through the effective Mach number $\mathcal{M}$. Note that 100 ks observation is sufficient to detect the resonant scattering signal even if gas motions are present. (Adapted from Zhuravleva et al. 2013)

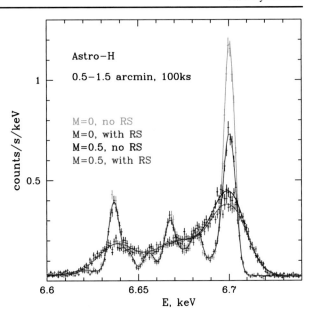

To make fully use of this effect, two lines of the same ion are needed, with different oscillator strength. Then from the radial profile of their line ratio (combined with information about the temperature and density profile of the cluster obtained from spectral modeling), the optical depth of the scattering line can be determined. This optical depth depends on the ratio of the turbulent to thermal energy density in the gas. Therefore this method directly probes the turbulence. Since the core of the cluster is typically much brighter than the outskirts, the optical depth is particularly sensitive to the radial component of the velocity field and small-scale eddies, while large-scale coherent motions tend to shift the line energy, but do not affect the optical depth (Zhuravleva et al. 2011).

One of the most promising ions to apply this method is Fe XXV. The 1s–2p and 1s–3p transitions of this ions have high and low oscillator strength, respectively. While BeppoSAX and ASCA observations of the Perseus cluster indicated the possible presence of a strong scattering effect and hence low levels of turbulence (Molendi et al. 1998; Akimoto et al. 1999), subsequent observations with XMM-Newton showed that the Fe XXV 1s–3p line is contaminated with nickel line emission; correcting for that results in insignificant resonance scattering hence the presence of substantial turbulence, with characteristic speeds of 0.36 times the sound speed and hence a turbulent pressure of about 10 % of the thermal pressure (Gastaldello and Molendi 2004; Churazov et al. 2004). Future ASTRO-H mission will have sufficient sensitivity and energy resolution to unambiguously identify signatures of the resonant scattering in the Fe XXV triplet (Fig. 24) in a 100 ks observations.

Another interesting diagnostic ion is Fe XVII. This has a strong resonance line at 15 Å, and by comparing this line with other Fe XVII lines, Xu et al. (2002) discovered resonance scattering in the elliptical NGC4636 using RGS data. This work was extended by Werner et al. (2009) using five giant elliptical halos, and using in addition to the RGS spectra also high-resolution images from Chandra. In four of these systems, the 15 Å line is suppressed relative to the other Fe XVII lines. For NGC4636 this leads to a turbulent velocity of less than $100 \, \mathrm{km \, s^{-1}}$ or less than 5 % for the turbulent to total pressure ratio.

de Plaa et al. (2012) obtained much deeper RGS spectra for two of the ellipticals studied by Werner et al. (2009). Interestingly, in these two systems the amount of turbulence appears to be rather high: in NGC5044 turbulent pressure constitutes at least 40 % of the total (with turbulent velocities between 320–720 $km\,s^{-1}$) and in NGC5813 turbulence contributes only 15–45 % to the total pressure, with velocities between 140–540 $km\,s^{-1}$. However, de Plaa et al. (2012) also point out that the atomic physics of Fe XVII is by no means undisputed, and some of these results depend on the adopted atomic parameters. Fe XVII still remains one of the most difficult and controversial ions, despite its relatively frequent occurrence in X-ray spectra.

The profiles of spectral lines that are affected by resonance scattering will be different from those of non-resonant lines. Because the optical depth in the line core is higher than that in the line wings, the core will be suppressed relative to the wings (e.g. Gilfanov et al. 1987; Shang and Oh 2013). The characteristic depression at the center of the line could therefore be used as a proxy for resonant scattering when observing bright cluster cores.

Apart from the distortions of the line intensity and shape, the resonant scattering causes polarization of the scattered line flux (Sazonov et al. 2002) at the level of 10–15 %. The polarization plane is expected to be perpendicular to the direction towards the cluster center. Gas motions will affect both the degree and the direction of polarization (Zhuravleva et al. 2010). The changes in the polarization signal are in particular sensitive to the gas motions perpendicular to the line of sight. This opens a principal possibility of measuring *transverse* component of the velocity field, once sensitive X-ray polarimeters with good spectral resolution become available.

8 Summary and Conclusions

Between the end of the 90's and the beginning of the 2000's, the X-ray satellites Chandra and XMM-Newton have definitely proven that galaxy clusters are extremely exciting physics laboratory for the characterization of complex plasma processes. In the same years, the development of deep radio observations of clusters have confirmed the presence of a non-thermal intracluster component (cosmic rays and magnetic fields) in the ICM, whose physical properties seem to be strongly connected to the dynamical state and evolutionary history of clusters. As presented in the previous sections, in the next years a new generation of X-ray and radio telescopes will allow us to do a further step forward in the characterization of the MHD physical processes that we expect to be shaping the ICM properties, but for which we are not yet able to make precise measurements.

Great advancements are expected in the direct measurements of turbulence thanks to the increased kinematic resolution of the X-ray satellites ASTRO-H first and Athena then. This will have a relevant impact on both astrophysical and cosmological cluster studies. As described in this paper, turbulent plasma motions are expected to play an important role in the AGN energy dissipation at the center of galaxy clusters and, together with merger induced shocks, in the acceleration of intracluster cosmic rays and in the amplification of magnetic fields. The greatest impact in the characterization of the non-thermal cluster components is expected to come from new and future radio interferometers (SKA in particular and, before, its precursors and pathfinders, such as JVLA, LOFAR, MWA, LWA, ASKAP, MeerKAT, . . . ; see Norris et al. 2013, and references therein). All together, these instruments will cover a broad spectral range, from a few tens of MHz to, possibly, ~ 15 GHz. Broadband low-frequency observations at arcsec resolution will be key to systematic searches of steep-spectrum sources, from radio ghosts at the center of clusters, to diffuse radio sources, expected to be hosted in a big fraction of low-mass and minor merging clusters (Cassano et al. 2012). Both total intensity and radio polarization observations from SKA and its pathfinders will allow us to access the magnetic field intensity and structure, as well as the acceleration mechanisms responsible for electron acceleration, within the intra- and, possibly, inter-cluster volume up to $z \sim 1$ (e.g. Bonafede et al. 2015; Ferrari et al. 2015; Govoni et al. 2015; Vazza et al. 2015).

Since random turbulent motions are expected to provide a pressure support to the ICM (thus affecting the measure of cluster masses based on the assumption of hydrostatic equilibrium), a precise quantification of the turbulence pressure is not only interesting from the astrophysical point of view, but also crucial for improving the constraints on cosmological parameters through cluster number counts. To this respect, the eROSITA satellite, to be launched in 2016, will survey the sky with unprecedented sensitivity and is expected to detect about 10^5 galaxy clusters down to 5×10^{13} $M_\odot/h$ and with a median redshift $z \sim 0.35$ (Pillepich et al. 2012).[2] Later on, the wide field imager onboard of Athena is expected to conduct blind field surveys, allowing to detect and characterize the mass and dynamical state of clusters and groups up to and beyond $z = 1$. To be noted that important developments are ongoing also in the (sub-)mm domain, allowing deep improvements in the observations of the Sunyaev-Zel'dovich effect (SZE).[3]

The gamma-ray observations reviewed above have placed stringent limits on the pressure contained in protons of energies above 100 MeV with hard energy spectra. The constraints on the energy density in non-thermal particles with steep enough spectra dominated by particles of energies well below 100 MeV remain to be established with the new low energy gamma-ray experiments which are under development (see e.g. von Ballmoos et al. 2012; Lebrun et al. 2014).

We can conclude that the excellent synergy between future X-ray satellites, together with the huge developments in the new generation of radio and mm telescopes, will allow us a breakthroughs in our understanding of the evolutionary physics of the intracluster plasma and in the exploitation of galaxy clusters as tools for cosmology.

Acknowledgements   We would like to thank the referee for useful comments and the ISSI staff for providing an inspiring atmosphere favorable for intense discussions. We thank Hiroki Akamatsu for providing us with Fig. 21 before publication. W. Forman acknowledges support from NASA contract NASA-03060 that funds the Chandra HRC project, the Chandra archive grant AR1-12007X, and the NASA observing grant GO2-13005X. SRON is financially supported by NWO (the Netherlands Organization for Scientific Research).

References

M. Ackermann, M. Ajello, A. Albert et al., Search for cosmic-ray-induced gamma-ray emission in galaxy clusters. Astrophys. J. **787**, 18 (2014). doi:10.1088/0004-637X/787/1/18

P.A.R. Ade, N. Aghanim, C. Armitage-Caplan, M. Arnaud, M. Ashdown, F. Atrio-Barandela, J. Aumont, H. Aussel, C. Baccigalupi et al., Planck 2013 results. XXIX. The Planck catalogue of Sunyaev-Zeldovich sources. Astron. Astrophys. **571**, 29 (2014). doi:10.1051/0004-6361/201321523

H. Akamatsu et al., Astron. Astrophys. (2014, in press)

F. Akimoto, A. Furuzawa, Y. Tawara, K. Yamashita, Iron K-line mapping of clusters of galaxies with the resonance scattering effect. Astron. Nachr. **320**, 283 (1999)

---

[2]A flat cosmology with $\Omega_\Lambda = 1 - \Omega_m$ and $h = 0.701$ is assumed here, see Pillepich et al. (2012).

[3]The SZE is the change in the apparent brightness of the Cosmic Microwave Background radiation towards a cluster of galaxies due to inverse Compton interaction between CMB photons and intracluster electrons., which provide a complementary and powerful tool for detecting clusters and for identifying pressure substructures in their atmospheres (e.g. Korngut et al. 2011; Reichardt et al. 2013; Hasselfield et al. 2013; Ade et al. 2014).

S.W. Allen, R.J.H. Dunn, A.C. Fabian, G.B. Taylor, C.S. Reynolds, The relation between accretion rate and jet power in X-ray luminous elliptical galaxies. Mon. Not. R. Astron. Soc. **372**, 21–30 (2006). doi:10.1111/j.1365-2966.2006.10778.x

S.W. Allen, A.E. Evrard, A.B. Mantz, Cosmological parameters from observations of galaxy clusters. Annu. Rev. Astron. Astrophys. **49**, 409–470 (2011). doi:10.1146/annurev-astro-081710-102514

E. Amato, P. Blasi, A kinetic approach to cosmic-ray-induced streaming instability at supernova shocks. Mon. Not. R. Astron. Soc. **392**, 1591–1600 (2009). doi:10.1111/j.1365-2966.2008.14200.x

N. Asai, N. Fukuda, R. Matsumoto, Three-dimensional MHD simulations of X-ray emitting subcluster plasmas in cluster of galaxies. Adv. Space Res. **36**, 636–642 (2005). doi:10.1016/j.asr.2005.04.041

Y. Ascasibar, M. Markevitch, The origin of cold fronts in the cores of relaxed galaxy clusters. Astrophys. J. **650**, 102–127 (2006). doi:10.1086/506508

A. Baldi, S. Ettori, S. Molendi, I. Balestra, F. Gastaldello, P. Tozzi, An XMM-Newton spatially-resolved study of metal abundance evolution in distant galaxy clusters. Astron. Astrophys. **537**, 142 (2012). doi:10.1051/0004-6361/201117836

A.M. Beck, K. Dolag, H. Lesch, P.P. Kronberg, Strong magnetic fields and large rotation measures in protogalaxies from supernova seeding. Mon. Not. R. Astron. Soc. **435**, 3575–3586 (2013). doi:10.1093/mnras/stt1549

A.R. Bell, Turbulent amplification of magnetic field and diffusive shock acceleration of cosmic rays. Mon. Not. R. Astron. Soc. **353**, 550–558 (2004). doi:10.1111/j.1365-2966.2004.08097.x

V.S. Berezinsky, P. Blasi, V.S. Ptuskin, Clusters of galaxies as storage room for cosmic rays. Astrophys. J. **487**, 529–535 (1997)

S. Bertone, C. Vogt, T. Enßlin, Magnetic field seeding by galactic winds. Mon. Not. R. Astron. Soc. **370**, 319–330 (2006). doi:10.1111/j.1365-2966.2006.10474.x

J. Binney, G. Tabor, Evolving cooling flows. Mon. Not. R. Astron. Soc. **276**, 663 (1995)

L. Bîrzan, D.A. Rafferty, P.E.J. Nulsen, B.R. McNamara, H.J.A. Röttgering, M.W. Wise, R. Mittal, The duty cycle of radio-mode feedback in complete samples of clusters. Mon. Not. R. Astron. Soc. **427**, 3468–3488 (2012). doi:10.1111/j.1365-2966.2012.22083.x

P. Blasi, S. Colafrancesco, Cosmic rays, radio halos and nonthermal X-ray emission in clusters of galaxies. Astropart. Phys. **12**, 169–183 (1999). doi:10.1016/S0927-6505(99)00079-1

H. Boehringer, W. Voges, A.C. Fabian, A.C. Edge, D.M. Neumann, A ROSAT HRI study of the interaction of the X-ray-emitting gas and radio lobes of NGC 1275. Mon. Not. R. Astron. Soc. **264**, 25–28 (1993)

H. Böhringer, N. Werner, X-ray spectroscopy of galaxy clusters: Studying astrophysical processes in the largest celestial laboratories. Astron. Astrophys. Rev. **18**, 127–196 (2010). doi:10.1007/s00159-009-0023-3

H. Bohringer, P.E.J. Nulsen, R. Braun, A.C. Fabian, The interaction of the radio halo of M87 with the cooling intracluster medium of the Virgo cluster. Mon. Not. R. Astron. Soc. **274**, 67–71 (1995)

H. Böhringer, K. Matsushita, E. Churazov, Y. Ikebe, Y. Chen, The new emerging model for the structure of cooling cores in clusters of galaxies. Astron. Astrophys. **382**, 804–820 (2002). doi:10.1051/0004-6361:20011708

A. Bonafede, F. Govoni, L. Feretti, M. Murgia, G. Giovannini, M. Brüggen, Fractional polarization as a probe of magnetic fields in the intra-cluster medium. Astron. Astrophys. **530**, 24 (2011). doi:10.1051/0004-6361/201016298

A. Bonafede, F. Vazza, M. Brüggen, T. Akahori, E. Carretti, S. Colafrancesco, L. Feretti, C. Ferrari, G. Giovannini, F. Govoni, M. Johnston-Hollitt, M. Murgia, L. Rudnick, A. Scaife, V. Vacca, Unravelling the origin of large-scale magnetic fields in galaxy clusters and beyond, in *Proceedings of Advancing Astrophysics with the Square Kilometre Array*, PoS(AASKA14)105 (2015)

H. Bondi, On spherically symmetrical accretion. Mon. Not. R. Astron. Soc. **112**, 195 (1952)

R.G. Bower, A.J. Benson, R. Malbon, J.C. Helly, C.S. Frenk, C.M. Baugh, S. Cole, C.G. Lacey, Breaking the hierarchy of galaxy formation. Mon. Not. R. Astron. Soc. **370**, 645–655 (2006). doi:10.1111/j.1365-2966.2006.10519.x

M. Brüggen, A. Bykov, D. Ryu, H. Röttgering, Magnetic fields, relativistic particles, and shock waves in cluster outskirts. Space Sci. Rev. **166**, 187–213 (2012). doi:10.1007/s11214-011-9785-9

G. Brunetti, T.W. Jones, Cosmic rays in galaxy clusters and their nonthermal emission. Int. J. Mod. Phys. D **23**, 30007 (2014). doi:10.1142/S0218271814300079

G. Brunetti, A. Lazarian, Acceleration of primary and secondary particles in galaxy clusters by compressible MHD turbulence: From radio haloes to gamma-rays. Mon. Not. R. Astron. Soc. **410**, 127–142 (2011). doi:10.1111/j.1365-2966.2010.17457.x

G. Brunetti, G. Setti, L. Feretti, G. Giovannini, Particle reacceleration in the Coma cluster: Radio properties and hard X-ray emission. Mon. Not. R. Astron. Soc. **320**, 365 (2001). doi:10.1046/j.1365-8711.2001.03978.x

E. Bulbul, R.K. Smith, M. Loewenstein, A new method to constrain supernova fractions using X-ray observations of clusters of galaxies. Astrophys. J. **753**, 54 (2012). doi:10.1088/0004-637X/753/1/54

A.M. Bykov, Particle acceleration and nonthermal phenomena in superbubbles. Space Sci. Rev. **99**, 317–326 (2001)

A.M. Bykov, Nonthermal particles and photons in starburst regions and superbubbles. Astron. Astrophys. Rev. **22**, 22–77 (2014). doi:10.1007/s00159-014-0077-8

A.M. Bykov, I. Toptygin, Particle kinetics in highly turbulent plasmas (renormalization and self-consistent field methods). Phys. Usp. **36**, 1020–1052 (1993). doi:10.1070/PU1993v036n11ABEH002179

A.M. Bykov, Y.A. Uvarov, Electron kinetics in collisionless shock waves. Sov. Phys. JETP **88**, 465–475 (1999). doi:10.1134/1.558817

A.M. Bykov, K. Dolag, F. Durret, Cosmological shock waves. Space Sci. Rev. **134**, 119–140 (2008). doi:10.1007/s11214-008-9312-9

A.M. Bykov, D.C. Ellison, M. Renaud, Magnetic fields in cosmic particle acceleration sources. Space Sci. Rev. **166**, 71–95 (2012). doi:10.1007/s11214-011-9761-4

A.M. Bykov, D.C. Ellison, S.M. Osipov, A.E. Vladimirov, Magnetic field amplification in nonlinear diffusive shock acceleration including resonant and non-resonant cosmic-ray driven instabilities. Astrophys. J. **789**, 137 (2014). doi:10.1088/0004-637X/789/2/137

R. Cassano, G. Brunetti, R.P. Norris, H.J.A. Röttgering, M. Johnston-Hollitt, M. Trasatti, Radio halos in future surveys in the radio continuum. Astron. Astrophys. **548**, 100 (2012). doi:10.1051/0004-6361/201220018

S.K. Chakrabarti, R. Rosner, S.I. Vainshtein, Possible role of massive black holes in the generation of galactic magnetic fields. Nature **368**, 434–436 (1994). doi:10.1038/368434a0

E. Churazov, W. Forman, C. Jones, H. Böhringer, Asymmetric, arc minute scale structures around NGC 1275. Astron. Astrophys. **356**, 788–794 (2000)

E. Churazov, M. Brüggen, C.R. Kaiser, H. Böhringer, W. Forman, Evolution of buoyant bubbles in M87. Astrophys. J. **554**, 261–273 (2001). doi:10.1086/321357

E. Churazov, R. Sunyaev, W. Forman, H. Böhringer, Cooling flows as a calorimeter of active galactic nucleus mechanical power. Mon. Not. R. Astron. Soc. **332**, 729–734 (2002). doi:10.1046/j.1365-8711.2002.05332.x

E. Churazov, W. Forman, C. Jones, H. Böhringer, XMM-Newton observations of the Perseus cluster. I. The temperature and surface brightness structure. Astrophys. J. **590**, 225–237 (2003). doi:10.1086/374923

E. Churazov, W. Forman, C. Jones, R. Sunyaev, H. Böhringer, XMM-Newton observations of the Perseus cluster—II. Evidence for gas motions in the core. Mon. Not. R. Astron. Soc. **347**, 29–35 (2004). doi:10.1111/j.1365-2966.2004.07201.x

E. Churazov, S. Sazonov, R. Sunyaev, W. Forman, C. Jones, H. Böhringer, Supermassive black holes in elliptical galaxies: Switching from very bright to very dim. Mon. Not. R. Astron. Soc. **363**, 91–95 (2005). doi:10.1111/j.1745-3933.2005.00093.x

E. Churazov, I. Zhuravleva, S. Sazonov, R. Sunyaev, Resonant scattering of X-ray emission lines in the hot intergalactic medium. Space Sci. Rev. **157**, 193–209 (2010). doi:10.1007/s11214-010-9685-4

E. Churazov, A. Vikhlinin, I. Zhuravleva, A. Schekochihin, I. Parrish, R. Sunyaev, W. Forman, H. Böhringer, S. Randall, X-ray surface brightness and gas density fluctuations in the Coma cluster. Mon. Not. R. Astron. Soc. **421**, 1123–1135 (2012). doi:10.1111/j.1365-2966.2011.20372.x

L.L. Cowie, J. Binney, Radiative regulation of gas flow within clusters of galaxies—A model for cluster X-ray sources. Astrophys. J. **215**, 723–732 (1977). doi:10.1086/155406

D.J. Croton, V. Springel, S.D.M. White, G. De Lucia, C.S. Frenk, L. Gao, A. Jenkins, G. Kauffmann, J.F. Navarro, N. Yoshida, The many lives of active galactic nuclei: Cooling flows, black holes and the luminosities and colours of galaxies. Mon. Not. R. Astron. Soc. **365**, 11–28 (2006). doi:10.1111/j.1365-2966.2005.09675.x

L.P. David, P.E.J. Nulsen, B.R. McNamara, W. Forman, C. Jones, T. Ponman, B. Robertson, M. Wise, A high-resolution study of the hydra a cluster with Chandra: Comparison of the core mass distribution with theoretical predictions and evidence for feedback in the cooling flow. Astrophys. J. **557**, 546–559 (2001). doi:10.1086/322250

F. de Gasperin, E. Orrú, M. Murgia et al., M 87 at metre wavelengths: The LOFAR picture. Astron. Astrophys. **547**, 56 (2012). doi:10.1051/0004-6361/201220209

S. de Grandi, S. Molendi, Metal abundances in the cool cores of galaxy clusters. Astron. Astrophys. **508**, 565 (2009). doi:10.1051/0004-6361/200912745

J. de Plaa, The origin of the chemical elements in cluster cores. Astron. Nachr. **334**, 416 (2013). doi:10.1002/asna.201211870

J. de Plaa, N. Werner, A.M. Bykov, J.S. Kaastra, M. Méndez, J. Vink, J.A.M. Bleeker, M. Bonamente, J.R. Peterson, Chemical evolution in Sérsic 159-03 observed with XMM-Newton. Astron. Astrophys. **452**, 397 (2006). doi:10.1051/0004-6361:20053864

J. de Plaa, N. Werner, J.A.M. Bleeker, J. Vink, J.S. Kaastra, M. Méndez, Constraining supernova models using the hot gas in clusters of galaxies. Astron. Astrophys. **465**, 345 (2007). doi:10.1051/0004-6361:20066382

J. de Plaa, I. Zhuravleva, N. Werner, J.S. Kaastra, E. Churazov, R.K. Smith, A.J.J. Raassen, Y.G. Grange, Estimating turbulent velocities in the elliptical galaxies NGC 5044 and NGC 5813. Astron. Astrophys. **539**, 34 (2012). doi:10.1051/0004-6361/201118404

R. della Ceca, R.E. Griffiths, T.M. Heckman, J.W. MacKenty, ASCA observations of starbursting dwarf galaxies: The case of NGC 1569. Astrophys. J. **469**, 662 (1996). doi:10.1086/177813

B. Dennison, Formation of radio halos in clusters of galaxies from cosmic-ray protons. Astrophys. J. Lett. **239**, 93–96 (1980). doi:10.1086/183300

T. Di Matteo, S.W. Allen, A.C. Fabian, A.S. Wilson, A.J. Young, Accretion onto the supermassive black hole in M87. Astrophys. J. **582**, 133–140 (2003). doi:10.1086/344504

K. Dolag, A.M. Bykov, A. Diaferio, Non-thermal processes in cosmological simulations. Space Sci. Rev. **134**, 311–335 (2008). doi:10.1007/s11214-008-9319-2

J. Donnert, K. Dolag, H. Lesch, E. Müller, Cluster magnetic fields from galactic outflows. Mon. Not. R. Astron. Soc. **392**, 1008–1021 (2009). doi:10.1111/j.1365-2966.2008.14132.x

J. Donnert, K. Dolag, G. Brunetti, R. Cassano, A. Bonafede, Radio haloes from simulations and hadronic models—I. The Coma cluster. Mon. Not. R. Astron. Soc. **401**, 47–54 (2010). doi:10.1111/j.1365-2966.2009.15655.x

Y. Dubois, R. Teyssier, Magnetised winds in dwarf galaxies. Astron. Astrophys. **523**, 72 (2010). doi:10.1051/0004-6361/200913014

R.J.H. Dunn, A.C. Fabian, Investigating AGN heating in a sample of nearby clusters. Mon. Not. R. Astron. Soc. **373**, 959–971 (2006). doi:10.1111/j.1365-2966.2006.11080.x

D. Eckert, S. Molendi, F. Vazza, S. Ettori, S. Paltani, The X-ray/SZ view of the virial region. I. Thermodynamic properties. Astron. Astrophys. **551**, 22 (2013). doi:10.1051/0004-6361/201220402

D. Eckert, S. Molendi, M. Owers, M. Gaspari, T. Venturi, L. Rudnick, S. Ettori, S. Paltani, F. Gastaldello, M. Rossetti, The stripping of a galaxy group diving into the massive cluster A2142. Astron. Astrophys. **570**, 119 (2014). doi:10.1051/0004-6361/201424259

T.A. Enßlin, S. Heinz, Radio and X-ray detectability of buoyant radio plasma bubbles in clusters of galaxies. Astron. Astrophys. **384**, 27–30 (2002). doi:10.1051/0004-6361:20020207

T. Enßlin, C. Pfrommer, F. Miniati, K. Subramanian, Cosmic ray transport in galaxy clusters: Implications for radio halos, gamma-ray signatures, and cool core heating. Astron. Astrophys. **527**, 99 (2011). doi:10.1051/0004-6361/201015652

S. Ettori, A.C. Fabian, Chandra constraints on the thermal conduction in the intracluster plasma of A2142. Mon. Not. R. Astron. Soc. **317**, 57–59 (2000). doi:10.1046/j.1365-8711.2000.03899.x

A.C. Fabian, Cooling flows in clusters of galaxies. Annu. Rev. Astron. Astrophys. **32**, 277–318 (1994). doi:10.1146/annurev.aa.32.090194.001425

A.C. Fabian, Observational evidence of active galactic nuclei feedback. Annu. Rev. Astron. Astrophys. **50**, 455–489 (2012). doi:10.1146/annurev-astro-081811-125521

A.C. Fabian, P.E.J. Nulsen, Subsonic accretion of cooling gas in clusters of galaxies. Mon. Not. R. Astron. Soc. **180**, 479–484 (1977)

A.C. Fabian, J.S. Sanders, S. Ettori, G.B. Taylor, S.W. Allen, C.S. Crawford, K. Iwasawa, R.M. Johnstone, P.M. Ogle, Chandra imaging of the complex X-ray core of the Perseus cluster. Mon. Not. R. Astron. Soc. **318**, 65–68 (2000). doi:10.1046/j.1365-8711.2000.03904.x

A.C. Fabian, J.S. Sanders, C.S. Crawford, C.J. Conselice, J.S. Gallagher, R.F.G. Wyse, The relationship between the optical Hα filaments and the X-ray emission in the core of the Perseus cluster. Mon. Not. R. Astron. Soc. **344**, 48–52 (2003). doi:10.1046/j.1365-8711.2003.06856.x

A.C. Fabian, J.S. Sanders, G.B. Taylor, S.W. Allen, C.S. Crawford, R.M. Johnstone, K. Iwasawa, A very deep Chandra observation of the Perseus cluster: Shocks, ripples and conduction. Mon. Not. R. Astron. Soc. **366**, 417–428 (2006). doi:10.1111/j.1365-2966.2005.09896.x

L. Feretti, G. Giovannini, F. Govoni, M. Murgia, Clusters of galaxies: Observational properties of the diffuse radio emission. Astron. Astrophys. Rev. **20**, 54 (2012). doi:10.1007/s00159-012-0054-z

L. Ferrarese, D. Merritt, A fundamental relation between supermassive black holes and their host galaxies. Astrophys. J. Lett. **539**, 9–12 (2000). doi:10.1086/312838

C. Ferrari, F. Govoni, S. Schindler, A.M. Bykov, Y. Rephaeli, Observations of extended radio emission in clusters. Space Sci. Rev. **134**, 93–118 (2008). doi:10.1007/s11214-008-9311-x

C. Ferrari, A. Dabbech, O. Smirnov et al., Non-thermal emission from galaxy clusters: Feasibility study with SKA1, in *Proceedings of Advancing Astrophysics with the Square Kilometre Array*, PoS(AASKA14)105 (2015)

W. Forman, P. Nulsen, S. Heinz, F. Owen, J. Eilek, A. Vikhlinin, M. Markevitch, R. Kraft, E. Churazov, C. Jones, Reflections of active galactic nucleus outbursts in the gaseous atmosphere of M87. Astrophys. J. **635**, 894–906 (2005). doi:10.1086/429746

W. Forman, C. Jones, E. Churazov, M. Markevitch, P. Nulsen, A. Vikhlinin, M. Begelman, H. Böhringer, J. Eilek, S. Heinz, R. Kraft, F. Owen, M. Pahre, Filaments, bubbles, and weak shocks in the gaseous atmosphere of M87. Astrophys. J. **665**, 1057–1066 (2007). doi:10.1086/519480

Y. Fujita, T. Matsumoto, K. Wada, Strong turbulence in the cool cores of galaxy clusters: Can tsunamis solve the cooling flow problem? Astrophys. J. Lett. **612**, 9–12 (2004). doi:10.1086/424483

J.M. Gabor, R. Davé, K. Finlator, B.D. Oppenheimer, How is star formation quenched in massive galaxies? Mon. Not. R. Astron. Soc. **407**, 749 (2010). doi:10.1111/j.1365-2966.2010.16961.x

E. Gallo, R.P. Fender, G.G. Pooley, A universal radio-X-ray correlation in low/hard state black hole binaries. Mon. Not. R. Astron. Soc. **344**, 60–72 (2003). doi:10.1046/j.1365-8711.2003.06791.x

M. Gaspari, M. Ruszkowski, P. Sharma, Cause and effect of feedback: Multiphase gas in cluster cores heated by AGN jets. Astrophys. J. **746**, 94 (2012). doi:10.1088/0004-637X/746/1/94

F. Gastaldello, S. Molendi, Ni abundance in the core of the Perseus cluster: An answer to the significance of resonant scattering. Astrophys. J. **600**, 670–680 (2004). doi:10.1086/379970

K. Gebhardt, R. Bender, G. Bower, A. Dressler, S.M. Faber, A.V. Filippenko, R. Green, C. Grillmair, L.C. Ho, J. Kormendy, T.R. Lauer, J. Magorrian, J. Pinkney, D. Richstone, S. Tremaine, A relationship between nuclear black hole mass and galaxy velocity dispersion. Astrophys. J. Lett. **539**, 13–16 (2000). doi:10.1086/312840

S. Ghizzardi, M. Rossetti, S. Molendi, Cold fronts in galaxy clusters. Astron. Astrophys. **516**, 32 (2010). doi:10.1051/0004-6361/200912496

S. Giacintucci, M. Markevitch, G. Brunetti, J.A. ZuHone, T. Venturi, P. Mazzotta, H. Bourdin, Mapping the particle acceleration in the cool core of the galaxy cluster RX J1720.1+2638. Astrophys. J. **795**, 73 (2014). doi:10.1088/0004-637X/795/1/73

M.R. Gilfanov, R.A. Syunyaev, E.M. Churazov, Radial brightness profiles of resonance X-ray lines in galaxy clusters. Sov. Astron. Lett. **13**, 3 (1987)

F. Govoni, M. Murgia, H. Xu, H. Li, M.L. Norman, L. Feretti, G. Giovannini, V. Vacca, Polarization of cluster radio halos with upcoming radio interferometers. Astron. Astrophys. **554**, 102 (2013). doi:10.1051/0004-6361/201321403

F. Govoni, M. Murgia, H. Xu et al., Cluster magnetic fields through the study of polarized radio halos, in *Proceedings of Advancing Astrophysics with the Square Kilometre Array*, PoS(AASKA14)105 (2015)

Y.G. Grange, J. de Plaa, J.S. Kaastra, N. Werner, F. Verbunt, F. Paerels, C.P. de Vries, The metal contents of two groups of galaxies. Astron. Astrophys. **531**, 15 (2011). doi:10.1051/0004-6361/201016187

S.F. Gull, K.J.E. Northover, Bubble model of extragalactic radio sources. Nature **244**, 80–83 (1973). doi:10.1038/244080a0

X. Guo, L. Sironi, R. Narayan, Non-thermal electron acceleration in low mach number collisionless shocks. I. Particle energy spectra and acceleration mechanism. Astrophys. J. **794**, 153 (2014a). doi:10.1088/0004-637X/794/2/153

X. Guo, L. Sironi, R. Narayan, Non-thermal electron acceleration in low mach number collisionless shocks. II. Firehose-mediated Fermi acceleration and its dependence on pre-shock conditions. Astrophys. J. **797**, 47 (2014b). doi:10.1088/0004-637X/797/1/47

M. Hasselfield, M. Hilton, T.A. Marriage et al., The Atacama cosmology telescope: Sunyaev-Zel'dovich selected galaxy clusters at 148 GHz from three seasons of data. J. Cosmol. Astropart. Phys. **7**, 8 (2013). doi:10.1088/1475-7516/2013/07/008

D.C. Hines, J.A. Eilek, F.N. Owen, Filaments in the radio lobes of M87. Astrophys. J. **347**, 713–726 (1989). doi:10.1086/168163

J. Hlavacek-Larrondo, A.C. Fabian, A.C. Edge, H. Ebeling, J.S. Sanders, M.T. Hogan, G.B. Taylor, Extreme AGN feedback in the MAssive cluster survey: A detailed study of X-ray cavities at $z > 0.3$. Mon. Not. R. Astron. Soc. **421**, 1360–1384 (2012). doi:10.1111/j.1365-2966.2011.20405.x

D.A. Howell, Type Ia supernovae as stellar endpoints and cosmological tools. Nat. Commun. **2**, 350 (2011). doi:10.1038/ncomms1344

S. Ichimaru, Bimodal behavior of accretion disks—Theory and application to Cygnus X-1 transitions. Astrophys. J. **214**, 840–855 (1977). doi:10.1086/155314

N.A. Inogamov, R.A. Sunyaev, Turbulence in clusters of galaxies and X-ray line profiles. Astron. Lett. **29**, 791–824 (2003). doi:10.1134/1.1631412

K. Iwamoto, F. Brachwitz, K. Nomoto, N. Kishimoto, H. Umeda, W.R. Hix, F.K. Thielemann, Nucleosynthesis in Chandrasekhar mass models for type IA supernovae and constraints on progenitor systems and burning- front propagation. Astrophys. J. Suppl. Ser. **125**, 439 (1999). doi:10.1086/313278

F.C. Jones, J.R. Jokipii, M.G. Baring, Charged-particle motion in electromagnetic fields having at least one ignorable spatial coordinate. Astrophys. J. **509**, 238–243 (1998). doi:10.1086/306480

J.S. Kaastra, A.M. Bykov, N. Werner, Non-Maxwellian electron distributions in clusters of galaxies. Astron. Astrophys. **503**, 373–378 (2009). doi:10.1051/0004-6361/200912492

A.I. Karakas, Updated stellar yields from asymptotic giant branch models. Mon. Not. R. Astron. Soc. **403**, 1413 (2010). doi:10.1111/j.1365-2966.2009.16198.x

C.F. Kennel, J.P. Edmiston, T. Hada, A quarter century of collisionless shock research, in *Washington DC American Geophysical Union Geophysical Monograph Series*, vol. 34 (1985), pp. 1–36

A.A. Kepley, S. Mühle, J. Everett, E.G. Zweibel, E.M. Wilcots, U. Klein, The role of the magnetic field in the interstellar medium of the post-starburst dwarf irregular galaxy NGC 1569. Astrophys. J. **712**, 536–557 (2010). doi:10.1088/0004-637X/712/1/536

U. Keshet, M. Markevitch, Y. Birnboim, A. Loeb, Dynamics and magnetization in galaxy cluster cores traced by X-ray cold fronts. Astrophys. J. Lett. **719**, 74–78 (2010). doi:10.1088/2041-8205/719/1/L74

U. Klein, E. Hummel, D.J. Bomans, U. Hopp, The synchrotron halo and magnetic field of NGC 4449. Astron. Astrophys. **313**, 396–404 (1996)

P.M. Korngut, S.R. Dicker, E.D. Reese, B.S. Mason, M.J. Devlin, T. Mroczkowski, C.L. Sarazin, M. Sun, J. Sievers, MUSTANG high angular resolution Sunyaev-Zel'dovich effect imaging of substructure in four galaxy clusters. Astrophys. J. **734**, 10 (2011). doi:10.1088/0004-637X/734/1/10

A.V. Kravtsov, S. Borgani, Formation of galaxy clusters. Annu. Rev. Astron. Astrophys. **50**, 353–409 (2012). doi:10.1146/annurev-astro-081811-125502

P.P. Kronberg, H. Lesch, U. Hopp, Magnetization of the intergalactic medium by primeval galaxies. Astrophys. J. **511**, 56–64 (1999). doi:10.1086/306662

S.M. Lea, The dynamics of the intergalactic medium in the vicinity of clusters of galaxies. Astrophys. J. **203**, 569–580 (1976). doi:10.1086/154113

F. Lebrun, R. Terrier, P. Laurent et al., The gamma cube: A novel concept of gamma-ray telescope, in *Society of Photo-Optical Instrumentation Engineers (SPIE) Conference Series*, vol. 9144 (2014). doi:10.1117/12.2055951

M. Lyutikov, Magnetic draping of merging cores and radio bubbles in clusters of galaxies. Mon. Not. R. Astron. Soc. **373**, 73–78 (2006). doi:10.1111/j.1365-2966.2006.10835.x

M. Machacek, A. Dosaj, W. Forman, C. Jones, M. Markevitch, A. Vikhlinin, A. Warmflash, R. Kraft, Infall of the elliptical galaxy NGC 1404 into the fornax cluster. Astrophys. J. **621**, 663–672 (2005). doi:10.1086/427548

M. Markevitch, Chandra observation of the most interesting cluster in the universe, in *The X-Ray Universe 2005*, ed. by A. Wilson. ESA Special Publication, vol. 604 (2006), p. 723

M. Markevitch, A. Vikhlinin, Shocks and cold fronts in galaxy clusters. Phys. Rep. **443**, 1–53 (2007). doi:10.1016/j.physrep.2007.01.001

M. Markevitch, T.J. Ponman, P.E.J. Nulsen, M.W. Bautz, D.J. Burke, L.P. David, D. Davis, R.H. Donnelly, W.R. Forman, C. Jones, J. Kaastra, E. Kellogg, D.W. Kim, J. Kolodziejczak, P. Mazzotta, A. Pagliaro, S. Patel, L. Van Speybroeck, A. Vikhlinin, J. Vrtilek, M. Wise, P. Zhao Chandra, Observation of Abell 2142: Survival of dense subcluster cores in a merger. Astrophys. J. **541**, 542–549 (2000). doi:10.1086/309470

M. Markevitch, A. Vikhlinin, P. Mazzotta, Nonhydrostatic gas in the core of the relaxed galaxy cluster A1795. Astrophys. J. Lett. **562**, 153–156 (2001). doi:10.1086/337973

M. Markevitch, A. Vikhlinin, W.R. Forman, A. High, Resolution picture of the intracluster gas, in *Matter and Energy in Clusters of Galaxies*, ed. by S. Bowyer, C.Y. Hwang. Astronomical Society of the Pacific Conference Series, vol. 301 (2003), p. 37

C.L. Martin, The impact of star formation on the interstellar medium in dwarf galaxies. II. The formation of galactic winds. Astrophys. J. **506**, 222–252 (1998). doi:10.1086/306219

F. Matteucci, F. Calura, Early chemical enrichment of the universe and the role of very massive population III stars. Mon. Not. R. Astron. Soc. **360**, 447 (2005). doi:10.1111/j.1365-2966.2005.08908.x

P. Mazzotta, M. Markevitch, A. Vikhlinin, W.R. Forman, L.P. David, L. van Speybroeck Chandra, Observation of RX J1720.1+2638: A nearly relaxed cluster with a fast-moving core? Astrophys. J. **555**, 205–214 (2001). doi:10.1086/321484

M. McDonald, M. Bayliss, B.A. Benson et al., A massive, cooling-flow-induced starburst in the core of a luminous cluster of galaxies. Nature **488**, 349–352 (2012). doi:10.1038/nature11379

M. McDonald, B.A. Benson, A. Vikhlinin et al., The growth of cool cores and evolution of cooling properties in a sample of 83 galaxy clusters at $0.3 < z < 1.2$ selected from the SPT-SZ survey. Astrophys. J. **774**, 23 (2013). doi:10.1088/0004-637X/774/1/23

B.R. McNamara, P.E.J. Nulsen, Heating hot atmospheres with active galactic nuclei. Annu. Rev. Astron. Astrophys. **45**, 117–175 (2007). doi:10.1146/annurev.astro.45.051806.110625

B.R. McNamara, M. Wise, P.E.J. Nulsen, L.P. David, C.L. Sarazin, M. Bautz, M. Markevitch, A. Vikhlinin, W.R. Forman, C. Jones, D.E. Harris, Chandra X-ray observations of the hydra a cluster: An interaction between the radio source and the X-ray-emitting gas. Astrophys. J. Lett. **534**, 135–138 (2000). doi:10. 1086/312662

B.R. McNamara, P.E.J. Nulsen, M.W. Wise, D.A. Rafferty, C. Carilli, C.L. Sarazin, E.L. Blanton, The heating of gas in a galaxy cluster by X-ray cavities and large-scale shock fronts. Nature **433**, 45–47 (2005). doi:10.1038/nature03202

G. Miley, The structure of extended extragalactic radio sources. Annu. Rev. Astron. Astrophys. **18**, 165–218 (1980). doi:10.1146/annurev.aa.18.090180.001121

E.T. Million, N. Werner, A. Simionescu, S.W. Allen, P.E.J. Nulsen, A.C. Fabian, H. Böhringer, J.S. Sanders, Feedback under the microscope—I. Thermodynamic structure and AGN-driven shocks in M87. Mon. Not. R. Astron. Soc. **407**, 2046–2062 (2010). doi:10.1111/j.1365-2966.2010.17220.x

S. Molendi, G. Matt, L.A. Antonelli, F. Fiore, R. Fusco-Femiano, J. Kaastra, C. Maccarone, C. Perola, How abundant is iron in the core of the Perseus cluster? Astrophys. J. **499**, 608–613 (1998)

R. Mushotzky, M. Loewenstein, K.A. Arnaud, T. Tamura, Y. Fukazawa, K. Matsushita, K. Kikuchi, I. Hatsukade, Measurement of the elemental abundances in four rich clusters of galaxies. I. Observations. Astrophys. J. **466**, 686 (1996). doi:10.1086/177541

R. Narayan, I. Yi, Advection-dominated accretion: A self-similar solution. Astrophys. J. Lett. **428**, 13–16 (1994). doi:10.1086/187381

R.P. Norris, J. Afonso, D. Bacon et al., Radio continuum surveys with square kilometre array pathfinders. Publ. Astron. Soc. Aust. **30**, 20 (2013). doi:10.1017/pas.2012.020

P. Nulsen, E. Roediger, Sloshing cold fronts & cluster G-modes, in *SnowCLUSTER 2013, Physics of Galaxy Clusters* (2013), p. 68

G.A. Ogrean, M. Brüggen, R.J. van Weeren, A. Burgmeier, A. Simionescu, No shock across part of a radio relic in the merging galaxy cluster ZwCl 2341.1+0000? Mon. Not. R. Astron. Soc. **443**, 2463–2474 (2014). doi:10.1093/mnras/stu1299

H. Omma, J. Binney, G. Bryan, A. Slyz, Heating cooling flows with jets. Mon. Not. R. Astron. Soc. **348**, 1105–1119 (2004). doi:10.1111/j.1365-2966.2004.07382.x

F.N. Owen, J.A. Eilek, N.E. Kassim, M87 at 90 centimeters: A different picture. Astrophys. J. **543**, 611–619 (2000). doi:10.1086/317151

A. Pedlar, H.S. Ghataure, R.D. Davies, B.A. Harrison, R. Perley, P.C. Crane, S.W. Unger, The radio structure of NGC1275. Mon. Not. R. Astron. Soc. **246**, 477 (1990)

C.B. Peres, A.C. Fabian, A.C. Edge, S.W. Allen, R.M. Johnstone, D.A. White, A ROSAT study of the cores of clusters of galaxies—I. Cooling flows in an X-ray flux-limited sample. Mon. Not. R. Astron. Soc. **298**, 416–432 (1998). doi:10.1046/j.1365-8711.1998.01624.x

J.R. Peterson, A.C. Fabian, X-ray spectroscopy of cooling clusters. Phys. Rep. **427**, 1–39 (2006). doi:10. 1016/j.physrep.2005.12.007

V. Petrosian, A. Bykov, Y. Rephaeli, Nonthermal radiation mechanisms. Space Sci. Rev. **134**, 191–206 (2008). doi:10.1007/s11214-008-9327-2

A. Pillepich, C. Porciani, T.H. Reiprich, The X-ray cluster survey with eRosita: Forecasts for cosmology, cluster physics and primordial non-Gaussianity. Mon. Not. R. Astron. Soc. **422**, 44–69 (2012). doi:10. 1111/j.1365-2966.2012.20443.x

F. Pizzolato, N. Soker, On the nature of feedback heating in cooling flow clusters. Astrophys. J. **632**, 821–830 (2005). doi:10.1086/444344

S. Planelles, D.R.G. Schleicher, A.M. Bykov, Large-scale structure formation: From the first non-linear objects to massive galaxy clusters. Space Sci. Rev. (2014). doi:10.1007/s11214-014-0045-7

D.A. Prokhorov, E.M. Churazov, Counting gamma rays in the directions of galaxy clusters. Astron. Astrophys. **567**, 93 (2014). doi:10.1051/0004-6361/201322454

D.A. Rafferty, B.R. McNamara, P.E.J. Nulsen, M.W. Wise, The feedback-regulated growth of black holes and bulges through gas accretion and starbursts in cluster central dominant galaxies. Astrophys. J. **652**, 216–231 (2006). doi:10.1086/507672

S.W. Randall, W.R. Forman, S. Giacintucci, P.E.J. Nulsen, M. Sun, C. Jones, E. Churazov, L.P. David, R. Kraft, M. Donahue, E.L. Blanton, A. Simionescu, N. Werner, Shocks and cavities from multiple outbursts in the galaxy group NGC 5813: A window to active galactic nucleus feedback. Astrophys. J. **726**, 86 (2011). doi:10.1088/0004-637X/726/2/86

M.J. Rees, The origin and cosmogonic implications of seed magnetic fields. Q. J. R. Astron. Soc. **28**, 197–206 (1987)

M.J. Rees, M.C. Begelman, R.D. Blandford, E.S. Phinney, Ion-supported tori and the origin of radio jets. Nature **295**, 17–21 (1982). doi:10.1038/295017a0

C.L. Reichardt, B. Stalder, L.E. Bleem et al., Galaxy clusters discovered via the Sunyaev-Zel'dovich effect in the first 720 square degrees of the South pole telescope survey. Astrophys. J. **763**, 127 (2013). doi:10. 1088/0004-637X/763/2/127

T.H. Reiprich, K. Basu, S. Ettori, H. Israel, L. Lovisari, S. Molendi, E. Pointecouteau, M. Roncarelli, Outskirts of galaxy clusters. Space Sci. Rev. **177**, 195–245 (2013). doi:10.1007/s11214-013-9983-8

E. Roediger, R.P. Kraft, W.R. Forman, P.E.J. Nulsen, E. Churazov, Kelvin-Helmholtz instabilities at the sloshing cold fronts in the Virgo cluster as a measure for the effective intracluster medium viscosity. Astrophys. J. **764**, 60 (2013). doi:10.1088/0004-637X/764/1/60

D. Romano, A.I. Karakas, M. Tosi, F. Matteucci, Quantifying the uncertainties of chemical evolution studies. II. Stellar yields. Astron. Astrophys. **522**, 32 (2010). doi:10.1051/0004-6361/201014483

M. Rossetti, D. Eckert, S. De Grandi, F. Gastaldello, S. Ghizzardi, E. Roediger, S. Molendi, Abell 2142 at large scales: An extreme case for sloshing? Astron. Astrophys. **556**, 44 (2013). doi:10.1051/0004-6361/201321319

D. Ryu, H. Kang, E. Hallman, T.W. Jones, Cosmological shock waves and their role in the large-scale structure of the universe. Astrophys. J. **593**, 599–610 (2003). doi:10.1086/376723

J.S. Sanders, A.C. Fabian, R.K. Smith, Constraints on turbulent velocity broadening for a sample of clusters, groups and elliptical galaxies using XMM-Newton. Mon. Not. R. Astron. Soc. **410**, 1797–1812 (2011). doi:10.1111/j.1365-2966.2010.17561.x

K. Sato, K. Tokoi, K. Matsushita, Y. Ishisaki, N.Y. Yamasaki, M. Ishida, T. Ohashi, Type Ia and II supernovae contributions to metal enrichment in the intracluster medium observed with Suzaku. Astrophys. J. **667**, 41 (2007). doi:10.1086/522031

S.Y. Sazonov, E.M. Churazov, R.A. Sunyaev, Polarization of resonance X-ray lines from clusters of galaxies. Mon. Not. R. Astron. Soc. **333**, 191–201 (2002). doi:10.1046/j.1365-8711.2002.05390.x

S.Y. Sazonov, J.P. Ostriker, R.A. Sunyaev, Quasars: The characteristic spectrum and the induced radiative heating. Mon. Not. R. Astron. Soc. **347**, 144–156 (2004). doi:10.1111/j.1365-2966.2004.07184.x

S.Y. Sazonov, J.P. Ostriker, L. Ciotti, R.A. Sunyaev, Radiative feedback from quasars and the growth of massive black holes in stellar spheroids. Mon. Not. R. Astron. Soc. **358**, 168–180 (2005). doi:10.1111/j.1365-2966.2005.08763.x

P. Schuecker, A. Finoguenov, F. Miniati, H. Böhringer, U.G. Briel, Probing turbulence in the Coma galaxy cluster. Astron. Astrophys. **426**, 387–397 (2004). doi:10.1051/0004-6361:20041039

K.M. Schure, A.R. Bell, L. O'C Drury, A.M. Bykov, Diffusive shock acceleration and magnetic field amplification. Space Sci. Rev. **173**, 491–519 (2012). doi:10.1007/s11214-012-9871-7

C. Shang, S.P. Oh, Disentangling resonant scattering and gas motions in galaxy cluster emission line profiles. Mon. Not. R. Astron. Soc. **433**, 1172–1184 (2013). doi:10.1093/mnras/stt790

A. Simionescu, S.W. Allen, A. Mantz, N. Werner, Y. Takei, R.G. Morris, A.C. Fabian, J.S. Sanders, P.E.J. Nulsen, M.R. George, G.B. Taylor, Baryons at the edge of the X-ray-brightest galaxy cluster. Science **331**, 1576 (2011). doi:10.1126/science.1200331

A. Simionescu, N. Werner, O. Urban, S.W. Allen, A.C. Fabian, J.S. Sanders, A. Mantz, P.E.J. Nulsen, Y. Takei, Large-scale motions in the Perseus galaxy cluster. Astrophys. J. **757**, 182 (2012). doi:10.1088/0004-637X/757/2/182

A. Stroe, R.J. van Weeren, H.T. Intema, H.J.A. Röttgering, M. Brüggen, M. Hoeft, Discovery of spectral curvature in the shock downstream region: CIZA J2242.8+5301. Astron. Astrophys. **555**, 110 (2013). doi:10.1051/0004-6361/201321267

G. Tabor, J. Binney, Elliptical galaxy cooling flows without mass drop-out. Mon. Not. R. Astron. Soc. **263**, 323 (1993)

P.A. Thomas, A.C. Fabian, K.A. Arnaud, W. Forman, C. Jones, The prevalence of cooling flows in early-type galaxies. Mon. Not. R. Astron. Soc. **222**, 655–672 (1986)

E.R. Tittley, M. Henriksen, Cluster mergers, core oscillations, and cold fronts. Astrophys. J. **618**, 227–236 (2005). doi:10.1086/425952

R.A. Treumann, Fundamentals of collisionless shocks for astrophysical application, 1. Non-relativistic shocks. Astron. Astrophys. Rev. **17**, 409–535 (2009). doi:10.1007/s00159-009-0024-2

T. Tsujimoto, K. Nomoto, Y. Yoshii, M. Hashimoto, S. Yanagida, F.K. Thielemann, Relative frequencies of type Ia and type II supernovae in the chemical evolution of the galaxy, LMC and SMC. Mon. Not. R. Astron. Soc. **277**, 945 (1995)

V. Vacca, M. Murgia, F. Govoni, L. Feretti, G. Giovannini, E. Orrú, A. Bonafede, The intracluster magnetic field power spectrum in Abell 665. Astron. Astrophys. **514**, 71 (2010). doi:10.1051/0004-6361/200913060

R.J. van Weeren, H.J.A. Röttgering, M. Brüggen, M. Hoeft, Particle acceleration on megaparsec scales in a merging galaxy cluster. Science **330**, 347 (2010). doi:10.1126/science.1194293

R.J. van Weeren, H.J.A. Röttgering, D.A. Rafferty et al., First LOFAR observations at very low frequencies of cluster-scale non-thermal emission: The case of Abell 2256. Astron. Astrophys. **543**, 43 (2012). doi:10.1051/0004-6361/201219154

E. Vangioni, J. Silk, K.A. Olive, B.D. Fields, Cosmic chemical evolution with an early population of intermediate-mass stars. Mon. Not. R. Astron. Soc. **413**, 2987 (2011). doi:10.1111/j.1365-2966.2011. 18372.x

F. Vazza, C. Ferrari, A. Bonafede, M. Brüggen, C. Gheller, R. Braun, S. Brown, Filaments of the radio cosmic web: Opportunities and challenges for SKA, in *Proceedings of Advancing Astrophysics with the Square Kilometre Array* PoS(AASKA14)105 (2015)

A. Vikhlinin, M. Markevitch, S.S. Murray, A moving cold front in the intergalactic medium of A3667. Astrophys. J. **551**, 160–171 (2001a). doi:10.1086/320078

A. Vikhlinin, M. Markevitch, S.S. Murray Chandra, Estimate of the magnetic field strength near the cold front in A3667. Astrophys. J. Lett. **549**, 47–50 (2001b). doi:10.1086/319126

H.J. Völk, F.A. Aharonian, D. Breitschwerdt, The nonthermal energy content and gamma-ray emission of starburst galaxies and clusters of galaxies. Space Sci. Rev. **75**, 279–297 (1996). doi:10.1007/BF00195040

P. von Ballmoos, J. Alvarez, N. Barrière et al., A DUAL mission for nuclear astrophysics. Exp. Astron. **34**, 583–622 (2012). doi:10.1007/s10686-011-9286-6

N. Werner, H. Böhringer, J.S. Kaastra, J. de Plaa, A. Simionescu, J. Vink, XMM-Newton high-resolution spectroscopy reveals the chemical evolution of M 87. Astron. Astrophys. **459**, 353 (2006a). doi:10. 1051/0004-6361:20065678

N. Werner, J. de Plaa, J.S. Kaastra, J. Vink, J.A.M. Bleeker, T. Tamura, J.R. Peterson, F. Verbunt, XMM-Newton spectroscopy of the cluster of galaxies 2A 0335+096. Astron. Astrophys. **449**, 475 (2006b). doi:10.1051/0004-6361:20053868

N. Werner, F. Durret, T. Ohashi, S. Schindler, R.P.C. Wiersma, Observations of metals in the intra-cluster medium. Space Sci. Rev. **134**, 337 (2008). doi:10.1007/s11214-008-9320-9

N. Werner, I. Zhuravleva, E. Churazov, A. Simionescu, S.W. Allen, W. Forman, C. Jones, J.S. Kaastra, Constraints on turbulent pressure in the X-ray haloes of giant elliptical galaxies from resonant scattering. Mon. Not. R. Astron. Soc. **398**, 23–32 (2009). doi:10.1111/j.1365-2966.2009.14860.x

N. Werner, A. Simionescu, E.T. Million, S.W. Allen, P.E.J. Nulsen, A. von der Linden, S.M. Hansen, H. Böhringer, E. Churazov, A.C. Fabian, W.R. Forman, C. Jones, J.S. Sanders, G.B. Taylor, Feedback under the microscope-II. Heating, gas uplift and mixing in the nearest cluster core. Mon. Not. R. Astron. Soc. **407**, 2063–2074 (2010). doi:10.1111/j.1365-2966.2010.16755.x

N. Werner, O. Urban, A. Simionescu, S.W. Allen, A uniform metal distribution in the intergalactic medium of the Perseus cluster of galaxies. Nature **502**, 656–658 (2013). doi:10.1038/nature12646

H. Xu, S.M. Kahn, J.R. Peterson, E. Behar, F.B.S. Paerels, R.F. Mushotzky, J.G. Jernigan, A.C. Brinkman, K. Makishima, High-resolution observations of the elliptical galaxy NGC 4636 with the reflection grating spectrometer on board XMM-Newton. Astrophys. J. **579**, 600–606 (2002). doi:10.1086/342828

H. Xu, H. Li, D.C. Collins, S. Li, M.L. Norman, Evolution and distribution of magnetic fields from active galactic nuclei in galaxy clusters. I. The effect of injection energy and redshift. Astrophys. J. **725**, 2152–2165 (2010). doi:10.1088/0004-637X/725/2/2152

H. Xu, H. Li, D.C. Collins, S. Li, M.L. Norman, Evolution and distribution of magnetic fields from active galactic nuclei in galaxy clusters. II. The effects of cluster size and dynamical state. Astrophys. J. **739**, 77 (2011). doi:10.1088/0004-637X/739/2/77

H. Xu, F. Govoni, M. Murgia, H. Li, D.C. Collins, M.L. Norman, R. Cen, L. Feretti, G. Giovannini, Comparisons of cosmological magnetohydrodynamic galaxy cluster simulations to radio observations. Astrophys. J. **759**, 40 (2012). doi:10.1088/0004-637X/759/1/40

I.V. Zhuravleva, E.M. Churazov, S.Y. Sazonov, R.A. Sunyaev, W. Forman, K. Dolag, Polarization of X-ray lines from galaxy clusters and elliptical galaxies—A way to measure the tangential component of gas velocity. Mon. Not. R. Astron. Soc. **403**, 129–150 (2010). doi:10.1111/j.1365-2966.2009.16148.x

I.V. Zhuravleva, E.M. Churazov, S.Y. Sazonov, R.A. Sunyaev, K. Dolag, Resonant scattering in galaxy clusters for anisotropic gas motions on various spatial scales. Astron. Lett. **37**, 141–153 (2011). doi:10. 1134/S1063773711010087

I. Zhuravleva, E. Churazov, R. Sunyaev, S. Sazonov, S.W. Allen, N. Werner, A. Simionescu, S. Konami, T. Ohashi, Resonant scattering in the Perseus cluster: Spectral model for constraining gas motions with Astro-H. Mon. Not. R. Astron. Soc. **435**, 3111–3121 (2013). doi:10.1093/mnras/stt1506

I. Zhuravleva, E. Churazov, A.A. Schekochihin, S.W. Allen, P. Arévalo, A.C. Fabian, W.R. Forman, J.S. Sanders, A. Simionescu, R. Sunyaev, A. Vikhlinin, N. Werner, Turbulent heating in galaxy clusters brightest in X-rays. Nature **515**, 85–87 (2014). doi:10.1038/nature13830

J.A. ZuHone, M. Markevitch, R.E. Johnson, Stirring up the pot: Can cooling flows in galaxy clusters be quenched by gas sloshing? Astrophys. J. **717**, 908–928 (2010). doi:10.1088/0004-637X/717/2/908

J.A. ZuHone, M. Markevitch, D. Lee, Sloshing of the magnetized cool gas in the cores of galaxy clusters. Astrophys. J. **743**, 16 (2011). doi:10.1088/0004-637X/743/1/16

J.A. ZuHone, M. Markevitch, G. Brunetti, S. Giacintucci, Turbulence and radio mini-halos in the sloshing cores of galaxy clusters. Astrophys. J. **762**, 78 (2013a). doi:10.1088/0004-637X/762/2/78

J.A. ZuHone, M. Markevitch, M. Ruszkowski, D. Lee, Cold fronts and gas sloshing in galaxy clusters with anisotropic thermal conduction. Astrophys. J. **762**, 69 (2013b). doi:10.1088/0004-637X/762/2/69

J.A. ZuHone, M.W. Kunz, M. Markevitch, J.M. Stone, V. Biffi, The effect of anisotropic viscosity on cold fronts in galaxy clusters. Astrophys. J. **798**, 90 (2015). doi:10.1088/0004-637X/798/2/90

DOI 10.1007/978-1-4939-3547-5_6
Reprinted from *Space Science Reviews* Journal, DOI 10.1007/s11214-014-0062-6

Supernova Remnants Interacting with Molecular Clouds: X-Ray and Gamma-Ray Signatures

**Patrick Slane · Andrei Bykov · Donald C. Ellison ·
Gloria Dubner · Daniel Castro**

Received: 8 March 2014 / Accepted: 17 June 2014 / Published online: 9 July 2014
© Springer Science+Business Media Dordrecht 2014

Abstract The giant molecular clouds (MCs) found in the Milky Way and similar galaxies
play a crucial role in the evolution of these systems. The supernova explosions that mark
the death of massive stars in these regions often lead to interactions between the supernova
remnants (SNRs) and the clouds. These interactions have a profound effect on our understanding of SNRs. Shocks in SNRs should be capable of accelerating particles to cosmic
ray (CR) energies with efficiencies high enough to power Galactic CRs. X-ray and γ-ray
studies have established the presence of relativistic electrons and protons in some SNRs and
provided strong evidence for diffusive shock acceleration as the primary acceleration mechanism, including strongly amplified magnetic fields, temperature and ionization effects on
the shock-heated plasmas, and modifications to the dynamical evolution of some systems.

P. Slane (✉)
Harvard-Smithsonian Center for Astrophysics, 60 Garden St., Cambridge, MA 02138, USA
e-mail: slane@cfa.harvard.edu

A. Bykov
Ioffe Institute for Physics and Technology, 194021 St. Petersburg, Russia
e-mail: byk@astro.ioffe.ru

A. Bykov
St. Petersburg State Polytechnical University, St. Petersburg, Russia

D.C. Ellison
Physics Department, North Carolina State University, Box 8202, Raleigh, NC 27695, USA
e-mail: don_ellison@ncsu.edu

G. Dubner
Instituto de Astronomía y Física del Espacio (IAFE), UBA-CONICET, CC 67, Suc. 28,
1428 Buenos Aires, Argentina
e-mail: gdubner@iafe.uba.ar

D. Castro
MIT-Kavli Center for Astrophysics and Space Research, 77 Massachusetts Avenue, Cambridge,
MA 02139, USA
e-mail: castro@mit.edu

 Springer

Because protons dominate the overall energetics of the CRs, it is crucial to understand this hadronic component even though electrons are much more efficient radiators and it can be difficult to identify the hadronic component. However, near MCs the densities are sufficiently high to allow the γ-ray emission to be dominated by protons. Thus, these interaction sites provide some of our best opportunities to constrain the overall energetics of these particle accelerators. Here we summarize some key properties of interactions between SNRs and MCs, with an emphasis on recent X-ray and γ-ray studies that are providing important constraints on our understanding of cosmic rays in our Galaxy.

Keywords Supernova remnants · Molecular clouds · X-rays · Gamma-rays

1 Introduction

Within star-forming galaxies, cold molecular material forms complex structures spanning a large range of spatial scales and densities, from giant molecular clouds (MCs) spanning tens of parsecs down to compact self-gravitating cores from which stars may form. For the most massive stars, their short lifetimes will not provide sufficient time to stray far from the molecular environments in which they were born before they undergo supernova explosions. We thus expect to find many supernova remnants (SNRs) associated with core-collapse supernovae to evolve in regions with dense MCs. Interactions with MCs can strongly impact the properties and long-term evolution of SNRs, and the distinct signatures of SNR/MC interactions can place strong constraints on this process, and on the properties of the evolving plasma. These interactions can drive shocks into the molecular material, providing distinct signatures of the interaction, and can play a role in triggering new star formation. In addition, the presence of dense gas from MC regions can play a crucial role in revealing the presence of energetic ions accelerated by SNR shocks, through the production of γ-rays.

The evolution of an SNR is arguably most readily investigated through X-ray observations. The rapid shocks heat the ambient medium and supernova (SN) ejecta to temperatures of $\sim 10^7$ K, producing thermal bremsstrahlung and line emission from which the density, ionization state, and abundances of the shocked gas can be constrained. Comparison of these derived properties with models for SNR evolution produce some of the most detailed information about the progenitor systems and their environments. At the same time, efficient particle acceleration often occurs in fast SNR shocks, resulting in X-ray synchrotron radiation in the compressed and amplified magnetic fields. The relativistic protons and electrons produced in this acceleration process can produce γ-ray emission. In particular, emission associated with the decay of neutral pions produced in proton-proton collisions can be significantly enhanced in environments with high densities, such as those in which SNRs are interacting with MCs. Thus, X-ray and γ-ray signatures from SNRs carry particularly important information on interactions with MCs.

2 Observational Signatures of SNR/MC Interactions

Resolved studies of SNRs and their associated clouds are most easily carried out for systems within the Milky Way, though with the increasing capacity of new observational facilities, the studies are rapidly extending to our neighbor galaxies, the Magellanic Clouds. In our Galaxy, molecular gas clouds account for less than one percent of the volume of the interstellar medium (ISM), yet as they are the densest part of the medium, they represent roughly one-half of the total gas mass interior to the Solar circle. This mass is distributed predominantly along the spiral arms and within a narrow midplane with a scale-height $Z \sim 50$–75 pc, much thinner than the atomic or ionized gas (Cox 2005). Most of the molecular gas ($\sim$90 %) appears to be in massive structures distributed in large clumps (Giant Molecular Clouds, GMCs), filaments and condensed clumps (which are the cradle for future stars). As mentioned above, most core-collapse SNe are located close to molecular concentrations, their birth places. Therefore a large percentage of the few hundreds of observed Galactic SNRs is expected to be physically related to, and maybe interacting with, MCs.

However, to unambiguously establish whether an SNR is physically associated with a MC, removing confusion introduced by unrelated gas along the line of sight is not trivial. Several distinct criteria can be used to demonstrate a possible SNR/MC interaction. Morphological traces along the periphery of the SNRs such as arcs of gas surrounding parts of the SNR or indentations in the SNR outer border encircling dense gas concentrations can be observed in images of SNRs (see Sect. 3.1.1). Usually such features indicate that a dense external cloud is disturbing an otherwise spherically symmetric shock expansion. These initial signatures need to be confirmed with more convincing, though more rare, features like broadenings, wings, or asymmetries in the molecular line spectra (Frail and Mitchell 1998), high ratios between lines of different excitation state (Seta et al. 1998), detection of near infrared H_2 or [Fe II] lines (e.g., Reach et al. 2005), peculiar infrared colors (Reach et al. 2006; Castelletti et al. 2011), or the presence of OH (1720) MHz masers (e.g., Frail et al. 1996; Green et al. 1997; Claussen et al. 1997; Hewitt et al. 2008), the most powerful tool to diagnose SNR/MC interactions. The fulfillment of one or more of these different criteria can be used to propose the existence of "definitive", "probable" or "possible" physical interactions between an SNR and a neighboring cloud.

Once the association between an SNR and a molecular feature is firmly established, it serves to provide an independent estimate for the distance to the SNR through the observed Doppler shift of the line centroid and by applying a circular rotation model for the Galaxy.[1] Based on a combination of different techniques, Jiang et al. (2010) presented a list of $\sim$70 Galactic SNRs suggested to be physically interacting with neighboring MCs, of which 34 cases are confirmed on the basis of simultaneous fulfillment of various criteria, 11 are probable, and 25 are classified as possible and deserve more studies.

The basic tool used to investigate cases of SNR/MC interactions is the survey of the interstellar medium in a field around the SNR using different spectral lines, from the cold, atomic hydrogen emitting at λ 21 cm to the dense shielded regions of molecular hydrogen (that constitutes at least 99 % of the molecular gas in the Milky Way) emitting in the millimeter and infrared ranges. The most widely used proxy to track down molecular gas is CO. This molecule has a nonzero dipole moment, and radiates much more efficiently than the abundant H_2 (with no dipole moment) and can be detected easily. However, such an indirect tracer can be biased, since there can still be CO-dark H_2 gas. Recent studies conducted with the Herschel Space Observatory revealed that the reservoir of molecular gas in our Galaxy can be hugely underestimated when it is traced with traditional methods (Pineda et al. 2013). Through the detection of ionized carbon [C II] at 158 μm one can trace the envelope of evolved clouds as well as clouds that are in transition between atomic and molecular. This might help to solve the question about the real proximity between clouds and the CR accelerator in some SNRs.

---

[1]The drawback of this method is that Galactic circular rotation is an approximation only valid for gas close to the Galactic disk, and for sources located inside the solar circle there is an ambiguity of two different distances producing the same radial velocity.

In what follows we discuss different tools for exploring interactions between SNRs and MCs.

2.1 HI Emission

Since the discovery of the λ 21 cm line of atomic hydrogen more than 60 years ago, it has been shown to be the perfect tool for investigating the distribution and kinematics of the interstellar medium, revealing Galactic arms, large concentrations, arcs, and bubbles. It is, then, the basic observational resource to explore the environs of SNRs searching for candidate structures with which the SNRs may be interacting. However, because of the high abundance and ubiquity, confusion with unrelated gas along the line of sight dominates, and the detection of an HI candidate structure needs to be confirmed with other indicators that reinforce the hypothesis of association. The HI can be studied either in emission or in absorption, the latter being a very effective tool for constraining the distance to the SNR. In addition, the HI mapping of large fields around SNRs is an excellent tool to explore the history of the precursor star, for example by detecting large wind-blown bubbles around the SNRs, and to estimate the gas density of the medium where the blast wave expands.

Numerous HI studies around SNRs have been carried out using single-dish and interferometric radio telescopes. Some examples are the results presented on the Lupus Loop (Colomb and Dubner 1982), G296.5+10.0 (Dubner et al. 1986), Vela (Dubner et al. 1998b), W50-SS 443 (Dubner et al. 1998a), Tycho (Reynoso et al. 1999), SN 1006 (Dubner et al. 2002), W28 (Velázquez et al. 2002), Puppis A (Reynoso et al. 2003), Kes 75 (Leahy and Tian 2008), IC 443 (Lee et al. 2008), and several Southern Galactic SNRs (Koo et al. 2004). Recent observations also indicate the existence of shells and super-shells around SNRs (Park et al. 2013), which are particularly useful as they permit the reconstruction of the history of the SNR progenitor, providing hints on the nature of "dark" γ-ray sources (e.g., Gabányi et al. 2013).

2.2 Molecular Emission

As mentioned above CO studies are the most widely used tool to analyze distribution and kinematics of cold, dense clouds with high molecular content. ^{12}CO in its different excitation states and ^{13}CO lines have been surveyed over most of the sky and public data are available from the classical survey "The Milky Way in Molecular Clouds" by Dame et al. (2001),[2] the "FCRAO CO Survey of the Outer Galaxy" by Heyer et al. (1998),[3] the "Galactic Ring Survey" by Jackson et al. (2006),[4] or the new "MOPRA Southern Galactic Plane CO Survey" by Burton et al. (2013), among others. These public databases are an excellent starting point to search for possible SNR/MC interactions. Additionally, dedicated studies using different facilities and in different molecular transitions have been conducted towards many Galactic SNRs. The SNR IC443 is a textbook case to analyze shock chemistry, from the early work by Denoyer and Frerking (1981), to tens of works investigating the chemical and physical transformations introduced by the strong SNR shocks on the surrounding MCs (e.g., White et al. 1987; Burton et al. 1988; Wang and Scoville 1992; van Dishoeck et al. 1993). More recently, Castelletti et al. (2011) showed a comparison between very high energy γ-ray emission as detected with VERITAS (Acciari et al. 2009),

---

[2]http://www.cfa.harvard.edu/mmw/MilkyWayinMolClouds.html.

[3]http://www.astro.umass.edu/~fcrao/telescope/2quad.html.

[4]http://www.bu.edu/galacticring/.

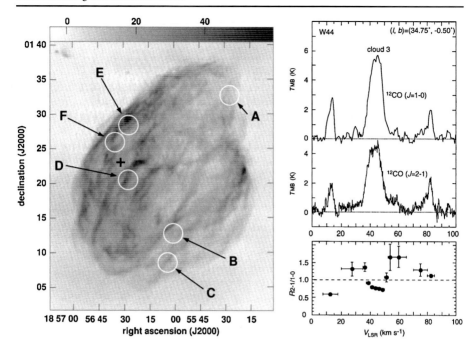

Fig. 1 *Left*: VLA image of W44. *Circles* indicate the positions of observed OH maser emission, and *cross* indicates the position from which the CO profile at the right is taken. *Right*: CO line profile from Cloud C in field of W44, illustrating broad wings produced by an interaction between the cloud and the SNR. [*Left panel* from Hoffman et al. 2005. *Right panel* from Seta et al. 1998. Reproduced by permission of the AAS]

with ^{12}CO $J = 1 - 0$ integrated emission (Zhang et al. 2010) revealing for the first time the excellent concordance between emissions in both regimes.

An important study carried out by Seta et al. (1998) proposed that an enhanced ^{12}CO($J = 2 - 1$)/^{12}CO($J = 1 - 0$) ratio in the line wings is a clear signature of physical interaction. Ratios are ≥ 3 in IC443, and higher than ~ 1 in W44, which also shows broadened line profiles indicating disrupted molecular material (Fig. 1, right).

Huang and Thaddeus (1986) carried out a search for possible SNR/MC associations in the outer Galaxy, within a limited region of the northern hemisphere, finding about 13 possible associations based on spatial coincidences (out of 26 SNRs investigated). More recently, Chen et al. (2014) reported CO observations towards a number of SNRs possibly interacting with MCs.

2.3 Masers

OH (1720 MHz) masers have long been recognized as signposts for SNR/MC interactions (e.g., Frail et al. 1994). This maser line is inverted through collisions with H_2 behind non-dissociative C-type shocks propagating into MCs (Frail and Mitchell 1998; Reach and Rho 1999). The conditions required for the maser formation are rather narrow, with temperatures in the range 50–125 K, densities between $n = 10^3$–10^5 cm^{-3}, and OH column densities of 10^{16}–10^{17} cm^{-2} (Lockett et al. 1999). The large column requirement exceeds what can be produced by the slow shocks, indicating that an additional contribution is required to dissociate water into OH (Wardle 1999), possibly from X-rays from the hot SNR shell or from cosmic rays that may have been accelerated by the SNR. As the physical conditions needed to pump this maser are so strict, it has to be noted that their presence is sufficient to demonstrate interaction, but its absence does not rule it out.

To date, OH (1720 MHz) masers have been found in $\sim$24 SNRs, or 10 % of the known SNRs in our Galaxy (Brogan et al. 2013). As summarized by Brogan et al. (2013), SNR OH (1720 MHz) masers have a large maser spot size, narrow and simple line profiles, low levels of circular polarization (less than 10 %), and low magnetic field strength.

The location of masers show, in general, very good coincidence with density/shock tracers. In the case of W44 (Frail et al. 1996; Claussen et al. 1997; Hoffman et al. 2005), there is a strong correlation between the morphology of the molecular gas and the relativistic gas traced by synchrotron emission at centimeter wavelengths.

In addition to providing a clear indication of SNR/MC interactions, OH (1720 MHz) masers also permit an independent estimate for the kinematic distances to the clouds, and thus for the remnants. Also, and very important, they provide the only means of directly observing the magnetic field strength using the Zeeman effect (Brogan et al. 2000, 2013).

2.4 Radiative Shocks

The evolution of supernova shells colliding with MCs differs from SNRs expanding in a pre-supernova wind or of those expanding in a homogeneous uniform ISM (Chevalier 1999). Observations have revealed that MCs are complex structures with a hierarchical structure of dense clumps embedded in an inter-clump gas of moderate density $\sim$10 H atoms cm^{-3}. The volume filling factor of the dense clumps is typically a few percent and the clump mass ranges from a fraction of $M_\circ$ to thousands of $M_\circ$. The total mass in the clumps can be comparable or more than the mass of the inter-clump gas.

The dense MC environment results in an early end of the adiabatic phase of the SNR expansion. An SNR leaves the adiabatic phase and enters the *pressure driven snowplow* (PDS) stage which eventually results in the formation of a radiative shell at an age estimated by Cioffi et al. (1988) to be $\tau_{sf} = 3.6 \times 10^4 E_{51}^{3/14} n_{10}^{-4/7} \zeta_m^{-5/14}$ years, where E_{51} is the kinetic energy of supernova shell in units of 10^{51} erg, n_{10} is the ambient number density in units of 10 cm^{-3}, and ζ_m is a metallicity factor of order 1 for solar abundances. Note that the transition of the SNR to the PDS stage is expected to occur somewhat earlier than τ_{sf}. The shock radius and velocity at the beginning of the PDS stage are $R_{pds} = 5.2 E_{51}^{2/7} n_{10}^{-3/7} \zeta_m^{-1/7}$ pc and $v_{pds} = 573 E_{51}^{1/14} n_{10}^{1/7} \zeta_m^{3/14}$ km s^{-1} correspondingly. Chevalier (1999) noted that a power law expansion in time is a reasonable approximation for the radiative shell over a range (5–50) τ_{pds}.

Radiative shocks in evolved SNRs are characterized by the efficient cooling of the post-shock plasma by line radiation from the dense shell behind the collisionless shock transition. The structure of the radiative shock depends on the line opacity which, for radiative SNR shocks, typically allows the escape of some optical and fine-structure infrared (IR) lines of abundant ions providing observational diagnostics of the shocks (see e.g., Raymond 1979; Hollenbach and McKee 1989; Gnat and Sternberg 2009; Bykov et al. 2013a).

In the case of core-collapse supernovae in such MC environments, the remnant becomes radiative at a radius of $\lesssim$6 pc, forming a shell that is magnetically supported and whose structure can be described by a self-similar solution (Chevalier 1999). The expected structure of the radiative shell is illustrated in the left panel in Fig. 2. The interaction of the radiative shell with an ensemble of molecular clumps of different sizes, masses, and cloud magnetization in the model results in both J and C-type shocks with a range of velocities. In some cases, OH maser emission may be associated with shocked molecular clumps.

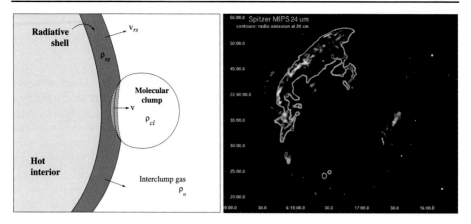

Fig. 2 *Left*: The sketch illustrating the interaction of the radiative supernova shell with a clumpy MC as described by Chevalier (1999). *Right*: *Spitzer MIPS* 24 μm image of IC 443 SNR with 1.4 GHz radio contours from Bykov et al. (2008). The radiative shock is located in the northeast part of the remnant

For IC 443, for example, Snell et al. (2005) detected the shocked clumps B, C, and G in H_2O, ^{13}CO, and C I line emission with *Submillimeter Wave Astronomy Satellite (SWAS)* detectors. They concluded that, to explain these observations, different type shocks with a range of velocities is likely required. Recently, Lee et al. (2012) found observational evidence for shocked clumps by mapping the southern part of IC 443 with ^{12}CO $J = 1 - 0$ and $HCO + J = 1 - 0$ lines.

Radiative shocks are observed in the northeast part of the SNR IC 443 where it is very likely interacting with an HI cloud. Spectrophotometry of [O III], [N II], and [S II] optical line emission performed by Fesen and Kirshner (1980) is found to be consistent with a radiative shock propagating in an inhomogeneous ISM. Electron densities up to 500 cm^{-3} were derived from the [S II] lines measurements. Kokusho et al. (2013) presented the near-IR [Fe II] 1.257 μm and [Fe II] 1.644 μm line maps ($30' \times 35'$) of IC 443 made with *IRSF/SIRIUS*. They found that [Fe II] filamentary structures exist all over the remnant, not only in the ionic shock NE-shell, but also in a molecular shock shell and a central region inside the shells.

Earlier, *Two Micron All Sky Survey (2MASS)* images of the entire IC 443 remnant in near-IR J (1.25 μm), H (1.65 μm), and Ks (2.17 μm) were analyzed by Rho et al. (2001) who revealed some clear morphological differences between the northeastern and southern parts. The J- and H-band *2MASS* emission from the NE rim was attributed mostly to different ionic fine-structure lines, where the H-band is often dominated by [Fe II] 1.644 μm line emission. The *2MASS* emission in the K_s-band with 2.12 μm H_2 molecular line emission, indicated a shocked molecular ridge spanning the southern region due to interaction of the remnant with the adjacent MC.

In the right panel of Fig. 2 we show the *Spitzer MIPS* 24 μm image of IC 443 SNR taken from Bykov et al. (2008) with 1.4 GHz radio contours. The 24 μm emission is likely dominated by [Fe II] 26 μm line emission rather than by the heated dust. This was first established by Oliva et al. (1999) who showed, using *ISO-SWS* spectroscopy, that most of the 12 μm and 25 μm band *IRAS* flux is accounted for by the line emission from [Ne II] and [Fe II] rather than dust emission. Apart from the [Fe II] 26 μm line, the other fine-structure far-IR lines [O I] 63.2 μm and [C II] 157.7 μm are predicted to be bright in the radiative shock models of Raymond (1979) and Hollenbach and McKee (1989). Indeed, Reach and

Rho (1999) estimated the luminosity of W44 in [O I] 63 μm line emission to be about 4×10^{36} erg s^{-1}, which Chevalier (1999) argued was consistent with the radiative shock models.

For the densities inferred for IC 443, the radiative shell is expected to form when the forward shock velocity drops below $v_{pds} \sim 600$ km s^{-1}. For shock speeds $> v_{pds}$, efficient diffusive shock acceleration (DSA) can occur and a substantial fraction of the shock ram pressure can be transferred into relativistic particles and turbulent magnetic fields which may be amplified to values well above those expected from adiabatic compression alone. The filamentary line emission region in the NE part of IC 443 is clearly correlated with the bright 1.4 GHz radio emission shell. This is an indication of the particle acceleration process in the shock. In contrast to the value of $v_{pds} \sim 600$ km s^{-1}, the radiative shock velocity in the NE shell of IC 443 was estimated to be about 100 km s^{-1}, while W44 is likely expanding at 150 km s^{-1}. It must be noted however, that these estimates were based on line emissivity derived in radiative shock models which did not account for efficient CR acceleration in the shocks. Particle acceleration, and the escape of high-energy CRs, result in higher compression and lower post-shock temperature at a given shock velocity than predicted ignoring DSA. The effect of CR acceleration on the line luminosity and ratios was discussed by Bykov et al. (2013a) who found a rather strong sensitivity of the line ratios to the extra shock compression effect. This implies the need for self-consistent models of radiative shocks including the efficient production of a non-thermal components.

3 X-Rays from SNR/MC Interactions

As a young SNR evolves, it compresses and heats the surrounding medium to X-ray emitting temperatures. Where the medium is extremely dense, however, the shock velocity is slow and the temperature of the shocked plasma can be considerably lower. In addition, the expansion of the remnant can be significantly impeded, leading to a distorted morphology. The high density can lead to rapid ionization, and can also drive a strong reverse shock into the ejecta. The X-ray characteristics thus provide important signatures of SNR/MC interactions.

3.1 Morphological Effects

In the Sedov phase of evolution, the radius of an SNR at a given age scales as $R_{SNR} \propto n_0^{-1/5}$. Despite this weak dependence, variations in density of a factor of five will produce significant changes in the remnant size of about 40 % and can easily cause noticeable modifications to the remnant shape. Molecular cloud environments are characterized by significant variations in density, and the clouds themselves have densities nearly 10^2–10^3 times higher than that of the typical ISM. SNR evolution in such environments can thus lead to deviations from the spherical morphology expected under ideal evolution in a uniform medium.

In addition, the *apparent* morphologies of SNRs can depend upon the density and distribution of foreground material, through energy-dependent absorption. The transmission of the ISM to X-rays for example, is $e^{-\sigma N_H}$ where σ is the photoelectric absorption coefficient and N_H is the column density of gas. Hydrogen itself does not produce significant absorption at X-rays energies. Rather, N_H acts as a tracer of other atoms with K-shell and L-shell transition energies in the soft X-ray band ($\sim$0.1–10 keV), given some relative abundance distribution relative to H. This has two significant effects on X-ray images of SNRs. First, for large values of N_H associated with large distances through the ISM, the observed X-ray emission will be faint. Second, the presence of MCs along the line of sight to an SNR will

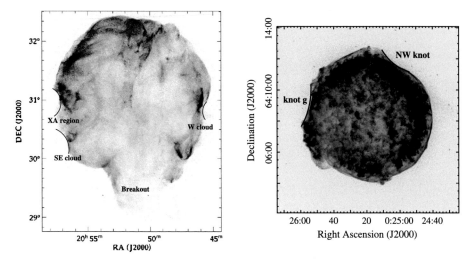

Fig. 3 *Left*: ROSAT HRI image of Cygnus Loop. Distinct deformation of the shell is seen at positions of cloud interactions indicated by arcs. *Right*: *Chandra* image of Tycho's SNR. Arcs in the east (so-called "knot g") and northwest indicate positions of cloud interactions

produce excess absorption. This can result in complete absorption from particular regions of the SNR, changing its apparent morphology. Because the absorption coefficient increases rapidly at low energies (e.g., Morrison and McCammon 1983), comparison of the morphology in low and high energy bands can reveal evidence for foreground MCs. Equivalently, spectral analysis of different regions of the SNR can reveal excess absorption of soft photons in the regions along the line of sight to MCs.

3.1.1 Shell Deformation

The complex CSM/ISM structure surrounding many SNRs, often characterized by significant large-scale density gradients, can result in very significant deviations from a circular morphology (e.g., Lopez et al. 2009). Direct interactions with dense clouds can also modify the morphology of SNRs, causing distinct deformation of the shell-like structure. In Fig. 3 (left) we show the ROSAT HRI image of the Cygnus Loop (Levenson et al. 1999), a nearby middle-aged SNR. The SNR shell is roughly circular, but significant deviations from circularity are seen in four regions. In the south, there is an apparent breakout associated with evolution into a low-density region. In the west, and in two positions in the east, there are very distinct arc-like deformations known as the W knot, the SE knot, and knot XA. The SE knot represents an encounter of the SNR blast wave with a protrusion from a large cloud. X-ray emission interior to radiative filaments in this region appears to originate from the reverse shock produced in the interaction (Graham et al. 1995). Knot XA appears to correspond to a dense, clumpy region resulting from an interaction with the wall of a cavity swept out by a precursor wind (McEntaffer and Brantseg 2011).

Also shown in Fig. 3 is a *Chandra* image of Tycho's SNR (right). While the overall morphology of this young, ejecta-dominated, historical remnant (SN 1572) is fairly circular, there are obvious depressions in the northwestern and eastern regions (indicated with arcs in the figure). Density estimates based on the IR emission from the remnant shell reveal dramatic increases in these regions (Williams et al. 2013), and X-ray proper motion measurements show that the expansion rate is lower in these regions than for adjacent regions

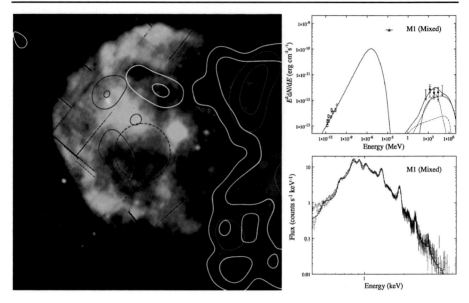

Fig. 4 *Left*: *XMM-Newton* image of CTB 109 (*cyan*) with *Spitzer* MIPS 24 μm image of adjacent MC (*red*). The *white* (*blue*) contours correspond to the CO line emission as observed by the CGPS, at velocity −51 (−54) km s⁻¹, and the levels correspond to 0.5 and 1.5 K. The 95 % confidence radius of the centroid of the associated Fermi-LAT source is shown as a *dashed red circle*. *Right*: Broadband emission model for CTB 109 (*top*) and thermal X-ray emission fit to predictions of broadband model (*bottom*). [Figures from Castro et al. 2012. Used by permission of the AAS]

of the remnant (Katsuda et al. 2010). These observations are consistent with the interpretation that Tycho is encountering dense clouds in the east and northwest, and optical studies of knot g, located in the eastern limb depression, show direct evidence of radiative shocks from the interaction.

A much more dramatic example of an SNR/MC interaction is that of CTB 109. As shown in Fig. 4 (left), the X-ray emission (cyan) is characterized by a half-shell morphology, with the western half of the SNR completely missing. CO measurements reveal a massive MC in this western region (white contours) from which IR emission is also seen (red image, from Spitzer). Radio observations at 1420 MHz show the same half-shell morphology (Kothes et al. 2002), confirming that the missing X-ray emission in the west is not produced by excess absorption. Rather, upon encountering the massive cloud, the SNR shock has apparently completely stalled.

A different approach was very recently adopted by Miceli et al. (2014), where the authors demonstrated the connection between shock-cloud interactions and particle acceleration in the southwestern limb of the historic remnant of SN 1006 based on the comparison of X-ray data with HI data. Figure 5 shows the X-ray image of the SW part of SN 1006 in the 0.3–2 keV band with HI contours superimposed. Exactly at the position where an indentation is observed in the X-ray (and radio) limb, there is an HI cloud. The existence of enhanced density was also confirmed by other means by Winkler et al. (2014). Several lines of evidence indicate that at this site particle acceleration is highly efficient and there is shock/cloud interaction, thus making this site a very promising source of γ-ray hadronic emission likely to be detectable with the *Fermi* LAT in the near future.

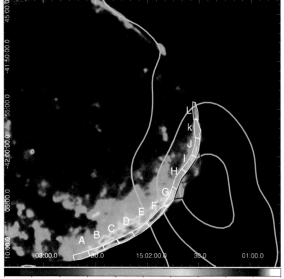

Fig. 5 *XMM-Newton* image of the southwestern portion of SN 1006 in the 0.3–2 keV band with HI column density plotted in contours showing the presence of a neutral gas cloud perfectly fitting the concavity in the SNR limb (figure from Miceli et al. 2014)

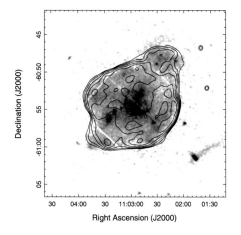

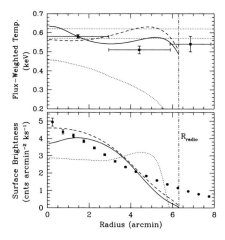

Fig. 6 *Left*: *XMM-Newton* image of MSH 11–61A with radio contours from MOST. The centrally-dominant emission is thermal in nature, and characterizes this as a MM SNR. *Right*: Brightness (*bottom*) and temperature (*top*) distributions for MSH 11–61A along with predictions from cloudy ISM (*solid*, *dashed*) and thermal conduction (*dotted*) models (after Slane et al. 2002). *Horizontal lines* in upper panel indicate mean temperature range

3.1.2 Mixed Morphology

A distinct subclass of remnants, known as mixed morphology (MM) SNRs, is characterized by a typical shell-like radio morphology contrasted by a centrally-bright X-ray morphology in which the central emission is thermal in nature (Fig. 6, left). At present, there are ∼40 known MM SNRs (see summary by Vink 2012), but the nature of the central X-ray emission is poorly understood. While abundance determinations from X-ray spectra indicate evidence for the presence of ejecta in some such remnants (e.g., Shelton et al. 2004; Lazendic and

Slane 2006; Bocchino et al. 2009; Pannuti et al. 2010; Yamaguchi et al. 2012), the estimated mass of X-ray emitting material in the central regions is generally much too high to be composed primarily of ejecta.

The spectral properties in MM SNRs provide significant constraints on the evolution of these systems. The plasma appears to be nearly isothermal in many systems, quite in contrast to the temperature profile expected in the Sedov phase of evolution. In addition, recent studies have shown that the plasma is overionized in several MM SNRs (see Sect. 3.2.2). This may result from early evolution through a dense wind profile created during the late phase of a red supergiant (or perhaps Wolf–Rayet) progenitor. In this scenario, the plasma is quickly ionized as the shock passes through the dense wind, but rapidly cools when the shock breaks through into the lower density surroundings, leaving the plasma in a higher ionization state than expected for its temperature (Moriya 2012). It is not clear exactly how such a scenario also leads to a centrally-bright morphology at later stages, nor have detailed models for such a scenario been carried out to confirm that a sufficient amount of material is shocked at this early stage to persist as an observable overionized feature in late stages of evolution.

Early models for MM SNRs centered primarily on two scenarios. In one, evolution to the radiative phase in which the shell temperature drops to low temperatures allows interstellar absorption to reduce the observed emission from the shell, while thermal conduction while the remnant is young and hot results in the transport of heat and bulk material into the central regions, smoothing out the temperature profile and increasing the emission from the remnant center (Cox et al. 1999). Application of this model to W44 is able to reproduce some general features, but fails to fully explain both the temperature and brightness distributions observed for this remnant (Shelton et al. 1999).

A second scenario centers on evolution in a surrounding medium filled with dense clouds (White and Long 1991). Upon being overtaken by the expanding SNR, the clouds evaporate through saturated conduction in the hot remnant interior, slowly increasing the central emission. For different combinations of the cloud evaporation timescale and the ratio of the mass in clouds to that in the intercloud material, significantly peaked brightness profiles can be obtained. Applying this model to the emission from MSH 11–61A, Slane et al. (2002) found that reasonable agreement could be obtained for both the radial brightness and temperature distributions (Fig. 6, right), but that the required evaporation timescale ($\sim$10–40 times the age of the remnant) appears much longer than expected. Application of the thermal-conduction/radiative-phase scenario was unable to reproduce the temperature and brightness distributions.

Importantly, the vast majority of MM SNRs appear to be interacting with MCs, and nearly half are observed to produce γ-ray emission, providing an additional clue as to the conditions that lead to the observed properties. At this stage, though, MM SNRs remain poorly understood. At the same time, increases in our knowledge of the abundance, ionization, and thermal properties of the shocked central plasma offer promise for constraining more complete models of these systems.

3.2 Spectral Signatures

The temperature, composition, and ionization state of the shocked gas in an SNR all depend crucially on the properties of the medium into which the remnant evolves. In particular, dense environments associated with the presence of MCs can produce spectral signatures that reveal the SNR/MC interactions. Such signatures are readily observed in a host of individual remnants.

3.2.1 Temperature and Absorption Variations

As an SNR sweeps up excessive amounts of material in the dense regions around a MC, the shock velocity drops. If the MC interaction is confined to only discrete regions in the SNR, this can result in temperature variations in the shocked plasma. X-ray studies of W51C (Koo et al. 2002, 2005) reveal a ~20 % temperature decrease in the northern regions where the SNR is interacting with a MC. Spectral modeling also indicates an increase in the column density, N_H, in this region, suggesting that a portion of the MC lies between the remnant and observer.

X-ray studies of 3C 391 also reveal N_H variations associated with a MC interaction, here in the northwestern regions of the remnant (Chen and Slane 2001). The increase of $\sim 5 \times 10^{21}$ cm^{-2} indicates a molecular cloud density of $\langle N(H_2) \rangle \sim 100 l_{pc}$ cm^{-3} where l_{pc} is the depth, in pc, of the MC region residing in front of the SNR.

3.2.2 Ionization Signatures

Due to the high densities for SNRs encountering MCs, one expects the shocked plasma to quickly reach ionization equilibrium. This is indeed observed in some remnants. For example, along the northeast limb of W28, Nakamura et al. (2014) find that the plasma is in collisional ionization equilibrium (CIE), with signs of density variations from region to region. In the central regions of the remnant, radiative recombination continuum (RRC) features of He-like Si and S are observed (Sawada and Koyama 2012), indicating a recombining plasma. Further, excess emission near the Kα lines of He-like Mg and Ne indicate different ionization temperatures for different elements.

Such overionization states are also observed in several other SNRs for which MC interactions occur. IC 443 shows an enhanced intensity ratio for H-like to He-like Si, indicating an overionized plasma (Kawasaki et al. 2002), and also distinct RRC features (Yamaguchi et al. 2009), for example. RRC features are also observed for W44, which also shows enhanced emission from H-like Si (Uchida et al. 2012), and from W49B (Ozawa et al. 2009). Interestingly, the nature of such overionized plasmas may actually be only indirectly associated with the presence of MCs. Although thermal conduction has been suggested as a mechanism by which heat flow from the hot SNR interior to regions with cold clouds could bring lower the electron temperature to values below the current ionization temperature of the ions, most calculations indicate that this process is too slow to operate efficiently in the typical lifetime of an SNR (e.g., Uchida et al. 2012). Instead, early evolution through a dense medium such as that from a stellar wind may have created a high ionization state, with subsequent rapid cooling as the remnant expands adiabatically into a low density cavity leading to a temperature that is lower than the ionization temperature (e.g., Yamaguchi et al. 2009; Uchida et al. 2012; Moriya 2012). Maps of the ionization state in W49B indicate higher states of overionization in the central and western regions (Miceli et al. 2010; Lopez et al. 2013), supporting the notion that rapid expansion in the direction opposite that of a dense MC in the east is responsible for the cooling.

Given the above scenario, it would seem that the most direct connection between recombining plasma in SNRs and their association with MCs is simply that the proposed early-phase evolution through a dense stellar wind implies a massive progenitor star, and the remnants of such stars are often found near the molecular cloud complexes from which they formed. However, it also appears that the overionized SNRs are all of the mixed morphology class. Whether or not the overall temperature, ionization, and brightness properties of this class require specific contributions from MC interactions remains unknown, at present, but is an area of active study.

3.3 Abundances and Nonthermal Emission

In evolved remnants that have undergone considerable interactions with MCs, the total swept-up mass can be very large. If the shock velocity in the dense interaction regions is low, much of this material will be too cool to produce X-rays. However, in many systems the X-ray emitting mass, M_x, is also large. Modeling of the emission from W44 indicates roughly $100M_\odot$ of hot gas in the remnant interior (Shelton et al. 2004), for example. The roughly solar abundances for this swept-up material will thus act to severely dilute the enhanced abundances of any (much smaller) ejecta component. Nonetheless, traces of ejecta appear common in many mixed morphology remnants (e.g., Slane et al. 2002; Shelton et al. 2004; Lazendic and Slane 2006; Bocchino et al. 2009), virtually all of which appear to be interacting with molecular clouds. In IC 443, a distinct ring-like structure of hot ($\sim$1.4 keV) plasma with significantly enhanced abundances of Mg and Si is observed in the vicinity of a MC interaction region, suggesting that a strong reverse shock produced in this interaction with dense material has produced enhanced ejecta emission (Troja et al. 2008).

Evidence for ejecta in IC 443 also exists in the form of compact knots of X-ray emitting material (Bocchino and Bykov 2003). The observed hard spectra and enhanced abundances suggest that these may be fast-moving knots of ejecta traveling into dense molecular material and producing Kα emission accompanied by nonthermal bremsstrahlung (NTB) emission from shock-accelerated electrons (Bykov 2002). Knots with similar spectral properties are observed in Kes 69 and are coincident with CO emission from a nearby MC (Bocchino et al. 2012), reinforcing this interpretation. Discrete X-ray knots directly along the SNR/MC interaction region in 3C 391 are also observed (Chen et al. 2004), but while the inferred density for these knots is high, consistent with structures being driven into clouds, the abundances do not show strong evidence for metal enhancements. We note that a complete analysis of the composition of cold, dense, metal-rich ejecta from X-ray observations needs to also account for the resonant photo-absorption effect.

Regions of *low* abundance plasma have also been observed in some regions of SNRs interacting with MCs. The Fe abundance appears to be subsolar in some bright knots in W44. The abundances of other elements in these knots appear to be enhanced, possibly indicating a scenario in which ejecta fragments are traveling through MC material in which some Fe remains condensed onto dust grains (Shelton et al. 2004). In W28, the emission from the bright northeast rim that is adjacent to a MC is best described by a high density plasma in CIE with subsolar abundances of refractory elements (Nakamura et al. 2014), but in this case the abundances of volatile elements are also subsolar, complicating any interpretation of condensation onto dust grains in MC material. Similar depletions of N, O, and Ne are observed in the XA region of the Cygnus Loop (McEntaffer and Brantseg 2011).

4 Gamma-Rays from SNR/MC Interactions

4.1 Particle Acceleration in SNRs

Particle acceleration in SNRs has long been suggested as a major contributor to the cosmic ray population, at least up to the knee in the spectrum at $\sim$10^{15} eV. The generally assumed mechanism is DSA, where some particles scatter off of self-generated turbulence and are returned to the shock region multiple times. In nonlinear DSA, a non-negligible fraction of the electrons and ions at the forward shock (FS) can reach ultrarelativistic energies. Diffusive shock acceleration has received considerable attention, and specific predictions of the nonlinear theory include (e.g., Blandford and Eichler 1987; Malkov and Drury 2001; Bykov et al. 2013b): (i) accelerated particles obtain enough pressure to modify the shock structure, with the shock developing an extended upstream precursor; (ii) the overall shock compression ratio can increase above the Rankine–Hugoniot value of four for strong shocks, while simultaneously the subshock compression, which is mainly responsible for heating the unshocked plasma, decreases below the Rankine–Hugoniot value; (iii) the particle power law expected from test-particle DSA becomes concave, with the highest energy particles developing a spectrum harder than the test-particle power law; (iv) a significant fraction of the highest energy particles can escape from the shock, adding to the nonlinear nature of the mechanism; and (v) the production of superthermal particles goes hand-in-hand with the production of magnetic turbulence, and strong magnetic field amplification can occur in high Mach number shocks.

While there is no direct evidence for the specific mechanism of DSA in the forward and reverse shocks of SNRs, there is overwhelming evidence for the production of relativistic particles, either ions or electrons, at the forward shocks in a number of remnants. There is also clear evidence for magnetic field amplification, and for the modification of the plasma hydrodynamics, predicted by efficient DSA, in several remnants (e.g., Patnaude and Fesen 2007; Uchiyama et al. 2007; Cassam-Chenaï et al. 2008; Uchiyama and Aharonian 2008) This, combined with the direct evidence for efficient DSA in spacecraft observations of heliospheric shocks (e.g., Ellison et al. 1990) and confirmation of fundamental aspects of the theory from particle-in-cell (PIC) simulations (e.g., Ellison et al. 1993; Sironi et al. 2013), has added to a general acceptance of the mechanism for cosmic ray production in SNRs.

Despite the progress made in understanding DSA and verifying some of its basic predictions, important open questions remain about the maximum particle energy E_{max}, the acceleration efficiency, the electron-to-proton ratio K_{ep} for the injected particles, and the eventual escape of cosmic rays from the acceleration region. As an example, local cosmic ray measurements, as well as theoretical expectations, suggest that shocks put far more energy in protons and heavier ions than in electrons. Yet the nonthermal emission from most astrophysical sources is dominated by radiation from relativistic electrons. While this can be understood, in part, because relativistic electrons radiate more efficiently than protons, the underlying energy budget of a source cannot be determined until constraints on the hadronic component are obtained and the accelerated electron-to-proton ratio determined. This is important for cosmic rays and for all sources where DSA is expected to occur.

Molecular clouds play a particularly important role in this regard because shocks in the high density MC environment will predominately produce γ-rays by hadronic interactions rather than leptonic ones (e.g., Aharonian et al. 1994).

4.2 Gamma-Ray Production

Gamma-ray production from such relativistic particles can proceed through inverse-Compton (IC) scattering of ambient photons by the energetic electrons, nonthermal bremsstrahlung (NTB) from collisions between relativistic electrons and ambient material, and the decay of neutral pions formed in collisions of energetic protons with surrounding nuclei. Figure 7 presents a simulation of the broadband spectrum produced by an SNR undergoing efficient acceleration of electrons and ions. The magenta curve represents synchrotron emission from the relativistic electrons, and the dotted blue curve corresponds to IC emission produced from that same electron population up-scattering photons from the cosmic microwave background (CMB). The dashed green curve represents NTB from the

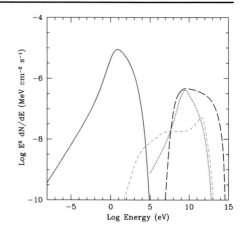

Fig. 7 Simulated broadband spectrum from SNR undergoing efficient DSA of electrons and protons. The *solid magenta curve* represents synchrotron emission, the *dotted blue curve* is IC emission, the *dashed green curve* corresponds to NTB, and the *long-dashed curve* represents the π^0-day emission

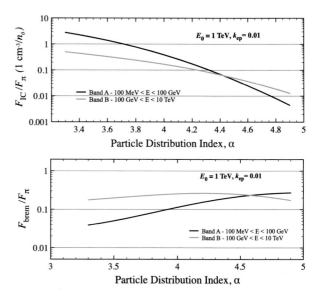

Fig. 8 IC-to-π^0-decay (*top*) and NTB-to-π^0-decay (*bottom*) flux ratios as a function of the particle momentum distribution spectral index, α. The *black curve* is for the photon energy band 100 MeV to 100 GeV, and the *green line* indicates the flux ratio in the 100 GeV to 10 TeV range. The exponential particle energy cutoff, for both electrons and protons, has been fixed at 1 TeV, and the electron-to-proton ratio is $K_{ep} = 0.01$. [From Castro et al. 2013. Reproduced by permission of the AAS]

relativistic electrons interacting with ambient material, and the long-dashed red curve corresponds to γ-rays from the decay of neutral pions produced by collisions of the relativistic proton component with ambient nuclei. (The curvature of the particle spectra associated with DSA is particularly evident in the synchrotron and IC emission.)

The emission from both NTB and π^0-decay scales with the ambient density n_0. As a result, in high density environments such as those encountered in SNR/MC interactions, significant γ-ray emission is expected if the SNR has been an active particle accelerator. For $K_{ep} \sim 10^{-2}$, as measured locally, the π^0-decay emission will dominate (see Fig. 8), making such γ-ray emission an important probe of the hadronic component of the particles accelerated by the SNR.

An important additional consideration for γ-ray production is the local photon energy density. Contributions from starlight as well as IR emission from local dust can increase the IC emission and change the spectral shape of this component due to the different effective temperatures of these photon components. These contributions are highly dependent upon galactocentric radius (e.g., Strong et al. 2000). Moreover, IR emission produced by the SNRs themselves can contribute significantly to the IC γ-ray emission (e.g., Morlino and Caprioli 2012; Slane et al. 2014).

Modeling of the broadband spectra from such SNRs, in order to ascertain the nature of the γ-ray emission, is complicated and has led to mixed interpretations, making the evidence for ion acceleration controversial in some cases. However, in a growing number of cases, γ-ray emission from some SNRs known to be interacting with MCs seems to clearly require a significant component from pion decay. We discuss emission from several of these sources below.

4.3 Gamma-Ray Observations of SNRs

To date, studies have identified γ-ray emission from $\sim$25 SNRs, the majority of which are interacting with MCs. Of these, the evidence for energetic hadrons as the source of these γ-ray s is compelling for more than 50 % of these based on energetic arguments and/or broadband spectral modeling. *Fermi* LAT observations of W51C reveal a spectrum that is consistent with π^0-decay, with dominant NTB ruled out unless $K_{ep} \gg 0.01$, and IC emission ruled out on energetic grounds (Abdo et al. 2009). W44 and IC 443 show clear evidence of a kinematic "pion bump" in their spectra, firmly establishing the presence of energetic ions in these remnants (Abdo et al. 2010; Giuliani et al. 2011; Ackermann et al. 2013). Gamma-ray emission from CTB 109 (Fig. 4, left) has been detected with the *Fermi* LAT (Castro et al. 2012), and modeling of the broadband emission (Fig. 4, upper right) indicates that the γ-ray emission arises from approximately equal contributions from IC scattering and π^0-decay, a result that is strongly constrained by the observed ionization state of the thermal X-ray emission (Fig. 4, lower right).

Castro and Slane (2010) carried out *Fermi* LAT studies of a set of four SNRs (G349.7+0.2, CTB 37A, 3C 391, and G8.7-0.1) known to be interacting with MCs based on observations of hydroxyl (OH) maser emission at 1720 MHz, and showed that all four were sources of GeV γ-ray emission. Based on the assumption that the γ-ray emission is dominated by π^0-decay produced in the compressed shell of the SNR, they derived a lower limit on the density of the γ-ray emitting material for each remnant, and compared this with the density inferred from X-ray measurements. They found that the density inferred from γ-ray measurements exceeds that from X-ray measurements by a factor of 20 or more. Subsequent studies of W41, MSH 17-39, and G337.7-0.1 (Castro et al. 2013) as well as Kes 79 (Auchettl et al. 2014) reveal γ-ray spectra indicative of hadronic emission, with leptonic scenarios requiring total electron energies in excess of 10^{51} erg or $K_{ep} \gg 0.01$. Similar discrepancies between densities inferred from γ-ray and X-ray measurements are obtained for these sources. The magnetic fields implied by the radio emission in these studies is typically much larger than expected from the compressed ISM, suggesting evidence for the magnetic field amplification expected in efficient particle acceleration in SNR shocks.

A plausible explanation for the observed discrepancy between n_x and n_γ is that these SNRs, by evolving in the presence of MCs, have swept up clumps of dense material. The SNR blast wave for each remnant has presumably evolved primarily through the low-density interclump medium, heating this material to X-ray emitting temperatures and accelerating particles through DSA. The proton component of these accelerated particles then interacts with both the dense clumps and the interclump gas, thus sampling a much higher effective density than that of the X-ray emitting gas (Fig. 9). A similar scenario has been

Fig. 9 Schematic diagram of SNR/MC interaction in which postshock region in SNR contains dense clumps within shocked interclump material. The low-density shocked interclump gas emits X-rays. Protons acceleration by the SNR encounter all of gas in the shell, and γ-rays are produced primarily through collisions with the dense clumps

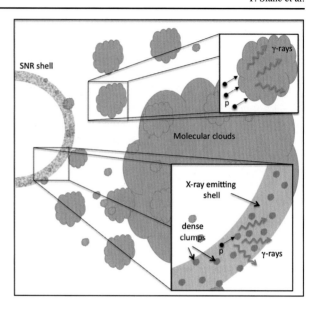

proposed by (Inoue et al. 2012) to explain the complete lack of thermal X-ray emission from RX J1713.7-3946, whose observed GeV and TeV γ-ray emission is otherwise required to be dominated by IC emission (Ellison et al. 2010; Ellison et al. 2012). Additional modifications of this picture of dense clumps embedded in the SNR shell invokes reacceleration of cosmic rays trapped in the cold molecular material rather than trapped particles in the shell that have been accelerated by the SNR (e.g., Uchiyama et al. 2010).

An alternative scenario that can explain γ-rays from SNR/MC interactions, and also $n_\gamma/n_x \gg 1$, is that of escaping CRs interacting with external MCs (e.g., Gabici et al. 2009). Here, as indicated in Fig. 9 (left), there is no connection between n_x and n_γ because the X-rays are produced in the SNR shell while the γ-rays are from the MC. Such a scenario has been suggested to explain the γ-rays from the source MAGIC J0616+225 as the result of energetic particles that have escaped from IC 443 and are interacting with dense clouds (Torres et al. 2008), for example, and discrete TeV sources outside the remnant W28 have been suggested to originate from particles escaping the SNR and interacting with adjacent clouds (Aharonian et al. 2008). The expected γ-ray spectrum in this scenario is complicated, however. Because the escaping particles are distributed around E_{max}, the resulting π^0-decay spectrum is peaked at higher energies than that for the shell (Zirakashvili and Ptuskin 2008; Ellison and Bykov 2011). However, because E_{max} is expected to decrease with time, the time-accumulated spectrum of particles interacting with a remote cloud will depend on the age of the SNR and the energy-dependent diffusion process by which the particles are transported. For simple assumptions about the time evolution and particle diffusion, this could lead to power law (Gabici et al. 2009) or broken power law (Ohira et al. 2011) particle distributions at the remote cloud site. The observed γ-ray spectra from W51C, W28, W44 and IC 443 show distinctly different slopes, leading to questions as to whether these differences are intrinsic to the acceleration or the result of complex evolution, propagation, and interaction with remote clouds.

5 Conclusions

Observations of SNR/MC interactions provide unique information on the distances to SNRs and on the physics of shock/cloud interactions. X-ray observations, in particular, probe the state of the shocked gas and provide diagnostics of the interaction through absorption, ionization, temperature, and general morphology measurements. For SNRs that are, or have been, active particle accelerators, these interactions with dense cloud material can lead to crucial measurements of emission from the hadronic component of their relativistic particle spectra, providing crucial information on acceleration efficiency, maximum energies, and particle escape in these systems.

Recent studies of γ-ray emission from numerous SNRs demonstrate that those interacting with MCs show strong evidence of relativistic protons. For some, X-ray measurements of the flux or ionization state of the thermal plasma provide constraints that allow us to determine what fraction of the γ-ray emission arises from this hadronic component. For a significant number of the remnants interacting with MCs, the inferred density of the γ-ray-emitting material is considerably higher than that determined from X-ray measurements, suggesting a complex emission environment with dense clumps embedded in hot interclump gas, or perhaps significant γ-ray emission associated with escaping cosmic rays interacting with external clouds. Considerable additional observations and modeling efforts are required to better understand the postshock regions of SNR/MC interactions, as well as the evolution and transport of the escaping particles. Of particular importance are simulations of SNR evolution into dense, cloudy environments to study the formation of clumps in the postshock region, as well as modeling of realistic particle acceleration with escape and energy-dependent diffusion to predict the spectrum of SNR cosmic rays interacting with nearby MCs. Equally important is progress in observations and modeling of supernova ejecta interactions with a complex circumstellar medium formed by the winds of massive stars at the different evolution stages of the massive stars. On the observational side, more sensitive X-ray and γ-ray observations are needed to provide improved spectra of SNRs interacting with MCs. Improved angular resolution at γ-ray energies, such as may be expected in future facilities like CTA, is particularly important. In combination with current studies of plasma instabilities, magnetic field amplification, and the reacceleration of relic cosmic rays, the outlook for reaching a quantitative understanding of the role SNRs play in producing Galactic cosmic rays is extremely positive.

Acknowledgements　　The authors would like to thank Tom Dame and Herman Lee for important discussions contributing to this paper. We also thank ISSI and its staff for supporting this effort and for creating a stimulating environment for the meeting at which this work was initiated. POS acknowledges support from NASA Contract NAS8-03060. AMB was supported in part by RAS Presidium 21 and OFN 15 programs.

References

A.A. Abdo, M. Ackermann, M. Ajello, L. Baldini, J. Ballet et al., Fermi LAT discovery of extended gamma-ray emission in the direction of supernova remnant W51C. Astrophys. J. Lett. **706**, 1–6 (2009)

A.A. Abdo, M. Ackermann, M. Ajello, L. Baldini, J. Ballet et al., Gamma-ray emission from the shell of supernova remnant W44 revealed by the Fermi LAT. Science **327**, 1103 (2010)

V.A. Acciari, E. Aliu, T. Arlen, T. Aune, M. Bautista et al., Observation of extended very high energy emission from the supernova remnant IC 443 with VERITAS. Astrophys. J. Lett. **698**, 133–137 (2009). doi:10.1088/0004-637X/698/2/L133

M. Ackermann, M. Ajello, A. Allafort, L. Baldini, J. Ballet et al., Detection of the characteristic pion-decay signature in supernova remnants. Science **339**, 807–811 (2013)

F.A. Aharonian, L.O. Drury, H.J. Voelk, GeV/TeV gamma-ray emission from dense molecular clouds overtaken by supernova shells. Astron. Astrophys. **285**, 645–647 (1994)

F. Aharonian, A.G. Akhperjanian, A.R. Bazer-Bachi, B. Behera, M. Beilicke et al., Discovery of very high energy gamma-ray emission coincident with molecular clouds in the W 28 (G6.4-0.1) field. Astron. Astrophys. **481**, 401–410 (2008). doi:10.1051/0004-6361:20077765

K. Auchettl, P. Slane, D. Castro, Fermi-LAT observations of supernova remnants Kesteven 79. Astrophys. J. **783**, 32 (2014). doi:10.1088/0004-637X/783/1/32

R. Blandford, D. Eichler, Particle acceleration at astrophysical shocks: a theory of cosmic ray origin. Phys. Rep. **154**, 1–75 (1987)

F. Bocchino, A.M. Bykov, XMM-Newton study of hard X-ray sources in IC 443. Astron. Astrophys. **400**, 203–211 (2003). doi:10.1051/0004-6361:20021858

F. Bocchino, M. Miceli, E. Troja, On the metal abundances inside mixed-morphology supernova remnants: the case of IC 443 and G166.0+4.3. Astron. Astrophys. **498**, 139–145 (2009). doi:10.1051/0004-6361/200810742

F. Bocchino, A.M. Bykov, Y. Chen, A.M. Krassilchtchikov, K.P. Levenfish, M. Miceli, G.G. Pavlov, Y.A. Uvarov, X. Zhou, A population of isolated hard X-ray sources near the supernova remnant Kes 69. Astron. Astrophys. **541**, 152 (2012). doi:10.1051/0004-6361/201219005

C.L. Brogan, D.A. Frail, W.M. Goss, T.H. Troland, OH Zeeman magnetic field detections toward five supernova remnants using the VLA. Astrophys. J. **537**, 875–890 (2000). doi:10.1086/309068

C.L. Brogan, W.M. Goss, T.R. Hunter, A.M.S. Richards, C.J. Chandler, J.S. Lazendic, B.-C. Koo, I.M. Hoffman, M.J. Claussen, OH (1720 MHz) masers: a multiwavelength study of the interaction between the W51C supernova remnant and the W51B star forming region. Astrophys. J. **771**, 91 (2013). doi:10.1088/0004-637X/771/2/91

M.G. Burton, T.R. Geballe, P.W.J.L. Brand, A.S. Webster, Shocked molecular hydrogen in the supernova remnant IC 443. Mon. Not. R. Astron. Soc. **231**, 617–634 (1988)

M.G. Burton, C. Braiding, C. Glueck, P. Goldsmith, J. Hawkes, D.J. Hollenbach, C. Kulesa, C.L. Martin, J.L. Pineda, G. Rowell, R. Simon, A.A. Stark, J. Stutzki, N.J.H. Tothill, J.S. Urquhart, C. Walker, A.J. Walsh, M. Wolfire, The Mopra southern galactic plane CO survey. Publ. Astron. Soc. Aust. **30**, 44 (2013). doi:10.1017/pasa.2013.22

A.M. Bykov, X-ray line emission from supernova ejecta fragments. Astron. Astrophys. **390**, 327–335 (2002). doi:10.1051/0004-6361:20020694

A.M. Bykov, A.M. Krassilchtchikov, Y.A. Uvarov, H. Bloemen, F. Bocchino, G.M. Dubner, E.B. Giacani, G.G. Pavlov, Isolated X-ray-infrared sources in the region of interaction of the supernova remnant IC 443 with a molecular cloud. Astrophys. J. **676**, 1050–1063 (2008). doi:10.1086/529117

A.M. Bykov, M.A. Malkov, J.C. Raymond, A.M. Krassilchtchikov, A.E. Vladimirov, Collisionless shocks in partly ionized plasma with cosmic rays: microphysics of non-thermal components. Space Sci. Rev. **178**, 599–632 (2013a). doi:10.1007/s11214-013-9984-7

A.M. Bykov, A. Brandenburg, M.A. Malkov, S.M. Osipov, Microphysics of cosmic ray driven plasma instabilities. Space Science Reviews (2013b)

G. Cassam-Chenaï, J.P. Hughes, E.M. Reynoso, C. Badenes, D. Moffett, Morphological evidence for azimuthal variations of the cosmic-ray ion acceleration at the blast wave of SN 1006. Astrophys. J. **680**, 1180–1197 (2008)

G. Castelletti, G. Dubner, T. Clarke, N.E. Kassim, High-resolution radio study of SNR IC 443 at low radio frequencies. Astron. Astrophys. **534**, 21 (2011). doi:10.1051/0004-6361/201016081

D. Castro, P. Slane, Fermi large area telescope observations of supernova remnants interacting with molecular clouds. Astrophys. J. **717**, 372–378 (2010)

D. Castro, P. Slane, D.C. Ellison, D.J. Patnaude, Fermi-LAT observations and a broadband study of supernova remnant CTB 109. Astrophys. J. **756**, 88 (2012). doi:10.1088/0004-637X/756/1/88

D. Castro, P. Slane, A. Carlton, E. Figueroa-Feliciano, Fermi-LAT observations of supernova remnants interacting with molecular clouds: W41, MSH 17-39, and G337.7-0.1. Astrophys. J. **774**, 36 (2013)

Y. Chen, P.O. Slane, ASCA observations of the thermal composite supernova remnant 3C 391. Astrophys. J. **563**, 202–208 (2001). doi:10.1086/323886

Y. Chen, Y. Su, P.O. Slane, Q.D. Wang, A Chandra ACIS view of the thermal composite supernova remnant 3C 391. Astrophys. J. **616**, 885–894 (2004). doi:10.1086/425152

Y. Chen, B. Jiang, P. Zhou, Y. Su, X. Zhou, H. Li, X. Zhang, Molecular environments of supernova remnants, in *IAU Symposium*, vol. 296, ed. by A. Ray, R.A. McCray (2014), pp. 170–177. Provided by the SAO/NASA Astrophysics Data System. doi:10.1017/S1743921313009423.

R.A. Chevalier, Supernova remnants in molecular clouds. Astrophys. J. **511**, 798–811 (1999). doi:10.1086/306710

D.F. Cioffi, C.F. McKee, E. Bertschinger, Dynamics of radiative supernova remnants. Astrophys. J. **334**, 252–265 (1988). doi:10.1086/166834

M.J. Claussen, D.A. Frail, W.M. Goss, R.A. Gaume, Polarization observations of 1720 MHz OH masers toward the three supernova remnants W28, W44, and IC 443. Astrophys. J. **489**, 143 (1997). doi:10.1086/304784

F.R. Colomb, G. Dubner, Neutral hydrogen associated with southern supernova remnants. II—Lupus Loop. Astron. Astrophys. **112**, 141–148 (1982)

D.P. Cox, The three-phase interstellar medium revisited. Annu. Rev. Astron. Astrophys. **43**, 337–385 (2005). doi:10.1146/annurev.astro.43.072103.150615

D.P. Cox, R.L. Shelton, W. Maciejewski, R.K. Smith, T. Plewa, A. Pawl, M. Różyczka, Modeling W44 as a supernova remnant in a density gradient with a partially formed dense shell and thermal conduction in the hot interior. I. The analytical model. Astrophys. J. **524**, 179–191 (1999)

T.M. Dame, D. Hartmann, P. Thaddeus, The Milky Way in molecular clouds: a new complete CO survey. Astrophys. J. **547**, 792–813 (2001). doi:10.1086/318388

L.K. Denoyer, M.A. Frerking, Some new results on shock chemistry in IC 443. Astrophys. J. Lett. **246**, 37–40 (1981). doi:10.1086/183548

G.M. Dubner, F.R. Colomb, E.B. Giacani, 1410 MHz continuum and H I line observations towards the SNR G296.5+10.0 and nearby sources—evidences of two SNRs tunneling through the interstellar medium. Astron. J. **91**, 343–353 (1986). doi:10.1086/114015

G.M. Dubner, M. Holdaway, W.M. Goss, I.F. Mirabel, A high-resolution radio study of the W50-SS 433 system and the surrounding medium. Astron. J. **116**, 1842–1855 (1998a). doi:10.1086/300537

G.M. Dubner, A.J. Green, W.M. Goss, D.C.-J. Bock, E. Giacani, Neutral hydrogen in the direction of the VELA supernova remnant. Astron. J. **116**, 813–822 (1998b). doi:10.1086/300466

G.M. Dubner, E.B. Giacani, W.M. Goss, A.J. Green, L.-Å. Nyman, The neutral gas environment of the young supernova remnant SN 1006 (G327.6+14.6). Astron. Astrophys. **387**, 1047–1056 (2002). doi:10.1051/0004-6361:20020365

D.C. Ellison, A.M. Bykov, Gamma-ray emission of accelerated particles escaping a supernova remnant in a molecular cloud. Astrophys. J. **731**, 87 (2011). doi:10.1088/0004-637X/731/2/87

D.C. Ellison, E. Moebius, G. Paschmann, Particle injection and acceleration at earth's bow shock—comparison of upstream and downstream events. Astrophys. J. **352**, 376–394 (1990)

D.C. Ellison, J. Giacalone, D. Burgess, S.J. Schwartz, Simulations of particle acceleration in parallel shocks: direct comparison between Monte Carlo and one-dimensional hybrid codes. J. Geophys. Res. **98**, 21085 (1993)

D.C. Ellison, D.J. Patnaude, P. Slane, J. Raymond, Efficient cosmic ray acceleration, hydrodynamics, and self-consistent thermal X-ray emission applied to supernova remnant RX J1713.7-3946. Astrophys. J. **712**, 287–293 (2010)

D.C. Ellison, P. Slane, D.J. Patnaude, A.M. Bykov, Core-collapse model of broadband emission from SNR RX J1713.7-3946 with thermal X-rays and gamma rays from escaping cosmic rays. Astrophys. J. **744**, 39 (2012)

R.A. Fesen, R.P. Kirshner, Spectrophotometry of the supernova remnant IC 443. Astrophys. J. **242**, 1023–1040 (1980). doi:10.1086/158534

D.A. Frail, G.F. Mitchell, OH (1720 MHz) masers as signposts of molecular shocks. Astrophys. J. **508**, 690–695 (1998). doi:10.1086/306452

D.A. Frail, W.M. Goss, V.I. Slysh, Shock-excited maser emission from the supernova remnant W28. Astrophys. J. Lett. **424**, 111–113 (1994)

D.A. Frail, W.M. Goss, E.M. Reynoso, E.B. Giacani, A.J. Green, R. Otrupcek, A survey for OH (1720 MHz) maser emission toward supernova remnants. Astron. J. **111**, 1651 (1996)

K.É. Gabányi, G. Dubner, E. Giacani, Z. Paragi, S. Frey, Y. Pidopryhora, Very long baseline interferometry search for the radio counterpart of HESS J1943+213. Astrophys. J. **762**, 63 (2013). doi:10.1088/0004-637X/762/1/63

S. Gabici, F.A. Aharonian, S. Casanova, Broad-band non-thermal emission from molecular clouds illuminated by cosmic rays from nearby supernova remnants. Mon. Not. R. Astron. Soc. **396**, 1629–1639 (2009). doi:10.1111/j.1365-2966.2009.14832.x

A. Giuliani, M. Cardillo, M. Tavani, Y. Fukui, S. Yoshiike et al., Neutral pion emission from accelerated protons in the supernova remnant W44. Astrophys. J. Lett. **742**, 30 (2011)

O. Gnat, A. Sternberg, Metal-absorption column densities in fast radiative shocks. Astrophys. J. **693**, 1514–1542 (2009). doi:10.1088/0004-637X/693/2/1514

J.R. Graham, N.A. Levenson, J.J. Hester, J.C. Raymond, R. Petre, An X-ray and optical study of the interaction of the Cygnus Loop supernova remnant with an interstellar cloud. Astrophys. J. **444**, 787–795 (1995). doi:10.1086/175651

A.J. Green, D.A. Frail, W.M. Goss, R. Otrupcek, Continuation of a survey of OH (1720 MHz) maser emission towards supernova remnants. Astron. J. **114**, 2058 (1997)

J.W. Hewitt, F. Yusef-Zadeh, M. Wardle, A survey of hydroxyl toward supernova remnants: evidence for extended 1720 MHz maser emission. Astrophys. J. **683**, 189–206 (2008). doi:10.1086/588652

M.H. Heyer, C. Brunt, R.L. Snell, J.E. Howe, F.P. Schloerb, J.M. Carpenter, The five college radio astronomy observatory CO survey of the outer Galaxy. Astrophys. J. Suppl. Ser. **115**, 241 (1998). doi:10.1086/313086

I.M. Hoffman, W.M. Goss, C.L. Brogan, M.J. Claussen, The OH (1720 MHz) supernova remnant masers in W44: MERLIN and VLBA polarization observations. Astrophys. J. **627**, 803–812 (2005). doi:10.1086/430419

D. Hollenbach, C.F. McKee, Molecule formation and infrared emission in fast interstellar shocks. III—results for J shocks in molecular clouds. Astrophys. J. **342**, 306–336 (1989). doi:10.1086/167595

Y.-L. Huang, P. Thaddeus, Molecular clouds and supernova remnants in the outer Galaxy. Astrophys. J. **309**, 804–821 (1986). doi:10.1086/164649

T. Inoue, R. Yamazaki, S.-i. Inutsuka, Y. Fukui, Toward understanding the cosmic-ray acceleration at young supernova remnants interacting with interstellar clouds: possible applications to RX J1713.7-3946. Astrophys. J. **744**, 71 (2012)

J.M. Jackson, J.M. Rathborne, R.Y. Shah, R. Simon, T.M. Bania, D.P. Clemens, E.T. Chambers, A.M. Johnson, M. Dormody, R. Lavoie, M.H. Heyer, The Boston university-five college radio astronomy observatory galactic ring survey. Astrophys. J. Suppl. Ser. **163**, 145–159 (2006). doi:10.1086/500091

B. Jiang, Y. Chen, J. Wang, Y. Su, X. Zhou, S. Safi-Harb, T. DeLaney, Cavity of molecular gas associated with supernova remnant 3C 397. Astrophys. J. **712**, 1147–1156 (2010)

S. Katsuda, R. Petre, J.P. Hughes, U. Hwang, H. Yamaguchi, A. Hayato, K. Mori, H. Tsunemi, X-ray measured dynamics of Tycho's supernova remnant. Astrophys. J. **709**, 1387–1395 (2010)

M.T. Kawasaki, M. Ozaki, F. Nagase, K. Masai, M. Ishida, R. Petre, ASCA observations of the supernova remnant IC 443: thermal structure and detection of overionized plasma. Astrophys. J. **572**, 897–905 (2002). doi:10.1086/340383

T. Kokusho, T. Nagayama, H. Kaneda, D. Ishihara, H.-G. Lee, T. Onaka, Large-area [Fe II] line mapping of the supernova remnant IC 443 with the IRSF/SIRIUS. Astrophys. J. Lett. **768**, 8 (2013). doi:10.1088/2041-8205/768/1/L8

B.-C. Koo, J.-H. Kang, N.M. McClure-Griffiths, HI 21 cm emission line study of southern galactic supernova remnants. J. Korean Astron. Soc. **37**, 61–77 (2004)

B.-C. Koo, J.-J. Lee, F.D. Seward, An ASCA study of the W51 complex. Astron. J. **123**, 1629–1638 (2002). doi:10.1086/339179

B.-C. Koo, J.-J. Lee, F.D. Seward, D.-S. Moon, Chandra observations of the W51C supernova remnant. Astrophys. J. **633**, 946–952 (2005). doi:10.1086/491468

R. Kothes, B. Uyaniker, A. Yar, The distance to supernova remnant CTB 109 deduced from its environment. Astrophys. J. **576**, 169–175 (2002). doi:10.1086/341545

J.S. Lazendic, P.O. Slane, Enhanced abundances in three large-diameter mixed-morphology supernova remnants. Astrophys. J. **647**, 350–366 (2006). doi:10.1086/505380

D.A. Leahy, W.W. Tian, The distance of the SNR Kes 75 and PWN PSR J1846-0258 system. Astron. Astrophys. **480**, 25–28 (2008). doi:10.1051/0004-6361:20079149

J.-J. Lee, B.-C. Koo, M.S. Yun, S. Stanimirović, C. Heiles, M. Heyer, A 21 cm spectral and continuum study of IC 443 using the very large array and the Arecibo telescope. Astron. J. **135**, 796–808 (2008). doi:10.1088/0004-6256/135/3/796

J.-J. Lee, B.-C. Koo, R.L. Snell, M.S. Yun, M.H. Heyer, M.G. Burton, Identification of ambient molecular clouds associated with galactic supernova remnant IC 443. Astrophys. J. **749**, 34 (2012). doi:10.1088/0004-637X/749/1/34

N.A. Levenson, J.R. Graham, S.L. Snowden, The Cygnus Loop: a soft-shelled supernova remnant. Astrophys. J. **526**, 874–880 (1999)

P. Lockett, E. Gauthier, M. Elitzur, OH 1720 megahertz masers in supernova remnants: C-shock indicators. Astrophys. J. **511**, 235–241 (1999)

L.A. Lopez, E. Ramirez-Ruiz, C. Badenes, D. Huppenkothen, T.E. Jeltema, D.A. Pooley, Typing supernova remnants using X-ray line emission morphologies. Astrophys. J. Lett. **706**, 106–109 (2009)

L.A. Lopez, S. Pearson, E. Ramirez-Ruiz, D. Castro, H. Yamaguchi, P.O. Slane, R.K. Smith, Unraveling the origin of overionized plasma in the galactic supernova remnant W49B. Astrophys. J. **777**, 145 (2013). doi:10.1088/0004-637X/777/2/145

M.A. Malkov, L.O. Drury, Nonlinear theory of diffusive acceleration of particles by shock waves. Rep. Prog. Phys. **64**, 429–481 (2001)

R.L. McEntaffer, T. Brantseg, Chandra imaging and spectroscopy of the eastern XA region of the Cygnus Loop supernova remnant. Astrophys. J. **730**, 99 (2011). doi:10.1088/0004-637X/730/2/99

M. Miceli, F. Bocchino, A. Decourchelle, J. Ballet, F. Reale, Spatial identification of the overionized plasma in W49B. Astron. Astrophys. **514**, 2 (2010). doi:10.1051/0004-6361/200913713

M. Miceli, F. Acero, G. Dubner, A. Decourchelle, S. Orlando, F. Bocchino, Shock-cloud interaction and particle acceleration in the southwestern limb of SN 1006. Astrophys. J. Lett. **782**, 33 (2014). doi:10.1088/2041-8205/782/2/L33

T.J. Moriya, Progenitors of recombining supernova remnants. Astrophys. J. Lett. **750**, 13 (2012)

G. Morlino, D. Caprioli, Strong evidence for hadron acceleration in Tycho's supernova remnant. Astron. Astrophys. **538**, 81 (2012)

R. Morrison, D. McCammon, Interstellar photoelectric absorption cross sections, 0.03–10 keV. Astrophys. J. **270**, 119–122 (1983). doi:10.1086/161102

R. Nakamura, A. Bamba, M. Ishida, R. Yamazaki, K. Tatematsu, K. Kohri, G. Pühlhofer, S.J. Wagner, M. Sawada, X-ray spectroscopy of the mixed morphology supernova remnant W28 with XMM-Newton. ArXiv e-prints (2014)

Y. Ohira, K. Murase, R. Yamazaki, Gamma-rays from molecular clouds illuminated by cosmic rays escaping from interacting supernova remnants. Mon. Not. R. Astron. Soc. **410**, 1577–1582 (2011). doi:10.1111/j.1365-2966.2010.17539.x

E. Oliva, D. Lutz, S. Drapatz, A.F.M. Moorwood, ISO-SWS spectroscopy of IC443 and the origin of the IRAS 12 and 25 MU M emission from radiative supernova remnants. Astron. Astrophys. **341**, 75–78 (1999)

M. Ozawa, K. Koyama, H. Yamaguchi, K. Masai, T. Tamagawa, Suzaku discovery of the strong radiative recombination continuum of iron from the supernova remnant W49B. Astrophys. J. Lett. **706**, 71–75 (2009). doi:10.1088/0004-637X/706/1/L71

T.G. Pannuti, J. Rho, K.J. Borkowski, P.B. Cameron, Mixed-morphology supernova remnants in X-rays: isothermal plasma in HB21 and probable oxygen-rich ejecta in CTB 1. Astron. J. **140**, 1787–1805 (2010). doi:10.1088/0004-6256/140/6/1787

G. Park, B.-C. Koo, S.J. Gibson, J.-h. Kang, D.C. Lane, K.A. Douglas, J.E.G. Peek, E.J. Korpela, C. Heiles, J.H. Newton, H I shells and supershells in the I-GALFA H I 21 cm line survey. I. Fast-expanding H I shells associated with supernova remnants. Astrophys. J. **777**, 14 (2013). doi:10.1088/0004-637X/777/1/14

D.J. Patnaude, R.A. Fesen, Small-scale X-ray variability in the Cassiopeia a supernova remnant. Astron. J. **133**, 147–153 (2007). doi:10.1086/509571

J.L. Pineda, W.D. Langer, T. Velusamy, P.F. Goldsmith, A Herschel [C ii] galactic plane survey. I. The global distribution of ISM gas components. Astron. Astrophys. **554**, 103 (2013). doi:10.1051/0004-6361/201321188

J.C. Raymond, Shock waves in the interstellar medium. Astrophys. J. Suppl. Ser. **39**, 1–27 (1979). doi:10.1086/190562

W.T. Reach, J. Rho, Excitation and disruption of a giant molecular cloud by the supernova remnant 3C 391. Astrophys. J. **511**, 836–846 (1999). doi:10.1086/306703

W.T. Reach, J. Rho, T.H. Jarrett, Shocked molecular gas in the supernova remnants W28 and W44: near-infrared and millimeter-wave observations. Astrophys. J. **618**, 297–320 (2005). doi:10.1086/425855

W.T. Reach, J. Rho, A. Tappe, T.G. Pannuti, C.L. Brogan, E.B. Churchwell, M.R. Meade, B. Babler, R. Indebetouw, B.A. Whitney, A Spitzer space telescope infrared survey of supernova remnants in the inner Galaxy. Astrophys. J. **131**, 1479–1500 (2006). doi:10.1086/499306

E.M. Reynoso, P.F. Velázquez, G.M. Dubner, W.M. Goss, The environs of Tycho's supernova remnant explored through the H I 21 centimeter line. Astron. J. **117**, 1827–1833 (1999). doi:10.1086/300814

E.M. Reynoso, A.J. Green, S. Johnston, G.M. Dubner, E.B. Giacani, W.M. Goss, Observations of the neutral hydrogen surrounding the radio-quiet neutron star RX J0822-4300 in Puppis A. Mon. Not. R. Astron. Soc. **345**, 671–677 (2003). doi:10.1046/j.1365-8711.2003.06978.x

J. Rho, T.H. Jarrett, R.M. Cutri, W.T. Reach, Near-infrared imaging and [O I] spectroscopy of IC 443 using two micron all sky survey and infrared space observatory. Astrophys. J. **547**, 885–898 (2001). doi:10.1086/318398

M. Sawada, K. Koyama, X-Ray observations of the supernova remnant W28 with Suzaku. I. spectral study of the recombining plasma. Publ. Astron. Soc. Jpn. **64**, 81 (2012)

M. Seta, T. Hasegawa, T.M. Dame, S. Sakamoto, T. Oka, T. Handa, M. Hayashi, J.-I. Morino, K. Sorai, K.S. Usuda, Enhanced CO $J = 2 - 1 / J = 1 - 0$ ratio as a marker of supernova remnant-molecular cloud interactions: the cases of W44 and IC 443. Astrophys. J. **505**, 286–298 (1998). doi:10.1086/306141

R.L. Shelton, K.D. Kuntz, R. Petre, Chandra observations and models of the mixed-morphology supernova remnant W44: global trends. Astrophys. J. **611**, 906–918 (2004). doi:10.1086/422352

R.L. Shelton, D.P. Cox, W. Maciejewski, R.K. Smith, T. Plewa, A. Pawl, M. Różyczka, Modeling W44 as a supernova remnant in a density gradient with a partially formed dense shell and thermal conduction in the hot interior. II. The hydrodynamic models. Astrophys. J. **524**, 192–212 (1999)

L. Sironi, A. Spitkovsky, J. Arons, The maximum energy of accelerated particles in relativistic collisionless shocks. Astrophys. J. **771**, 54 (2013)

209

P. Slane, R.K. Smith, J.P. Hughes, R. Petre, An X-ray study of the supernova remnant G290.1-0.8. Astrophys. J. **564**, 284–290 (2002)

P. Slane, S.-H. Lee, D.C. Ellison, D.J. Patnaude, J.P. Hughes, K.A. Eriksen, D. Castro, S. Nagataki, A CR-hydro-NEI model of the structure and broadband emission from Tycho's SNR. Astrophys. J. **783**, 33–42 (2014)

R.L. Snell, D. Hollenbach, J.E. Howe, D.A. Neufeld, M.J. Kaufman, G.J. Melnick, E.A. Bergin, Z. Wang, Detection of water in the shocked gas associated with IC 443: constraints on shock models. Astrophys. J. **620**, 758–773 (2005). doi:10.1086/427231

A.W. Strong, I.V. Moskalenko, O. Reimer, Diffuse continuum gamma rays from the Galaxy. Astrophys. J. **537**, 763–784 (2000). doi:10.1086/309038

D.F. Torres, A.Y. Rodriguez Marrero, E. de Cea Del Pozo, MAGIC J0616+225 as delayed TeV emission of cosmic rays diffusing from the supernova remnant IC 443. Mon. Not. R. Astron. Soc. **387**, 59–63 (2008). doi:10.1111/j.1745-3933.2008.00485.x

E. Troja, F. Bocchino, M. Miceli, F. Reale, XMM-Newton observations of the supernova remnant IC 443. II. Evidence of stellar ejecta in the inner regions. Astron. Astrophys. **485**, 777–785 (2008). doi:10.1051/0004-6361:20079123

H. Uchida, K. Koyama, H. Yamaguchi, M. Sawada, T. Ohnishi, T.G. Tsuru, T. Tanaka, S. Yoshiike, Y. Fukui, Recombining plasma and hard X-ray filament in the mixed-morphology supernova remnant W 44. Publ. Astron. Soc. Jpn. **64**, 141 (2012)

Y. Uchiyama, F.A. Aharonian, Fast variability of nonthermal X-ray emission in Cassiopeia A: probing electron acceleration in reverse-shocked ejecta. Astrophys. J. Lett. **677**, 105–108 (2008). doi:10.1086/588190

Y. Uchiyama, F.A. Aharonian, T. Tanaka, T. Takahashi, Y. Maeda, Extremely fast acceleration of cosmic rays in a supernova remnant. Nature **449**, 576–578 (2007). doi:10.1038/nature06210

Y. Uchiyama, R.D. Blandford, S. Funk, H. Tajima, T. Tanaka, Gamma-ray emission from crushed clouds in supernova remnants. Astrophys. J. Lett. **723**, 122–126 (2010). doi:10.1088/2041-8205/723/1/L122

E.F. van Dishoeck, D.J. Jansen, T.G. Phillips, Submillimeter observations of the shocked molecular gas associated with the supernova remnant IC 443. Astron. Astrophys. **279**, 541–566 (1993)

P.F. Velázquez, G.M. Dubner, W.M. Goss, A.J. Green, Investigation of the large-scale neutral hydrogen near the supernova remnant W28. Astron. J. **124**, 2145–2151 (2002). doi:10.1086/342936

J. Vink, Supernova remnants: the X-ray perspective. Astron. Astrophys. Rev. **20**, 49 (2012). doi:10.1007/s00159-011-0049-1

Z. Wang, N.Z. Scoville, Strongly shocked interstellar gas in IC 443. I—high-resolution molecular observations. Astrophys. J. **386**, 158–169 (1992). doi:10.1086/171001

M. Wardle, Enhanced OH in C-type shock waves in molecular clouds. Astrophys. J. Lett. **525**, 101–104 (1999)

G.J. White, R. Rainey, S.S. Hayashi, N. Kaifu, Molecular line observations of IC 443—the interaction of a molecular cloud and an interstellar shock. Astron. Astrophys. **173**, 337–346 (1987)

R.L. White, K.S. Long, Supernova remnant evolution in an interstellar medium with evaporating clouds. Astrophys. J. **373**, 543–555 (1991)

B.J. Williams, K.J. Borkowski, P. Ghavamian, J.W. Hewitt, S.A. Mao, R. Petre, S.P. Reynolds, J.M. Blondin, Azimuthal density variations around the rim of Tycho's supernova remnant. Astrophys. J. **770**, 129 (2013)

P.F. Winkler, B.J. Williams, S.P. Reynolds, R. Petre, K.S. Long, S. Katsuda, U. Hwang, A high-resolution X-ray and optical study of SN 1006: asymmetric expansion and small-scale structure in a type Ia supernova remnant. Astrophys. J. **781**, 65 (2014). doi:10.1088/0004-637X/781/2/65

H. Yamaguchi, M. Ozawa, K. Koyama, K. Masai, J.S. Hiraga, M. Ozaki, D. Yonetoku, Discovery of strong radiative recombination continua from the supernova remnant IC 443 with Suzaku. Astrophys. J. Lett. **705**, 6–9 (2009). doi:10.1088/0004-637X/705/1/L6

H. Yamaguchi, M. Tanaka, K. Maeda, P.O. Slane, A. Foster, R.K. Smith, S. Katsuda, R. Yoshii, Elemental abundances in the possible type Ia supernova remnant G344.7-0.1. Astrophys. J. **749**, 137 (2012). doi:10.1088/0004-637X/749/2/137

Z. Zhang, Y. Gao, J. Wang, CO observation of SNR IC 443. Sci. China, Phys. Mech. Astron. **53**, 1357–1369 (2010). doi:10.1007/s11433-010-4010-5

V.N. Zirakashvili, V.S. Ptuskin, Diffusive shock acceleration with magnetic amplification by nonresonant streaming instability in supernova remnants. Astrophys. J. **678**, 939–949 (2008). doi:10.1086/529580

DOI 10.1007/978-1-4939-3547-5_7
Reprinted from *Space Science Reviews* Journal, DOI 10.1007/s11214-015-0186-3

The Heliosphere: What Did We Learn in Recent Years and the Current Challenges

M. Opher[1]

Received: 25 June 2015 / Accepted: 10 July 2015
© Springer Science+Business Media Dordrecht 2015

Keywords Interstellar medium · Heliosphere

As the Sun moves through the interstellar medium it carves a bubble called the heliosphere. A fortunate confluence of missions has provided a treasury of data that will likely not be repeated for decades The measurements in-situ by the Voyager spacecrafts, combined with the all-sky images of the heliospheric boundaries by the Interstellar Boundary Explorer and CASSINI missions have transformed our understanding of heliosphere. In particular one of the first surprises was that both Voyager spacecrafts found no evidence for the acceleration of the anomalous cosmic rays (ACRs) at the termination shock as expected for approximately 25 years. Another challenge are the energetically particles intensities and the plasma flows that are dramatically different at Voyager 1 and 2. There are several other observations that are key challenges to the heliospheric models that indicate that the nature of the heliosheath (the region where the solar wind is subsonic) is much more complex than thought, such as (a) Why the azimuthal magnetic flux is not conserved along the Voyager 1 trajectory? (b) What causes the flow stagnation region seen at Voyager 1? (c) What causes the unexpected observation of a depletion-region beginning in 2012 at Voyager 1? These observations point to the need to move past the standard description of the heliosphere. In this paper, I will review the state-of-the art of our understanding of this "new heliosphere". In late 2012 Voyager 1 observed several events that indicated a magnetic connectivity between the heliosheath magnetic field and the interstellar medium magnetic field; where the energetic particles of the heliosheath leaked out while the galactic cosmic rays penetrated the heliosheath. With the radio observations confirming densities of the interstellar medium, there is a consensus that Voyager 1 left the heliosphere in September 2012 and entered the

Note by the editor: This paper was meant to be part of the topical volume on 'Multi-Scale Structure Formation and Dynamics in Cosmic Plasmas', Volume 188, 2015, edited by A. Balogh, A. Bykov, J.P. Eastwood, and J.S. Kaastra.

✉ M. Opher
  mopher@bu.edu

[1] Department of Astronomy, Boston University, Boston, MA, USA

interstellar medium. We will review as well our current understanding of the nature of the heliopause. The knowledge gain from the edge of the heliosphere will have consequences for other astrospheres and astrosheaths where the magnetic nature of the winds could be much more complex that previously thought.

1 Introduction

"O Mar sem fim sera grego ou troiano, o mar sem fim sera português; The Ocean that has an end will be greek or trojan; the Ocean without an end will be portugues", Fernando Pessoa.

As the Sun moves through the interstellar medium its solar wind carves a bubble called the *heliosphere*. The solar wind is supersonic with wind speeds around 400–800 km/s and as it approaches the edge of the *heliosphere* it goes through a shock, called the termination shock (TS). Beyond the termination shock the solar wind is subsonic as it approaches the very edge that separates the solar wind domain from the interstellar medium (ISM), called the *heliopause* (HP). The HP is thought a tangential discontinuity where the pressure of the solar wind equalizes the pressure from the ISM. The structure of the heliosphere with its different components is seen in Fig. 1.

The *termination shock* (*TS*) marks the boundary where the supersonic solar wind decelerates to slower subsonic speeds. The *heliopause* (*HP*) is the boundary separating the hot solar wind and the colder, denser interstellar plasma and is often considered as the boundary of the heliosphere. The region between the TS and HP with decelerated compressed hot solar wind is called the *heliosheath*. The interstellar medium is disturbed by the interaction with the heliosphere. Depending on the properties of the local interstellar medium a *bow shock* or *bow wave* forms in the interstellar plasma in front of the heliosphere.

There are several basic features of the very nature of heliosphere that are still not well understood. These aspects stem from the very "shape" of the heliosphere; the extent of its tail; the nature of the heliosheath; the structure of the interstellar medium just ahead of it. Both the in-situ measurements by Voyager spacecrafts and the remote energetic neutral atoms (ENA) maps from IBEX and CASSINI help us solve some of the problems but brought many more puzzles. These missions will continue to unravel more surprises and help us constrain some of the models. However only with a revisit of this region with a modern instrumentation; we will be able to shade light on the very fundamental aspects of our home within the galaxy, the heliosphere.

The shape of the heliosphere and the structure of the interface are determined by various physical processes. Interstellar hydrogen atoms penetrating into the heliosphere interact with the solar wind protons in a charge exchange process creating an energetic population of ions called *pick up ions*. Early theoretical studies (Baranov and Malama 1993) predicted that the charge exchange process decelerates the solar wind and pushes the heliosphere boundary toward the Sun.

Both solar wind and interstellar medium are magnetized and the magnetic field is one of the key elements determining the structure of the outer heliosphere. A tilted interstellar magnetic field distorts the shape of the heliosphere producing the asymmetry of the TS and HP (Fig. 2). The B_{ISM} distort the heliosphere pushing the southern side closer to the Sun. The heliospheric asymmetry was confirmed by the crossing of the TS by Voyager 2, 10 AU closer to the Sun than V1 (Stone et al. 2008), although part of the asymmetry could be due to time-dependent effects (as argued by works such as Pogorelov et al. 2009).

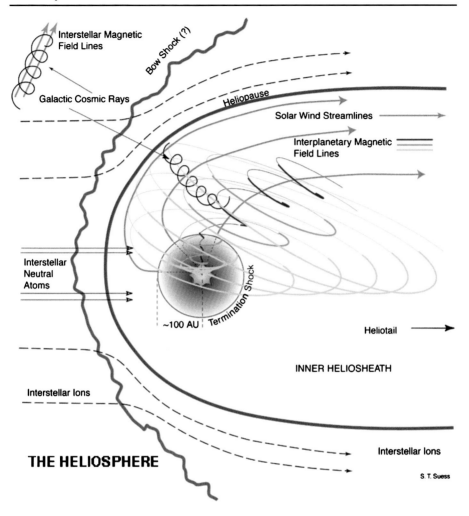

Fig. 1 Structure of the region where solar wind interacts with the ISM. SOURCE: Jet Propulsion Laboratory (1999a), courtesy of Steven T. Suess

The very nature and direction of the magnetic field ahead of the heliosphere is being debated. In order to explain the observed heliospheric asymmetries seen by Voyager (Opher et al. 2009; Izmodenov et al. 2009) suggest a strong interstellar magnetic field with the strength of ~ 4 µG and north-south component producing a tilt angle ~ 10–$20°$ relative to the interstellar flow direction v_{ISM} (in respect to the Sun). Anther constrain on the B_{ISM} is the deflection of the H atoms with respect to the He atoms (Lallement et al. 2005, 2010) that constrain the plane B–V of the $B_{ISM} - v_{ISM}$ to be in what is referred as the "Hydrogen Deflection Plane" ($60°$ from the ecliptic plane). In 2009, the Interstellar Boundary Explorer (IBEX) revealed that the energetic neutral atoms maps produced a ribbon of higher intensity around energies ~ 1 keV (McComas et al. 2009). There is an ongoing debate where the ribbon is produced and by which mechanism; although generally it seem to be organized by the direction where the radial component of B_{ISM} goes to zero ($B_{ISM} * r = 0$). Works that try to fit the IBEX ribbon by mechanisms that produce them outside the Heliopause (e.g.,

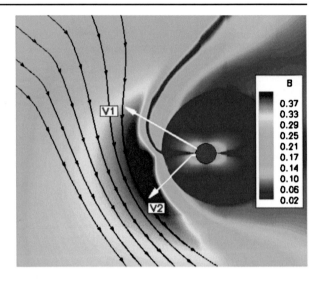

Fig. 2 Tilted interstellar magnetic field (*black curves*) creates asymmetric heliosphere. Trajectories by the *white arrows* of Voyager 1 and 2 are shown by *white arrows* (Opher et al. 2006)

Heerikhuisen et al. 2010); use a direction where the title angle is larger $\sim$30–40° relative to the interstellar flow direction and intensity not exceeding 3.5 µG with a B–V plane that differ from the HDP plane by 20°. These mechanisms assume secondary charge exchange in which the energetic neutral atoms (ENAs) charge exchange in the plasma outside the heliopause creating pick-up ions (PUIs). It is assumed that these PUIs will retain a ring-beam distribution with the velocity component along the interstellar magnetic field for sufficient time until they are charge exchanged again. These "secondary" ENAs will be enhanced in the locations where $B_{ISM} * r = 0$. This debate can only be solved as the Voyager mission or a future one will adventure farther into the interstellar medium ahead of the Heliosphere.

Another debate is the extent with which the influence of the heliosphere in the local interstellar medium and how B_{ISM} drape around the heliosphere; if as an ideal draping (i.e., draping on a surface without interacting with the surface itself) or mediated by another process (such as temporal instabilities; or reconnection). The expected direction of B_{ISM} implies that the interstellar magnetic field is not parallel to the solar Parker spiral magnetic field, which has an east-west direction. The models predicted the dramatic rotation of the magnetic field direction after the heliopause crossing. However when Voyager 1 crossed the HP at the distance of $\sim$120 AU in August 2012 observations revealed completely unexpected behavior of the magnetic field. The magnetic field magnitude increased from 1 µG in the heliosheath to $\sim$4 µG outside the HP but there was almost no change in the direction of the magnetic field. These data sparked a search for physical processes responsible for such behavior of the magnetic field at the heliosphere boundary. Recent work (Opher and Drake 2013) suggested that the draping of the interstellar magnetic field B_{ISM} around the HP is strongly affected by the solar wind magnetic field. As it approaches the heliopause B_{ISM} twists and acquires the east-west component. The physical reasons of such interaction of the heliospheric and interstellar magnetic fields remain to be understood. Some recent works argue that the observed direction of B_{ISM} outside the HP can be explained by draping around an ideal surface (Isenberg et al. 2015). Others explain the change in direction by temporal instabilities (Krimigis et al. 2013; Florinski et al. 2015; Pogorelov et al. 2014).

Another recent debate is the very shape of the heliosphere and the extent of its tail. The long accepted view of the shape of the heliosphere is that it is a comet-like object (Parker 1961; Baranov and Malama 1993) with a long tail opposite to the direction in which the

Fig. 3 Two-lobe structure of the heliosphere. *Yellow surface* shows the heliopause surface. *Grey curves* show the solar magnetic field lines, *red curves*—interstellar magnetic field lines (Opher et al. 2015)

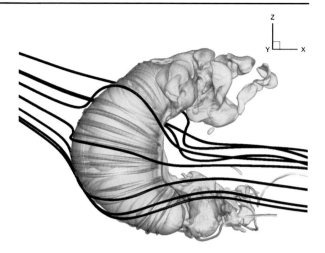

solar system moves through the local interstellar medium (ISM). The solar magnetic field at a large distance from the Sun is azimuthal, forming a spiral (the so called "Parker spiral") as a result of the rotation of the Sun. The traditional picture of the heliosphere as a cometary-like structure comes from the assumption that even though the solar wind becomes subsonic at the termination shock as it flows down the tail, it is able to stretch the solar magnetic field. Opher et al. (2015) argued based on MHD simulations, that the twisted magnetic field of the Sun confines the solar wind plasma and drives jets to the north and south very much like some astrophysical jets (Fig. 3).

The two lobes are formed by the magnetic tension of the solar azimuthal magnetic field that in the heliosheath resist being stretched by the subsonic flows. The ratio between the force stretching the magnetic field due to the flows and the magnetic tension (hoop stress) resisting the stretch is given by $F_{streatch}/F_{tension} \sim P_{ram}/2P_B$ where P_{ram} is the ram pressure and P_B is the magnetic pressure. In the heliosheath this ratio is <1 so the magnetic tension (hoop stress) is sufficient to resist the stretching by the flows and can collimate jets. The result is a tail divided in two separate plasmas confined by the solar magnetic field.

For our local interstellar medium (ISM), the pressure is not strong enough to force the two lobes in a single tail. For astrospheres where the ISM ram pressure is strong enough the two lobes might join in a unique tail.

In the heliosheath the plasma pressure is generally much higher than the magnetic pressure, so it might seem surprising that the magnetic field controls the formation and structure of the jets. As shown in Drake et al. (2015) the overall structure of the heliosheath is controlled by the solar magnetic field even in the limit in which the ratio of the plasma to magnetic field pressure, $\beta = 8\pi P/B^2$, is large. The tension of the solar magnetic field produces a drop in the total pressure between the termination shock (TS) and the heliopause. This same pressure drop accelerates the plasma flow downstream of the TS into the North and South directions to form two collimated jets.

Other magnetospheres (such as Earth, Siscoe et al. 2004, and Saturn, Zieger et al. 2010; Jia et al. 2012) exhibit a two-lobe structure. These structures are not related to the phenomena discussed in this paper, but to reconnection of the down-tail component of the draped solar magnetic field that produces a dominant midtail x-line. The key ingredient here is the solar magnetic field that confines and collimates the solar wind and the ISM pressure that maintains the separation of the two lobes in the tail.

Astrophysical jets around massive black holes are thought to originate from Keplerian accretion disks and are driven by centrifugal forces (Blandford and Payne 1982). However, the jets in the case of the heliosphere are driven downstream of the termination shock similar to what was proposed for the Crab Nebula (Chevalier and Luo 1994; Lyubarsky 2002). In this region of subsonic flows, the magnetic tension (hoop) force is strong enough to collimate the wind. The tension force is also the primary driver of the outflow. (Fig. 3, Opher et al. 2015).

The overall two-lobe structure is consistent with the ENA images from IBEX that for the first time mapped the heliotail. Such images show two lobes (McComas et al. 2013) with an excess of low energy ENA (<1 keV) and a deficit at higher energy (>2 keV) around the solar equator. The ENA images from Cassini (Krimigis et al. 2009; Dialynas et al. 2013; at much higher energies, 5–55 keV) revealed intensities that were comparable in the direction of the nose and tail. The observers therefore concluded that the heliosphere might be "tailless" because the emission from these high-energy ENAs is believed to come from the heliosheath. The two-lobe heliosphere is in fact almost "tailless" with the distance down the tail to the ISM between the lobes being nearly equal to the distance toward the nose. McComas et al. (2013) interpreted the ENA tail measurements as a result of a slower wind to the fact that the Sun has been sending out fast solar wind near its poles and slower wind near its equator. With additional ENA measurements through an extended solar cycle it will be possible to distinguish between the two scenarios.

2 Termination Shock

Both Voyager 1 and 2 (V1 and V2) are now beyond the TS, V1 most likely beyond the HP, although there are works that disagree that V1 is beyond the HP (Fisk and Gloeckler 2013; 2014; McComas and Schwadron 2012; Schwadron and McComas 2013a).

The main disagreement stems from the magnetic field measurements that indicate that the magnetic field as measured by V1 didn't change direction across that boundary. We will come back to that later on when we discuss the HP. V2 is the only spacecraft (among the Voyagers) that carry a working plasma instrument (although the plasma flows in the RT plane can be inferred from the particle anisotropies from V1—Richardson and Decker 2014).

With the crossing of TS by V2 that carried the working plasma instrument, it become clear that the TS was not a just a one-fluid MHD perpendicular shock as previously expected. One of the surprises was that the heliosheath plasma temperature was much colder, by an order of magnitude than expected if all the energy upstream was transferred to the plasma thermal population (Richardson et al. 2008) (Fig. 4). The measurements downstream the TS are consistent with 80 % of the energy transferred to the suprathermal population, the pick up ions (Gloeckler et al. 2005; Zank et al. 1996). Pick-up ions, are not measured by Voyager spacecrafts. *In fact there is a gap in energy between the thermal plasma (at energies <1 keV) to 40 keV*, the lowest energy measured by the LECP instrument. It is also possible that electrons played an important role in the energy budget stilling part of the energy downstream (Zieger et al. 2015). Again there is a gap between what the plasma instrument measure (∼ eV) to the lowest energies at LECP (30 keV). It is possible that hot electrons play an important role in the TS crossing and downstream in the heliosheath thermodynamics (Chalov and Fahr 2013; Chashei and Fahr 2013; 2014). This can only be resolved with a new visit to that region with proper instrumentation that bridge the gap in those energies; i.e., able to measure the suprathermal PUI population from 1 keV–40 keV and energetic

Fig. 4 From Richardson et al. (2008). The V2 data measured at TS (crosses), "in comparison with V2 data measured at Neptune's inbound bow shock crossing (*diamonds*). The solar wind parameters upstream of Neptune are normalized to those upstream of the TS; the timescales are identical. The solar wind speed (**a**; Neptune data divided by 1.3) at the bow shock fell by a factor of four but at the termination shock the speed decreased by a factor of only two. The density (**b**; Neptune data divided by five) at the bow shock increased by a factor of four, but at the termination shock by a factor of two. The major difference is in the temperature (**c**; Neptune data divided by two): at the bow shock it increased by a factor of 100, but at the termination shock by a factor of only ten. The differences between these two shocks are probably caused by the greater abundance of pickup ions at the TS

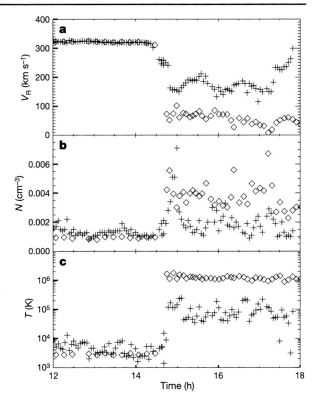

electrons in the same energy. Only then we will be able to definitively probe the structure of the TS and the thermodynamics of the HS.

3 Heliosheath

As Voyager 1 and 2 adventured into the region where the solar wind is subsonic, the heliosheath (HS) it became clear that there are several observations that challenge our understanding of that region. While global models advanced rapidly in sophistication in the last decade, these models are still not able to predict self-consistently the flows, fields and particles behavior in the HS. Furthermore, none of the current standard global models predict the very thin HS (~30–40 AU) implying that Voyager 1 (V1) did indeed cross the HP.

There are several observations that are key challenges to the heliospheric models:

(a) The flows at V1 and Voyager 2 (V2) are very different; (b) the presence of a flow stagnation region seen at Voyager 1; (c) the V1 observations suggest that the magnetic flux in the HS is not conserved; (d) the fact that the Anomalous Cosmic Ray (ACR) spectrum roll out well into the heliosheath; (e) the thin heliosheath; and (f) different behavior of energetic particles at V1 and 2; including dropouts of ~1 MeV electrons and the most energetic ACRs at V2.

One of the biggest puzzles is why the flows in the heliosheath are so different at V1 and 2 (Fig. 5). After six years in the sheath, V2 flow magnitudes remain high, near 150 km/s, while V1 flows dropped to zero after 2010 and are sometimes negative. In fact, all the components of the speed at V1 became small in 2010 (Krimigis et al. 2011). Current global models

Fig. 5 Very different flows on board of V1 (*left panel*) and 2 (*right panel*). V1 doesn't measure directly the flows. They are inferred. The velocity components in V1 are calculated from measurements of 53–85 keV ion intensities. The components that V1 is able to extract are in the RT plane in the R–T–N heliographic polar coordinates in which the transverse (+T) direction is that of planetary motion around the Sun and +R is the radial direction relative to the Sun. Panel (**b**) shows the VR components as well as density N; temperature T and the RT = tan −1(VR/VT) and RN = tan −1(VR/VT) angles on V2

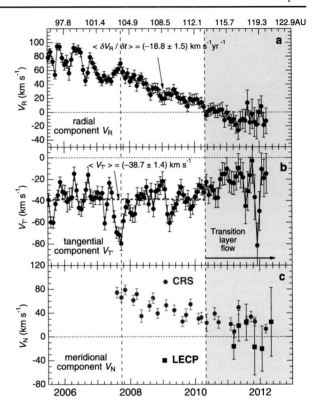

don't correctly predict the observed flows at V1 and V2 either in magnitude or direction. All current models (Opher et al. 2009; Ratkiewicz and Grygorczuk 2008; Izmodenov et al. 2009; Pogorelov et al. 2007) predict the HS flows will slowly turn to the flanks and to the poles as the Voyagers move deeper into the sheath. Instead, the V2 flows turn much more rapidly in the transverse direction than in the normal direction. Is the HP flatter than we thought or are we missing something else?

In particular, the zero values of radial flow at V1 pose a challenge to the models, since in current models the flow rotates parallel to the HP and the radial component gradually decreases asymptotically (not abruptly) to zero, and it should become zero only at the HP itself.

There have been recent suggestions that the flows can be explained by the gradients in pressure as shown by the integrated pressure flux of PUIs (McComas and Schwadron 2014)

Another puzzle comes from the magnetic field. We expect that from flux conservation $B_T V_R R \sim$ const. However, when V_R decreased at V1 the magnitude of B_T didn't increase as expected (Richardson et al. 2013) (Fig. 6). Even as V_R went to zero B_T stayed around 0.1–0.2 nT (Burlaga and Ness 2012). (The exact conservation is $B_T V_\perp L \sim const.$, where $V_\perp = \sqrt{V_R^2 + V_N^2}$, and L is the separation between streamlines.) The non-conservation of magnetic flux cannot be explained by solar cycle variations of the solar wind and magnetic field intensity (Michael et al. 2015).

After the crossing of the TS by V1 and then by V2, one of the first surprises was that both Voyager spacecrafts found no evidence for the acceleration of the anomalous cosmic rays (ACRs) at the TS, as expected for approximately 25 years (Fisk et al. 1974). The expectation

 🦊 Springer

Fig. 5 (*Continued*)

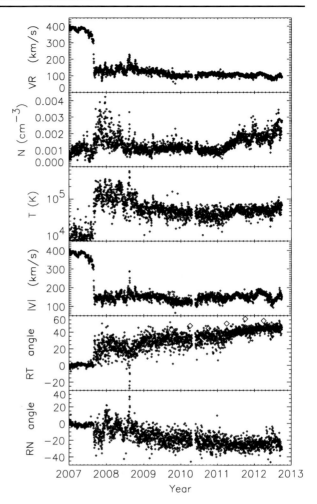

was that the ACRs were accelerated at the largest shock in the heliosphere, the TS. The ACR intensities not only didn't peak at the shock, but their intensity kept increasing as the spacecraft moved deeper into the sheath (Decker et al. 2010; Cummings et al. 2008). This finding generated several hypotheses for the ACRs acceleration mechanisms and locations: in the flanks of the shock (McComas and Schwadron 2006); in "hot spots" in a turbulent TS (Kota 2010; Guo et al. 2010); deep in the sheath; by reconnection (Lazarian and Opher 2009; Drake et al. 2010); or by turbulence processes also deep in the hot HS (Fisk and Gloeckler 2009).

Another mystery comes from the different behavior of energetic particles at V1 and V2 (Fig. 7). The particles at V2 show variations of intensity of more than three orders of magnitude correlated with periods when the spacecraft was in and out of the sector region (as indicated by Wilcox data) (Hill et al. 2014), while the intensities at V1 remained steady. When V2 is in the sector region the intensities are substantially higher then when it is in the unipolar region. There is more than a three order of magnitude energy range (highest energies not shown) over which ions and electrons vary coherently with the passage of the temporally varying spatial structure, the edges of the sector region.

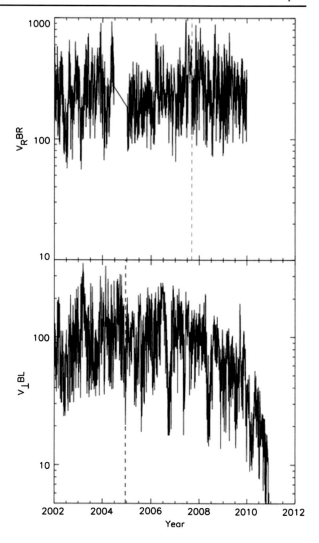

Fig. 6 What happened to the missing azimuthal magnetic flux at Voyager 1? The magnetic flux observed at V2 (*top*) and V1 (*bottom*). B is normalized by the values at 1 AU, V is in km/s and L is in AU. The *vertical dashed lines* show the TS locations (from Richardson et al. 2013)

There is also the problem of the HS thickness. Most models predict a thickness of ~50 AU even after accounting for time dependence (Richardson and Wang 2010; Provornikova et al. 2014). Models that include both the thermal and suprathermal components, such as pick-up ions (e.g., Malama et al. 2006) predict some reduction in the thickness. But these models still don't match the observed heliosheath thickness of 27 AU.

Which other aspects of the nature of the HS are we missing in our models that could thin the HS?

To solve these puzzles, in the last couple of year have been several suggestions for additional effects such as reconnection in the sector region (the region where the solar magnetic field reverses polarity) and near the HP, turbulence, and time dependent effects.

Reconnection within the sector region (as suggested by Opher et al. 2011; Drake et al. 2010) explains the ACR spectrum rolling over well into the HS by acceleration from reconnection. It can also explain the dropout of particles on V2; while particle were steady at V1 by different transport properties within a reconnected sector region – given that V2 was

Fig. 7 Temporal variations of the latitudinal boundaries of the sectored HS and V1 and 2 energetic ion and electron intensities, where V2 shows a clear correspondence to the sector configuration (Hill et al. 2014)

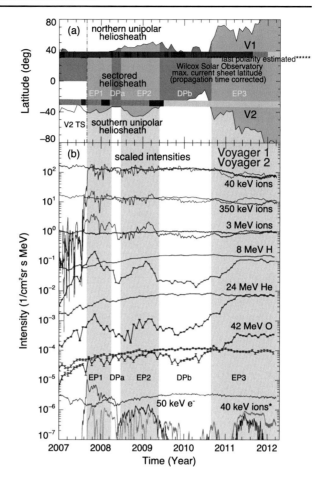

in an out of the sector while V1 was immersed within it throughout its trajectory (Opher et al. 2011). Reconnection can also explain the missing azimuthal magnetic flux at V1 and potentially the flow stagnation region seen at V1 (Opher et al. 2012).

Reconnection within the sector region is a new regime of reconnection different than any other location in the heliosphere; is where plasma β (ratio of thermal to magnetic pressure) is high (while usually reconnection occurs in regions of low plasma β) and the guide field is zero (anti-symmetric reconnection). In that regime (Opher et al. 2011; Schoeffler et al. 2013) the magnetic islands are very elongated and the magnetic profile is similar to the sector (Fig. 8). This poses a challenge to the magnetometer on Voyager 1 and 2 that is tuned to strong field for the strong fields of the outer planets and not for the week fields of the heliosheath. The uncertainty on the magnetometer on V1 is 0.03 nT in each component and on V2 0.05 nT while the average field intensity in the HS is 0.1 nT.

We need a way to extract energy from the HS. Is reconnection within the sector region (as suggested by Opher et al. 2011; Drake et al. 2010) sufficient (Fig. 8)? Perhaps the HS has a strong turbulent component (as suggested by Fisk and Gloeckler 2013)? Are temporal effects such as instabilities (Pogorelov et al. 2012; Florinski et al. 2015) or other non-ideal MHD effects important? Most likely instabilities such as Rayleigh–Taylor instability won't be present because of the stabilization effect of the interstellar magnetic field. Izmodenov et al.

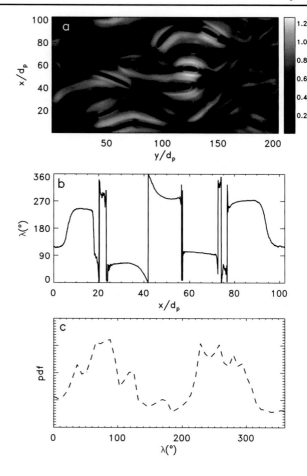

Fig. 8 Multi-current system in its final stage after reconnection ceased. Elongated magnetic islands are formed with intense magnetic field at the walls. The *bottom panels* show the magnetic field magnitude, direction, and distribution function for a cut through the model results (Opher et al. 2011)

(2009) suggests that electron thermal conduction can thin significantly the heliosheath; in the limiting case where the thermal conduction is very effective the heliosheath was thinned to 32 AU.

4 Heliopause

Between May and August 2012 there was a series of puzzling events. The cosmic ray flux increased rapidly in May. Then in August the intensity of particles that were accelerated in the heliosphere (from ∼30 keV to MeV) decreased to background levels (intensity decreases of a factor of ∼1000). At the same time the galactic cosmic rays intensity again increased, this time to the highest level ever observed. The magnetic field magnitude simultaneously increased (Fig. 9). This transition had been dubbed the "heliocliff". One of the expected signatures of the crossing of the HP, was that the magnetic field direction would significantly change. This is expected because the solar magnetic field just inside the HP is azimuthal, or east-west, on average (called the "Parker field"), while the magnetic field in the interstellar medium (derived from several indirect indicators) is widely believed to be inclined significantly to the east-west direction (Izmodenov et al. 2009; Opher et al. 2009;

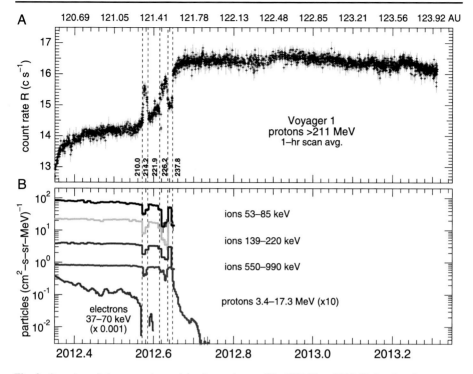

Fig. 9 Overview of the energetic particle observations at V1, 2012.35 to 2013.40 showing the contrary behavior of GCRs and lower-energy particles (from Krimigis et al. 2013)

Pogorelov et al. 2009; Ratkiewicz and Grygorczuk 2008). The absence of a significant rotation in the direction of the magnetic field at the times of dropouts of energetic particles were initially interpreted as indicating that V1 was still in the HS (Burlaga et al. 2013; Krimigis et al. 2013; Stone et al. 2013; Fisk and Gloeckler 2013; McComas and Schwadron 2012) although some models suggested the contrary (Swisdak et al. 2013).

However, in September of 2013 the plasma wave team announced the detection of 2–3 kHz plasma waves, so the plasma densities indicated V1 was in the interstellar medium (ISM) (Gurnett et al. 2013), although not all agree (Fisk and Gloeckler 2013; McComas and Schwadron 2012, 2013a).

If V1 were beyond the HP, then why is the magnetic field outside the HP still within ~20° of the Parker spiral direction (Burlaga and Ness 2014) and thus very different from the B direction expected deeper in the ISM? Could this difference be due to the shape of the HP and magnetic draping geometry, magnetohydrodynamic (MHD) instabilities, temporal aspects, or not having really crossed HP? Opher and Drake (2013) propose that, regardless of the direction in the ISM, near the HP the field twists to the Parker direction (Fig. 11). Not all modelers agree and this question is being hotly debated. Some argue that *ideal* draping, i.e., draping on a surface without communication between the solar and interstellar magnetic field can account for that (e.g., Grygorczuk et al. 2014). Do other aspects such as reconnection or turbulence play a role in this local rotation? The implications of understanding the behavior of the magnetic field ahead of the HP has consequences not only for what V2 will encounter as it approaches and crosses the HP, but for what V1 will see as it adventures farther away from the HP into the ISM.

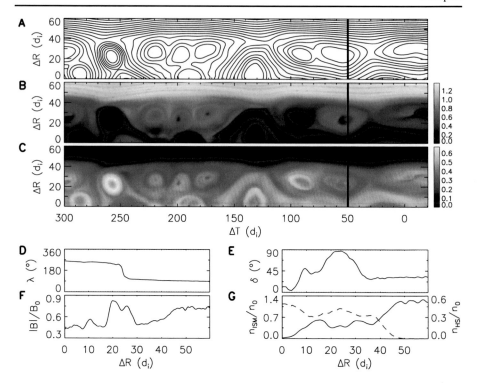

Fig. 10 Structure of the HP and adjacent LISM and HS at late time from the PIC simulation. In the R/T plane in (**A**) the magnetic field lines and in (**B**) and (**C**) the number density n_{LISM} (n_{HS}) of particles originally in the LISM (HS). Panels (**D**)–(**G**) are cuts along the vertical line in panels (**A**)–(**C**). In (**D**) λ is the angle of B in the R–T plane with respect to the R direction. In (**E**) δ is the angle between B and the R–T plane. In (**F**), the magnitude of B and, in (**G**), the number density n_{LISM} (*solid*) and the number density n_{HS} (*dashed red*) (from Swisdak et al. 2013)

Swisdak et al. (2013) based on particle-in-cell simulations, based on cuts through the MHD model at V1's location, suggest that the sectored region of the HS produces large-scale magnetic islands that reconnect with the interstellar magnetic field while mixing the local interstellar medium (LISM) and HS plasma. Cuts across the simulation reveal multiple, anti-correlated jumps in the number densities of LISM and HS particles at magnetic separatrices where there is essentially no magnetic field rotation (Fig. 10). The absence of rotation at these dropouts is consistent with the V1 observations. In this model (Swisdak et al. 2013) the authors argue that V1 had crossed the HP at the end of July 2012. Soon after this paper was published, the Voyager team reached the conclusion that V1 was in the interstellar space based on the detection of radio emissions (Gurnett et al. 2013).

Other works proposed that the HP dropouts can be explained by MHD reconnection predicting islands structures before the crossing of the HP (Strumik et al. 2013).

It is debated within the community if similar structure should be expected or will be seen when V2 will cross the HP. In any case the plasma instrument on board of V2 will only be sensitive to the thermal component. In order to sort out the different scenarios this region should be revisited with sensitive magnetometer and a particle instrument covering the suprathermal populations especially in the gap between 1 keV–40 keV.

Fig. 11 View at the nose of the heliosphere from the interstellar medium towards the Sun. The nose of the HP is shown in the *yellow* iso-surface (defined by $\ln T = 11.9$–12). The *gray field lines* are the B_{ISM} wrapping and twisting around the HP (Opher et al. 2013)

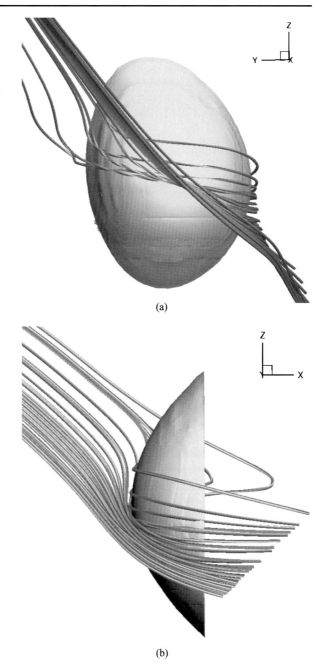

(a)

(b)

5 ENA Observations of the Global Structure of the Heliosphere

Another way to probe the global structure of the heliosphere is through energetic neutral atoms (ENAs). Both Interstellar Boundary Explorer (IBEX) and CASSINI/INCA instruments mapped that region in different energy ranges. IBEX is a small explorer mission

Fig. 12 From Schwadron et al. (2014). Pressure of plasma protons that form observed ENAs integrated over line-of-sight (LOS) as observed by *IBEX* and referenced to the inertial frame *IBEX*-Hi measurements from *IBEX*-Hi (from 0.7 to 4.3 keV) from 0.7 to 4.3 keV

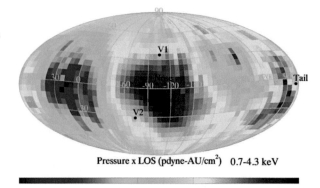

Pressure x LOS (pdyne-AU/cm²)   0.7-4.3 keV

| 10 | 15 | 20 | 25 | 30 |

that revolves around the Earth returning ENAs images in the range 0.2 keV–4.3 keV range (McComas et al. 2009). CASSINI/INCA measures ENAs in much higher energy range (~5.4–55 keV) (Krimigis et al. 2009). Both spacecrafts measured unexpected features, IBEX, a so called- "ribbon" around 1 keV energies and CASSINI/INCA a so- called "belt" around 4–13 keV. The IBEX "ribbon" seem to be organized by the interstellar magnetic field $B_{ISM} \cdot R = 0$ (or the location where the radial component of B_{ISM} is zero) and prompted a series of papers trying to explain its origin. In any case these data demonstrated as well as the Voyager heliospheric asymmetries that the heliosphere is strongly affected by the interstellar magnetic field.

All the different theories have pros and cons when compared to the data as summarized by McComas et al. (2011). Because of the ordering with $B_{ISM} \cdot R = 0$ most proposed mechanisms are outside the heliosphere in the outer heliosheath; Some proposed mechanisms make use of secondary charge exchange (e.g., Heerikhuisen et al. 2010); magnetic mirror (Chalov et al. 2010); etc. However there is an issue of scattering and stability of the pick-up ions (PUIs) in the local ISM (Florinski et al. 2010) so more recent mechanisms use some kind of a trapping mechanism (Schwadron and MComas 2013b). Chalov et al. (2010), similarly to Heerikhuisen et al. (2010) consider the pick-up ion (PUI) population that is produced by charge exchange between interstellar protons and heliospheric ENAs in the case of negligible scattering. They consider the motion of these PUIs around B_{ISM} that gets compressed outside the heliopause. The regions of the strong magnetic field can be considered as magnetic mirrors. The location of the magnetic mirrors is where the PUIs spend a considerable longer time (since their parallel speeds are small) so are the ideal places for a production of ENAs. Therefore the positions of the magnetic mirrors (where the radial component of B_{ISM} is zero) are the location of the IBEX ribbon.

Very few works tackled the origin of the CASSINI belt that seem to organize itself in a "belt" in a location similar but not equal to the IBEX ribbon.

Recently, the IBEX team separated the distributed flux emission from the ribbon (McComas et al. 2013; Schwadron et al. 2014) (Fig. 12). The distributed flux emission gives a global view of the structure of the heliosphere since it's believed to be produced in the inner heliosheath. In particular the tail emission seem to be organized by a two-lobe structure. The ENA tail observations (McComas et al. 2013) reveal two lobes at high latitudes and depletion in low latitudes of high energy ENAs (~4 keV) while in low energies (~0.7 keV) the tail appears as two separate enhancements in low latitudes. McComas et al. suggested that these observations resulted from the spatial separation of slow and fast winds. The ENA images from Cassini (Krimigis et al. 2009;

Dialynas et al. 2013; at 5–55 keV) revealed intensities that were comparable in the direction of the nose and tail. The observers therefore concluded that the heliosphere might be "tailless" because the emission from these high-energy ENAs is believed to come from the heliosheath. The two-lobe heliosphere is almost "tailless" with the distance down the tail to the ISM between the lobes being comparable to the distance to the ISM at the nose.

Moreover the ENA emissions show strong time variations (McComas and Schwadron 2014; Dialynas et al. 2013) that need to be explained.

Finally an interesting complement is the low energy ENAs that are order of magnitude higher than models predict. The low energy ENAs (measured by IBEX-Lo) struggle with signal-to-noise ratio so the statistics is poor. The low energy ENAs could indicate that additional heating has to be occurring within the heliosheath (Opher et al. 2013) or that there are additional pick-up ion population that is important outside the HP (Desai et al. 2014).

6 Conclusions

As described above there are several challenges to our understanding of the nature of the heliosheath, the heliopause and even the very global structure of the heliosphere. To really understand the nature of the heliosheath and help resolve between the different scenarios, a new visit to that region is necessary with proper instrumentation, with high sensitivity magnetometer and energetic particle instrument that bridge the gap in those energies; i.e., able to measure the suprathermal PUI population from 1 keV–40 keV and energetic electrons in the same energy.

The problem that the IBEX team faced (similar problem with CASSINI/INCA) is the sensitivity of the instruments requiring 3 years to be able, for example, to separate the tail emission from the rest, or the distributed flux. In the meantime, with new proposed missions such as IMAP that propose a much high sensitivity ENAs we will be able to constrain further the global structure of the heliosphere.

References

V.B. Baranov, Yu.G. Malama, Model of the solar wind interaction with the local interstellar medium—numerical solution of self-consistent problem. J. Geophys. Res. **98**(A9), 15157–15163 (1993), ISSN:0148–0227

R.D. Blandford, D.G. Payne, Hydromagnetic flows from accretion discs and the production of radio jets. Mon. Not. R. Astron. Soc. **199**, 883–903 (1982)

L.F. Burlaga, N.F. Ness, *Magnetic Field Fluctuations Observed in the Heliosheath by Voyager. 1 at* $114 \pm 2\,AU$ *During 2010* (2012)

L.F. Burlaga, N.F. Ness, E.C. Stone, Magnetic field observations as voyager 1 entered the heliosheath depletion region. Science **341**, 147 (2013)

L.F. Burlaga, N.F. Ness, Interstellar magnetic fields observed by voyager 1 beyond the heliopause. Astrophys. J. **784**, 146 (2014), 14 pp.

S.V. Chalov, D.B. Alexashov, D. McComas, V.V. Izmodenov, Y.G. Malama, N. Schwadron, Scatter-free pickup ions beyond the heliopause as a model for the interstellar boundary explorer ribbon. Astrophys. J. Lett. **716**(2), L99–L102 (2010)

S.V. Chalov, H.J. Fahr, The role of solar wind electrons at the solar wind termination shock. Mon. Not. R. Astron. Soc. Lett. **433**(1), L40–L43 (2013)

I.V. Chashei, H.J. Fahr, On the electron temperature downstream of the solar wind termination shock. Ann. Geophys. **31**(7), 1205–1212 (2013)

I.V. Chashei, H.J. Fahr, On solar-wind electron heating at large solar distances. Sol. Phys. **289**(4), 1359–1370 (2014)

R.A. Chevalier, D. Luo, Magnetic shaping of planetary nebulae and other stellar wind bubbles. Astrophys. J. **421**(1), 225–235 (1994). Part 1 (ISSN 0004-637X)

A.C. Cummings, E.C. Stone, F.B. McDonald, B.C. Heikkila, N. Lal, W.R. Webber, Anomalous cosmic rays in the heliosheath, in *Particle Acceleration and Transport in the Heliosphere and Beyond: 7th Annual International Astrophysics Conference. AIP Conference Proceedings*, vol. 1039 (2008), pp. 343–348

R.B. Decker, S.M. Krimigis, E.C. Roelof, M.E. Hill, Variations of low-energy ion distributions measured in the heliosheath, in *Pickup Ions Throughout the Heliosphere and Beyond: Proceedings of the 9th Annual International Astrophysics Conference. AIP Conference Proceedings*, vol. 1302 (2010), pp. 51–57

M.I. Desai et al., Energetic neutral atoms measured by the interstellar boundary explorer (IBEX): evidence for multiple heliosheath populations. Astrophys. J. **780**(1), 98 (2014), 11 pp.

J.F. Drake, M. Swisdak, M. Opher, A Model of the Heliosphere with Jets. Astrophys. J. Lett. **808**, L44 (2015)

K. Dialynas, S.M. Krimigis, D.G. Mitchell, E.C. Roelof, R.B. Decker, A three-coordinate system (Ecliptic, galactic, ISMF) spectral analysis of heliospheric ENA emissions using Cassini/INCA measurements. Astrophys. J. **778**(1), 40 (2013), 13 pp.

J.F. Drake, M. Opher, M. Swisdak, J.N. Chamoun, A magnetic reconnection mechanism for the generation of anomalous cosmic rays. Astrophys. J. **709**(2), 963–974 (2010)

L.A. Fisk, B. Kozlovsky, R. Ramaty, An interpretation of the observed oxygen and nitrogen enhancements in low-energy cosmic rays. Astrophys. J. **190**, L35 (1974)

L.A. Fisk, G. Gloeckler, The acceleration of anomalous cosmic rays by stochastic acceleration in the heliosheath. Adv. Space Res. **43**(10), 1471–1478 (2009)

L.A. Fisk, G. Gloeckler, The global configuration of the heliosheath inferred from recent voyager 1 observations. Astrophys. J. **776**, 79 (2013)

L.A. Fisk, G. Gloeckler, On whether or not voyager 1 has crossed the heliopause. Astrophys. J. **789**(1), 41 (2014), 9 pp.

V. Florinski, G.P. Zank, J. Heerikhusen, Q. Hu, I. Khazanov, Stability of a pickup ion ring-beam population in the outer heliosheath: implications for the IBEX ribbon. Astrophys. J. **719**(2), 1097–1103 (2010)

V. Florinski, E.C. Stone, A.C. Cummings, J.A. le Roux, Energetic particle anisotropies at the heliospheric boundary. II. transient features and rigidity dependence. Astrophys. J. **803**(1), 47 (2015), 8 pp.

G. Gloeckler, L.A. Fisk, L.J. Lanzerotti, Acceleration of solar wind and pickup ions by shocks, in *Connecting Sun and Heliosphere (Proc. Solar Wind 11/SOHO 16 Conf.)*, ed. by B. Fleck, T.H. Zurbuchen, H. Lacoste (ESA, Noordwijk, 2005), pp. 107–112

F. Guo, J.R. Jokipii, J. Kota, Particle acceleration by collisionless shocks containing large-scale magnetic-field variations. Astrophys. J. **725**(1), 128–133 (2010)

D.A. Gurnett, W.S. Kurth, L.F. Burlaga, N.F. Ness, In situ observations of interstellar plasma with voyager 1. Science, **341**, 1489 (2013)

J. Grygorczuk, A. Czechowski, S. Grzedzielski, Why are the magnetic field directions measured by voyager 1 on both sides of the heliopause so similar? Astrophys. J. Lett. **789**(2), L43 (2014), 4 pp.

J. Heerikhuisen, N.V. Pogorelov, G.P. Zank, G.B. Crew, P.C. Frisch, H.O. Funsten, P.H. Janzen, D.J. McComas, D.B. Reisenfeld, N.A. Schwadron, Pick-up ions in the outer heliosheath: a possible mechanism for the interstellar boundary EXplorer ribbon. Astrophys. J. Lett. **708**(2), L126–L130 (2010)

M.E. Hill, R.B. Decker, L.E. Brown, J.F. Drake, D.C. Hamilton, S.M. Krimigis, M. Opher, Dependence of energetic ion and electron intensities on proximity to the magnetically sectored heliosheath: voyager 1 and 2 observations. Astrophys. J. **781**, 94 (2014)

P.A. Isenberg, T.G. Forbes, E. Möbius, Draping of the interstellar magnetic field over the heliopause: a passive field model. Astrophys. J. **805**(2), 153 (2015), 10 pp.

V.V. Izmodenov et al., Kinetic-gasdynamic modeling of the heliospheric interface. Space Sci. Rev. **146**, 329 (2009)

X. Jia, K.C. Hansen, T.I. Gombosi et al., Magnetospheric configuration and dynamics of Saturn's magnetosphere: a global MHD simulation. J. Geophys. Res. **117**, A05225 (2012)

J. Kota, Particle acceleration at near-perpendicular shocks: the role of field-line topology. Astrophys. J. **723**(1), 393–397 (2010)

S. Krimigis et al., Imaging the interaction of the heliosphere with the interstellar medium from Saturn with cassini. Science **326**, 971 (2009)

S.M. Krimigis, E.C. Roelof, R.B. Decker, M.E. Hill, Zero outward flow velocity for plasma in a heliosheath transition layer. Nature **474**(7351), 359–361 (2011)

S.M. Krimigis et al., Search for the exit: voyager 1 at heliosphere's border with the galaxy. Science **341**, 144 (2013)

R. Lallement et al., Deflection of the interstellar neutral hydrogen flow across the heliospheric interface. Science **307**(5714), 1447–1449 (2005)

R. Lallement et al., The interstellar H flow: updated analysis of SOHO/SWAN data, in *Twelfth International Solar Wind Conference. AIP Conference Proceedings*, vol. 1216 (2010), pp. 555–558

Y.E. Lyubarsky, On the structure of the inner Crab Nebula. Mon. Not. R. Astron. Soc. **329**(2), L34–L36 (2002)

A. Lazarian, M. Opher, Model of acceleration of anomalous cosmic rays by reconnection in the heliosheath. Astrophys. J. **703**, 8 (2009)

Y.G. Malama, V.V. Izmodenov, S.V. Chalov, Modeling of the heliospheric interface: multi-component nature of the heliospheric plasma. Astron. Astrophys. **445**(2), 693–701 (2006)

D.J. McComas, N.A. Schwadron, An explanation of the voyager paradox: particle acceleration at a blunt termination shock. Geophys. Res. Lett. **33**(4), L04102 (2006)

D.J. McComas et al., Global observations of the interstellar interaction from the Interstellar Boundary Explorer (IBEX). Science **326**, 959 (2009)

D.J. McComas et al., IBEX observations of heliospheric energetic neutral atoms: current understanding and future directions. Geophys. Res. Lett. **38**(18), L18101 (2011)

D.J. McComas, N.A. Schwadron, Disconnection from the termination shock: the end of the voyager paradox. Astrophys. J. **758**, 19 (2012)

D.J. McComas, M.A. Dayeh, H.O. Funsten, G. Livadiotis, N.A. Schwadron, The heliotail revealed by the interstellar boundary explorer. Astrophys. J. **771**(2), 77 (2013), 9 pp.

D.J. McComas, N.A. Schwadron, Plasma flows at voyager 2 away from the measured suprathermal pressures. Astrophys. J. Lett. **795**(1), L17 (2014), 3 pp.

A. Michael, M. Opher, E. Provornikova, J. Richardson, G. Toth, Magnetic flux conservation in the heliosheath including solar cycle variations of magnetic field intensity. Astrophys. J. Lett. **803**, L6 (2015)

M. Opher, E.C. Stone, P.C. Liewer, The effects of a local interstellar magnetic field on voyager 1 and 2 observations. Astrophys. J. **640**(1), L71–L74 (2006)

M. Opher et al., A strong, highly-tilted interstellar magnetic field near the solar system. Nature **462**, 1036 (2009)

M. Opher, J.F. Drake, M. Swisdak, K.M. Schoeffler, J.D. Richardson, R.B. Decker, G. Toth, Is the magnetic field in the heliosheath laminar or a turbulent sea of bubbles? Astrophys. J. **734**(1), 71 (2011), 10 pp.

M. Opher, J.F. Drake, M. Velli, G. Toth, R. Decker, Near the boundary of the heliosphere: a flow transition region. Astrophys. J. **751**, 80 (2012)

M. Opher, J.F. Drake, On the rotation of the magnetic field across the heliopause. Astrophys. J. Lett. **778**, L26 (2013)

M. Opher, C. Prested, D.J. McComas, N. Schwadron, J. Drake, Probing the nature of the heliosheath with the neutral atom spectra measured by IBEX in the voyager 1 direction. Astrophys. J. Lett. **776**, L32 (2013)

M. Opher, J.F. Drake, B. Zieger, T.I. Gombosi, Magnetized jets driven by the Sun: the structure of the heliosphere revisited. Astrophys. J. Lett. **800**(2), L28 (2015), 7 pp.

E.N. Parker, Astrophys. J. **134**, 20 (1961)

N.V. Pogorelov, E.C. Stone, V. Florinski, G.P. Zank, Termination shock asymmetries as seen by the voyager spacecraft: the role of the interstellar magnetic field and neutral hydrogen. Astrophys. J. **668**(1), 611–624 (2007)

N.V. Pogorelov, S.N. Borovikov, G.P. Zank, T. Ogino, Three-dimensional features of the outer heliosphere due to coupling between the interstellar and interplanetary magnetic fields. III. The effects of solar rotation and activity cycle. Astrophys. J. **696**, 1478 (2009)

N.V. Pogorelov, S.N. Borovikov, G.P. Zank, L.F. Burlaga, R.A. Decker, E.C. Stone, Radial velocity along the voyager 1 trajectory: the effect of solar cycle. Astrophys. J. Lett. **750**(1), L4 (2012), 6 pp.

N.V. Pogorelov, S.N. Borovikov, J. Heerikhuisen, T.K. Kim, G.P. Zank, in *Int. Conf. Numerical Modeling of Space Plasma Flows*, ed. by N.V. Pogorelov, E. Audit, G.P. Zank. ASP Conf. Ser., vol. 488 (ASP, San Francisco, 2014), p. 8th, 167

E. Provornikova, M. Opher, V.V. Izmodenov, J.D. Richardson, G. Toth, Plasma flows in the heliosheath along the voyager 1 and 2 trajectories due to effects of the 11 yr solar cycle. Astrophys. J. **794**(1), 29, (2014), 9 pp.

R. Ratkiewicz, J. Grygorczuk, Orientation of the local interstellar magnetic field inferred from voyagers' positions. Geophys. Res. Lett. **35**, L23105 (2008)

J.D. Richardson, J.C. Kasper, C. Wang, J.W. Belcher, A.J. Lazarus, Cool heliosheath plasma and deceleration of the upstream solar wind at the termination shock. Nature **454**(7200), 63–66 (2008)

J.D. Richardson, C. Wang, Plasma near the heliosheath: observations and interpretations. Astrophys. J. Lett. **711**(1), L44–L47 (2010)

J.D. Richardson, L.F. Burlaga, R.B. Decker, J.F. Drake, N.F. Ness, M. Opher, Magnetic flux conservation in the heliosheath. Astrophys. J. Lett. **762**(1), L14 (2013), 4 pp.

J.D. Richardson, R.B. Decker, Voyager 2 observations of plasmas and flows out to 104 AU. Astrophys. J. **792**(2), 126, (2014), 5 pp.

K.M. Schoeffler, J.F. Drake, M. Swisdak, K. Knizhnik, The role of pressure anisotropy on particle acceleration during magnetic reconnection. Astrophys. J. **764**(2), 126, (2013), 8 pp.

N.A. Schwadron, D.J. McComas, Is voyager 1 inside an interstellar flux transfer event? Astrophys. J. Lett. **778**(2), L33 (2013a), 5 pp.

N.A. Schwadron, D.J. MComas, Spatial retention of ions producing the IBEX ribbon. Astrophys. J. **764**(1), 92, (2013b), 11 pp.

N.A. Schwadron et al., Separation of the ribbon from globally distributed energetic neutral atom flux using the first five years of IBEX observations. Astrophys. J. Suppl. Ser. **215**(1), 13, (2014), 18 pp.

G. Siscoe, J. Raeder, A.J. Ridley, Transpolar potential saturation models compared. J. Geophys. Res. **109**, A09203 (2004)

E.C. Stone et al., An asymmetric solar wind termination shock. Nature **454**(7200), 71–74 (2008)

E.C. Stone et al., Voyager 1 observes low-energy galactic cosmic rays in a region depleted of heliospheric ions. Science **341**, 150 (2013)

M. Swisdak, J. Drake, M. Opher, A porous, layered heliopause. Astrophys. J. Lett. **774**, L8 (2013)

M. Strumik, A. Czechowski, S. Grzedzielski, W.M. Macek, R. Ratkiewicz, Small-scale local phenomena related to the magnetic reconnection and turbulence in the proximity of the heliopause. Astrophys. J. Lett. **773**(2), L23 (2013), 5 pp.

G. Zank, H. Pauls, I. Cairns, G. Webb, Interstellar pickup ions and quasi- perpendicular shocks: implications for the termination shock and interplanetary shocks. J. Geophys. Res. **101**, 457–477 (1996)

B. Zieger, K.C. Hansen, T.I. Gombosi, D.L. DeZeeuw, Periodic plasma escape from the mass-loaded Kronian magnetosphere. J. Geophys. Res. **115**, A08208 (2010)

B. Zieger, M. Opher, G. Toth, Constraining the Pick-up Ion Parameters at a Double Heliospheric Termination Shock. J. Geophys. Res. (2015, submitted)

DOI 10.1007/978-1-4939-3547-5_8
Reprinted from *Space Science Reviews* Journal, DOI 10.1007/s11214-014-0111-1

Structures in the Outer Solar Atmosphere

**L. Fletcher · P.J. Cargill · S.K. Antiochos ·
B.V. Gudiksen**

Received: 13 June 2014 / Accepted: 20 October 2014 / Published online: 28 November 2014
© Springer Science+Business Media Dordrecht 2014

Abstract The structure and dynamics of the outer solar atmosphere are reviewed with emphasis on the role played by the magnetic field. Contemporary observations that focus on high resolution imaging over a range of temperatures, as well as UV, EUV and hard X-ray spectroscopy, demonstrate the presence of a vast range of temporal and spatial scales, mass motions, and particle energies present. By focusing on recent developments in the chromosphere, corona and solar wind, it is shown that small scale processes, in particular magnetic reconnection, play a central role in determining the large-scale structure and properties of all regions. This coupling of scales is central to understanding the atmosphere, yet poses formidable challenges for theoretical models.

Keywords Sun · Corona · Hard X-rays

L. Fletcher (✉)
SUPA School of Physics and Astronomy, University of Glasgow, Glasgow, G12 8QQ, UK
e-mail: lyndsay.fletcher@glasgow.ac.uk

P.J. Cargill
Space and Atmospheric Physics, The Blackett Laboratory, Imperial College, London, SW7 2BZ, UK

P.J. Cargill
School of Mathematics and Statistics, University of St Andrews, St Andrews, Fife, KY16 9SS, UK

S.K. Antiochos
NASA Goddard Space Flight Center, Greenbelt, MD, 20771, USA

B.V. Gudiksen
Institute of Theoretical Astrophysics, University of Oslo, P.O. Box 1029, Blindern, 0315, Oslo, Norway

B.V. Gudiksen
Center of Mathematics for Applications, University of Oslo, P.O. Box 1053, Blindern, 0316, Oslo, Norway

1 Introduction

The structure and dynamics of the outer solar atmosphere, defined as extending from the base of the chromosphere into interplanetary space, is a topic of central importance for understanding solar activity. The fundamental questions have been well known for decades: why do the chromosphere and corona have temperatures considerably in excess of the solar photosphere and why do dynamic phenomena such as flares and coronal mass ejections (CMEs) occur? Only the solar magnetic field can provide the required energy to account for these phenomena.

Since the results from the Skylab observatory became available four decades ago, it has been recognized that, as a consequence of the magnetic field, the chromosphere/corona system is highly structured. The chromosphere is a complicated region that extends above the photosphere to a temperature of around 30000 K and in which the density falls to of order 10^{11}–10^{12} cm^{-3} (Sect. 2). Above this is the corona, with a temperature > 1 MK, considered to be either magnetically closed with field lines forming confined structures usually referred to as coronal loops (Sects. 3 and 4), or magnetically open (the source of high-speed solar wind) with field lines extending into interplanetary space (Sect. 5). There are a variety of magnetically closed structures, the most conspicuous being bright active regions (ARs), the site of intense EUV and X-ray emission. The rest of the magnetically closed corona is termed the quiet Sun, having a smaller level of emission. Figure 1 shows a typical coronal image with the different regions highlighted.

At the photosphere, the Sun's vector magnetic field can be measured routinely, and comprises many discrete sources ranging from small flux elements (10^{17} Mx with a scale of 100 km) to large sunspots (10^{22} Mx) that form a power law distribution over several decades (Thornton and Parnell 2011). The photospheric field strength is roughly its equipartition value, between one and two kG. The relative importance of the magnetic field as one rises through the outer atmosphere is seen by considering the value of the plasma beta ($\beta = 8\pi p/B^2$). At the photosphere, β is of order unity. Above the photosphere the pressure falls rapidly due to the small pressure scale height and so the photospheric field elements expand to fill all space in the upper chromosphere and corona. (Note that at this time, chromospheric field measurements are not made routinely Lagg et al. 2004, Metcalf et al. 2005, Leka et al. 2012.) A consequence is that β decreases throughout the chromosphere and becomes $\ll 1$, perhaps 0.1 in quiet regions and 0.01 in regions of very strong field, until well

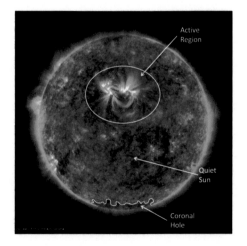

Fig. 1 An image from SDO (adapted from Cargill and de Moortel 2011), showing Fe IX emission at around 1 MK. The main features are the dark regions (magnetically open coronal holes: Sect. 4), bright magnetically closed active regions associated with sunspot groups, brightenings associated with small bipolar field regions (X-ray bright points) and "the rest": the quiet Sun

✷ Springer

out in the corona ($> 2R_s$) when the expanding solar wind becomes important. This implies that between the middle chromosphere and outer corona, magnetic forces dominate over pressure ones. Plasma also moves with the magnetic field due to magnetic flux freezing, so that a convenient description of these regions is in terms of magnetic field topology and the motion of magnetic field lines. A further consequence of this magnetic connectivity is that any complexity in the photospheric field due to the hierarchy of structures and associated motions of the solar surface must map in some way into the corona.

One difficulty is the interconnection of the various regions (for example coronal plasma must originate in the chromosphere), and the small scales (sub-resolution) required for the relevant processes that release magnetic energy. In all regions under discussion, effective Reynolds numbers are large, whether based on viscosity, Coulomb or anomalous resistivity (including Hall effects), or Pedersen conductivity, implying small scales. In particular, the magnetic reconnection process that is widely believed to be an essential mechanism for magnetic energy dissipation, and is the subject of much of this paper, requires scales $\ll$ 1 km, yet has consequences on global scales which CAN be observed. This poses major challenges, especially for theoretical modelling.

In keeping with the cross-disciplinary nature of the workshop, this paper is aimed at the non-specialist. It is not a comprehensive review of the subject, rather it focuses on specific aspects in which considerable progress is being made at this time. The structure is traditional, beginning with the chromosphere, then the closed corona, and dynamic phenomenon therein, and leading on to the solar wind.

2 The Chromosphere

In the past the solar chromosphere was seen as a thin layer in the solar atmosphere with relatively little influence, except for the formation of three prominent spectral lines: Ca II H, K and H_α, and the last of these gave the chromosphere its name, as it is largely H_α that produces the red ring visible in a full solar eclipse. This view of the chromosphere is that it is a layer of the solar atmosphere, roughly 0.5–1.5 Mm thick, with a temperature between 2500 and 25000 K. It is sandwiched between the photosphere and corona, with another thin layer of only a few hundred kilometers, called the transition region (Sect. 3), separating chromosphere and the corona.

However, increasingly sophisticated ground- and space-based observations have made it clear that the chromosphere is not a flat thin layer of constant depth, but rather a highly time-varying undulating layer of varying thickness. Indeed a precise definition of the chromosphere is challenging. The original definition in terms of H_α is not consistent with other definitions that use, for example, a specific temperature range. Such a temperature definition was often based on static one-dimensional calculations of an atmosphere where the aim was to reproduce the intensity of as many spectral lines as possible. One dimensional planar models of the chromosphere are now widely thought to be inadequate, so a present view is that the chromosphere is a highly dynamic layer where many important transitions in physical quantities and processes happen.

2.1 Structure

The structure of the chromosphere is determined by the dynamics of the photosphere (whose scales are also present in the chromosphere), its magnetic field, and the influence of motions on that field. The photosphere is dominated by the convective motions, which are organized on all scales from 1 Mm to at least 30 Mm in convective cells, where the sizes of the convective cells are inversely proportional to the velocity of the plasma they contain, and their lifetimes are proportional to the square of their size, with the smallest granules having a lifetime of roughly 5 minutes, while the largest have lifetimes of more than a day. The velocity pattern is generated by a superposition of granules of different sizes, and it sweeps magnetic flux into concentrated collections of all sizes from large sunspots, all the way down to the smallest concentrations with scales of less than 100 km. As discussed in the Introduction, the magnetic field undergoes a major change through the chromosphere, with a transition from plasma dominated dynamics ($\beta > 1$) to magnetic field dominated dynamics ($\beta \ll 1$). In addition, the small chromospheric pressure scale height implies a large gradient in the sound speed which creates an almost impassable barrier for all magnetosonic (compressional) waves generated in the photosphere. Such waves turn into shock fronts, are reflected by the large wave-speed gradient, and can deliver a substantial amount of energy to the chromospheric plasma.

Other complications that arise are (i) that radiation of all wavelengths becomes decoupled from the plasma and can leave the Sun without being further absorbed or scattered. The decoupling is dependent on wavelength and so the details of the radiative transfer are important when trying to understand the energetics of the chromosphere and (ii) that the plasma changes from being very weakly ionized at the photosphere to fully ionized at the base of the corona. This has significant implications for both transport and dissipation of currents. These disparities in fundamental physical properties over such a small distance, as well as the intrinsic time-dependence, make any sort of modelling very difficult.

When defined as a temperature interval, the chromosphere is split into two different parts. The lower is dominated by waves generated by the photospheric granules. Because the wave speed has such a large gradient, the waves shock in the lower chromosphere creating a pattern of colliding shock fronts. That pattern is prominent in the quiet sun where the magnetic field is generally weak and can be seen in the central panel of Fig. 2 taken with a filter centered on the Ca II H line. The same region observed in the H_α line appears dramatically different (see the right panel of Fig. 2), and since the prominent H_α line is the reason for naming the chromosphere, the part of the chromosphere showing the interacting shock fronts is now named the "Clapotisphere" (Rutten 2007) by some. (For comparison a white light image is shown in the left panel of Fig. 2.) An observation of the solar chromosphere in H_α is dominated by the structure of the magnetic field. Now the chromosphere looks like the fur of a haired animal, with dark and bright hairs with a width down to the resolution limit of our present instruments. The thin structures are due to the magnetic field's ability to isolate the plasma across the field direction making each strand a separate one-dimensional atmosphere. This very different look of the chromosphere is due to the change in the plasma β, so that the lower chromosphere is dominated by the plasma, while the upper region where H_α is produced is dominated by the magnetic field.

2.2 Energetics

The energy balance in the solar chromosphere is quite subtle. Thermal conduction plays little role for two reasons. Since most of the chromosphere is at a temperature greater than the photosphere, there is no heat conduction into the chromosphere from below. On the other hand the corona has a temperature in excess of 1 MK, and energy is conducted downwards to the cooler regions (Sect. 3). The heat flux is strongly temperature-dependent: $F_c = \kappa_0 T^{5/2} \nabla T$ where $\kappa_0 = 10^{-6}$ ergs cm^{-1} s^{-1} K$^{-7/2}$. One can estimate the average F_c in the chromosphere by simply taking an average scale for the temperature gradients of 1 Mm and an average temperature of 10^4 K, which yields a heat flux F_c of order 1 erg cm^{-2} s^{-1},

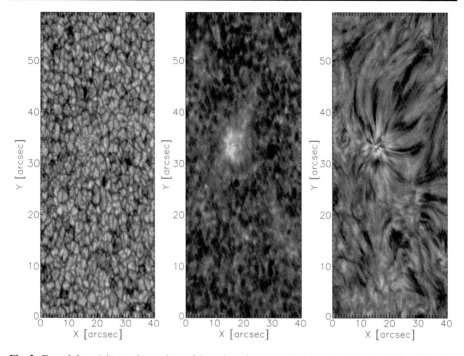

Fig. 2 From *left* to *right* are observations of the quiet solar atmosphere in white light, Ca II H and H_α. *The axes* are in units of arcsec, corresponding to roughly 725 km on the sun. The observations are made with the Swedish 1-meter solar telescope at La Palma, by Luc Rouppe van der Voort on June 18, 2006

a completely negligible value. (The importance of this downward heat flux for coronal structure is discussed in Sect. 3.) The large outward radiative flux from the photosphere could also potentially heat the chromosphere, but the chromosphere is generally optically thin and does not absorb a significant part of the incident radiative spectrum. In fact radiation is the main coolant of the chromosphere, so there is no way radiation can sustain its temperature.

The two remaining possibilities are the dissipation of shock waves initiated in the solar photosphere, and the dissipation of magnetic energy. The magnetic energy arises because the photospheric mass motions inevitably lead to magnetic perturbations, for example in Alfven or magnetosonic waves. Both of these mechanisms transport energy from the photospheric kinetic energy reservoir, which can be estimated to be roughly 1000 ergs cm^{-3} by assuming a mass density of 2×10^{-7} g cm^{-3} and a typical velocity of 1 km/s. The velocities can deliver an amount of work over a granular length scale and over a granular lifetime of 3×10^8 ergs cm^{-2} s^{-1}. The kinetic energy available in the photosphere is then able deliver more than enough energy to heat the chromosphere (estimated to be 3–14×10^6 ergs cm^{-2} s^{-1}, see references in Fossum and Carlsson 2006) and the corona (estimated to be roughly 10^6 ergs cm^{-2} s^{-1}, at least for the quiet Sun). However, it turns out that the sonic portion of the shock waves does not contain enough energy to maintain the chromosphere's temperature (Fossum and Carlsson 2005, 2006; Carlsson et al. 2007) even in the quiet sun. This leaves the options of heating either via pure Alfvén waves, the magnetic part of magnetosonic waves, by the reconnection of oppositely directed magnetic field components, or by other forms of current dissipation.

The injection of Poynting flux (**S**) into the chromosphere and corona happens primarily in the photosphere. For a perfectly conducting plasma, one can write this as:

$$\mathbf{S} = B^2\mathbf{V}/4\pi - (\mathbf{V}\cdot\mathbf{B})\mathbf{B}/4\pi \tag{1}$$

where **V** and **B** are the velocity and magnetic field vectors. Breaking these down into components in the plane of the photosphere and in the vertical direction, the first term corresponds to direct injection of magnetic flux into the corona. Such emerging flux is recognized as a key component of coronal activity, in particular large flares (Sect. 4). The second term is commonly discussed in terms of fast and slow photospheric motions. For present purposes, assume that the typical magnetic concentration in the photosphere has a magnetic field strength of 1 kG and that the typical velocity is roughly 1 km/s. Then, for motions in the plane of the photosphere, and no emerging flux, one finds $S_z = \frac{1}{8\pi}|B|^2 u_t \sin(2\phi)$ where u_t is the horizontal component of the velocity antiparallel to the horizontal magnetic field component. This is the stressing of the field by the horizontal motions, and will be discussed in greater detail in Sect. 3. The maximum value of S_z is roughly 4×10^8 ergs cm^{-2} s^{-1}. Of course the filling factor for the magnetic field is quite small but the amount of energy available where there is a magnetic field is very large.

At some point the magnetic field will be stressed sufficiently that the associated free magnetic energy will be converted into heat either through the dissipation of electric currents, through magnetic reconnection events or through dissipation of waves and shocks. As it evolves, the field will seek a minimum energy state, and for low-β plasma this will be a potential field, or perhaps a force-free field characterized by field-aligned electric currents. Note that the most readily accessible minimum energy state need not be potential (Bareford et al. 2010). The force-free state is present in the low corona, but not necessarily in the plasma-dominated lower chromosphere. However, if we assume that the chromosphere also has a force-free magnetic field, and the magnetic resistivity is constant, then the magnetic energy that may be converted into heat simply scales with the available magnetic energy, i.e. as B^2. All else being equal, the release of energy should decrease with height, and so the heating should be large in the chromosphere on the basis of this argument alone. The theoretical dissipation scale in the outer solar atmosphere is on the order of a few meters based on either resistive diffusion or collisionless processes and a typical temperature of 10^6 K and density of 10^9 cm^{-3}, so it would appear that the dissipation happens in a very small volume with high velocities as a consequence. However, the topological change occurring in reconnection also facilitates energy release over much larger scales through relaxation of stressed fields, particle acceleration, shocks etc. (Dungey 1953; Petschek 1964). The dissipation of magnetic energy as the main heating mechanism in the chromosphere means that there are relevant scales from just a few tens of meters to tens of megameters. If the dissipation happens in the same way in the chromosphere as it does in the corona, it happens in a bursty fashion, so the relevant time scales are seconds for the individual dissipation events, to tens of hours for the stressing of the magnetic field on large scales.

2.3 Modern Analysis of the Chromosphere

Major strides forward have been taken in recent years using space-based observations and MHD modelling. Given the short spatial and temporal scales, observations with short exposure times are needed, as well as high spectral resolution from everywhere in the 2D field of view with good spatial resolution. These are challenging requirements. The Hinode and Interface Region Imaging Spectrograph (IRIS) missions were both launched with the aim of understanding the outer solar atmosphere: implications for the corona from Hinode results are discussed in Sect. 3. Hinode has imagers with passbands at the wavelengths of the strong chromospheric spectral lines, so as to understand the linkage between corona, TR and chromosphere. IRIS is focused on the chromosphere, with instruments designed to understand the chromosphere and lower transition region.

One of the discoveries of Hinode has been the very long thin spicule-like structures at chromospheric temperatures that penetrate many megameters into the hot corona and that of which a subgroup seem to be ballistically driven by shock waves, while another subgroup seems to be longer lived and disappear while they still penetrate far into the corona (Type 2 spicules: De Pontieu et al. 2007). We have for some time known that the dynamics of the chromosphere and transition region produce correlations between the intensities of spectral lines formed in the chromosphere, while there is much less, if any, correlation between the intensity of spectral lines formed in the chromosphere and the transition region. We also know that there seems to be a correlation between the non-thermal width of spectral lines and their intensity. These correlations can only be reproduced by models that are able to catch the dynamics on all scales, and so far that has not been done.

As we have made clear, modelling the chromosphere is challenging, but is now being undertaken successfully by some groups (Martínez-Sykora et al. 2012). The chromosphere cannot be modeled analytically, so multi-dimensional numerical 3-D MHD models are needed. Even then there are problems converting model (or simulated) time- and length-scales to those inferred from the observations. Such models need to include radiative transfer and scattering and require a resolution which is high enough that the dynamical scales can be resolved. Further physics has been modeled such as non-equilibrium hydrogen ionization (Leenaarts et al. 2007) and the effect of a large fraction of neutral atoms in the atmosphere, resulting in a generalized Ohm's law (Martínez-Sykora et al. 2012), both of which have an effect on chromospheric energetics. However not all of the physics can be included in the same numerical simulation. Even though the models have led to new discoveries and a deeper understanding of the solar chromosphere, it is questionable whether the simulations actually capture all of the small scale dynamics present.

The chromosphere is hiding the answers to a number of important questions. The most important might be how the mass transfer from the photosphere to the corona happens, but another and most likely connected question is what type of magnetic heating happens in the chromosphere, the transition region and corona: are they the same and will we be able to identify the initiation of large flares by observing the chromosphere? The last major question is what produces the so called first ionization potential (FIP) effect (Laming 2004) in which the observed abundances of metals with low first ionization potential seems to be higher in the corona than in the photosphere, and a satisfactory physical explanation for that has not yet been put forward. The obvious conclusion is that it is a electromagnetic effect, but how exactly and why relatively more metals with low ionization potential are present in the corona is unknown.

3 The Non-flaring Closed Corona

The general properties of the non-flaring, closed corona are well known. The basic structures are magnetic loops that connect regions of opposite surface magnetic polarity. Loops can be considered as mini-atmospheres with plasma properties determined by the energy input. Reale (2010) provides a useful summary. (It should however be stressed that when one sees, for example, a coronal loop in EUV emission, one is looking at a collection of magnetic field lines that happen to be illuminated, not at an isolated bundle of field in a surrounding plasma.) AR loops have typical scales of 25–120 Mm, the temperature at the peak of the emission is around $10^{6.6}$ K (e.g. Warren et al. 2012), and they are bright because they are relatively dense (a few 10^9 cm^{-3}). The quiet Sun is cooler and less bright and exhibits larger structures in excess of 100 Mm such as streamers (Sect. 5) and interconnecting loops.

A major difficulty in understanding the corona is the determination of its magnetic field. In general, the Zeeman effect is not detectable due to the thermal broadening of EUV emission lines. An exception are coronagraph observations of the outer corona using Fe lines in the visible or IR (e.g. Lin et al. 2000, 2004). It is to be hoped that the Daniel K Inouye Solar Telescope (DKIST) will permit improved observations. Though not done regularly, the field strength can be constrained by microwave maps of the primarily gyrosynchrotron radiation, together with optically thin EUV temperature and density diagnostics (e.g. Brosius et al. 2002). Future microwave arrays such as the Frequency Agile Solar Radiotelescope (FASR Bastian 2004) and its pathfinder, the Extended Owens Valley Solar Array (EOVSA) will do this with high spatial, temporal and spectral resolution, and in principle the variations of magnetic energy can be deduced, although strong (few hundred G) fields are required.

Theoretical concepts can also be used to deduce coronal magnetic field properties. One approach is coronal seismology which relies on conjectured properties of oscillating coronal magnetic loops (e.g. Nakariakov et al. 1999; Nakariakov and Verwichte 2005; Tomczyk et al. 2007). Given a period of loop oscillation, a loop length and an estimate of the density, the magnetic field magnitude can be obtained. Typical values are 10–20 G. Force-free reconstruction of the coronal field from photospheric measurements (e.g. Schrijver et al. 2006; Metcalf et al. 2008) are widely used. This is a difficult task, beset with observational and computational difficulties, and solutions are non-unique (De Rosa et al. 2009). More reliable extrapolation results will be possible when the chromospheric vector magnetic field is measured. However, unlike seismological methods, a 3D field can be constructed. Also useful are EUV and X-ray images and movies, although they provide no information on field magnitude. Since the field and plasma are frozen, and strong thermal conduction leads to nearly-isothermal conditions in the high corona, plasma structures are very often taken to be an indicator of field geometry. This in turn provides information on both large-scale structures such as separatrices (Sect. 4) and the small-scale structure present within a large-scale field: for example Brooks et al. (2012) identified distinct loops with scales of 1". Results from the Hi-C rocket flight suggests that field topology on sub-arc second scales can be inferred from images and movies (Cirtain et al. 2013).

3.1 Energy Requirements and Atmospheric Thermal Structure

The energy requirements of the various coronal regions date to Withbroe and Noyes (1977) and are 10^7, 10^6 and a few 10^5 ergs cm^{-2} s^{-1} for ARs, coronal holes and quiet Sun respectively. The Poynting flux at the base of the chromosphere is described in Sect. 2. Of the hydromagnetic waves generated by the small-scale motions in the photosphere, only the Alfvén wave can reach the corona, and these have an upward Poynting flux of $\rho V_A \delta V^2$ where ρ is the mass density, V_A the Alfven speed and δV the wave amplitude. For slow (e.g. granular) motions (< 1 km/s) the Poynting flux is $V B_t B_a / 4\pi$, where B_a and B_t are magnetic field components perpendicular to, and in the plane of the photosphere respectively. If one takes typical values: $B_a = 150$ G, $B_t = 50$ G, $V = 1$ km/s, $\delta V = 30$ km/s, the Poynting flux for both Alfvén waves and slow injection is a few 10^7 ergs cm^{-2} s^{-1}, so in principle sufficient energy is injected.

How a corona forms as a consequence of heating is well understood, and involves the small but important transition region (TR) between corona and chromosphere. The behavior of the corona/TR system is governed by three energy transfer processes: thermal conduction from the corona to the lower atmosphere, optically thin radiation to space and mass flows between the upper chromosphere, TR and corona. Vesecky et al. (1979) noted that the TR can be defined as the region above the top of the chromosphere and below the location where thermal conduction changes from being an energy loss to an energy gain. In a static loop, the TR structure is determined roughly by a balance between a downward heat flux from the corona and radiation. Indeed the radiation from the TR is of order twice that from the corona (Cargill et al. 2012), despite its small size ($< 10\%$ of a typical loop).

We can use this to understand how a corona forms. Consider a low density coronal loop that is heated rapidly. Heat is conducted from the corona, but the TR is not dense enough to radiate it away. The only way for the TR and chromosphere to respond is by an enthalpy flux into the corona, increasing the density until a steady state is reached (e.g. Antiochos and Sturrock 1978). Turning the heating off has the opposite effect. As the corona cools, the conduction is not strong enough to power the TR radiation. Instead an enthalpy flux is set up to power the TR and the corona drains (e.g. Bradshaw and Cargill 2010). A simple physically motivated account of these processes is in Klimchuk et al. (2008) and Cargill et al. (2012).

It is very important to understand that the TR is not at a fixed location or temperature: these adapt in response to the conditions within any loop. Thus for a hot AR loop, Fe IX emission is from the TR, as is seen with the moss. For a cooler loop (1 MK peak temperature say), plasma only below 0.5 MK is TR. Misconceptions of the TR persist in the literature with reference to TR plasma or TR lines or in the TR. Such definitions are meaningless without understanding the overall context of the loop for which they are made.

3.2 Magnetic Field Equilibrium and Stability

It is, in principle, straightforward to develop MHD models of individual coronal loops and the more general large-scale coronal structures such as streamers, arcades, prominences etc, and examples can be found in Sects. 4 and 5. With any such magnetic field, two questions are important: (i) does an equilibrium exist for given boundary conditions and (ii) if it does, is it linearly (and non-linearly) stable? Seen in movies, the corona is in continual motion, but large eruptive flares are rare, so most of the time the magnetic field configuration is in, or close to, a state of equilibrium, or undergoing a weak instability that does not destroy the global field configuration. So perhaps the real question is not: why do large flares occur? Rather it is: why do they not occur more often? It is in fact quite difficult to eject material from the Sun (Sect. 4).

One reason for this stability lies in the line-tying of the magnetic field lines at the photosphere. The high density there means that magnetic field lines cannot move in response to a perturbation originating in the corona. For example, in a cylindrical arcade extending over π in the corona, the most unstable ($m = 1$) mode cannot arise, and the other modes are absolutely stable over a wide range of equilibria (Hood 1983; Cargill et al. 1986). For a cylindrical loop tied at both ends, the threshold for the (destructive) kink instability increases by of order 50% over that expected in the laboratory (e.g. Hood and Priest 1979). In both these geometries global resistive instabilities such as the tearing mode are suppressed except for the most eccentric field conditions. Also, proof of linear instability does not mean that the pre-instability field structure is entirely destroyed since the instability can saturate (non-linearly) at a low level. This is important because the vast majority of flares are not associated with a CME, so require very significant energy release but not the destruction of the field configuration.

3.3 Coronal Heating: MHD Aspects

Coronal heating has been divided for many decades into studies of wave heating (Alfvén waves generated at the photosphere) and heating by small-scale coronal reconnection. Alfvén wave heating has been discussed extensively (Nakariakov and Verwichte 2005; Klimchuk 2006) and requires structuring in the atmosphere and/or magnetic field to get dissipation by either phase mixing (e.g. Browning and Priest 1984) or resonance absorption (e.g. Ionson 1978). We will not discuss this further but note that Ofman et al. (1998) pointed out that resonance absorption became an impulsive heating process when feedback from the chromosphere was included. A different approach to wave heating is due to van Ballegooijen et al. (2011), Asgari-Targhi and van Ballegooijen (2012), and Asgari-Targhi et al. (2013). They argued that the generation of Alfvén waves at two loop footpoints, with different wave properties at each, would lead to a turbulent cascade as the counter-propagating waves interacted, eventually reaching dissipation scales. The heating is highly time and spatially dependent, with bursts of energy being released on top of a low background level of dissipation.

The consequences for the coronal magnetic field of slow photospheric motions have been studied extensively. For AR heating, the conditions imposed by this scenario are quite severe (Parker 1988), with a ratio B_t/B_a of order 0.25–0.4 being required. Here B_t is a typical field strength in a direction between the loop footpoints and B_a the field in the direction parallel to the photosphere, where we assumed that a curved loop has been straightened out. This implies that the coronal field must resist any desire to dissipate before a stressed condition is reached. The energy released if the dissipation returns B_t to zero is of order 10^{23}–10^{26} ergs, which led to the term nanoflares. It takes in excess of 10^4 secs to build up the energy in a nanoflare, with implications discussed shortly. Thus, in this scenario, the corona is maintained at its temperature by a swarm of nanoflares, each occurring in a small volume (Parker 1988; Cargill 1994). Theoretical evidence for nanoflare heating with such conditions on B_t/B_a is not yet convincing. Dahlburg et al. (2005, 2009) and Bowness et al. (2013) have shown that large values of shear are possible, but computational limitations are a concern in transferring their results to the real corona.

There is a burgeoning body of work that treats the corona as a global system in an MHD simulation, including a chromosphere, and attempts to impose realistic photospheric motions (e.g. Gudiksen and Nordlund 2005; Bingert and Peter 2013). While these models do produce something that looks like a corona, the temperatures are low (in part because a large enough simulation is unfeasible), heating occurs near the base of any loop, and is attributed to ohmic heating. However, it may be this ohmic heating may be an artefact of numerical resolution. The level of energy release at say an x-point is rather small: this was pointed out by Dungey as long ago as 1953 (see also Cargill 2014a). Magnetic reconnection is a facilitator of energy release: shocks, waves, particle acceleration, turbulence away from a reconnection site are far more important.

Addressing the real dissipation processes is difficult, and requires more local models which should be seen as complementary to the global ones. As an example, consider the nonlinear evolution of the kink instability (Browning et al. 2008; Hood et al. 2009). The energy release here is more appropriate for a microflare or a nanoflare storm than AR heating, but it makes an important point. Figure 3 shows the current and velocity at three stages of the instability. First (left panel), regions form where oppositely-directed field components are

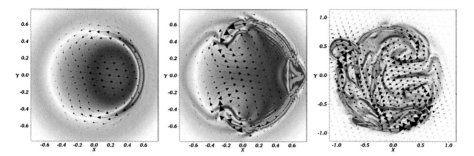

Fig. 3 Evolution of a kink instability. *The three plots* show current magnitude (*colored background*) and velocity vectors (*arrows*) at three different times. Note the formation of very fine-scale structure in the final plot. From Hood et al. (2009)

pushed together, initiating forced magnetic reconnection. A single current sheet evolves and then begins to fragment (2nd panel). Finally, a wide range of small-scale structures emerges (3rd panel) which then dissipate, leading to a magnetic field structure that is stable to the kink mode.

While it is tempting to invoke the language of turbulence here, the numerical grid does not permit the construction of a meaningful distribution of scales. But what is clear is the evolution of a smooth magnetic field into a very fragmented one through the instability. An important aspect is that the main energy release is not at the obvious current sheet, but may be due to slow shocks driven by the vortical flow, with secondary heating in the rest of the turbulent structure (Bareford and Hood 2013). These dissipating structures need to be resolved properly to actually understand what is going on. So calling everything 'ohmic heating' is a little lazy and possibly misleading.

Another common approach is use the reduced MHD approximation which assumes $B_t \ll B_a$. A number of papers Rappazzo et al. (2008, 2010), Dahlburg et al. (2012) have demonstrated that dissipation occurs readily for a range of photospheric motions with discrete heating events occurring. However, it is impossible for RMHD models to account for AR heating. The required value of B_t/B_a is beyond the viability of the model. Indeed, one could say that dissipation occurs too easily in these models: this may be due to assumptions about how the coronal field responds to photospheric motions. For example, adjustments to neighboring equilibrium states are forbidden, so that dynamic evolution must occur instead.

3.4 Deducing Properties of Heating Mechanisms

In the absence of direct observations of coronal magnetic field dynamics, inference of heating processes relies on the interpretation of images and spectra. In both cases, it is important to have comprehensive temperature coverage and data from a wide range of emission lines can provide information on plasma parameters such as mass motions, density etc. using spectral techniques. The results have been mixed. Density-sensitive line pairs (i.e. same element, same ionization state, same temperature, different transition) can provide an absolute measurement of the electron density. Combined with an estimate of the density from an (assumed unresolved) image, this provides a handle on the filling factor: the ratio of the volume radiating to the actual volume. In turn, this can provide information on the fundamental scales associated with coronal heating. Cargill and Klimchuk (1997) obtained scales

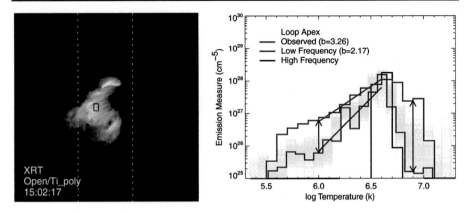

Fig. 4 *On the left*: AR loops seen on the disk by the Hinode X-ray telescope (XRT). *The small rectangle in the center corresponds to loops in the AR core that were analyzed by the Hinode EIS instrument (from Warren et al. 2011). On the right is the emission measure as a function of temperature from EIS. The red line is the observed value, and the blue and black correspond to low and high frequency nanoflares as defined by Warren et al. (2011)*

of <100 km. Future high-quality spectroscopic data has the potential to provide such information.

Analysis of coronal mass motions provides another possible diagnostic. An important aspect of magnetic reconnection, at least in a simple form, is the prediction of high-speed reconnection jets which have been detected at magnetopause and solar wind reconnection sites. In the corona, these should lead to Doppler shifts or non-thermal broadening (if multiple sites are convolved: e.g. Cargill 1996) well in excess of 100 km/s. While individual jets have been well observed over many years (see, for example, Innes et al. 1997, Klimchuk 1998), and more recently coronal manifestations of spicules (McIntosh and De Pontieu 2009; Peter 2010), line broadening in active regions of the magnitude predicted is not seen (Doschek 2012). Reasons may include: the temperature of the initial jet is not measured, the jet interacts with the surrounding magnetic field and thermalizes rapidly or the initial jet density is small (as would be the case in a low filling factor corona), with negligible emission measure (Zirker 1993; Klimchuk 1998) or a nanoflare may be too small (e.g. Testa et al. 2013).

At the present time, the most promising approach by which progress is being made in understanding coronal heating is through emission measure analysis that evaluates the dependence of the emission on temperature. Figure 4 shows an image of an active region core (Warren et al. 2011) and other AR loops have been analyzed by a number of workers (Warren et al. 2011, 2012; Tripathi et al. 2011; Schmelz and Pathak 2012). Using many spectral lines from the Hinode EIS instrument, as well as data from SDO and Hinode, they were able to construct the $EM(T)$ profile over a wide temperature range. One popular heating model, that due to relatively infrequent nanoflares that involve slow injection of energy followed by rapid dissipation, predicts that $EM(T) \sim T^2$ as shown in Fig. 5 (Cargill 1994; Cargill and Klimchuk 2004). Warren et al. (2011) found $EM \sim T^{3.1}$ and concluded that infrequent nanoflares were not heating the corona, instead that steady heating was responsible. Subsequent parameter studies showed that there were some ARs that did fit the original prediction, but that many did not.

The real question here is: what is the frequency of the nanoflares? The key parameter is the ratio of the time taken for a loop to cool from its initial heated state to below

Fig. 5 The EM-T profile from a single nanoflare obtained from the EBTEL model (Cargill et al. 2012)

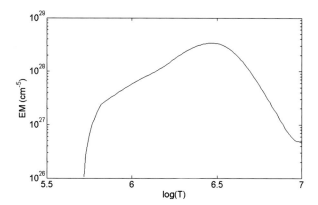

1 MK, to the recurrence rate of the nanoflare along a heated sub-loop. The former is of order 1000–3000 secs, scales with the loop length (Cargill et al. 1995), and is almost entirely independent of the nanoflare energy (Cargill 2014b). Recently it has been shown that these AR observations cannot be reproduced if the nanoflare recurrence time is much under 500 s, and it needs to lie in the range 500–2000 s. In addition, the nanoflare energy distribution should take on the form of a power law, with the delay time between each nanoflare proportional to the energy of the second nanoflare. If all these conditions are met, then the AR results can be accounted for (Cargill 2014b).

This result poses questions for coronal heating by nanoflares. Namely, it is impossible to power the corona if the nanoflare involves the complete relaxation of the field to a near-potential state since, as was noted above, the time to rebuild the nanoflare energy is long ($> 10^4$ s). Instead, nanoflares must involve a small relaxation of the field to a slightly less stressed state, which then permits a rapid rebuilding. How this happens in the framework of MHD is unclear.

A second important clue from contemporary data concerns hot coronal plasma. Any impulsive heating process leads to a plasma component that lies well above the peak of the emission measure, of order $10^{6.7}$–$10^{6.9}$ K, as seen on the right hand side of Fig. 5. Detection of this would be a powerful result. This has been accomplished by several workers at this time (Reale et al. 2009; Testa et al. 2011; Testa and Reale 2012). Unfortunately characterizing the physical properties of such plasma is very difficult. Low emission measures, ionization non-equilibrium, the difficult nature of thermal conduction and presence of non-Maxwellian distributions make this an ongoing challenge.

3.5 Discussion

The last few years have seen remarkable progress in the understanding of the properties of the hot non-flaring corona. This has come about through the extensive data bases of the Hinode and SDO missions, modelling, both hydrodynamic and MHD, and the ability to forward model observables from these models. Yet major puzzles remain. The time-dependence of coronal heating mechanisms remains unclear, and associated with this is a lack of clarity of the plasma environment in which heating occurs (tenuous $< 10^8$ cm^{-3} or dense: a few 10^9 cm^{-3}). This in turn is likely to feed into the extent of any hot plasma component discussed above and/or the presence of accelerated particles in the non-flaring corona. The lack of detection of the latter in particular is a puzzle since magnetic reconnection is well known as a good facilitator of particle acceleration.

Progress on these issues requires analysis of current data, modelling, and innovative uses of new data sources such as the Interface Region Imaging Spectrograph (IRIS) and the Nuclear Spectroscopic Telescope Array (NuSTAR) telescope. One viewpoint is that the smoking gun of coronal heating may lie in the presence or absence of these hot and energetic components. Unfortunately for the small nanoflare energies now being discussed, detection is likely to be even more difficult and may require long integrations over multiple sources to build up a real signal. An additional problem is that while nanoflares are often discussed as mini-flares, it seems most unlikely that the efficient acceleration occurring in flares that leads to 30 % of the energy going into accelerated particles persists in nanoflares. Much smaller fractions seem likely, enhancing the detection problem.

4 Flares and Eruptions

4.1 Introduction: Flare and CME Properties

Solar flares are abrupt and dramatic outbursts of radiation in the solar atmosphere, almost always taking place in magnetic active regions, and spanning a range of scales in size and energy from the smallest (currently) observable events called 'microflares' with energies of 10^{26} ergs, to the largest great flares with energy of up to a few $\times 10^{32}$ ergs (e.g. Hannah et al. 2011). Flares involve the acceleration of large numbers of electrons and ions up to mildly relativistic energies, a minority of which have access to the interplanetary magnetic field. The remainder are contained in the closed magnetic structures of the lower solar atmosphere. Particle acceleration characterizes the time of the main energy release in solar flares, and is a major focus of theoretical and observational attention. However, the flare is defined by its radiation burst, the majority of which is produced in the optical and UV parts of the spectrum (Woods et al. 2004) and emitted primarily by the solar chromosphere and photosphere in concentrated sources known as 'footpoints' or 'ribbons' depending on whether they have a point-like or a linear morphology. This emission is generated when the lower solar atmosphere is heated during a flare, and collisional heating by particles probably plays a considerable role in this, in addition to producing high energy radiations; hard X-rays (HXRs) from the electrons and γ-rays from the ions. Other mechanisms, e.g. damping of Alfvén waves, may also be important in heating the lower atmosphere at depths where particles cannot readily penetrate (Russell and Fletcher 2013) The corona radiates mostly in extreme UV and soft X-rays during a flare, but the total energy is a small fraction of the bolometric radiated energy (Emslie et al. 2012). The coronal emission is organized into beautiful, well-defined arcades, called flare loops. A composite image of a well-observed flare is shown in Fig. 6.

Flares take place in a variety of magnetic structures. Characteristic of large flares are active region filaments or filament channels, which are configurations including concave-upwards magnetic fields capable of supporting cool, dense plasma typically close to the solar surface and roughly parallel with the magnetic polarity inversion line. Smaller flares can take place in single or small groups of loops. The instability mechanisms are likely quite different in each. The magnetic reconfiguration that takes place in a flare can also lead to the ejection of magnetized plasma into the heliosphere, as a coronal mass ejection (CME) as shown in Fig. 7. CMEs typically travel around 500 km s^{-1} with a total mass (measured from Thomson scattering) of around 10^{15} g. Statistically, more energetic flares are more likely to be associated with a CME (Yashiro et al. 2006). A CME's total energy has been estimated at half an order of magnitude greater than that radiated in a flare (Emslie et al. 2012) but a flare

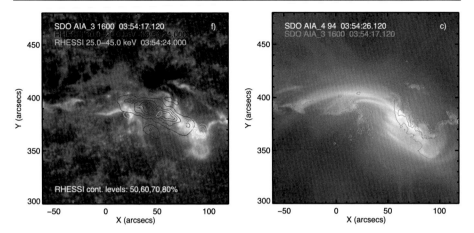

Fig. 6 *Two images* from a flare on 9th March 2012. *The LH panel* shows UV emission from the chromosphere, dominated by a number of flare ribbons. *Cyan intensity contours* are hard X-ray footpoints, and *red intensity contours* are thermal emission with a strong coronal component. *The RH panel* shows hot coronal emission at around 10 MK structured into quite well-organized loops. *Orange contours* show the ribbon locations. From Simões et al. (2013)

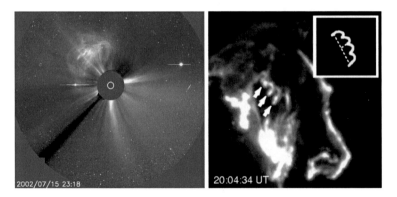

Fig. 7 A CME observed by the LASCO instrument on SOHO (*left-hand panel*), and a UV snapshot of the eruption during its related flare (*right-hand panel*), as reported by Gary and Moore (2004). The flare clearly shows an erupting twisted structure—with individual twists indicated by *the arrows*. Twist is also visible in the CME. Note, this is an extreme example of twist in an eruption; fewer turns are normally visible

has a much higher energy density. The median time between the flare HXR peak and CME peak acceleration is around one minute (Berkebile-Stoiser et al. 2012). The overall picture is of a dramatic re-organization of coronal magnetic structures which results in both upward-going and downward-going energy fluxes, with downward-going energy channeled by the (strong) magnetic field, resulting in heating and particle acceleration. A secondary response to both of these is expansion of chromospheric material into closed coronal field, usually termed 'chromospheric evaporation' and visible clearly in spectroscopic observations. This results in the bright EUV-emitting flare loops.

Large flares occur almost exclusively in solar active regions hosting sunspots; for example Dodson and Hedeman (1970) found that 7 % of large Hα flares occur in regions with small or no spots, while 82 % of X1 flares were observed to occur in magnetically complex

"$\beta\gamma\delta$" spots by Sammis et al. (2000). The fastest CMEs are associated with active region flares (Yashiro et al. 2005; Bein et al. 2012), but CMEs can also occur with the eruption of quiescent prominences outside active regions without an accompanying flare. The magnetic field in an active region, which is readily observed emerging and developing at the solar photosphere, is the primary agent imposing the structure, and determining the evolution, of solar flares and CMEs at launch (the solar wind dynamical pressure and frictional stresses also plays a role in later CME evolution). The heart of the flare problem is to work out how energy stored in large-scale and organized coronal magnetic structures is released and transferred to scales at which it can be picked up by individual electrons and ions, resulting in heating and acceleration.

4.2 Flare Morphology, Magnetic Structure and Magnetic Topology

There is a tendency for flares to occur close to the magnetic polarity inversion line in complicated, sunspot groups with a large amount of free magnetic energy. Flaring spot groups also exhibit rapid evolution in their strong-field regions: for example, Schrijver (2007) finds that the total unsigned magnetic flux within 15 Mm of the polarity inversion line increases by around 20 % in the 2 days prior to a major flare. The bulk of the free energy is associated with elongated and twisted (i.e. current-carrying) magnetic fields—called flux ropes—concentrated within a few 1000 km of the magnetic polarity inversion line (e.g. Sun et al. 2012). It is not yet possible to probe the current-carrying field in detail, but extrapolations of the magnetic field into the corona indicate that the bulk current-carrying structures shift in position during a flare, and change twist. While the field carrying most of the energy for the flare is organized into a relatively simple flux-rope structure on scales of a thousand km or so, there are doubtless smaller-scale structures embedded within this which may be critical for the flare triggering and evolution. It is estimated that at least 40 % of CMEs are flux-rope eruptions (Vourlidas et al. 2013), the remainder appearing more like jets and outflows.

Though the energy for the flare appears to be stored in a substantial volume of the corona, it is focused on release into a small number of compact patches, known as footpoints and ribbons. The footpoints are sites of the most energetic emission in optical and also in HXRs (which indicate high number densities of non-thermal electrons) and tend to have a point-like morphology. They are a subset of narrow, well-defined and elongated UV flare ribbons (Hudson et al. 2006).

The coronal magnetic field is rooted in a complicated photospheric field, and attempts to describe this call on notions of magnetic topology, which describe how one part of the field is linked to another. A simple example of this applied to the late evolution of a solar flare is the so-called 'CSHKP' model (after the primary authors, Carmichael 1964; Sturrock 1966; Hirayama 1974; Kopp and Pneuman 1976), which provides a plausible explanation for the appearance and the slow spreading apart of ribbons in the late phase of a flare in terms of the interface between post-reconnection closed loops and pre-reconnection unconnected field. Because it is a steady, two-dimensional view its applicability is restricted to the flare gradual phase, and it is used here to indicate why magnetic topology is of concern in understanding observed structures. In the CSHKP model (Fig. 8) reconnection between oppositely directed magnetic field occurs in the corona. Reconnected field retracts from the reconnection region, both upwards and downwards, allowing free energy to be liberated from the larger-scale magnetic structure as the field relaxes to a lower energy state. The liberated energy, ducted along the magnetic field, heats the lower atmosphere near the ends of the just-reconnected field on opposite sides of the polarity inversion line. The next set of field lines brought into

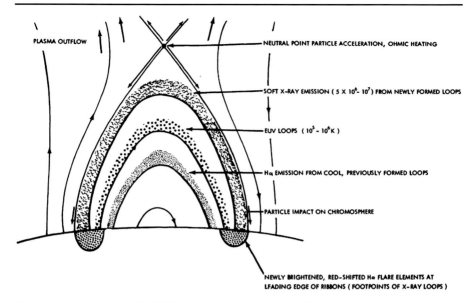

Fig. 8 A rendering of the 2-D CSHKP model by Svestka et al. (1980) provides an illustration of the link between field topology and flare evolution. The field lines extending to the photosphere from the neutral point (X-point) are separatrices which, translated into the 3rd dimension, become separatrix surfaces. Energy is delivered to the lower atmosphere at or near these surfaces, and field advecting into the X-point from the outer regions results in the spreading of flare ribbons

the reconnection region are rooted somewhat further from the polarity inversion, so that the location of heating moves outwards. Ribbons would arise from the extension of this 2-D model in an invariant direction, with their narrowness reflecting both the ducting of the energy along the field, and the rapid cooling of the lower atmosphere by radiation.

In a 2.5D model like CSHKP, the interface between closed and open field is an example of a separatrix surface. In more realistic topologies, separatrix surfaces curve around and intersect with one another at 'separator' field lines. The identification of such flux domains and intersections as locations of particular significance in a flare was made by Sweet (1969), and developed by many authors. An extensive review can be found in Longcope (2005). Particularly relevant for flares is the work of Hénoux and Somov (1987) who discussed how coronal currents could be generated and flow along coronal separators and Demoulin et al. (1992) who developed an early observationally motivated 3D model of coronal separators and separatrix surfaces (see Fig. 9). The calculated positions of the intersections of separatrix surfaces (or, more realistically in the case of continuous photospheric field distributions, quasi-separatrix layers, e.g. Priest and Démoulin 1995) with the photosphere agree well with the observed positions of flare ribbons (e.g. Mandrini et al. 1991). Quasi-separatrix layers are locations where coronal 'slip-running' reconnection may occur, leading to brightenings running along ribbons (Masson et al. 2009). Separators may also be critical structures, though this is by no means so clear. Theoretically, the minimum energy state of the corona consistent with a given photospheric boundary (extrapolated from discrete magnetic charges) is one in which the currents are concentrated along separator field lines while the rest of the field is current-free—the 'minimum current corona' model of Longcope (1996). This could be seen as the state to which a corona evolves under slow driving. Observationally Metcalf et al. (2003) and Des Jardins et al. (2009) identify some HXR footpoints with the photospheric ends of separators, which move when magnetic flux is transferred between magnetic

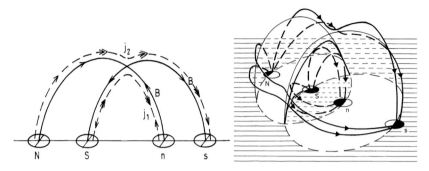

Fig. 9 This cartoon from Hénoux and Somov (1987) shows on the lefthand side a simple arrangement of 4 magnetic sources in a line, with *the single arrowheads* indicating the field and *the double arrowheads* indicating the current. *The solid lines* indicate the 2-D separatrix field lines separating 4 domains of different connectivity (including the exterior domain) and intersecting in an X-point. *The righthand side* shows this in 3-D; the separatrix lines are now domed separatrix surfaces, and the X-point is extended into a separator field line. This figure begins to suggest the structural complexity that can be arrived at from multiple magnetic sources in 3-D

domains as reconnection proceeds. There are relatively few studies in the literature dealing explicitly and carefully with the magnetic and topological evolution of a flaring active region and the primary (HXR) emission sites within it, so more case studies and equally importantly a proper synthesis of these case studies to identify common traits are necessary to understand the significance of the locations of these strongest energy release sites. This will have to involve detailed study of the peculiarities of each event before a coherent picture can be established—a very time-consuming process.

We note that, as expected locations of strong currents, separators, separatrix surfaces and other singular (or quasi-singular) structures have frequently been proposed as sites of particle acceleration. We return to this in Sect. 4.6.

4.3 Magnetic Structure, Complexity, and Flaring

Substantial effort has been expended on seeking relationships between various properties of sunspot groups and photospheric magnetic fields, and the flare productivity of a region. The level of complexity in the active region photospheric field is generally reflected in the complexity of the spatial arrangement of sunspot umbrae and penumbrae (McIntosh 1990). We do know for certain that large, complicated sunspot groups which are rapidly evolving tend to be flare-productive, but it is likely to be difficult to pick apart the properties of an active region that leads to a propensity to flare—for example rapid emergence (tending to lead to strong magnetic gradients and high shear), a high total flux, or complexity. Each of these is linked to a possible scenario for flare occurrence. Rapid emergence would lead to the development of strong current sheets at the interface with pre-existing coronal field which has been proposed as a trigger (Hirayama 1974). Strong shear or twisting flows may also indicate the rapid emergence of a flux rope (Manchester et al. 2004), or of a structure carrying a high current (Magara and Longcope 2003). A complex field implies a complex topology, with multiple 'stress points'—the nulls, separators and QSLs discussed previously at which reconnection is likely to happen

Both global properties (e.g. total magnetic flux and current) and local properties (e.g. variations in the magnetic shear around the neutral line) are important for flaring, but it is surprising how little can be said about flare productivity from photospheric observations:

'the state of the photospheric magnetic field at any given time has limited bearing on whether [a] region will be flare productive' (Leka and Barnes 2007). Flaring behavior also depends on more subtle properties of the magnetic structure. Perhaps the 'disconnect' between photospheric and coronal magnetic fields implied by the change from non-force free to force free across the chromosphere means that photospheric fields will never hold the key to flare prediction. Perhaps it is necessary to look in detail at the distribution and driving of topological structures of the magnetic field. Or perhaps it is necessary to understand microphysical properties—for example the development of resistivity in a coronal current sheet—which might never be accessible to our observational or theoretical tools, leaving us in the disheartening situation that the time and exact location of flare onset are determined by plasma properties of which we have only a weak observational grasp.

4.4 The Role of Magnetic Structures in Flare and Eruption Onset

The basic flare structure is a magnetic field carrying strong electrical currents, line-tied at a high inertia photosphere and embedded in a magnetized corona which supports currents with sub-photospheric origins (see Kuijpers et al. 2014, this issue). The manner in which a flare or eruption starts and develops depends both on the intrinsic stability of the strong-current structure and on how it interacts with the surrounding magnetic field, and given the enormous variety of configurations that can arise it may well be that there is not a unique mechanism.

Flares can be eruptive or 'confined' (not exhibiting a CME), and one can imagine all sorts of reasons for the distinction. For example, internal reconnections in a loop being twisted at its base and supporting internal currents could lead to flaring energy release inside the loop without any associated ejection, via a (non-erupting) kink instability (Liu and Alexander 2009). Similar ideas are proposed for micro-flaring (e.g. Bareford et al. 2010). Eruption on the other hand requires rapid inflation of a closed field, e.g. by photospheric current injection (thought unlikely on the timescale of a flare) or by removing or weakening overlying magnetic structure, or by allowing material on closed field access to open field via reconnection. The flare and eruption is thought likely to start with slow evolution towards an MHD instability, followed by the instability and reconnection permitting the magnetic field to re-configure as the MHD instability dictates. In support of this, filaments are often observed to start to rise slowly, indicating the loss of equilibrium of a previously stable magnetic system, before the main flare thermal or non-thermal radiation is detectable (Kahler et al. 1988; Sterling and Moore 2005).

Theory tends to consider two broad classes of coronal structure that can become unstable: sheared arcades and flux ropes. In the former class, a set of magnetic loops rooted on either side of a polarity inversion line are driven by photospheric shear flows, inflating the field until it erupts (Mikic and Linker 1994). During the eruption, reconnection between and underneath the loops of the arcade result in the formation of a twisted flux rope which is expelled into space, and the flare occurs as the coronal field re-organizes behind this. In the latter class, the initial configuration is modeled as a flux-rope that has either emerged bodily from underneath the photosphere into an overlying arcade field that stabilizes it, providing a magnetic tension force to counteract the hoop force in the flux rope (Titov and Démoulin 1999) or has been formed in situ as reconnection happens during the emergence of a sheared arcade (Amari et al. 2003). Eruption happens (e.g. Fig. 10) when the flux rope is perturbed by some critical amount leading to one of the classical MHD instabilities of a twisted flux tube (Török and Kliem 2005; Kliem and Török 2006; Fan and Gibson 2007).

Interaction of the stressed magnetic structure carrying the free energy for the flare with its environment is central in understanding what drives what. For example, in a simple geometry

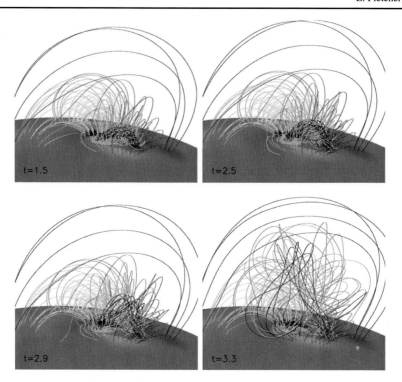

Fig. 10 Snapshots from an MHD simulation by Fan (2011) of a solar eruption that occurred on 13 December 2006. The fieldlines drawn show a compact flux rope (*blue/green lines*) embedded in a larger-scale potential field. Just prior to the snapshots shown the flux rope has emerged rapidly through the lower boundary (photosphere) and this is followed by a period of slow emergence, during which the eruption occurs because the rope fails to reach a static equilibrium with its surroundings

with a rope emerging into an overlying arcade, the flux rope becoming kink-unstable can force the field aside and burst through, with reconnection between flux rope and overlying field resulting from the MHD instability (Fan and Gibson 2007). On the other hand in the 'magnetic breakout' model (Antiochos et al. 1999) it is reconnection between the energy-storing sheared structure and the surroundings that destabilizes the system and leads to the eruption. Schmieder et al. (2013) argue that all of the proposed mechanisms can play a role, but favor the torus instability (Kliem and Török 2006) that occurs when the outwards Lorentz force of a current-carrying ring or partial ring of magnetic field (the 'hoop' force) is unbalanced by forces exerted by an external or 'strapping' field. The reasons given by Schmieder et al. (2013) for favoring this instability are that simulations of other types of instability have failed to produce eruptions, and also that observationally the occurrence or otherwise of an eruption seems to depend on the decay of the magnetic field with height: the torus instability is known to require an environment where the overlying field with height more rapidly than some critical value (Török and Kliem 2007).

Adding another complicating factor, the global coronal magnetic field on its largest scales can have an influence on the onset of flares. With high time-resolution and coverage of the complete solar disk from the Atmospheric Imaging Assembly on the Solar Dynamics Observatory it has been established that an event in one active region can be directly (causally) connected to an event in another region (Schrijver et al. 2013). Though not common, it does imply that the structure and stability of the global solar coronal magnetic field, and not just that of the region where the flare occurs, may play a role in enabling or preventing a flare and CME.

4.5 Flares at Different Scales and Relationship to Coronal Heating

Flare parameters (e.g. total thermal and non-thermal energy, peak power) follow power-law distributions, suggesting at least some similarity of structure across physical size scales (Hannah et al. 2011). It is clear that small flares (i.e. microflares, of the smallest GOES classification) do share many characteristics with larger events, such as spatial morphology and production of non-thermal particles (Lin et al. 2001; Hannah et al. 2008). Individual coronal 'nanoflares' have not been observed though a response in the lower atmosphere may have been (Testa et al. 2013) and so it is yet to be established whether the low-energy end of the observed flare distribution continues smoothly into these proposed coronal heating events. There is also as yet no evidence that the process that heats the non-flaring corona produces accelerated electrons with the high non-thermal energy content that characterizes larger flares (Hannah et al. 2010), though this may merely be a detector sensitivity issue.

As mentioned in Sect. 4.4 the kink instability in a strand subjected to twisting motions about its own axis is a model for confined flares and a model of coronal strands wrapped about one another is also proposed as a mechanism for heating the corona (Parker 1988). The idea of storing energy for a flare in a realistic corona by Parker-like random shuffling of its photospheric footpoints has also been investigated by Bingert and Peter (2013) and by Dimitropoulou et al. (2011) who incorporated a sandpile-like release. Statistical distributions of flare populations can be obtained in such models, as can the bursty behavior characterizing individual events. In both geometries it is the formation of current sheets and either Joule heating or component reconnection that leads to the coronal heating and energy dissipation, but whereas coronal heating requires a quasi-continuous transfer of energy from field to particles, flares require instead that energy is stored for some time, and released intermittently, in large events. It is not clear why the same twisting or shuffling process should have such different outcomes, but the character of the braiding may be an important factor in determining whether a continuous and space-filling heating arises, rather than a more flare-like intermittent behavior (Wilmot-Smith et al. 2011).

4.6 Magnetic Structures and Flare Particle Acceleration

The main energy release phase of a solar flare is characterized by intense bursts of radiation generated by accelerated electrons and ions, i.e. bremsstrahlung hard X-rays, and nuclear line and continuum γ-radiation. The bremsstrahlung radiation is observed to originate primarily in the solar chromosphere, but coronal HXR sources occur frequently as well (Krucker et al. 2008). Imaging of nuclear γ-radiation is much more difficult but a lower atmosphere location has been identified in a small number of strong flares and, curiously, it is not always consistent with the location of the non-thermal HXRs (Hurford et al. 2006). There has been no direct imaging of nuclear γ-ray sources in the corona, but γ-ray observations from flares with chromospheric footpoints over the limb clearly show evidence for large populations of accelerated coronal ions (Ryan 2000). Detections *in situ* in space, and via radio emission, give another view of the particle population and its access to the heliosphere.

There are different roles for magnetic structure in particle acceleration. Charged particles are accelerated by electric fields which can arise from time-varying magnetic fields in MHD waves or turbulence, or in magnetic reconnection regions. Models involving acceleration of

Fig. 11 A flare cartoon from Vlahos (1994) showing multiple sites of particle acceleration within a complex coronal magnetic structure. Though the sites are shown here distributed randomly through the corona, they must exist within the overall large-scale magnetic organization revealed by flare observations and field reconstructions

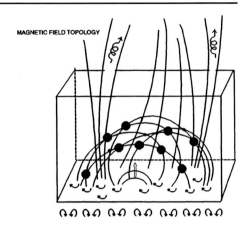

particles in the electric field set up in a reconnecting singular structure all face a problem, which is exacerbated as the dimension number of the structure reduces from current sheets (or separatrix layers) to separator lines, to nulls. If this accelerator is physically separate from the chromospheric location where the radiation appears, a very large number of electrons per second is required to explain HXR observations—on the order of 10^{36} electrons per second (Holman et al. 2003) for a few minutes. In a (pre-flare) corona of typical density 10^9 electrons cm^{-3} a large volume of coronal plasma per second needs to be 'processed' through the reconnecting structure. A single large current sheet of $\sim$ 30000 km on a side, with an external Alfvén speed of 1000 km s^{-1} could provide this number flux, but any other single reconnecting structure one can imagine, e.g. an X-line, cannot (Hannah and Fletcher 2006).

Acceleration throughout a large coronal volume, in turbulence or by multiple interactions with many smaller current sheets as shown in Fig. 11, is often proposed (e.g. Turkmani et al. 2005; Cargill et al. 2012), though such 'volumetric' acceleration still does not fully address the number requirements of a coronal acceleration model. The required electron rate and typical pre-flare coronal density implies the equivalent of $\sim 10^{27}$ cm^{-3} of corona being emptied of all electrons each second. The separation of impulsive phase HXR footpoints, typically 20–60 arcseconds or 15000–45000 km on the surface of the Sun (Saint-Hilaire et al. 2008) suggests a coronal volume involved of a few $\times 10^{27}$–10^{28} cm^{-3}, which will therefore need to be replenished during the flare. It may be possible to do this by return flows in the electric-field-free parts of the plasma. If a means can be found to increase the bremsstrahlung yield per electron (e.g. acceleration/re-acceleration in the radiating source) the demands on electron number or supply can be reduced.

Disordered small-scale field structures must exist within a large-scale organizing structure, defined by large-scale field stresses before the flare, and ordered post-flare loops afterwards. Individual flare HXR lightcurves show bursty behavior that can be characterized as fractal in time (McAteer et al. 2007), but inspection of flare images show that in space the HXR sources undergo rather ordered motions, associated with the evolution of the large-scale magnetic field (Grigis and Benz 2005). Whatever is happening in the corona, it is not completely random in an individual flare event. The observed power-law distributions in the properties of large numbers of flares have led to notions of self-organized criticality (SOC) being applied to individual flares (e.g Vlahos et al. 1995), but the observed distributions show only that the ensemble of active regions over some large portion of a solar cycle exhibits SOC-like scaling (Lu and Hamilton 1991). It remains to be seen whether SOC ideas can be applied successfully also to an individual flare in a single active region.

4.7 Discussion

Understanding flares and eruptions requires a deep knowledge of coronal magnetic structures. The free energy of a flare is stored in non-potential magnetic fields in the low-β corona, which emerge from below the photosphere, and the current-carrying magnetic flux rope is therefore a basic ingredient of the active corona. The current paths through the corona are determined by its topological structure and may be complex. Topology and topological changes also determine the post-flare states that can be accessed (and thus how much free energy can be released), and whether or not a magnetic restructuring will lead to an ejection. Within the larger scale structure, smaller structures must exist or form to allow energy to be transferred to scales at which both thermal plasma and non-thermal particles can pick it up. In more than 40 years of study the theory and observation of many aspects of flares and ejections have become highly refined but the answers to basic questions, such as identifying the conditions that precede a flare, or identifying how and where flare non-thermal particles are accelerated, still elude us.

5 The Open Solar Corona

The study of the Sun's open magnetic field corona began with the Parker (1958) theory of the steady, spherically symmetric solar wind. Parker argued that if the solar atmosphere were static with a temperature of order 1 MK out to several solar radii, then the resulting gas pressure at infinity would be finite. Consequently, the corona cannot be static; instead, it must expand outward to interstellar space as a steady, supersonic outflow, as was confirmed a few years later by the Mariner II observations (Neugebauer and Snyder 1962). However, it is known that the true corona and wind exhibit diverse structure, as shown in the eclipse image in Fig. 12. Note that this image was taken near solar minimum; at solar maximum the structure is even more complex. This spatial complexity is primarily due to the distribution of magnetic flux at the photosphere, which has enormous structure throughout the solar cycle. Given that the flux distribution at the photosphere is constantly evolving via flux emergence, cancellation, and the multi-scale photospheric motions, the large-scale corona of Fig. 12 must be fully dynamic and unlikely to be in a true steady state. If the photospheric flux evolution is slow, however, then a quasi-steady approximation may be valid. This is the fundamental assumption for most present-day modeling of the large-scale solar/heliospheric magnetic field.

5.1 Structure and Dynamics of the Open-Field Corona

As with the magnetically closed solar atmosphere, the structure and associated dynamics of the magnetically open corona are due to the effects of the Sun's magnetic field: the freezing of plasma and field implies that the structure seen in Fig. 12 traces out the magnetic field. Figure 12 also shows two primary types of structure: the finite-length arcs or loops discussed in Sect. 3 above and the semi-infinite rays associated with open magnetic field lines that extend from the solar surface out beyond the edge of the image, along which the solar wind must flow. While these open field lines extend outward to the heliopause at ~ 140 AU, the large-scale properties of the heliosphere, as well as the distribution of open and closed flux, are governed by the magnetic structure and dynamics at the photosphere. For example, the most widely used model for the global coronal magnetic field is the Potential Field Source Surface (PFSS) model, which assumes that in the low corona, below some radius $R_s \sim 2R_\odot$,

Fig. 12 White light image of the solar eclipse of August 1, 2008 (Pasachoff et al. 2009)

the field is potential, and purely radial at R_s (Altschuler and Newkirk 1969; Schatten et al. 1969; Hoeksema 1991). For production models such as the Wang–Sheeley–Arge (Arge and Pizzo 2000) the source surface radius is held fixed, so that the only real input to the PFSS model is the observed normal flux at the photosphere. Given the extreme simplicity of the PFSS model and the errors inherent in the input data, the model does surprisingly well at reproducing the observed large-scale distribution of open and closed field at the Sun (Riley et al. 2006), at least, during solar minimum when the photospheric flux is not changing too rapidly. The reason is that $\beta \ll 1$ in the low corona, and the field is believed to be close to potential except at filament channels, which lie in the innermost regions of the closed field, such as seen in the helmets at the NW and SE (upper right and lower left) limbs in Fig. 12. Fully 3D MHD models also show that the open flux structure is determined predominantly by the photospheric flux distribution (Riley et al. 2006).

In fact there are two types of open field regions in the corona. The most obvious are the large-scale, long-lived open regions corresponding to coronal holes as seen in EUV (Fig. 1) which evolve quasi-statically. The second are the boundaries of these coronal holes, which are likely to be fully dynamic and have a scale of order a supergranule at the photosphere. Note that if we assume a purely radial extrapolation, this scale corresponds to an angular width of order 5 degrees or so in the heliosphere, which is roughly the width of the streamer stalks observable in Fig. 12.

Accompanying the spatial spectrum of photospheric flux is a spectrum of temporal scales ranging from days for the emergence and disappearance of active region flux to minutes for the elemental flux of the so-called magnetic carpet (Schrijver et al. 1997). In addition, there exists a complex of photospheric motions such as the granular and supergranular flows that continuously drive the coronal fieldlines at their photospheric footpoints. All this activity at the photosphere and interior is imprinted onto the corona and heliosphere. Small-scale

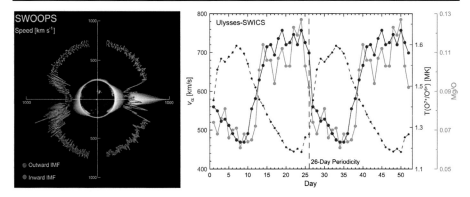

Fig. 13 *Left Panel*: Solar wind speed versus latitude from McComas et al. (2008). *The inset* consists of a SOHO/EIT image imbedded in a SOHO/LASCO C2 image. *Right Panel*: Solar Variation of solar wind speed (*red*), charge state (*blue*) and composition (*green*) for a time period during the 1995, from Geiss et al. (1995)

photospheric motions are expected to result in upward propagating Alfvén waves on field lines near the Sun's poles, such as seen in Fig. 12. These waves have long been proposed as the source of the energy and momentum that powers the wind (Cranmer 2012), and are likely to be a major source of solar wind turbulence (Verdini et al. 2010). Indeed some authors have argued that the supergranular structure can be seen directly in the heliosphere as well-defined fluxtubes (Borovsky 2008); but there is debate over this result (Greco et al. 2009).

A key question that is only now starting to be explored in detail is the effect of the photospheric motions on field lines near the interface between open and closed flux, such as the boundaries of the various streamers in Fig. 12. Topologically, the open-closed interface is a separatrix surface of zero width (Lau and Finn 1990), but this holds only for a true steady-state. Dynamically, the width of the interface will be set by the time-scale for establishing a steady wind, generally a day or so, which for typical photospheric speeds ~ 1 km/s corresponds to a spatial scale of ~ 30000 km, the scale of a supergranule. For time/spatial scales longer than this the open-closed interface can be considered to evolve quasi-steadily, well approximated by, for example, a sequence of PFSS solutions. On smaller scales, however, the interface must be fully dynamic and the opening and closing of field lines calculated explicitly.

5.2 Fast and Slow Solar Wind

In parallel with the two types of open field regions in the corona, it has long been known that there are two distinct types of solar wind: the fast and slow. This can be seen in the left panel of Fig. 13 (McComas et al. 2008), which shows the speed of the wind as a function of heliospheric latitude between 1992 and 1998. These measurements were obtained during solar minimum, when the large-scale photospheric magnetic field is predominantly dipolar and exhibits the least global structure and evolution. Note, however, that due to the constant emergence and cancellation of the small-scale dipoles of the magnetic carpet, the small-scale field of the chromosphere and corona is always far from a steady state. It is evident from the figure that the wind at high latitudes is fast, speeds >500 km/s, and fairly steady, while that at low latitudes is slow <500 km/s, with large variability. Note that the slow wind surrounds the region at the ecliptic where streamer stalks and the heliospheric current sheet (HCS) are located.

This extreme dichotomy of the two winds suggests that they have different sources at the Sun and is best examined by measuring the compositions of the two winds, because the plasma composition directly connects the wind material to its source at the Sun (Zurbuchen et al. 2002). Indeed speed is not a valid discriminator of the two types of solar wind. The wind from small equatorial coronal holes is often observed to be slow, $V \sim 500$ km/s, but yet it has the spatial, temporal, and compositional signatures of the fast wind (Zhao et al. 2009).

To demonstrate this, the right panel of Fig. 13 shows a superposed epoch analysis of solar wind composition for ten solar rotations during 1992–1993 when Ulysses was periodically sampling fast and slow wind during a solar rotation (Geiss et al. 1995). The red curve plots the alpha particle speed, which is similar to that of protons, and shows the characteristic smooth gradient where the fast stream outruns the slow and a sharp gradient where the fast runs into the slow. The blue curve shows the freeze-in temperature of the plasma back near the Sun as derived from charge state ratios of Oxygen ions. There is a clear correlation with wind speed; the slow wind originates from markedly hotter plasma than the fast. The green curve plots the ratio of Magnesium, an element with low first ionization potential (FIP) to Oxygen, which has high FIP. Note that the ratio is consistently a factor of 3 or so higher in the slow wind than in the fast. The FIP ratio of the slow wind is very similar to that measured for the closed corona, while that of the fast wind is similar to that measured for the photosphere.

From these and many other observations of the wind and corona, the differences between the two winds can be summarized as follows:

(1) Spatial Properties: The fast wind is predominantly found at high latitudes near solar minimum and, consequently, is believed to originate from long-lived coronal holes, as can be seen from Fig. 13. The slow wind is found at low latitudes and surrounds the HCS (heliospheric current sheet) (Burlaga et al. 2002). The HCS is always observed to be embedded in slow wind, which is sometimes observed to extend as much as 30° or more from the HCS.

(2) Temporal Properties: The fast wind has predominately steady speed and composition (Geiss et al. 1995; von Steiger et al. 1995; Zurbuchen 2007), similar to the quasi-steady wind of (Parker 1958) but with a significant amount of additional physics such as momentum deposition, turbulence, and kinetic effects (Verdini and Velli 2007). The slow wind, on the other hand, is strongly variable in both speed and composition (Zurbuchen 2007).

(3) Composition: The ionic composition of the fast wind implies a freeze-in temperature at its source of $T \sim 1.2$ MK, typical of coronal holes, and an elemental abundance close to that of the photosphere (von Steiger et al. 2001). The slow wind has a freeze-in $T \sim 1.5$ MK and an abundance close to that of the closed corona (Geiss et al. 1995). Many have argued that composition rather than speed is the physical property that defines the wind, because the wind from small, long-lived coronal holes can be slow (~ 500 km/s) yet has "fast" wind composition (Zhao et al. 2009).

At present the source of the fast wind at the Sun is generally accepted to be coronal holes. Both imaging and in situ data support this conclusion. Almost immediately after the initial discovery of coronal holes by Skylab, it was inferred that these are the source of the so-called high speed streams (Zirker 1977). Since then, many in situ observations of high-speed streams have indicated that they are magnetically connected back to coronal holes on the disk. The Ulysses results are especially definitive (McComas et al. 2008). There is no doubt from Fig. 13 that when Ulysses is over a polar coronal hole it sees fast wind. Furthermore, direct spectroscopic imaging of coronal holes appears to show outflows along network boundaries (Hassler et al. 1999), exactly as would be expected for the source of the fast wind. This result has been further confirmed and elaborated by Tu et al. (2005) who used the spectroscopic date to determine the height at which the fast wind starts to flow in the network fluxtubes. The main questions regarding the fast wind, therefore, are not as to its source region on the Sun, but as to the actual mechanism for its acceleration.

5.2.1 Models for the Sources of the Slow Wind

Unlike the fast wind, the source of the slow wind remains as one of the outstanding questions in solar/heliospheric physics. Three general types of theories have been proposed for the location of slow wind origin at the Sun. Given its observed location in the heliosphere and its association with the HCS, as evident in Fig. 13, all three theories involve the closed field regions in some manner. The three theories differ most strongly in the location of the source regions at the Sun and in the role of dynamics.

The Expansion Factor Model   Perhaps the most straightforward theory for the slow wind is the so-called expansion factor model in which the slow wind originates from coronal hole regions, just like the fast wind, but only from open flux tubes near the boundary of the coronal hole with the closed flux region (Suess 1979; Withbroe 1988; Wang and Sheeley 1990). The model was originally described in terms of a single parameter, the expansion factor, defined as the ratio of the area of a flux tube at the PFSS source surface to its area at the solar surface. The basic idea underlying the model is that open flux tubes deep inside a coronal hole expand outward approximately radially, whereas flux tubes very near the boundary expand super-radially due to the presence of the closed flux. For example, the open flux that connects to the Y-point at the top of a helmet streamer has, in principle, infinite expansion. This streamer topology is discussed in detail below. It is well known that a large expansion factor can lead to slower velocities in the usual steady-state wind equations (Holzer and Leer 1980).

A problem with the original expansion factor model is that it predicts that the wind from the vicinity of pseudostreamers, which are also discussed in detail below, should be fast (Wang et al. 2007), but instead this wind is observed to be slow (Riley et al. 2006; Riley and Luhmann 2012). Consequently, the expansion factor model has been generalized substantially in recent years to consider the effects of the detailed variations of flux tube geometry on solar wind properties. The speed and other properties of the wind are sensitive to the locations of the heat and momentum deposition in an open flux tube, which in turn can vary greatly with flux tube geometry. In fact, Cranmer et al. (2007) have shown that non-steady solar wind solutions exist even for completely time-independent flux tube geometry and energy/momentum input. Consequently, the observed differences between the two winds may arise solely from the geometrical difference between open-field flux tubes near some open-closed boundary and flux tubes deep in the coronal-hole interior. It is clear from this discussion that, at least, for the expansion factor models, the exact topology of the open-closed boundary and neighboring open flux is essential for understanding the slow wind.

The Interchange Model   Another prominent theory for the sources of the slow wind is the interchange model (Fisk et al. 1998; Fisk 2005), in which the small-scale dynamics of the photospheric magnetic field (e.g., the magnetic carpet) play the central role. This model can be thought of as the exact opposite of the expansion factor model in that the slow wind source is the closed field region and dynamics are all-important. The key idea underlying the interchange model is that open flux is conserved and diffuses via reconnection throughout the corona, even into the apparently closed field regions (Fisk 2003). Unlike the expansion factor model, which assumes a quasi-steady field, the magnetic field is inherently dynamic and the slow wind is postulated to escape from the closed-field region via continuous interchange reconnection between open and closed flux. The interchange model has obvious advantages in accounting for slow wind observations: it naturally produces a continuously variable wind with closed-field plasma composition, located around the HCS but with large extent. The primary challenge for the model is to verify that interchange reconnection induced by photospheric dynamics does, indeed, enable open flux to penetrate deep into closed-field regions. The simulations, to date, have found that the open-closed boundary remains smooth and topologically well defined, even during interchange reconnection (Edmondson et al. 2009; Linker et al. 2011). On the other hand, these calculations lacked the topological complexity of observed photospheric flux distributions; consequently, it remains to be seen whether interchange reconnection with sufficiently complex magnetic topology at the open-closed boundary can produce an effective diffusion of the open flux deep into the closed. We note that for this model, as well, the open-closed boundary topology plays a central role.

The Streamer Top/S-Web Model　Given its associations with the HCS, its variability, and especially its composition, the conjecture that the slow wind is due to the release of closed field plasma onto open field lines, as in the interchange model above, seems promising. Many authors have argued that this release would naturally occur at the open-closed boundary of streamer tops (Suess et al. 1996; Einaudi et al. 1999; Endeve et al. 2004; Rappazzo et al. 2005). This streamer-top model can be thought of as intermediate between the expansion factor and interchange in that the location of the slow wind source is at a boundary region between open and closed, as in the expansion factor, but dynamics are essential, as in the interchange. The release of closed field plasma at streamer tops can account for all three properties of the slow wind, except for the observation that it can extend up 30° from the HCS. As discussed above, the expected angular width of the interface region due to supergranular flow is only 5° or so, which is too small to explain the observations. In recent years, however, it has been realized that for the photospheric flux distributions observed on the Sun, the open-closed boundary involves topological structures such as pseudo-streamers, which add much more topological complexity to the open field than expected from only the streamer belt. This topological complexity, referred to as the S-Web model (Antiochos et al. 2011, 2012) results in the existence of open flux located far from the HCS in the heliosphere, but mapping very near the open-closed boundary back in the corona. Interchange reconnection of such flux with closed field would readily release closed field plasma far from the HCS in the heliosphere.

　　A key point here is that the topological complexity of the open-closed boundary invoked by the S-Web is present in the purely steady models, such as a PFSS solution. It is not an assumption of the S-Web model. The invoked dynamics to release closed plasma is, indeed, an assumption, but the large-scale magnetic topology is an inherent feature resulting from the observed photospheric flux distributions and, hence, must be included in any model of the corona and heliosphere. It is clear, therefore, that a starting point for any theory of the origins of the slow wind and of the corona-heliosphere connection, in general, is an understanding of the topology open-closed boundary. In the following section we describe the salient features of this topology.

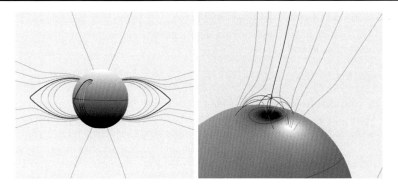

Fig. 14 *Left Panel*: Source surface model for a helmet streamer topology. *Black line* on the photosphere is the polarity inversion line at the equator and the field lines in the corona are *colored*: *red* for closed, *green* for open, and *blue* for separatrix boundary. *Right Panel*: Source surface model for parasitic polarity region embedded in polar coronal hole of *left panel*. *The blue separatrix lines* converge onto a 3D null point and define a fan surface and two spines, one open and one closed

5.3 The Open-Closed Boundary and Its Extension into the Heliosphere

For a completely general photospheric field, the open-closed boundary can have arbitrary topological complexity, but for the flux distributions actually observed on the Sun, there are only three types of structures whose boundaries are of importance: streamers, plumes/jets, and pseudo-streamers (originally defined by Hundhausen 1972 as plasma sheets), and all are evident in Fig. 12. The helmets and stalks on the lower left and right side are streamers; the bright rays emanating from the two poles are plumes (some may be jets); and the bright helmet and stalk on the upper left is a pseudo-streamer. Note also that the three open-field structures appear to originate at different heights above the photosphere: the streamer stalks at R_s, the pseudo-streamer at $R_s/4$ (Wang et al. 2007), and the plumes essentially at the photosphere. The magnetic topology of these boundaries, and associated dynamics, especially magnetic reconnection, are the key to understanding the corona-heliosphere connection.

5.3.1 Streamers

Assume the simplest possible photospheric flux distribution; that due to a single dipole **d** located at Sun center. Then the source-surface magnetic field is given by $\mathbf{B} = -\nabla\phi$, where the potential ϕ is,

$$\phi = \mathbf{d} \cdot \mathbf{r}\left(R_s^3 - r^3\right)/(R_s r)^3 \tag{2}$$

and R_s is the radius of the source surface. This field is plotted on the left of Fig. 14. The topology consists of a closed arcade centered about the equator (red field lines) and two polar coronal holes (green open field lines). The boundary between the open and closed regions in the corona is a toroidal-like surface, (the helmet), that wraps around the Sun (blue lines). The intersection of this surface with the photosphere forms two closed curves encircling the Sun roughly latitudinally that define the boundary between open and closed flux at the photosphere; these are the coronal hole boundaries. Note that the open-closed boundary is a true separatrix in that the magnetic connectivity is discontinuous across this surface. Consequently, every field line on this surface must end up at a null point. For the simple potential field above, the nulls are of the X-type and form a closed circle that defines the apex of the helmet. These can be seen as the tips of the blue field lines. In the solar case the X-points are deformed by the solar wind flow into Y-points that form a circle, with the connected HCS emanating from the circle. This simple picture agrees well with the helmet streamers observable in Fig. 12.

5.3.2 Plumes/Jets

Let us now consider the simplest possible extension of the magnetic topology above by adding a small dipole $\mathbf{d}_1$ below the photosphere. Such a field would correspond to, for example, a single bipolar active region or a single bipole of the magnetic carpet. If the dipole occurs at low latitudes where the coronal field is closed, then it would not change the basic structure of the open-closed topology, only its detailed shape. However, if the dipole occurs in the open field region a new open-closed boundary must occur on the Sun, because a parasitic polarity region (polarity opposite to that of the coronal hole) will appear on the photosphere, and this parasitic flux must be closed. The resulting topology is shown in the right hand panel of Fig. 14. This field is calculated using a source surface model in which a contribution ϕ_1 due to the small dipole is added to the potential above:

$$\phi_1 = \mathbf{d}_1 \cdot (\mathbf{r} - \mathbf{r}_d)/|\mathbf{r} - \mathbf{r}_d|^3 - R_s r_d^3 (\mathbf{d}_1 \cdot (R_s^2 \mathbf{r} - r^2 \mathbf{r}_d) / |r_d^2 \mathbf{r} - R_s^2 \mathbf{r}_d|^3 \tag{3}$$

where $\mathbf{r}_d$ is the location of the new dipole. The new open-closed topology is that of the well-known embedded bipole consisting of, in the photosphere: a closed circular curve defining the boundary between the small closed-field region and the surrounding open, in the corona: a dome-like fan surface with a single 3D null point near the apex of the dome (blue lines in Fig. 14), and in the heliosphere: a single line, the spine, emanating outward from the null (Antiochos 1998). There is also a downward spine, but this is simply part of the closed-field system.

Adding photospheric motions to this topology is expected to produce continuous reconnection at the null and fan surface between the closed flux and the surrounding coronal hole open flux. In this case the release of closed field plasma can occur very low in the corona, < 10000 km, depending on the size of the dipole. This type of interchange reconnection has been proposed as the mechanism for a broad range of observed phenomena, including coronal jets (Pariat et al. 2009), plumes (DeForest and Gurman 1998), and even the quasi-steady fast wind itself (Axford and McKenzie 1992). Since the magnetic carpet is ubiquitous throughout the Sun, we expect coronal holes to be riddled with small closed parasitic polarity regions. Figure 12 shows the presence of numerous plumes in the polar hole regions, each plume is believed to have a small parasitic polarity region at its base.

There remain questions, however, in reconciling the model with in situ measurements. Heliospheric measurements of coronal hole wind, the fast wind, do not show evidence for plume structure; nor do they show evidence of closed field plasma that has been released by interchange reconnection. As implied by Fig. 13, the wind from coronal holes appears to be uniformly fast wind, with no evidence of the variability or composition of the slow. There seems to be a disconnect, therefore, between the heliospheric data and coronal observations and with the dynamic interface model. This disconnect is one of the major puzzles in coronal/heliospheric physics. The Solar Probe Plus and Solar Orbiter Missions will hopefully resolve this puzzle by measuring the wind much closer to its origin at the Sun.

5.3.3 Plasma Sheets/Pseudostreamers

The topology of the third bright open-field structure in Fig. 12, the plasma sheet or pseudostreamer visible on the North-East limb, can be thought of as an intermediate case between the streamer and plume topology. A high-lying Y-null circle characterizes streamers,

Fig. 15 Source surface model for a plasma sheet topology. As in Fig. 4, *the black elliptical curve* at the photosphere is the PIL for the parasitic polarity region, which is inside the coronal hole. *The green lines* are open, and the intersections of the three pairs of heavy *blue lines* defines the locations of the three coronal nulls

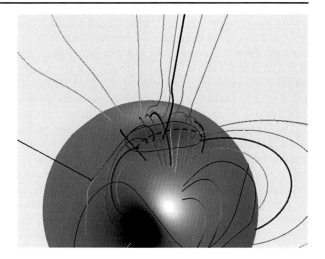

whereas a very low-lying X-point characterizes plumes. For a plasma sheet/pseudostreamer the corresponding coronal singularity is a finite line segment. In principle, this could be an actual null-line, but such lines are topologically unstable, so the singular line segment is due instead to a separator connecting two or more null points (Lau and Finn 1990). It is well known from studies of the Earth's magnetosphere that separators are equivalent to null lines in terms of sites for reconnection (Greene 1988).

Figure 15 shows one source surface model realization of a plasma sheet topology, but we emphasize that this is not unique. Plasma sheets can form from both finite width (Antiochos et al. 2007, 2011) and singular width (Titov et al. 2011) structures in the corona. Here we will describe only the key features of the singular width case so as to more easily make the comparison with plumes and streamers. The most intuitive picture for the plasma sheet magnetic structure is to start with the embedded bipole of Fig. 14, and then elongate the parasitic polarity region by simply adding more dipoles aligned in a row below the solar surface. This was the procedure used to obtain Fig. 15. If the parasitic polarity region is elongated sufficiently, the coronal null in Fig. 14 splits up to form three nulls connected by separator lines. These three points are located at the intersection of the thick blue lines in Fig. 15. Connecting these points are two field lines, the separators, indicated by the red dashed lines. Since these are singular lines it is impossible to find them with a graphics program, so they are simply drawn by hand.

The topology of Fig. 15 may appear to resemble that of a plume/jet above, but the nature of the singularity in the heliosphere is very different. In the plume topology above, the singularity is a single line, but for a plasma sheet it is a fan-like surface that emanates from the central null point. The central null has as its separatrix fan a vertical surface that is part open and part closed, while the two side nulls share the dome-like separatrix surface surrounding the parasitic polarity region as their fans. The intersection of the vertical fan of the central null and the dome fan of the side nulls defines the two separator lines where reconnection can occur.

We conclude, therefore, that the open-closed topology of a plasma sheet consists of in the photosphere: a closed elliptical curve defining the extent of the parasitic polarity flux, in the corona: a dome-like surface with multiple 3D null points and separator lines connecting them, and in the heliosphere: a surface of finite angular extent that connects down to the central null and separator lines. Again, adding photospheric dynamics to this picture implies interchange reconnection all along the separator lines in the corona and the release of closed field plasma into the heliosphere. Note that the release in this case is intermediate to the two cases above, it is not along a single line as in a plume, or along a full planar sheet as in a streamer/HSC, but along a surface that forms only a finite arc in the heliosphere. Furthermore, the height of the release is expected to be intermediate to plumes and streamers in that to have a sufficiently elongated geometry, the parasitic polarity region must be fairly large, but will not be larger that a streamer (Wang et al. 2007).

The question remains, however, if pseudostreamers will be a source of slow wind observed in the heliosphere. Plumes apparently are not, but streamers definitely are. This question requires detailed calculations, which have yet to be performed, of the interchange dynamics and closed plasma release for the topology of Fig. 15. Another important question is the amount of slow wind expected from such structures and whether it can account for all the slow wind observed. It may seem that multiple null points with separator lines would form only rarely on the Sun, but in fact, they are fairly common (Titov et al. 2012). As a result of the so-called "rush to the poles" of trailing polarity flux during the solar cycle, long tongues of opposite polarity often cut into coronal hole regions, leading to structures such as shown in Fig. 15.

For the observed solar flux distributions the magnetic topology is more complex than described above in that there is inevitably an interaction between the open-closed separatrix surfaces defining the coronal hole and the separatrices of the parasitic polarity region. Although not easily apparent, this interaction is present in the field of Fig. 15. Note that the dark blue lines defining the side nulls are closed, whereas the line defining the central null is open; consequently, the dome fan surface of the parasitic polarity must intersect the open-closed surface of the coronal hole. As a result the heliospheric separatrix surface emanating from the central null and separators takes the shape of an arch whose footpoints lie on the HCS. This is a key result. It implies that regions of slow wind in the heliosphere generally connect to the HCS, but can extend to high latitudes, in agreement with observations. Furthermore, the number of such arches can be so large for an observed photospheric flux distribution that they form a dense web, the S-web, in the heliosphere (Antiochos et al. 2011, 2012; Crooker et al. 2002).

5.4 Discussion

We conclude from the discussions above that the nature of the boundary between open and closed field in the corona plays the dominant role in determining the properties of the solar wind, in particular the slow wind. The large-scale geometry of this boundary is determined by the distribution of magnetic flux at the photosphere, but since this boundary is a singular surface, its dynamics are determined by its detailed topology and by magnetic reconnection across this topological singularity. Fortunately, there are only three important cases to consider, streamers, plumes, and plasma sheets, each of which can be seen in eclipse or coronagraph images. The Sun and Heliosphere, therefore, constitute an amazing example of a multi-scale coupled physical system in that structure at the photosphere on the scale of a solar radius ($\sim 10^{11}$ cm) couples to reconnection dynamics in the corona at kinetic scales (~ 100 cm), which then produce the plasma structures that we observe out at 1 AU ($\sim 10^{13}$ cm)!

Another major challenge remains the acceleration mechanism for the fast wind. Since there have been many recent reviews of progress on this problem (Hansteen and Velli 2012), we have only touched upon it here. Two general types of mechanisms have been studied: Alfvén waves (Cranmer et al. 2007) and reconnection as in the plume model above (Axford and McKenzie 1992; Fisk 2003). Most of the work has focused on the wave model, and there are now highly detailed calculations as to how waves generated by photospheric motions can couple nonlinearly to produce turbulent heating and acceleration on 1D open flux tubes (Cranmer 2012; van der Holst et al. 2014). The challenge is to extend this work to 3D and to include self-consistently the energy cascade down to the dissipation scale in the global models. There has been comparatively much less quantitative work on the reconnection model, because this inherently requires calculation of reconnection dynamics in a fully 3D topology. It appears inevitable, therefore, that further progress on both the slow and fast winds will require a much deeper understanding of multi-scale coupling and reconnection in the Sun's corona.

6 Conclusions

This paper has examined the properties of structures in the outer solar atmosphere. For reasons of space, the focus has been on a limited range of topics that have seen progress in recent years. It is clear that many factors have contributed to the major advances seen in the last decade or so. Of these, comprehensive spatial and spectral coverage of the outer solar atmosphere is probably the most significant since it has revealed the complexity of the magnetic and plasma structures, and their inherent dynamics. In turn, this has led to general acceptance that coupling between a hierarchy of scales is central to what is seen and that magnetic reconnection is the prime mechanism driving the observed global phenomena.

At this time, those studying the solar atmosphere have a greater range of data than previously. Putting these different data sources together has yielded outstanding insights, as we have discussed. However, we emphasize the need to undertake extensive parameter surveys: as an example we contrast the results discussed earlier of Warren et al. (2011) who studied a single AR, with those of Warren et al. (2012) who studied almost 20 ARs and reached different conclusions about the time-variability of coronal heating. Such extended studies may not have the excitement of a brand new result, but are essential for understanding the generality and breadth of the result. Unfortunately today due to the very richness of data, there is sometimes a tendency to show just one example and move on.

Numerical simulations also play a major role. For one-dimensional hydrodynamic models as discussed in Sect. 3, the important length and time scales are now being resolved adequately. For 3D MHD the situation is rather different in that one is dealing with transport coefficients orders of magnitude larger than exist in reality, and this impacts in particular magnetic reconnection studies. For forced (or driven) reconnection, one can argue that numerical reconnection handles the process adequately, at least in a time-averaged sense, so that confidence can be maintained in models of large scale eruptions. For smaller scales, and weakly stressed systems, such confidence may be misplaced.

With the launch of the IRIS satellite, the present rush of solar missions takes a pause, until Solar Orbiter in the latter part of this decade and then Solar Probe and Solar-C in the early 2020. The mission instrumentation is largely decided and in some cases under construction: Solar Orbiter will study the connection between the solar surface and solar wind at 0.3 AU, Solar Probe will fly closer to the Sun ($8.5R_s$), and Solar-C will fly next-generation coronal spectrometers and imagers, as well as measuring the magnetic field higher in the chromosphere. Beyond these, what else would be important to do? The solar flare problem is unsolved. We do not know how to accelerate the required number of particles. So a "son of RHESSI" with improved instrumentation, perhaps along the lines of the FOXSI rocket flight, and complementary imaging spectroscopy, is desirable.

Acknowledgements   We thank Andre Balogh for organizing this workshop and ISSI staff for their hospitality. The work of L. Fletcher has been supported by STFC grant ST/I001808/1 and that of S. Antiochos has been supported by the NASA TR&T and SR&T Programs.

References

M.D. Altschuler, G. Newkirk, Sol. Phys. **9**, 131 (1969). doi:10.1007/BF00145734

T. Amari, J.F. Luciani, J.J. Aly, Z. Mikic, J. Linker, Astrophys. J. **585**, 1073 (2003). doi:10.1086/345501

S.K. Antiochos, Astrophys. J. Lett. **502**, L181 (1998). arXiv:astro-ph/9806030. doi:10.1086/311507

S.K. Antiochos, C.R. DeVore, J.T. Karpen, Z. Mikić, Astrophys. J. **671**, 936 (2007). arXiv:0705.4430. doi:10.1086/522489

S.K. Antiochos, C.R. DeVore, J.A. Klimchuk, Astrophys. J. **510**, 485 (1999). arXiv:astro-ph/9807220. doi:10.1086/306563

S.K. Antiochos, J.A. Linker, R. Lionello, Z. Mikić, V. Titov, T.H. Zurbuchen, Space Sci. Rev. **172**, 169 (2012). doi:10.1007/s11214-011-9795-7

S.K. Antiochos, Z. Mikić, V.S. Titov, R. Lionello, J.A. Linker, Astrophys. J. **731**, 112 (2011). arXiv:1102.3704. doi:10.1088/0004-637X/731/2/112

S.K. Antiochos, P.A. Sturrock, Astrophys. J. **220**, 1137 (1978). doi:10.1086/155999

C.N. Arge, V.J. Pizzo, J. Geophys. Res. **105**, 10465 (2000). doi:10.1029/1999JA000262

M. Asgari-Targhi, A.A. van Ballegooijen, Astrophys. J. **746**, 81 (2012). doi:10.1088/0004-637X/746/1/81

M. Asgari-Targhi, A.A. van Ballegooijen, S.R. Cranmer, E.E. DeLuca, Astrophys. J. **773**, 111 (2013). arXiv:1306.6038. doi:10.1088/0004-637X/773/2/111

W.I. Axford, J.F. McKenzie, in *Solar Wind Seven Colloquium*, ed. by E. Marsch, R. Schwenn (1992), pp. 1–5

M. Bareford, A. Hood, in *AAS/Solar Physics Division Meeting*. AAS/Solar Physics Division Meeting, vol. 44 (2013), p. 200.02

M.R. Bareford, P.K. Browning, R.A.M. van der Linden, Astron. Astrophys. **521**, A70 (2010). arXiv:1005.5249. doi:10.1051/0004-6361/201014067

T.S. Bastian, Planet. Space Sci. **52**, 1381 (2004). doi:10.1016/j.pss.2004.09.015

B.M. Bein, S. Berkebile-Stoiser, A.M. Veronig, M. Temmer, B. Vršnak, Astrophys. J. **755**, 44 (2012). arXiv:1206.2144. doi:10.1088/0004-637X/755/1/44

S. Berkebile-Stoiser, A.M. Veronig, B.M. Bein, M. Temmer, Astrophys. J. **753**, 88 (2012). doi:10.1088/0004-637X/753/1/88

S. Bingert, H. Peter, Astron. Astrophys. **550**, A30 (2013). arXiv:1211.6417. doi:10.1051/0004-6361/201220469

J.E. Borovsky, J. Geophys. Res. **113**, A08110 (2008). doi:10.1029/2007JA012684

R. Bowness, A.W. Hood, C.E. Parnell, Astron. Astrophys. **560**, A89 (2013). doi:10.1051/0004-6361/201116652

S.J. Bradshaw, P.J. Cargill, Astrophys. J. **717**, 163 (2010). doi:10.1088/0004-637X/717/1/163

D.H. Brooks, H.P. Warren, I. Ugarte-Urra, Astrophys. J. Lett. **755**, L33 (2012). doi:10.1088/2041-8205/755/2/L33

J.W. Brosius, E. Landi, J.W. Cook, J.S. Newmark, N. Gopalswamy, A. Lara, Astrophys. J. **574**, 453 (2002). doi:10.1086/340923

P.K. Browning, C. Gerrard, A.W. Hood, R. Kevis, R.A.M. van der Linden, Astron. Astrophys. **485**, 837 (2008). doi:10.1051/0004-6361:20079192

P.K. Browning, E.R. Priest, Astron. Astrophys. **131**, 283 (1984)

L.F. Burlaga, N.F. Ness, Y.-M. Wang, N.R. Sheeley, J. Geophys. Res. **107**, 1410 (2002). doi:10.1029/2001JA009217

P. Cargill, I. de Moortel, Nature **475**, 463 (2011). doi:10.1038/475463a

P.J. Cargill, Astrophys. J. **422**, 381 (1994). doi:10.1086/173733

P.J. Cargill, Sol. Phys. **167**, 267 (1996). doi:10.1007/BF00146339

P.J. Cargill, in *Fundamental Space Plasma Problems: Jim Dungey at 90*, ed. by D.J. Southwood, S.W.H. Cowley (2014a)

P.J. Cargill, Astrophys. J. **784**, 49 (2014b). doi:10.1088/0004-637X/784/1/49

P.J. Cargill, S.J. Bradshaw, J.A. Klimchuk, Astrophys. J. **752**, 161 (2012). arXiv:1204.5960. doi:10.1088/0004-637X/752/2/161

P.J. Cargill, A.W. Hood, S. Migliuolo, Astrophys. J. **309**, 402 (1986). doi:10.1086/164612

P.J. Cargill, J.A. Klimchuk, Astrophys. J. **478**, 799 (1997)

P.J. Cargill, J.A. Klimchuk, Astrophys. J. **605**, 911 (2004). doi:10.1086/382526

P.J. Cargill, J.T. Mariska, S.K. Antiochos, Astrophys. J. **439**, 1034 (1995). doi:10.1086/175240

P.J. Cargill, L. Vlahos, G. Baumann, J.F. Drake, Å. Nordlund, Space Sci. Rev. **173**, 223 (2012). doi:10.1007/s11214-012-9888-y

M. Carlsson, V.H. Hansteen, B. de Pontieu, S. McIntosh, T.D. Tarbell, D. Shine, S. Tsuneta, Y. Katsukawa, K. Ichimoto, Y. Suematsu, T. Shimizu, S. Nagata, Publ. Astron. Soc. Jpn. **59**, 663 (2007). arXiv:0709.3462. doi:10.1093/pasj/59.sp3.S663

H. Carmichael, NASA Spec. Publ. **50**, 451 (1964)

J.W. Cirtain, L. Golub, A.R. Winebarger, B. de Pontieu, K. Kobayashi, R.L. Moore, R.W. Walsh, K.E. Korreck, M. Weber, P. McCauley, A. Title, S. Kuzin, C.E. Deforest, Nature **493**, 501 (2013). doi:10.1038/nature11772

S.R. Cranmer, Space Sci. Rev. **172**, 145 (2012). arXiv:1007.0954. doi:10.1007/s11214-010-9674-7

S.R. Cranmer, A.A. van Ballegooijen, R.J. Edgar, Astrophys. J. Suppl. Ser. **171**, 520 (2007). arXiv:astro-ph/0703333. doi:10.1086/518001

N.U. Crooker, J.T. Gosling, S.W. Kahler, J. Geophys. Res. **107**, 1028 (2002). doi:10.1029/2001JA000236

R.B. Dahlburg, G. Einaudi, A.F. Rappazzo, M. Velli, Astron. Astrophys. **544**, L20 (2012). arXiv:1208.2459. doi:10.1051/0004-6361/201219752

R.B. Dahlburg, J.A. Klimchuk, S.K. Antiochos, Astrophys. J. **622**, 1191 (2005). doi:10.1086/425645

R.B. Dahlburg, J.-H. Liu, J.A. Klimchuk, G. Nigro, Astrophys. J. **704**, 1059 (2009). doi:10.1088/0004-637X/704/2/1059

B. De Pontieu, S.W. McIntosh, M. Carlsson, V.H. Hansteen, T.D. Tarbell, C.J. Schrijver, A.M. Title, R.A. Shine, S. Tsuneta, Y. Katsukawa, K. Ichimoto, Y. Suematsu, T. Shimizu, S. Nagata, Science **318**, 1574 (2007). doi:10.1126/science.1151747

M.L. De Rosa, C.J. Schrijver, G. Barnes, K.D. Leka, B.W. Lites, M.J. Aschwanden, T. Amari, A. Canou, J.M. McTiernan, S. Régnier, J.K. Thalmann, G. Valori, M.S. Wheatland, T. Wiegelmann, M.C.M. Cheung, P.A. Conlon, M. Fuhrmann, B. Inhester, T. Tadesse, Astrophys. J. **696**, 1780 (2009). arXiv:0902.1007. doi:10.1088/0004-637X/696/2/1780

C.E. DeForest, J.B. Gurman, Astrophys. J. Lett. **501**, L217 (1998). doi:10.1086/311460

P. Demoulin, J.C. Henoux, C.H. Mandrini, Sol. Phys. **139**, 105 (1992). doi:10.1007/BF00147884

A. Des Jardins, R. Canfield, D. Longcope, C. Fordyce, S. Waitukaitis, Astrophys. J. **693**, 1628 (2009). doi:10.1088/0004-637X/693/2/1628

M. Dimitropoulou, H. Isliker, L. Vlahos, M.K. Georgoulis, Astron. Astrophys. **529**, A101 (2011). arXiv:1102.2352. doi:10.1051/0004-6361/201015569

H.W. Dodson, E.R. Hedeman, Sol. Phys. **13**, 401 (1970). doi:10.1007/BF00153560

G.A. Doschek, Astrophys. J. **754**, 153 (2012). doi:10.1088/0004-637X/754/2/153

J.W. Dungey, Philos. Mag. **44**, 725 (1953)

J.K. Edmondson, B.J. Lynch, S.K. Antiochos, C.R. De Vore, T.H. Zurbuchen, Astrophys. J. **707**, 1427 (2009). doi:10.1088/0004-637X/707/2/1427

G. Einaudi, P. Boncinelli, R.B. Dahlburg, J.T. Karpen, J. Geophys. Res. **104**, 521 (1999). doi:10.1029/98JA02394

A.G. Emslie, B.R. Dennis, A.Y. Shih, P.C. Chamberlin, R.A. Mewaldt, C.S. Moore, G.H. Share, A. Vourlidas, B.T. Welsch, Astrophys. J. **759**, 71 (2012). arXiv:1209.2654. doi:10.1088/0004-637X/759/1/71

E. Endeve, T.E. Holzer, E. Leer, Astrophys. J. **603**, 307 (2004). doi:10.1086/381239

Y. Fan, Astrophys. J. **740**, 68 (2011). arXiv:1109.3734. doi:10.1088/0004-637X/740/2/68

Y. Fan, S.E. Gibson, Astrophys. J. **668**, 1232 (2007). doi:10.1086/521335

L.A. Fisk, J. Geophys. Res. **108**, 1157 (2003). doi:10.1029/2002JA009284

L.A. Fisk, Astrophys. J. **626**, 563 (2005). doi:10.1086/429957

L.A. Fisk, N.A. Schwadron, T.H. Zurbuchen, Space Sci. Rev. **86**, 51 (1998). doi:10.1023/A:1005015527146

A. Fossum, M. Carlsson, Nature **435**, 919 (2005). doi:10.1038/nature03695

A. Fossum, M. Carlsson, Astrophys. J. **646**, 579 (2006). doi:10.1086/504887

G.A. Gary, R.L. Moore, Astrophys. J. **611**, 545 (2004). doi:10.1086/422132

J. Geiss, G. Gloeckler, R. von Steiger, Space Sci. Rev. **72**, 49 (1995). doi:10.1007/BF00768753

A. Greco, W.H. Matthaeus, S. Servidio, P. Chuychai, P. Dmitruk, Astrophys. J. Lett. **691**, L111 (2009). doi:10.1088/0004-637X/691/2/L111

J.M. Greene, J. Geophys. Res. **93**, 8583 (1988). doi:10.1029/JA093iA08p08583

P.C. Grigis, A.O. Benz, Astrophys. J. Lett. **625**, L143 (2005). arXiv:astro-ph/0504436. doi:10.1086/431147

B.V. Gudiksen, Å. Nordlund, Astrophys. J. **618**, 1020 (2005). arXiv:astro-ph/0407266. doi:10.1086/426063

I.G. Hannah, L. Fletcher, Sol. Phys. **236**, 59 (2006). doi:10.1007/s11207-006-0139-9

I.G. Hannah, H.S. Hudson, M. Battaglia, S. Christe, J. Kašparová, S. Krucker, M.R. Kundu, A. Veronig, Space Sci. Rev. **159**, 263 (2011). arXiv:1108.6203. doi:10.1007/s11214-010-9705-4

I.G. Hannah, H.S. Hudson, G.J. Hurford, R.P. Lin, Astrophys. J. **724**, 487 (2010). arXiv:1009.2918. doi:10. 1088/0004-637X/724/1/487

I.G. Hannah, S. Krucker, H.S. Hudson, S. Christe, R.P. Lin, Astron. Astrophys. **481**, L45 (2008). arXiv:0712.0369. doi:10.1051/0004-6361:20079019

V.H. Hansteen, M. Velli, Space Sci. Rev. **172**, 89 (2012). doi:10.1007/s11214-012-9887-z

D.M. Hassler, I.E. Dammasch, P. Lemaire, P. Brekke, W. Curdt, H.E. Mason, J.-C. Vial, K. Wilhelm, Science **283**, 810 (1999). doi:10.1126/science.283.5403.810

J.C. Hénoux, B.V. Somov, Astron. Astrophys. **185**, 306 (1987)

T. Hirayama, Sol. Phys. **34**, 323 (1974). doi:10.1007/BF00153671

J.T. Hoeksema, Adv. Space Res. **11**, 15 (1991)

G.D. Holman, L. Sui, R.A. Schwartz, A.G. Emslie, Astrophys. J. Lett. **595**, L97 (2003). doi:10.1086/378488

T.E. Holzer, E. Leer, J. Geophys. Res. **85**, 4665 (1980). doi:10.1029/JA085iA09p04665

A.W. Hood, Sol. Phys. **87**, 279 (1983). doi:10.1007/BF00224841

A.W. Hood, P.K. Browning, R.A.M. van der Linden, Astron. Astrophys. **506**, 913 (2009). doi:10.1051/ 0004-6361/200912285

A.W. Hood, E.R. Priest, Sol. Phys. **64**, 303 (1979). doi:10.1007/BF00151441

H.S. Hudson, C.J. Wolfson, T.R. Metcalf, Sol. Phys. **234**, 79 (2006). doi:10.1007/s11207-006-0056-y

A.J. Hundhausen, *Coronal Expansion and Solar Wind*. Physics and Chemistry in Space, vol. 5 (Springer, Berlin, 1972). doi:10.1007/978-3-642-65414-5

G.J. Hurford, S. Krucker, R.P. Lin, R.A. Schwartz, G.H. Share, D.M. Smith, Astrophys. J. Lett. **644**, L93 (2006). doi:10.1086/505329

D.E. Innes, B. Inhester, W.I. Axford, K. Wilhelm, Nature **386**, 811 (1997). doi:10.1038/386811a0

J.A. Ionson, Astrophys. J. **226**, 650 (1978). doi:10.1086/156648

S.W. Kahler, R.L. Moore, S.R. Kane, H. Zirin, Astrophys. J. **328**, 824 (1988). doi:10.1086/166340

B. Kliem, T. Török, Phys. Rev. Lett. **96**(25), 255002 (2006). arXiv:physics/0605217. doi:10.1103/ PhysRevLett.96.255002

J.A. Klimchuk, in *Solar Jets and Coronal Plumes*, ed. by T.-D. Guyenne. ESA Special Publication, vol. 421 (1998), p. 233

J.A. Klimchuk, Sol. Phys. **234**, 41 (2006). arXiv:astro-ph/0511841. doi:10.1007/s11207-006-0055-z

J.A. Klimchuk, S. Patsourakos, P.J. Cargill, Astrophys. J. **682**, 1351 (2008). arXiv:0710.0185. doi:10.1086/ 589426

R.A. Kopp, G.W. Pneuman, Sol. Phys. **50**, 85 (1976). doi:10.1007/BF00206193

S. Krucker, M. Battaglia, P.J. Cargill, L. Fletcher, H.S. Hudson, A.L. MacKinnon, S. Masuda, L. Sui, M. Tom-czak, A.L. Veronig, L. Vlahos, S.M. White, Astron. Astrophys. Rev. **16**, 155 (2008). doi:10.1007/ s00159-008-0014-9

J. Kuijpers, H.U. Frey, L. Fletcher, Space Sci. Rev. Online First (2014). doi:10.1007/s11214-014-0041-y

A. Lagg, J. Woch, N. Krupp, S.K. Solanki, Astron. Astrophys. **414**, 1109 (2004). doi:10.1051/0004-6361:20031643

J.M. Laming, Astrophys. J. **614**, 1063 (2004). arXiv:astro-ph/0405230. doi:10.1086/423780

Y.-T. Lau, J.M. Finn, Astrophys. J. **350**, 672 (1990). doi:10.1086/168419

J. Leenaarts, M. Carlsson, V. Hansteen, R.J. Rutten, Astron. Astrophys. **473**, 625 (2007). arXiv:0709.3751. doi:10.1051/0004-6361:20078161

K.D. Leka, G. Barnes, Astrophys. J. **656**, 1173 (2007). doi:10.1086/510282

K.D. Leka, D.L. Mickey, H. Uitenbroek, E.L. Wagner, T.R. Metcalf, Sol. Phys. **278**, 471 (2012). doi:10. 1007/s11207-012-9958-z

H. Lin, J.R. Kuhn, R. Coulter, Astrophys. J. Lett. **613**, L177 (2004). doi:10.1086/425217

H. Lin, M.J. Penn, S. Tomczyk, Astrophys. J. Lett. **541**, L83 (2000). doi:10.1086/312900

R.P. Lin, P.T. Feffer, R.A. Schwartz, Astrophys. J. Lett. **557**, L125 (2001). doi:10.1086/323270

J.A. Linker, R. Lionello, Z. Mikić, V.S. Titov, S.K. Antiochos, Astrophys. J. **731**, 110 (2011). doi:10. 1088/0004-637X/731/2/110

R. Liu, D. Alexander, Astrophys. J. **697**, 999 (2009). doi:10.1088/0004-637X/697/2/999

D.W. Longcope, Sol. Phys. **169**, 91 (1996). doi:10.1007/BF00153846

D.W. Longcope, Living Rev. Sol. Phys. **2**, 7 (2005). doi:10.12942/lrsp-2005-7

E.T. Lu, R.J. Hamilton, Astrophys. J. Lett. **380**, L89 (1991). doi:10.1086/186180

T. Magara, D.W. Longcope, Astrophys. J. **586**, 630 (2003). doi:10.1086/367611

W. Manchester IV, T. Gombosi, D. DeZeeuw, Y. Fan, Astrophys. J. **610**, 588 (2004). doi:10.1086/421516

C.H. Mandrini, P. Demoulin, J.C. Henoux, M.E. Machado, Astron. Astrophys. **250**, 541 (1991)

J. Martínez-Sykora, B. De Pontieu, V. Hansteen, Astrophys. J. **753**, 161 (2012). arXiv:1204.5991. doi:10. 1088/0004-637X/753/2/161

S. Masson, E. Pariat, G. Aulanier, C.J. Schrijver, Astrophys. J. **700**, 559 (2009). doi:10.1088/0004-637X/700/1/559

R.T.J. McAteer, C.A. Young, J. Ireland, P.T. Gallagher, Astrophys. J. **662**, 691 (2007). doi:10.1086/518086

D.J. McComas, R.W. Ebert, H.A. Elliott, B.E. Goldstein, J.T. Gosling, N.A. Schwadron, R.M. Skoug, Geophys. Res. Lett. **35**, L18103 (2008). doi:10.1029/2008GL034896

P.S. McIntosh, Sol. Phys. **125**, 251 (1990). doi:10.1007/BF00158405

S.W. McIntosh, B. De Pontieu, Astrophys. J. **707**, 524 (2009). arXiv:0910.5191. doi:10.1088/0004-637X/707/1/524

T.R. Metcalf, D. Alexander, H.S. Hudson, D.W. Longcope, Astrophys. J. **595**, 483 (2003). doi:10.1086/377217

T.R. Metcalf, M.L. De Rosa, C.J. Schrijver, G. Barnes, A.A. van Ballegooijen, T. Wiegelmann, M.S. Wheatland, G. Valori, J.M. McTiernan, Sol. Phys. **247**, 269 (2008). doi:10.1007/s11207-007-9110-7

T.R. Metcalf, K.D. Leka, D.L. Mickey, Astrophys. J. Lett. **623**, L53 (2005). doi:10.1086/429961

Z. Mikic, J.A. Linker, Astrophys. J. **430**, 898 (1994). doi:10.1086/174460

V.M. Nakariakov, L. Ofman, E.E. Deluca, B. Roberts, J.M. Davila, Science **285**, 862 (1999). doi:10.1126/science.285.5429.862

V.M. Nakariakov, E. Verwichte, Living Rev. Sol. Phys. **2**, 3 (2005). doi:10.12942/lrsp-2005-3

M. Neugebauer, C.W. Snyder, Science **138**, 1095 (1962). doi:10.1126/science.138.3545.1095-a

L. Ofman, J.A. Klimchuk, J.M. Davila, Astrophys. J. **493**, 474 (1998). doi:10.1086/305109

E. Pariat, S.K. Antiochos, C.R. DeVore, Astrophys. J. **691**, 61 (2009). doi:10.1088/0004-637X/691/1/61

E.N. Parker, Astrophys. J. **128**, 664 (1958). doi:10.1086/146579

E.N. Parker, Astrophys. J. **330**, 474 (1988). doi:10.1086/166485

J.M. Pasachoff, V. Rušin, M. Druckmüller, P. Aniol, M. Saniga, M. Minarovjech, Astrophys. J. **702**, 1297 (2009). arXiv:0907.1643. doi:10.1088/0004-637X/702/2/1297

H. Peter, Astron. Astrophys. **521**, A51 (2010). arXiv:1004.5403. doi:10.1051/0004-6361/201014433

H.E. Petschek, NASA Spec. Publ. **50**, 425 (1964)

E.R. Priest, P. Démoulin, J. Geophys. Res. **100**, 23443 (1995). doi:10.1029/95JA02740

A.F. Rappazzo, M. Velli, G. Einaudi, Astrophys. J. **722**, 65 (2010). arXiv:1003.3872. doi:10.1088/0004-637X/722/1/65

A.F. Rappazzo, M. Velli, G. Einaudi, R.B. Dahlburg, Astrophys. J. **633**, 474 (2005). arXiv:1002.3325. doi:10.1086/431916

A.F. Rappazzo, M. Velli, G. Einaudi, R.B. Dahlburg, Astrophys. J. **677**, 1348 (2008). arXiv:0709.3687. doi:10.1086/528786

F. Reale, Living Rev. Sol. Phys. **7**, 5 (2010). arXiv:1010.5927. doi:10.12942/lrsp-2010-5

F. Reale, P. Testa, J.A. Klimchuk, S. Parenti, Astrophys. J. **698**, 756 (2009). arXiv:0904.0878. doi:10.1088/0004-637X/698/1/756

P. Riley, J.A. Linker, Z. Mikić, R. Lionello, S.A. Ledvina, J.G. Luhmann, Astrophys. J. **653**, 1510 (2006). doi:10.1086/508565

P. Riley, J.G. Luhmann, Sol. Phys. **277**, 355 (2012). doi:10.1007/s11207-011-9909-0

A.J.B. Russell, L. Fletcher, Astrophys. J. **765**, 81 (2013). arXiv:1302.2458. doi:10.1088/0004-637X/765/2/81

R.J. Rutten, in *The Physics of Chromospheric Plasmas*, ed. by P. Heinzel, I. Dorotovič, R.J. Rutten. Astronomical Society of the Pacific Conference Series, vol. 368 (2007), p. 27. arXiv:astro-ph/0703637

J.M. Ryan, Space Sci. Rev. **93**, 581 (2000)

P. Saint-Hilaire, S. Krucker, R.P. Lin, Sol. Phys. **250**, 53 (2008). arXiv:1111.4247. doi:10.1007/s11207-008-9193-9

I. Sammis, F. Tang, H. Zirin, Astrophys. J. **540**, 583 (2000). doi:10.1086/309303

K.H. Schatten, J.M. Wilcox, N.F. Ness, Sol. Phys. **6**, 442 (1969). doi:10.1007/BF00146478

J.T. Schmelz, S. Pathak, Astrophys. J. **756**, 126 (2012). doi:10.1088/0004-637X/756/2/126

B. Schmieder, P. Démoulin, G. Aulanier, Adv. Space Res. **51**, 1967 (2013). arXiv:1212.4014. doi:10.1016/j.asr.2012.12.026

C.J. Schrijver, Astrophys. J. Lett. **655**, L117 (2007). doi:10.1086/511857

C.J. Schrijver, M.L. De Rosa, T.R. Metcalf, Y. Liu, J. McTiernan, S. Régnier, G. Valori, M.S. Wheatland, T. Wiegelmann, Sol. Phys. **235**, 161 (2006). doi:10.1007/s11207-006-0068-7

C.J. Schrijver, A.M. Title, A.A. van Ballegooijen, H.J. Hagenaar, R.A. Shine, Astrophys. J. **487**, 424 (1997)

C.J. Schrijver, A.M. Title, A.R. Yeates, M.L. DeRosa, Astrophys. J. **773**, 93 (2013). arXiv:1305.0801. doi:10.1088/0004-637X/773/2/93

P.J.A. Simões, L. Fletcher, H.S. Hudson, A.J.B. Russell, ArXiv e-prints (2013). arXiv:1309.7090

A.C. Sterling, R.L. Moore, Astrophys. J. **630**, 1148 (2005). doi:10.1086/432044

P.A. Sturrock, Nature **211**, 695 (1966). doi:10.1038/211695a0

S.T. Suess, Space Sci. Rev. **23**, 159 (1979). doi:10.1007/BF00173809

S.T. Suess, A.-H. Wang, S.T. Wu, J. Geophys. Res. **101**, 19957 (1996). doi:10.1029/96JA01458

X. Sun, J.T. Hoeksema, Y. Liu, T. Wiegelmann, K. Hayashi, Q. Chen, J. Thalmann, Astrophys. J. **748**, 77 (2012). arXiv:1201.3404. doi:10.1088/0004-637X/748/2/77

Z. Svestka, S.F. Martin, R.A. Kopp, in *Solar and Interplanetary Dynamics*, ed. by M. Dryer, E. Tandberg-Hanssen. IAU Symposium, vol. 91 (1980), pp. 217–221

P.A. Sweet, Annu. Rev. Astron. Astrophys. **7**, 149 (1969). doi:10.1146/annurev.aa.07.090169.001053

P. Testa, B. De Pontieu, J. Martínez-Sykora, E. DeLuca, V. Hansteen, J. Cirtain, A. Winebarger, L. Golub, K. Kobayashi, K. Korreck, S. Kuzin, R. Walsh, C. DeForest, A. Title, M. Weber, Astrophys. J. Lett. **770**, L1 (2013). arXiv:1305.1687. doi:10.1088/2041-8205/770/1/L1

P. Testa, F. Reale, Astrophys. J. Lett. **750**, L10 (2012). arXiv:1204.0041. doi:10.1088/2041-8205/750/1/L10

P. Testa, F. Reale, E. Landi, E.E. DeLuca, V. Kashyap, Astrophys. J. **728**, 30 (2011). arXiv:1012.0346. doi:10.1088/0004-637X/728/1/30

L.M. Thornton, C.E. Parnell, Sol. Phys. **269**, 13 (2011). doi:10.1007/s11207-010-9656-7

V.S. Titov, P. Démoulin, Astron. Astrophys. **351**, 707 (1999)

V.S. Titov, Z. Mikić, J.A. Linker, R. Lionello, S.K. Antiochos, Astrophys. J. **731**, 111 (2011). arXiv: 1011.0009. doi:10.1088/0004-637X/731/2/111

V.S. Titov, Z. Mikic, T. Török, J.A. Linker, O. Panasenco, Astrophys. J. **759**, 70 (2012). arXiv:1209.5797. doi:10.1088/0004-637X/759/1/70

S. Tomczyk, S.W. McIntosh, S.L. Keil, P.G. Judge, T. Schad, D.H. Seeley, J. Edmondson, Science **317**, 1192 (2007). doi:10.1126/science.1143304

T. Török, B. Kliem, Astrophys. J. Lett. **630**, L97 (2005). arXiv:astro-ph/0507662. doi:10.1086/462412

T. Török, B. Kliem, Astron. Nachr. **328**, 743 (2007). arXiv:0705.2100. doi:10.1002/asna.200710795

D. Tripathi, J.A. Klimchuk, H.E. Mason, Astrophys. J. **740**, 111 (2011). arXiv:1107.4480. doi:10.1088/0004-637X/740/2/111

C.-Y. Tu, C. Zhou, E. Marsch, L.-D. Xia, L. Zhao, J.-X. Wang, K. Wilhelm, Science **308**, 519 (2005). doi:10.1126/science.1109447

R. Turkmani, L. Vlahos, K. Galsgaard, P.J. Cargill, H. Isliker, Astrophys. J. Lett. **620**, L59 (2005). doi:10.1086/428395

A.A. van Ballegooijen, M. Asgari-Targhi, S.R. Cranmer, E.E. DeLuca, Astrophys. J. **736**, 3 (2011). arXiv:1105.0402. doi:10.1088/0004-637X/736/1/3

B. van der Holst, I.V. Sokolov, X. Meng, M. Jin, W.B. Manchester IV, G. Tóth, T.I. Gombosi, Astrophys. J. **782**, 81 (2014). arXiv:1311.4093. doi:10.1088/0004-637X/782/2/81

A. Verdini, M. Velli, Astrophys. J. **662**, 669 (2007). arXiv:astro-ph/0702205. doi:10.1086/510710

A. Verdini, M. Velli, W.H. Matthaeus, S. Oughton, P. Dmitruk, Astrophys. J. Lett. **708**, L116 (2010). arXiv:0911.5221. doi:10.1088/2041-8205/708/2/L116

J.F. Vesecky, S.K. Antiochos, J.H. Underwood, Astrophys. J. **233**, 987 (1979). doi:10.1086/157462

L. Vlahos, Space Sci. Rev. **68**, 39 (1994). doi:10.1007/BF00749115

L. Vlahos, M. Georgoulis, R. Kluiving, P. Paschos, Astron. **299**, 897 (1995)

R. von Steiger, R.F.W. Schweingruber, J. Geiss, G. Gloeckler, Adv. Space Res. **15**, 3 (1995)

R. von Steiger, T.H. Zurbuchen, J. Geiss, G. Gloeckler, L.A. Fisk, N.A. Schwadron, Space Sci. Rev. **97**, 123 (2001). doi:10.1023/A:1011886414964

A. Vourlidas, B.J. Lynch, R.A. Howard, Y. Li, Sol. Phys. **284**, 179 (2013). arXiv:1207.1599. doi:10.1007/s11207-012-0084-8

Y.-M. Wang, N.R. Sheeley Jr., Astrophys. J. **355**, 726 (1990). doi:10.1086/168805

Y.-M. Wang, N.R. Sheeley Jr., N.B. Rich, Astrophys. J. **658**, 1340 (2007). doi:10.1086/511416

H.P. Warren, D.H. Brooks, A.R. Winebarger, Astrophys. J. **734**, 90 (2011). arXiv:1009.5976. doi:10.1088/0004-637X/734/2/90

H.P. Warren, A.R. Winebarger, D.H. Brooks, Astrophys. J. **759**, 141 (2012). arXiv:1204.3220. doi:10.1088/0004-637X/759/2/141

A.L. Wilmot-Smith, D.I. Pontin, A.R. Yeates, G. Hornig, Astron. Astrophys. **536**, A67 (2011). arXiv:1111.1100. doi:10.1051/0004-6361/201117942

G.L. Withbroe, Astrophys. J. **325**, 442 (1988). doi:10.1086/166015

G.L. Withbroe, R.W. Noyes, Annu. Rev. Astron. Astrophys. **15**, 363 (1977). doi:10.1146/annurev.aa.15.090177.002051

T.N. Woods, F.G. Eparvier, J. Fontenla, J. Harder, G. Kopp, W.E. McClintock, G. Rottman, B. Smiley, M. Snow, Geophys. Res. Lett. **31**, L10802 (2004). doi:10.1029/2004GL019571

S. Yashiro, S. Akiyama, N. Gopalswamy, R.A. Howard, Astrophys. J. Lett. **650**, L143 (2006). arXiv:astro-ph/0609197. doi:10.1086/508876

S. Yashiro, N. Gopalswamy, S. Akiyama, G. Michalek, R.A. Howard, J. Geophys. Res. **110**, A12S05 (2005). doi:10.1029/2005JA011151

L. Zhao, T.H. Zurbuchen, L.A. Fisk, Geophys. Res. Lett. **36**, L14104 (2009). doi:10.1029/2009GL039181

J.B. Zirker, Rev. Geophys. Space Phys. **15**, 257 (1977). doi:10.1029/RG015i003p00257

J.B. Zirker, Sol. Phys. **148**, 43 (1993). doi:10.1007/BF00675534

T.H. Zurbuchen, Annu. Rev. Astron. Astrophys. **45**, 297 (2007). doi:10.1146/annurev.astro.45.010807.154030

T.H. Zurbuchen, L.A. Fisk, G. Gloeckler, R. von Steiger, Geophys. Res. Lett. **29**, 1352 (2002). doi:10.1029/2001GL013946

DOI 10.1007/978-1-4939-3547-5_9
Reprinted from *Space Science Reviews* Journal, DOI 10.1007/s11214-014-0050-x

What Controls the Structure and Dynamics of Earth's Magnetosphere?

J.P. Eastwood · H. Hietala · G. Toth · T.D. Phan · M. Fujimoto

Received: 15 January 2014 / Accepted: 25 April 2014 / Published online: 14 June 2014
© The Author(s) 2014. This article is published with open access at Springerlink.com

Abstract Unlike most cosmic plasma structures, planetary magnetospheres can be extensively studied in situ. In particular, studies of the Earth's magnetosphere over the past few decades have resulted in a relatively good experimental understanding of both its basic structural properties and its response to changes in the impinging solar wind. In this article we provide a broad overview, designed for researchers unfamiliar with magnetospheric physics, of the main processes and parameters that control the structure and dynamics of planetary magnetospheres, especially the Earth's. In particular, we concentrate on the structure and dynamics of three important regions: the bow shock, the magnetopause and the magnetotail. In the final part of this review we describe the current status of global magnetospheric modelling, which is crucial to placing in situ observations in the proper context and providing a better understanding of magnetospheric structure and dynamics under all possible input conditions. Although the parameter regime experienced in the solar system is limited, the plasma physics that is learned by studying planetary magnetospheres can, in principle, be translated to more general studies of cosmic plasma structures. Conversely, studies of cosmic plasma under a wide range of conditions should be used to understand Earth's

J.P. Eastwood (✉) · H. Hietala
The Blackett Laboratory, Imperial College London, London SW7 2AZ, UK
e-mail: jonathan.eastwood@imperial.ac.uk

G. Toth
Center for Space Environment Modeling, Dept. of AOSS, University of Michigan, Ann Arbor, MI 48109, USA

T.D. Phan
Space Sciences Laboratory, University of California, Berkeley 94720, USA

M. Fujimoto
Institute of Space and Astronautical Science, Japan Aerospace Exploration Agency, 3-1-1 Yoshinodai, Chuo-ku, Sagamihara, Kanagawa 252-5210, Japan

M. Fujimoto
Earth-Life Science Institute, Tokyo Institute of Technology, 2-12-1 Ookayama, Meguro, Tokyo 152-8551, Japan

magnetosphere under extreme conditions. We conclude the review by discussing this and summarizing some general properties and principles that may be applied to studies of other cosmic plasma structures.

Keywords Magnetosphere · Bow shock · Magnetopause · Magnetotail · Computer simulation

1 Introduction

Planetary magnetospheres are examples of cosmic plasmas whose structure and dynamics can be studied extensively in situ. Since the majority of cosmic plasmas are inaccessible to in situ measurements, which can be considered crucial in fully understanding the physics, the study of planetary magnetospheres must therefore play a vital role in exploring and understanding general principles of cosmic plasma structure formation and dynamics. In this review, we describe the basic structure and dynamics of the magnetosphere, focussing in particular on the bow shock, the magnetopause and the magnetotail. We will mainly discuss the Earth's magnetosphere, but also mention, where relevant, properties of other planetary magnetospheres. This review is derived from presentations given at an International Space Science Institute (ISSI) workshop on multi-scale structure formation in cosmic plasmas, and so our aim is to present this work in a manner useful for those new to the field, and those working in associated areas of plasma and astrophysics with little or no previous experience in magnetospheric physics. In particular we note that many of the physical processes important in magnetospheric physics are directly relevant for solar physics. Many books and monographs have been written on magnetospheric physics (e.g., Kivelson and Russell 1995; Baumjohann and Treumann 1996; Cravens 1997), and in each section we have also included citations to more detailed and specific review articles.

We begin by briefly reviewing the overall structure and configuration of the magnetosphere, as shown in Fig. 1. The cartoon is drawn with the Sun to the left, and the solar wind flowing from left to right. The Geocentric Solar Magnetospheric (GSM) coordinate system is often used when studying the magnetosphere, where the x direction points sunward, the x–z plane contains the dipole, and x–y–z is a right handed triple. As such, a cut in the meridional x–z GSM plane is shown. However, note that the cartoon is somewhat simplified because in reality the dipole is typically tilted with respect to the z axis (due to both the tilt of the Earth's rotational axis relative to the ecliptic plane, and the offset of the magnetic dipole axis from the rotation axis). The solar wind (yellow) first encounters the bow shock (dotted line), where it is slowed, compressed, heated and deflected around the blunt obstacle that is the magnetosphere. The solar wind is magnetised (its magnetic field is often referred to as the interplanetary magnetic field (IMF)) and largely behaves as an ideal plasma where the magnetic field is frozen in. In this cartoon the magnetic field is shown as pointing southward, opposite to the Earth's magnetic field at the dayside magnetopause, but in fact the solar wind magnetic field orientation constantly varies on timescales of minutes. The bow shock, where the solar wind is slowed and deflected around the magnetosphere, is both a particle accelerator and an important site for studying and understanding plasma energy conversion processes. In particular, the structure and properties of the bow shock strongly depend on the orientation of the solar wind magnetic field relative to the shock normal, and this is discussed in more detail in Sect. 2.

In Fig. 1, the magnetopause (dashed line) separates the shocked solar wind in the magnetosheath (orange) from the magnetospheric plasma. The low latitude boundary layer (red)

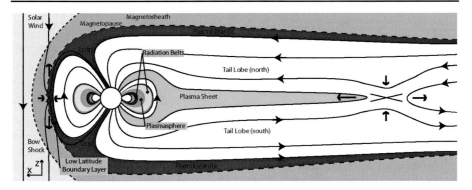

Fig. 1 A cartoon of the Earth's magnetosphere

forms on the dayside on the magnetospheric side. Since in general there is a change in the strength and orientation of the magnetic field (in common with the other magnetospheric boundaries), it is also referred to as the magnetopause current sheet. The magnetopause is not a perfect shield and can be 'open' or 'closed' depending on the surrounding plasma conditions. This implies that there can be mass transport across the magnetopause. For example, under southward IMF conditions, as is drawn in Fig. 1, magnetic reconnection occurs along an X-line extending out of the page, and creates plasma jets directed north and south along the magnetopause current sheet. However, although energy transfer is enhanced for southward IMF, the mass transport is enhanced when the IMF is northward. Thus, the structure and dynamics of the magnetopause are also strongly dependent on the solar wind conditions, and are discussed further in Sect. 3.

The dynamic pressure of the solar wind compresses the magnetosphere on the dayside, but on the nightside an extended tail is formed. The plasma mantle forms at the edge of the magnetotail and can, to some extent, be considered the tailward extension of the low latitude boundary layer. The tail lobes consist of magnetic flux that is connected to the planet at one end, and are largely devoid of plasma compared to the other magnetospheric regions. Magnetic reconnection also occurs at a tail X-line, returning plasma on closed field lines towards the planet (Dungey 1961). This is leads to the formation of the plasma sheet (green). Closer to Earth are located the plasmasphere (light blue) and the radiation belts (yellow/dark blue). We do not further discuss the inner magnetosphere, radiation belts or aurora here. The structure and dynamics of the magnetotail fundamentally depend on the manner in which plasma and magnetic flux enters the magnetotail. The magnetotail exhibits dynamic and bursty behaviour, controlled by reconnection, which corresponds to the storage and release of plasma and magnetic energy. This is described further in Sect. 4.

Whilst this cartoon is relatively simple and appears to suggest that the magnetosphere exists in a state of dynamic equilibrium, it should be understood that since the solar wind varies continuously, on time scales of minutes to hours, these boundaries and regions are not stationary and exhibit considerable dynamical behavior. In addition, there are various instabilities that create dynamics even with a steady solar wind. Computer simulations are crucial to placing in situ experimental observations in context, as well as revealing the large scale system dynamics. In the final part of this review we discuss global simulations of the magnetosphere, and describe how models are being developed and coupled in order to better capture the various physical processes at work.

1.1 Controlling Parameters

Before proceeding further, we briefly discuss the main parameters that control the structure and dynamics of planetary magnetospheres. In the subsequent sections it will be seen that different parameters are important for different regions.

Solar Wind Magnetic Field Strength and Orientation   The orientation of the solar wind (or interplanetary) magnetic field (IMF) is described in different ways depending on the context. For studies of the bow shock, the angle between the solar wind magnetic field and the local shock normal, θ_{BN}, is used. If $\theta_{BN} < 45°$, the shock is said to be quasi-parallel (i.e., the shock normal is approximately parallel to the magnetic field), whereas if $\theta_{BN} > 45°$, the shock is said to be quasi-perpendicular. For magnetospheric studies, the clock angle ($\theta = \tan^{-1}(B_y/B_z)$) is used to describe the orientation of the field relative to the dipole field. If the solar wind and magnetospheric magnetic fields are aligned at the nose of the magnetopause (clock angle $= 0$), the IMF is said to point northward. In contrast, if the clock angle $= 180°$, then the solar wind and magnetospheric fields are opposite, and the IMF is said to point southward. Finally, the cone angle ($\varphi = \cos^{-1}(B_x/|B|)$) is the angle between the IMF and the x direction. If the cone angle is small or close to $180°$, the IMF is said to be radial. To illustrate the typical properties of the IMF at Earth, Fig. 2(a) shows Wind satellite measurements of the solar wind magnetic field strength over the 11 year solar cycle (year averages) (Acuña et al. 1995). It can be seen that the strength of the IMF at Earth is of the order of 5 nT (although this masks considerable variability on short timescales), and was at a maximum value in 2003, around solar maximum.

Solar Wind Plasma Beta and Mach Number   Figure 2(b–d) shows Wind yearly averaged measurements of the solar wind velocity, density and temperature. Figure 2(e) shows the solar wind ion plasma beta (β_{SW}), the ratio of the thermal to magnetic pressure in the solar wind. This typically varies between 0.4 and 0.8, although again this masks considerable variability on short timescales. The solar wind Mach number is the ratio of the solar wind speed to a characteristic plasma wave speed (either Alfvén, or magnetosonic) M_A or M_{MS}— the solar wind Alfvén Mach number typically varies between 6 and 12, although again this masks considerable variability.

Planetary Rotation Rate and Plasma Sources   At Earth, the magnetosphere is largely driven by the solar wind, although ionospheric plasma (oxygen) is often present in the magnetosphere (e.g., Strangeway et al. 2000). Mercury's magnetosphere is even more strongly driven by the solar wind (e.g., Slavin et al. 2012b). At Jupiter and Saturn, Io and Enceladus are sources of plasma that load the magnetosphere internally (Bagenal et al. 2007; Dougherty et al. 2009). Furthermore, both Jupiter and Saturn are rapidly rotating, and so the interaction of Jupiter (in particular) with the solar wind is very different to that at Earth, since it is dominated by rotation and internal plasma, and the solar wind driving is thought to be rather weak (Vasyliunas 1983). As such, comparative magnetospheric studies are becoming increasingly important (e.g., Keiling et al. 2014).

2 The Bow Shock

2.1 Planetary Bow Shock Structure

The bow shocks of planetary magnetospheres are curved, and so even if the upstream conditions are uniform, the geometry of the shock changes across its surface as shown in Fig. 3.

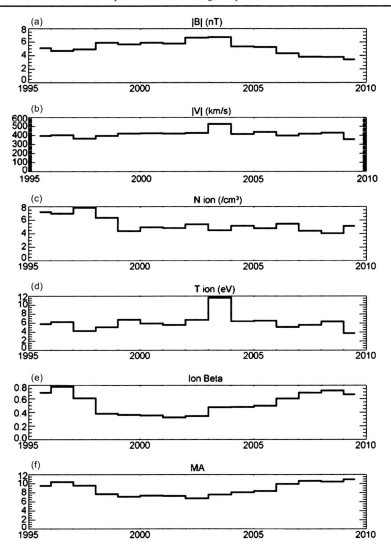

Fig. 2 Average properties of the solar wind at Earth as measured by the Wind satellite. The average value for each year between 1996 and 2009 is shown, essentially 1 solar cycle. *The top four panels* show the magnetic field strength, the solar wind speed and the solar wind density and temperature. *The bottom two panels* show the solar wind ion beta and the solar wind Alfvén Mach number

Here we briefly summarise some of the main features; many more specific reviews have been published (e.g., Tsurutani and Stone 1985; Fuselier 1994, 1995; Le and Russell 1994; Burgess 1995; Bale et al. 2005; Burgess et al. 2005; Eastwood et al. 2005b). In Fig. 3, the magnetic field is perpendicular to the shock normal in the lower part of the figure, and it is parallel to the shock normal in the upper part. As the solar wind reaches the shock, the tangent field line represents the first point of contact (here the magnetic field is exactly perpendicular). Just behind the tangent field line, electrons are able to stream back along the field into the upstream region; this is denoted by the electron foreshock boundary. The ions are not able to escape along the tangent field line, but can escape from quasi-perpendicular

Fig. 3 Geometry of the bow
shock and foreshock

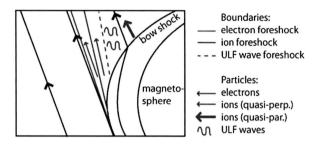

Boundaries:
—— electron foreshock
—— ion foreshock
- - - ULF wave foreshock

Particles:
⟵ electrons
⟵ ions (quasi-perp.)
⟸ ions (quasi-par.)
∿ ULF waves

geometries. The ion beam streaming from the quasi-perpendicular foreshock is typically cold, and is a field aligned beam. The combination of these beams and the solar wind core distribution is subject to the right hand resonant ion beam instability, generating ULF waves with characteristic periods of $\sim$30 s at Earth (e.g., Eastwood et al. 2005a). However, the waves are immediately convected anti-sunward into the quasi-parallel shock, and so the waves are not necessarily observed in conjunction with the distributions that created them.

Hotter and more energetic beams, known as diffuse distributions, are also observed in the foreshock. It was originally thought that these hot beams arose because of wave-particle interactions and disruption of the field-aligned beams. However, the diffuse distributions contain solar wind quantities of He^{2+}, whereas the field-aligned beams consist only of protons, and so they are not directly related (Fuselier 1995). In fact, diffuse distributions arise from the quasi-parallel shock. However, the curvature of the bow shock at Earth means that there is considerable cross talk and the quasi-perpendicular and parallel sections are strongly coupled. More generally, the magnetic geometry means that the quasi-perpendicular shock jump is thin (e.g., Bale et al. 2005) whereas the quasi-parallel shock is much more complex and extended (e.g., Burgess et al. 2005).

This curvature also means that the shock itself is quite spatially inhomogeneous, even when only the Rankine-Hugoniot jump conditions are considered. To illustrate this, Fig. 4 shows how the shock compression ratio (r) varies over the bow shock surface for different values of the upstream Mach number and plasma beta. These figures are generated with the IMF aligned to the upstream flow to better illustrate the effect of bow shock curvature, and we use the Merka et al. (2005) shock model. The model shape depends on M_A. We note that as the solar wind ram pressure increases (P_{dyn}), the magnetosphere compresses, and so the obstacle and the shock change in size. In this figure, M_A is varied only by varying the magnetic field strength, and P_{dyn} is constant. The plasma beta is varied independently by changing the solar wind temperature. High solar wind M_A suppresses the effects of curvature since in the high M_A limit $r \to 4$, and increasing the plasma beta decreases the compression ratio.

2.2 The Bow Shock/Foreshock as a Site for Particle Acceleration

Particle acceleration is intimately related to the foreshock; for example, Fermi acceleration is naturally expected to occur at parallel shocks. Fermi acceleration of a charged particle in space takes place when the particle finds itself in a situation analogous to a ball confined between two converging walls (Fermi 1949). Because the ULF waves described above are moving towards the shock front in the shock-rest-frame, the ions undergoing multiple interactions with upstream ULF waves, the shock itself, and downstream waves are in a situation similar to a ball sandwiched between converging walls. Indeed they are accelerated and higher energies are available for those ions that interact more. Diffusive shock acceleration

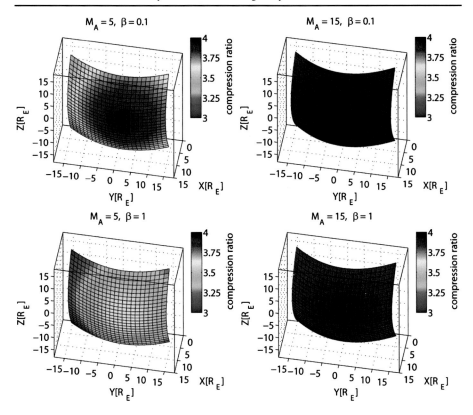

Fig. 4 The highly curved nature of planetary bow shocks means that there is significant variability in shock parameters over their surface, even when the upstream conditions are uniform

is thought to enable the acceleration of ions in the diffuse distribution to high energies and overall this physics is thought to lie at the heart of cosmic ray production in astrophysical shocks (for a recent review see e.g., Burgess et al. 2012).

The ion acceleration process described above has been studied in detail using hybrid code simulation, a technique to study MHD-scale phenomena including ion kinetic effects (e.g., Hellinger et al. 2002). Some modifications to the original idea were made through detailed analysis of the results. For example, while it had not been expected in the original theory, it was found that ions were accelerated upon interaction at a shock front in addition to the acceleration upon scattering by the wave in the upstream region (Sugiyama et al. 2001). The simulation results also nicely reproduced the features reported by observations at the terrestrial bow shock (Burgess et al. 2005).

The problem, however, is that the interaction time available at the terrestrial bow shock is very limited compared to those at huge astronomical objects. Detailed in-situ observations at the terrestrial bow shock, whilst certainly very useful in unveiling complicated plasma physical processes in some aspects, cannot definitely say what will happen when the interaction interval is highly extended. Because of their confidence in the investigation scheme, backed-up by an accumulation of successful observation-simulation benchmark results, Sugiyama (2011) has run a simulation for a much longer time in a much larger box than any other studies (and in the terrestrial situation). It is found that the accelerated population forms a power-law shaped energy distribution and the maximum energy of the power-law part increases in time with the power-law index being unchanged. The questions that remain to be addressed are: What is the key element that enables the expansion of the energy spectrum in time? Can the physics of the energy spectrum be understood within the test-particle approach, or are self-consistent ion-kinetic effects also playing a critical role? Another limitation of observational studies at the terrestrial bow shock is in the Mach number range that can be surveyed. The Alfven Mach number M_A at the terrestrial bow shock rarely exceeds ~20 while a much higher M_A is expected in astronomical scenes. Is there a substantial change in parallel shock physics and the associated ion acceleration process when M_A is so high? Once these three questions are answered by large-scale hybrid simulations, with regard to ion acceleration process at parallel shocks, a clear path bridging the gap between heliophysical and astrophysical shocks will be established.

Whilst ion acceleration leads to cosmic rays that are detected directly, in the astrophysical context electron acceleration is detected via radio and X-ray emission. Images of supernova remnant SN1006 show that this emission is inhomogeneous and spatially limited, and so it is of interest to establish whether this emission occurs in the parallel or perpendicular shock regime. At low Mach number shocks at Earth, electron acceleration occurs for perpendicular geometry as seen in the previous section (e.g., Krimigis 1992; Oka et al. 2006). More generally however, supernova remnant (SNR) Mach numbers are much higher. If electrons are accelerated by parallel shocks, then above a certain energy they will interact with ion generated waves and also undergo Fermi acceleration. One may then ask how the seed population arises—the so-called 'injection problem' and ideally one would wish to study this with in situ measurements. Whilst this cannot be investigated at Earth, recent observations have shown the existence of energetic electrons at Saturn's bow shock where $M_A \sim 100$, providing new insights into electron acceleration (Masters et al. 2013).

2.3 Time Variability: Interaction of Solar Wind Structure with the Bow Shock

In this subsection, we consider how discrete structure in the solar wind can control the structure and dynamics of the bow shock, with consequences for the overall dynamics of the magnetosphere itself. Whilst there are many interactions that can be mentioned, here we highlight two in particular: the interaction of a tangential discontinuity in the solar wind magnetic field with the bow shock, which can generate a Hot Flow Anomaly, and the interaction between an interplanetary shock and the bow shock, which leads to strong particle acceleration.

Hot Flow Anomalies (HFAs) are disruptions of the solar wind flow observed in the vicinity of the terrestrial bow shock and are caused by the interaction of a tangential discontinuity (TD) in the IMF with the bow shock (Schwartz et al. 1985; Thomsen et al. 1986), as illustrated in Fig. 5. Since the solar wind is frozen in, there is a convection electric field, which is perpendicular to the IMF. If the magnetic field connects to the shock on at least one side (Omidi and Sibeck 2007b), and if the solar wind convection electric field points into the TD on at least one side (Thomsen et al. 1993), ions specularly reflected at the shock are channeled back along the current sheet (Burgess 1989; Thomas et al. 1991). The backstreaming ion population is then heated and expands, driving weak shock waves (Fuselier et al. 1987; Lucek et al. 2004) and excavating the solar wind, so reducing the bulk density, and changing the plasma velocity moment. The core of the HFA typically contains hot plasma; $T_i \sim 100$ eV ($\sim10 \times$ greater than the ambient solar wind), accompanied by a significant deflection of the plasma flow away from the antisunward direction. In the core, the bulk plasma density is reduced, whereas in the edge

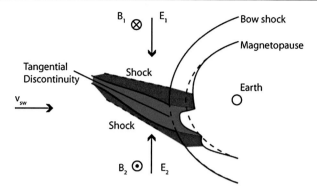

Fig. 5 A cartoon illustrating the formation of a Hot Flow Anomaly when a tangential discontinuity with appropriate magnetic field geometry intersects the bow shock. The hot plasma generated in the core of the HFA (*blue*) expands laterally resulting in regions of compressed plasma (*red*) at the edges of the HFA which can also develop shock waves on the outer edges (*dashed line*). The solar wind ram pressure is reduced inside the hot core region causing the magnetopause to dramatically expand. The HFA is attached to the TD that is convected by the solar wind, and so the point of contact tracks across the bow shock surface, whilst the HFA develops

regions the plasma density and magnetic field are enhanced. These features are typical of HFAs previously identified in the literature (Paschmann et al. 1988; Schwartz et al. 1988; Thomsen et al. 1988).

HFAs typically last a few minutes, and in dramatically disrupting the solar wind dynamic pressure, they have been observed to cause changes of more than 5 Earth radii (R_E) in the magnetopause location (Sibeck et al. 1998, 1999a; Jacobsen et al. 2009). The deformation of the magnetopause creates field-aligned currents that connect to the high-latitude ionosphere, and create ionospheric disturbances such as magnetic impulse events (Eastwood et al. 2008) and traveling convection vortices (Jacobsen et al. 2009). HFAs have also been seen to create transient magnetospheric ULF waves (Eastwood et al. 2011) and enhanced auroral emission (Fillingim et al. 2011). It is important to note that HFAs are generated by apparently ordinary discontinuities in the solar wind. It is therefore the kinetic physics of the bow shock that leads to a highly non-linear and global system response.

Interaction of an Interplanetary Shock with the Bow Shock   The shock-shock interaction is a well-established particle acceleration mechanism in astrophysical and space plasmas, but very difficult to study observationally using in situ measurements. Recently, the collision of an interplanetary (IP) shock with the Earth's bow shock on 10 August 1998 was identified as one of the rare events where detailed in situ observations of the different acceleration phases could be made (Hietala et al. 2011). This study used data from the ACE, Wind and Geotail spacecraft as shown in Fig. 6.

The interaction could be divided into 4 phases. ACE was magnetically connected to the IP shock but not to the bow shock, allowing the seed population to be characterized (phase 1). In phase 2, Wind observed several particle bursts coming from the bow shock direction at times when it was connected to the quasi-perpendicular bow shock. In phase 3, Wind became continuously connected to both shocks, and measured an increasing flux as well as two counter-streaming populations until the IP shock passed over it. Geotail was closer to Earth than Wind and was continuously connected to both shocks. It recorded the highest intensity at the IP shock crossing, but immediately after the crossing, Geotail also observed a burst of very high energy particles propagating sunwards (phase 4). Based on the velocity

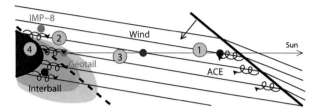

Fig. 6 Schematic picture of the shock-shock collision observed on August 9–10, 1998. The *black lines* depict the IMF. The *red paraboloid* represents the bow shock of the Earth, and the *pink region* its foreshock. The *red line* represents the IP shock, and the *red dashed line* indicates its position when it hit the bow shock. The *numbers* refer to the different phases of particle acceleration. Adapted from Hietala et al. (2012b)

dispersion of the burst and analysis of the geometry of the two shocks, Hietala et al. (2011) deduced that these particles had been released from the magnetic trap between the shocks as they collided.

Hietala et al. (2012b) used a global 2.5D test-particle simulation to further study particle acceleration in this event. They concentrated on the last phases (3 and 4) of the shock-shock interaction, when the shocks approached and passed through each other. Comparison of the simulation results with the observations of Wind and Geotail verified that the main features of the measurements can be explained by shock-shock interaction in this magnetic geometry, and are in agreement with the previous interpretation made by Hietala et al. (2011) of particle release.

2.4 Structure and Dynamics Downstream of the Bow Shock: High Speed Jets

Here we discuss recent progress in studying the structure and dynamics of the magnetosheath, and in particular the formation of High Speed Jets (HSJs), temporally and spatially localised regions of magnetosheath plasma which exhibit high dynamic pressure, kinetic energy density, and/or particle flux (Nemecek et al. 1998; Savin et al. 2008; Hietala et al. 2009; Amata et al. 2011; Archer and Horbury 2013; Plaschke et al. 2013). These studies show the dynamic pressure can be enhanced by up to ×15 the surrounding magnetosheath pressure, and in the subsolar region (i.e., the magnetosheath downstream of the nose of the bow shock) HSJs are super-magnetosonic 14.2 % of the time.

Studies show that HSJs have a size of ∼4000 km parallel to the flow and ∼1200–7000 km perpendicular to the flow, meaning that they exist on fluid scales (Hietala et al. 2012a; Archer and Horbury 2013; Plaschke et al. 2013). Thus, HSJs can locally distort the magnetopause (Hietala et al. 2009; Shue et al. 2009; Amata et al. 2011) generating both magnetopause surface waves and/or inner-magnetospheric compressional waves (Plaschke et al. 2009). As such the effects of the jets have been seen in the magnetosphere all the way to the ground (Dmitriev and Suvorova 2012; Hietala et al. 2012a; Archer et al. 2013).

HSJ occurrence seems to be controlled only by the IMF cone-angle (see Sect. 1); subsolar HSJs occur predominantly during intervals of low cone-angle (Suvorova et al. 2010; Archer and Horbury 2013; Plaschke et al. 2013). They tend to be found more often downstream of the quasi-parallel shock and closer to the bow shock than the magnetopause (Archer and Horbury 2013; Plaschke et al. 2013). As such, HSJs are thought to be connected to quasi-parallel shock geometry and ion foreshock processes. HSJs interact with the surrounding flow in the magnetosheath, and can slow down and become indistinguishable closer to the magnetopause, but still a significant number reach the magnetopause and impact the magnetosphere.

The main source of HSJs is not solar wind discontinuities (Archer and Horbury 2013; Plaschke et al. 2013). Instead, it appears to be related to ripples on the bow shock itself. Large scale structure in the foreshock is thought to steepen as it is convected earthward, leading to a dynamic and cyclical shock reformation process that would naturally introduce rippling of the shock surface (Omidi et al. 2005; Blanco-Cano et al. 2009). In this mechanism, the ripples change the angle at which the incoming flow meets the shock. At an appropriately inclined shock the downstream flow is compressed but not significantly decelerated (Hietala et al. 2009).

In a recent study, Hietala and Plaschke (2013), combining four years of THEMIS satellite data (Plaschke et al. 2013) with model calculations of bow shock ripples, and concentrating on the magnetosheath close to the shock during intervals when the IMF cone angle was small, found that 97 % of the observed jets could be produced by local ripples of the shock under the observed upstream conditions. The relevant parameters for the ripple-based jet formation mechanism were found to be the upstream M_A and β, in addition to the IMF orientation. They also found that the coherent jets formed a significant fraction of the high dynamic pressure tail of the magnetosheath flow distribution. This distribution is consistent with ripples that have a dominant amplitude to wavelength ratio of about 9 % (i.e. a wavelength of $\sim 1 R_E$ and an amplitude of $0.1 R_E$), present ~ 12 % of the time at any given location.

3 The Magnetopause

3.1 Basic Structure of the Magnetopause

As described in the introduction, the magnetopause marks the edge of the magnetosphere where the ram pressure of the shocked solar wind balances the magnetic pressure of the confined dipole. Here we briefly outline some of the essential results and features of magnetopause structure and dynamics. The magnetopause itself has been the subject of numerous more-detailed reviews and monographs (e.g., Lundin 1988; Song et al. 1995; Paschmann 1997; Sibeck et al. 1999b; Farrugia et al. 2001; De Keyser et al. 2005; Phan et al. 2005a; Lavraud et al. 2011; Hasegawa 2012), and the interested reader is invited to consult these articles and reviews for more information.

The magnetopause is not impermeable, and in reality solar wind plasma can and does cross the magnetopause onto closed magnetospheric field lines. This means that the basic structure of the magnetopause is more extended than might be expected, with the formation of what is known as the low-latitude boundary layer (LLBL), as shown in Fig. 1 in the introduction. The LLBL is a near-permanent feature earthward of the magnetopause containing mainly solar wind plasma and in fact a large fraction of plasma in the magnetosphere is of solar wind origin.

The structure and dynamics of both the magnetopause and the LLBL strongly depend on the solar wind magnetic field and plasma conditions. This is due to the fact that several different physical mechanisms operate at the magnetopause with some being more dominant than others depending on the boundary conditions. As such, the main goals of magnetopause research in the past several decades have been to reveal the key mechanisms operating therein, their consequences for transporting plasmas across the boundary, and what controls their occurrence or dominance. The key processes that drive magnetopause dynamics include *magnetic reconnection*, the *Kelvin-Helmholtz instability* and *diffusive entry*, and are illustrated in Fig. 7. These are general plasma physical processes that may also occur in other

Fig. 7 The three most important modes of plasma entry at the magnetopause are (**a**) magnetic reconnection, (**b**) diffusive entry and (**c**) the Kelvin-Helmholtz instability. Each of these is discussed in more detail in the text. (Sketches courtesy of M.S. Davis)

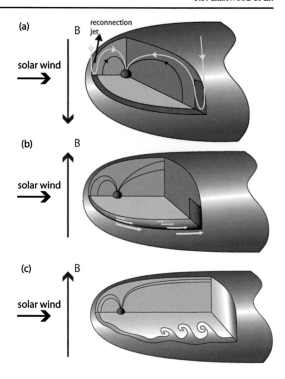

cosmic plasma structures, and so the knowledge gained in studying their properties at the magnetopause could be applicable in many other contexts. Finally, the manner in which solar wind plasma enters the magnetosphere directly influences magnetotail dynamics; this is discussed further in Sect. 4.

3.2 Magnetic Reconnection

Magnetic reconnection is in itself one of the most important phenomena that can occur in plasmas, and in the context of the magnetosphere is an important topic of study in its own right (for recent reviews see e.g., Eastwood et al. 2013b; Paschmann et al. 2013). It enables the formation of open field lines, connected to the Earth in one direction and into the solar wind in the other direction. The simplest configuration usually considered corresponds to southward IMF, which is shown in Fig. 7(a). Reconnection between the IMF (red field line) and the magnetospheric magnetic field (black field line) at the subsolar magnetopause, where the IMF is anti-parallel to the magnetospheric magnetic field, creates an 'open' field line (yellow). Note that reconnection also results in an equivalent mirror-image open field line formed below the ecliptic plane (this is not shown in the figure, but is visible in Fig. 1). Magnetic reconnection results in plasma jets confined to the magnetopause; observations of such jets are a crucial part of the overall experimental evidence confirming that reconnection does indeed occur (Paschmann et al. 1979; Sonnerup et al. 1981; Gosling et al. 1986). During southward IMF reconnection, a boundary layer is formed (shown in red), however this boundary is thin and essentially corresponds to the reconnection layer itself. There is not deep penetration of the solar wind across the dayside magnetopause, but rather the importance of the process lies in the fact that it enables transport of energy and magnetic flux through the cusps and into the magnetotail (Palmroth et al. 2003), as previously mentioned in Sect. 1. We now discuss some different aspects of magnetopause reconnection that impact both its structure and dynamics.

3.2.1 General IMF Orientation: Anti-parallel vs. Component Reconnection

The majority of cartoon descriptions of magnetic reconnection show anti-parallel magnetic field configurations. In reality, the IMF can of course assume any orientation. Furthermore, because the magnetopause is not flat, even if the IMF is anti-parallel to the magnetospheric magnetic field at one location, this does not guarantee that this is the case over the whole magnetopause. Historically, there has been debate as to whether purely anti-parallel reconnection (Crooker 1979) or component reconnection (Gonzalez and Mozer 1974; Sonnerup 1974; Cowley 1976) prevails at the magnetopause, since this influences where and when it can occur, and therefore limits the degree to which solar wind plasma can enter the magnetosphere via reconnection (e.g., Cowley and Owen 1989). As such there has been a considerable effort to determine the extent of the X-line over the magnetopause. Two-point in situ measurements first demonstrated the existence of a stable elongated X-line under southward IMF conditions (Phan et al. 2000). However since the magnetopause cannot yet be imaged, 'remote sensing' (in fact ions precipitating into the cusp from the dayside magnetopause) provides an alternative way to probe the magnetopause under a variety of solar wind conditions (e.g., Trattner et al. 2007a, 2007b). These results show that if there is a substantial B_y (i.e., out of the page in Fig. 1) then component reconnection is observed and the X-line is tilted, whereas if there is also a strong B_x (i.e., along the Sun-Earth line), then anti-parallel reconnection occurred. This has since been developed into the so-called maximum magnetic shear model (Trattner et al. 2012). A recent serendipitous 10 spacecraft conjunction using Cluster, THEMIS and Double-Star observations indicates from in situ data that a component reconnection X-line can form, extending across the dayside magnetopause (Dunlop et al. 2011a, 2011b). Ultimately, determining the extent of the X-line using in situ satellite data is very difficult but recent developments in X-ray imaging via solar wind charge exchange suggest that this is in fact technically possible, leading to a future mission proposal called AXIOM (Branduardi-Raymont et al. 2012).

3.2.2 Northward IMF

One may naively think that if the IMF points northward, parallel to the Earth's magnetic field at the subsolar magnetopause, then reconnection will not occur. In fact, the IMF is then anti-parallel to the magnetospheric field on the anti-sunward side of the cusp and so-called cusp reconnection can then occur (Dungey 1963). It is also possible that so-called dual-lobe reconnection could occur, i.e., cusp reconnection in both hemispheres simultaneously, leading to substantial plasma entry (Song and Russell 1992; Raeder et al. 1995, 1997; Li et al. 2005). This is related to an important and perhaps counterintuitive point regarding solar wind magnetosphere coupling which is that mass transport across the magnetopause appears to be enhanced for northward IMF (Palmroth et al. 2006). This results in the formation of a trapped layer of plasma inside the dayside magnetopause, which is observed in satellite data (e.g., Lavraud et al. 2005, 2006; Øieroset et al. 2005, 2008). However, it is still unclear how to precisely distinguish reconnection created boundary layer from a diffusion-created boundary layer during northward IMF; this is discussed further below.

3.2.3 Asymmetric Plasma Conditions and Reconnection Onset

Since reconnection at the magnetopause involves the high plasma density, low field strength magnetosheath and the low plasma density, high field strength magnetosphere, the boundary conditions are asymmetric and this changes both the physics of reconnection, and the structure of the central diffusion region surrounding the X-line (e.g., review by Eastwood et al. 2013b). In fact, in the central diffusion region the stagnation point and the X-line are not collocated; the stagnation point is displaced to the low mass flux side and the X-line is displaced to the high beta (strong field) side (Cassak and Shay 2007). If there is also a guide field, then the combination of this and asymmetry across the current sheet can in theory suppress reconnection entirely. This is the so-called beta-shear condition (Swisdak et al. 2003, 2010). This was first tested in the solar wind, where it was found that all reconnecting current sheets satisfied this condition (Phan et al. 2010) and led to the conclusion that beta-shear is indeed a necessary (although not sufficient condition) for reconnection to occur. It was subsequently tested and found to hold at the magnetopause with both reconnecting and non-reconnecting current sheets (Phan et al. 2013a). In this study, the fact that the condition divided the two groups of observations implies that the magnetopause current sheet is generally sufficiently thin to allow reconnection, and that the beta-shear condition is possibly the main mechanism controlling reconnection onset.

The fact that the solar wind plasma beta systematically varies through the heliosphere, increasing with heliocentric distance, has implications for the effect of reconnection on planetary magnetospheres. At Mercury, close to the Sun, the magnetosheath plasma beta is low and in theory reconnection can occur for a wide variety of IMF orientations; this has been investigated using MESSENGER spacecraft data (DiBraccio et al. 2013). In contrast, at the outer planets, e.g. Saturn, $\Delta\beta$ is larger and reconnection is more likely to be suppressed, which may explain the less frequent detections of local reconnection signatures there (Masters et al. 2012).

3.2.4 Plasma Heating

In releasing magnetic energy, reconnection creates jets of heated plasma. However, heating is not ubiquitous, and is also species dependent. For example, in the solar wind there is essentially no electron heating (Gosling et al. 2007), whereas at the magnetopause both ions and electrons are heated (Gosling et al. 1990). Recent reconnection measurements from the magnetotail show that reconnection preferentially heats ions (Eastwood et al. 2013a). The mechanisms by which reconnection heats the plasma jets are still not well understood, but recent work shows that the degree to which electrons are heated depends on the Alfvén speed of the inflowing plasma, and analysis of THEMIS observations of magnetopause reconnection show that 1.7 % of the inflowing energy is converted to electron heating (Phan et al. 2013b). This empirical model predicts that in the solar wind, electron heating of less than 1 eV is expected, which is not in practice detectable. In the magnetotail, electron heating is predicted to be of the order of 1 keV, which is also consistent with typical magnetotail properties.

3.2.5 Non Steady State Reconnection: Flux Transfer Events (FTEs)

Simple cartoons of magnetopause reconnection dynamics imply a smooth and continuous interaction, and evidence shows that on occasion reconnection can persist for long times, both for northward and southward IMF orientations (Frey et al. 2003; Phan et al. 2003,

2004). Nevertheless, there is a lot of experimental evidence showing that reconnection can be bursty and time dependent and generate significant structure. An important phenomenon in this regard is the Flux Transfer Event (FTE) (Russell and Elphic 1978; Wang et al. 2006; Fear et al. 2008). At the most basic level, an FTE may be thought of as a bulge propagating along the magnetopause, and a number of formation mechanisms have been proposed including patchy reconnection (Russell and Elphic 1978), bursty (i.e., time dependent) reconnection from a single X-line (Scholer 1988; Southwood et al. 1988), and the formation of a flux rope by multiple X-line reconnection (Lee and Fu 1985; Raeder 2006; Omidi and Sibeck 2007a). Whilst all mechanisms are possible, distinguishing between them is difficult (Fear et al. 2012). Recent work using multi-spacecraft analysis has demonstrated that FTEs can arise as a result of flux ropes formed by multiple X-line reconnection (Hasegawa et al. 2010; Øieroset et al. 2011; Zhong et al. 2013), and that these flux ropes can persist far along the tail magnetopause (Eastwood et al. 2012).

3.3 The Kelvin Helmholtz Instability

Across the flank magnetopause, the fast shocked solar wind flow relative to the stagnant magnetospheric plasma creates a large velocity shear which could lead to the Kelvin-Helmholtz Instability (KHI). The KHI in itself does not lead to plasma transport across the boundary. However, when the instability reaches its nonlinear stage it results in rolled-up vortices that contain local sharp gradients that are susceptible to microscale instabilities or reconnection that can then lead to plasma transport across the boundary (e.g., Nykyri and Otto 2001).

The KHI is a well-known instability in hydrodynamics. It also occurs in MHD although the magnetic field could have a stabilizing effect (Southwood 1968; Hasegawa 1975). Theory indicates that magnetic fields that have a component aligned with the plasma flow could stabilize the instability. For the equatorial magnetopause this means that the KHI would be favoured if the shocked solar wind magnetic field is either northward or southward oriented. Furthermore, Miura (1995) and Thomas and Winske (1993) simulated the effect of the IMF orientation on the KHI and found that even a small magnetic field rotation across the magnetopause could reduce the nonlinear growth of the instability. This means that the KHI is expected to occur at the low-latitude flank magnetopause only when the shocked solar wind magnetic field is aligned to the magnetospheric field at that location.

Observationally, evidence for the KHI linear stage (steepening of the waves along the magnetopause) as well as its nonlinear stage (rolled-up vortices) have been reported (Fairfield et al. 2000; Gustafsson et al. 2001; Hasegawa et al. 2004; Owen et al. 2004). The reported events tend to occur for northward rather than southward magnetic field conditions, in agreement with theoretical expectations, although observations of KHI under southward IMF conditions have recently been presented (Hwang et al. 2011). The most convincing evidence for the existence of rolled-up KH vortices leading to plasma transport across the magnetopause was obtained by the four Cluster spacecraft during a flank magnetopause crossing (Hasegawa et al. 2004). The spacecraft detected large flow and magnetic field vortices, in addition to the presence of shocked solar wind plasma on magnetospheric field lines embedded inside the vortices. These observations provide evidence that the KHI could play a major role in the transport of solar wind plasma into the magnetosphere from the flanks during northward solar wind magnetic field conditions when reconnection is not expected to be efficient at the low-latitude magnetopause and explain in part why mass transport across the magnetopause is enhanced during northward IMF.

3.4 Diffusive Plasma Entry

Cross-magnetic-field diffusion is a possible means for solar wind plasma to cross the magnetopause (e.g., Axford and Hines 1961; Sonnerup 1980). In collisionless plasmas, classical Spitzer diffusion is extremely weak and would have minimal impact on plasma transfer across the magnetopause. However, wave-particle interactions due to micro- or macro-instabilities could lead to anomalous diffusion (e.g., Lotko and Sonnerup 1995; Treumann et al. 1995; Johnson and Cheng 1997; Chaston et al. 2008). Expected signatures for diffusive solar wind entry creating the boundary layer include (1) boundary layer flows tangential to the magnetopause that are aligned with the direction of the shocked solar wind flow on the other side of the magnetopause, (2) smoothly varying velocity and density profiles across the boundary layer, (3) increasing boundary layer thickness with increasing distance from the subsolar point, and (4) boundary layer located on closed field lines. Such boundary layer characteristics have been reported primarily when the solar wind magnetic field points northward and at the low-latitude magnetopause (e.g., Eastman et al. 1985; Mitchell et al. 1987; Phan et al. 1997). Furthermore, there have also been observations showing that the outermost portion of the boundary layer (closest to the magnetopause itself) exhibits reconnection jet signatures with orientations vastly different from the shocked solar wind flow direction, while the inner part of the boundary layer displayed velocity profiles consistent with diffusion (Fujimoto et al. 1996; Phan et al. 2005b). These observations seem to suggest that the diffusion process is always present to create a boundary layer, but if the conditions are such that reconnection also occurs, reconnection acts to destroy the diffusion generated boundary layer.

4 The Magnetotail

4.1 Basic Structure of Earth's Magnetotail and Typical Properties

The interaction of the solar wind with the Earth's magnetic field leads to the stretching of field lines connected to the planet on the night side. In this section we briefly review some aspects of magnetotail structure and dynamics. Whilst our main focus is the Earth, we will also compare and contrast the behaviour of other planetary magnetotails in the solar system. The Earth's magnetotail and its properties and dynamics are an area of considerable active research and it is not possible to cover all aspects in a single review; for more information the reader should consult, for example, reviews by Walker et al. (1999), Nishida (2000), Russell (2001), Sergeev et al. (2012), Milan (2014) and Eastwood and Kiehas (2014).

Figure 1 shows that the magnetotail can be divided into two halves, with oppositely directed magnetic fields connecting to the two magnetic poles of the Earth. The magnetic field points Earthward in the northern lobe and tailward in the southern lobe. The lobes are separated by the magnetotail current sheet, which is enveloped by the plasma sheet (Ness 1965; Behannon 1968). The plasma sheet contains hot plasma, bounded on the north and south by plasma sheet boundary layers that contain beams and field-aligned flow. The density of the plasma sheet is typically about 0.1–1 cm^{-3} with a temperature of 1–10 keV (Baumjohann et al. 1989; Walker et al. 1999). The plasma sheet is predominantly a hydrogen plasma (e.g., Haaland et al. 2010), but can contain oxygen, especially during active times (e.g., Moore and Horwitz 2007). The locations of these regions have been mapped out by several different satellites (e.g., Christon et al. 1998).

The most well studied region of the magnetotail is the near-Earth magnetotail, which usually refers to the region up to $30R_{\mathrm{E}}$ downtail. The overall structure of the distant magnetotail has been mapped out by Geotail and ISEE-3 with the basic structure extending to at least $240R_E$ (Fairfield 1992; Christon et al. 1998). Further observations far downstream of the Earth in the solar wind indicate that signatures of the magnetotail may persist and extend to $3100R_E$ from the Earth (Intriligator et al. 1979). The magnetotail has a diameter of 50–$60R_E$ beyond 30–$50R_E$ downtail (Walker et al. 1999; Russell 2001), but whether or not the magnetotail is flattened, or has a circular cross section is still not well understood because of a lack of measurements, and variability in the solar wind conditions as we now discuss.

4.2 Correlation of Earth's Magnetotail Properties with Solar Wind Properties

The properties of the Earth's magnetotail depend on the solar wind (e.g., Borovsky et al. 1998 and references therein). In particular, the density of the plasma sheet is correlated with the solar wind density, the temperature of plasma sheet correlates with solar wind speed, and B_y in the plasma sheet is correlated with B_y in solar wind. The so-called 'penetration' of IMF B_y is a direct effect of magnetopause reconnection which also results in twisting of the tail (Cowley 1981). This happens because the external IMF B_y field exerts a torque on the internal, lobe field, such that the northern lobe twists towards dawn for a positive B_y and towards dusk for a negative B_y (Sibeck et al. 1985; Maezawa et al. 1997). The shape of the tail also varies in response to the orientation of the solar wind magnetic field. Observations show that the tail flaring angle (i.e., the angle between the magnetopause and the x-direction) depends on the B_z component of the IMF. If B_z is southward, then the flaring angle increases as the size of B_z increases, whereas for northward IMF the angle is fairly constant (Petrinec and Russell 1996).

4.3 Earth's Magnetotail: Southward IMF

4.3.1 Magnetospheric Convection

The basic dynamics of the Earth's magnetosphere for southward IMF were first outlined by Dungey (1961), and are illustrated in Fig. 8. Magnetic reconnection between the closed magnetospheric field (red line) and the solar wind magnetic field (blue) leads to the formation of 'open' field lines that are connected to the planet at one end (purple). Via the action of the solar wind, and the unbending of the kinked magnetic field, plasma is transported poleward, through the cusps and into the magnetotail. This leads to the accumulation of magnetic flux on open field lines in the tail. Magnetic reconnection across the tail current sheet returns plasma on closed field lines (red) back towards the nightside of the Earth. This basic pattern of magnetospheric convection is known as the Dungey cycle. In the context of the Earth's magnetosphere, based on lobe convection calculations (e.g. Haaland et al. 2007), the tail reconnection site may be understood as a distant X-line $\sim 100R_E$ downtail, and reconnection there is thought to lead to the formation of the plasma sheet (Walker et al. 1999), although the temporal and spatial scales of reconnection at the distant X-line are still not well understood.

4.3.2 Substorms

Whilst the concept of the open magnetosphere and the Dungey cycle is crucial to understanding the behaviour of the magnetosphere, it does not capture the fact that the magnetosphere can be highly dynamic. Continued dayside reconnection during an interval of

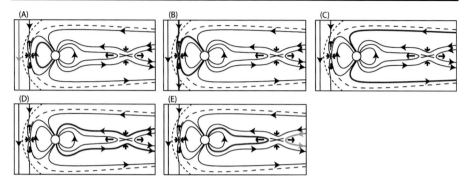

Fig. 8 Cartoon showing the progression of the Dungey cycle

Fig. 9 Cartoon showing the development of the near Earth neutral line and the formation of a plasmoid/flux rope which is released downtail once the near Earth neutral line enables reconnection between open field lines from the northern and southern lobe

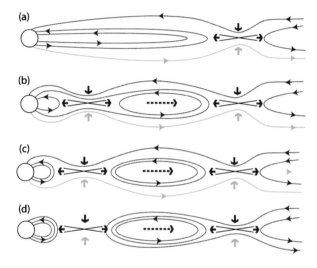

southward IMF leads to an accumulation of open flux in the magnetotail, and storage of energy in the lobe magnetic field. This accumulated energy undergoes periodic release in what is known as a magnetospheric substorm (Rostoker et al. 1980; Baker et al. 1996; Angelopoulos et al. 2008). This process involves the formation of a second neutral line closer to Earth, as shown in Fig. 9.

As the magnetic field strength in the lobe increases, the magnetotail current sheet thins. Whilst there is still some debate about the exact trigger, a reconnection X-line forms between $20–30R_E$ from the Earth (Nagai et al. 1998). Initially reconnection occurs on closed field lines, leading to the formation of a closed loop in this idealised cartoon (Hones 1977). This is a plasmoid, which is bounded by magnetic tension of the surrounding closed field lines connected at both ends to the Earth. Eventually, reconnection begins to process open field lines, at which point the plasmoid is released and can move downtail (Hones 1977; Baker et al. 1996; Slavin et al. 1999, 2002). In reality, if the reconnecting fields are not perfectly anti-parallel, then the plasmoid will in fact be a flux rope. The formation and properties of flux ropes in Earth's magnetotail is discussed in detail by Eastwood and Kiehas (2014).

The formation of the near Earth neutral line results in the Earthward injection of plasma and leads to bright auroral displays and associated disturbances to ground based magnetometers—the auroral substorm (Akasofu 1964). Since intervals of southward IMF occur naturally in the solar wind, substorms occur on a daily basis (Borovsky et al. 1993).

The dynamics of substorms are very complex and are the principal subject of the THEMIS mission (Angelopoulos 2008). For a comprehensive overview of recent progress in understanding the physics of substorms the reader is referred to the review by Sergeev et al. (2012). THEMIS, consisting of five satellites distributed in the magnetotail and a comprehensive network of ground-based imaging and magnetometers, has made great progress in establishing the timeline of events during substorms. Results appear to show that transport of magnetic flux (from reconnection) in the mid-magnetotail precedes auroral breakup (Liu et al. 2011), extending previous statistical investigations (Miyashita et al. 2009). The timing is consistent with tracking of distinct auroral features (streamers, or poleward boundary intensifications) which propagate equatorward with auroral breakup subsequently being observed (Nishimura et al. 2010). One issue regarding timing which remains to be resolved is that the time delay between reconnection onset and auroral breakup can be surprisingly short (Angelopoulos et al. 2008). This may be due to kinetic physics effects, and the transmission of kinetic Alfvén waves from the reconnection site at speeds considerably greater than the bulk outflow (Shay et al. 2011). Another major focus of activity for THEMIS has been the study of Dipolarization Fronts, thin interfaces which form at the leading edge of the reconnection jet (Sitnov et al. 2009). Satellite data shows that these fronts can propagate relatively long distances through the magnetotail whilst remaining coherent and, for example, are sites for particle acceleration (e.g., Runov et al. 2009; Khotyaintsev et al. 2011; Angelopoulos et al. 2013).

Given the dynamic behaviour of the magnetotail, it is perhaps not surprising that the instantaneous (or even the short-term average) reconnection rates at the dayside magnetopause and in the magnetotail need not match. As such, the amount of 'open' magnetic field constantly varies with time. This area of open field, the polar cap, therefore varies according to the balance between the rates of flux opening and closing on the dayside and nightside and is referred to as the Expanding/Contracting Polar Cap (ECPC) paradigm (e.g., Milan 2014). The size of the polar cap, i.e., the location of the open/closed field line boundary, can be approximated by the size of the auroral oval. This provides a means by which magnetotail loading and unloading can be monitored on a global level by analysing auroral observations (e.g., Milan 2009).

4.3.3 Storms

If the solar wind magnetic field is southward for a long period of time (hours), then a geomagnetic storm may develop (Gonzalez et al. 1994; Russell 2001). Storms are characterised by energization of the radiations belts and enhanced associated particle fluxes, which cause a significant depression in the equatorial magnetic field, measured by the Dst index for example. Storms result from strong, prolonged convection of plasma into the inner magnetosphere. The corresponding solar wind conditions tend to correspond to coherent solar wind structure such as Coronal Mass Ejections (CMEs) and Corotating Interaction Regions (CIRs) (Gonzalez et al. 1999), which can both provide long intervals of southward IMF (Forbes 2000; Tsurutani et al. 2006). CMEs are responsible for the largest geomagnetic disturbances, with the storm size depending on the solar wind speed, the field strength and the southward component of the magnetic field (Gosling et al. 1991; Richardson et al. 2001). The connection between storms and substorms is not straightforward, in the sense that a storm does not necessarily consist of a series of elemental substorms (Kamide et al. 1998), and in fact the dipolarization of the magnetotail by the enhanced ring current may inhibit substorm onset (Milan 2009).

4.4 Earth's Magnetotail: Northward IMF

During intervals of low geomagnetic activity, observations show that the plasma sheet can become significantly colder and denser (e.g., Terasawa et al. 1997 and references therein). This so-called cold dense plasma sheet (CDPS) has densities of ~ 1 cm^{-3}, and temperatures <1 keV (Li et al. 2005) and is found on closed field lines, with larger densities nearer the flanks than in the center of the magnetotail (e.g., Fujimoto et al. 1998). The CDPS does not contain cold O^{+} ions and so is not thought to be sourced from the ionosphere (Rème et al. 2001).

In fact, the plasma content of the CDPS increases during intervals of northward IMF with the best correlation occurring when the IMF parameters are averaged ~ 9 hours before the plasma sheet observations (Terasawa et al. 1997). As such, the CDPS is due to solar wind plasma entering under northward IMF conditions. Further work using in-situ data (Fujimoto et al. 2005), and maps derived from precipitating particles (Wing and Newell 2002) show that cold and dense populations appear along the tail flanks, suggesting entry through the flank low latitude boundary layer. As described in Sect. 3, it is thought that magnetopause plasma entry processes during northward IMF lead to the formation of the CDPS. In particular, the Kelvin-Helmholtz Instability will occur and enable the transfer of cold solar wind plasma into the magnetosphere at the flanks (Hasegawa et al. 2004). In fact the CDPS is a stable and persistent feature. In a case study examining the response of the magnetosphere to more than 32 hours of northward IMF, the CDPS was observed for more than 30 hours (Øieroset et al. 2005).

4.5 Other Planets

Several other solar system bodies also exhibit global magnetospheres and magnetotails. In particular, the magnetospheres of Mercury, Jupiter and Saturn have been extensively explored with orbiting spacecraft (MESSENGER, Galileo and Cassini respectively), and we now briefly mention some similarities and differences in the context of what we know about Earth's magnetotail. For a comprehensive review of solar system magnetotails, we refer the reader to the recent AGU Chapman Monograph on this subject (Keiling et al. 2014).

4.5.1 Mercury

The MESSENGER spacecraft, which entered orbit around Mercury in March 2011, has transformed our understanding of Mercury's magnetosphere. Whilst Mercury has a dipole magnetic field (Anderson et al. 2011), its magnetosphere is very small (Johnson et al. 2012), and is strongly driven by the solar wind (DiBraccio et al. 2013) (as mentioned in Sect. 1). Basic elements of this interaction were revealed during the first three flybys (Slavin et al. 2012a). During the first MESSENGER flyby, the solar wind magnetic field was largely northward, and the magnetotail, observed relatively close to the planet was quiescent. During the second flyby, the solar wind magnetic field was southward, and the magnetotail was highly dynamic, with evidence for numerous plasmoids. During the third flyby, the IMF orientation was variable, and MESSENGER observed dynamic loading and unloading corresponding to Hermian substorms.

4.5.2 Jupiter

Jupiter has the largest magnetosphere in the solar system and its properties act as a useful counterpoint to Earth (Bagenal et al. 2007). Its interaction with the solar wind is very different to that of the Earth's magnetosphere, and this is due to its strong magnetic field, its rapid rotation, and the presence of internal plasma sources (Delamere and Bagenal 2010). The first observations by Pioneer and Voyager revealed the presence of a strong internal plasma source, Io, which produces sulphur and oxygen atoms that subsequently ionise. This plasma corotates with Jupiter in a plasma disk, since centrifugal forces confine the plasma to a disk with the corresponding equatorial magnetic field stretched into a magnetodisk. As the plasma disk rotates, it is confined by the solar wind pressure on the dayside. On the nightside plasma-laden flux tubes can stretch out, and reconnect, leading to the ejection of a plasmoid down-tail. This plasmoid removes the internally generated plasma, and the cycle of loading and releasing was first presented by Vasyliunas (1983), now referred to as the Vasyliunas cycle.

A key question is then whether the solar wind plays any role in the dynamics of Jupiter's magnetotail. Observations made by Ulysses during a fly-by showed the presence of anti-sunward flow in the dawn magnetotail and solar wind type plasma in the high-latitude magnetosphere (Cowley et al. 1993). This suggested that the Dungey cycle and the Vasyliunas cycle could operate in tandem (Cowley et al. 2003), an idea further developed by Kivelson and Southwood (2005). However, the size of Jupiter's magnetosphere means that the timescale for the Dungey cycle appears to be much larger than the rotation rate and so it has been proposed that turbulent reconnection on the flanks intermittently creates small regions of open field (Delamere and Bagenal 2010).

4.5.3 Saturn

Saturn's magnetosphere has mainly been explored by the Cassini spacecraft (Dougherty et al. 2009), as well as by Pioneer 11, Voyager 1 and Voyager 2 during flybys. Like Jupiter, Saturn is a rapid rotator, with a relatively strong magnetic field. Cassini also showed that there is a significant internal plasma source: in this case, the moon Enceladus generates plasma, which also creates a magnetodisk. However, the dynamics of Saturn's magnetotail appear to involve a balance between rotational and solar wind driven effects (Cowley et al. 2005). The first evidence for magnetotail reconnection came during Saturn Orbit Insertion (Bunce et al. 2005), and subsequently plasmoids have been observed (Jackman et al. 2007, 2011).

5 Using Simulations to Accurately Capture Magnetosphere Structure and Dynamics

The magnetosphere occupies a large volume relative to typical plasma scale lengths, especially in the magnetotail, which is very extended. One could attempt to model the whole magnetosphere with a single plasma code that can properly represent all the relevant physical processes, such as non-thermal energy distribution, gradient and curvature drifts, multiple ion species, reconnection physics, etc. These processes cannot be captured with a simple magnetohydrodynamic (MHD) code, so one would need to use a particle-in-cell (PIC), a hybrid, or a Vlasov solver. The PIC code represents the kinetic effects with macro-particles. The macro-particles are much larger and heavier than the actual ions and electrons, and they represent an ensemble of particles with similar positions and velocity vectors. Hybrid codes are similar to PIC codes but they treat the electrons as a fluid to allow both larger grid cells and time steps and to reduce the computational costs. A Vlasov code solves the continuous Vlasov equations for the distribution function in a six dimensional (6D) phase space representing the distribution of particles in spatial coordinates and velocity space. All three of these techniques are computationally expensive, and so tend to be used more for studying phenomena locally (see e.g. Burgess and Scholer 2013 and Karimabadi et al. 2013 for recent reviews in the context of shock physics and reconnection respectively). To date, the hybrid code is the only one that has been used to represent the whole magnetosphere (Karimabadi et al. 2006, 2011), but even this simulation required about 10^5 CPU cores. Full PIC simulations have been used to model a part of the magnetosphere, for example dipolarization fronts in the tail in 2D (Sitnov and Swisdak 2011). Palmroth et al. (2013) used the Vlasiator code, a Vlasov solver, for a 3D magnetosphere simulation initially in a test particle mode in combination with a global MHD solver. Subsequently, Vlasiator has been used to simulate the entire dayside magnetospheric interaction, resulting in good reproduction of ion kinetic features in the foreshock and magnetosheath commonly observed by spacecraft, including backstreaming ion beams, wave generation and magnetosheath mirror modes (Pokhotelov et al. 2013).

At the time of writing, kinetic codes are still too expensive to be used for full 3D and time dependent magnetosphere simulations. The alternative is to use fluid codes that assume a thermal distribution of the particles. Further commonly used simplifications are to assume isotropy of the pressure tensor, use a single ion species, and to neglect the Hall term, explicit resistivity, heat conduction, and viscosity all of which leads to an ideal MHD description of the plasma. There are several MHD codes (e.g., BATS-R-US—Powell et al. 1999; Toth et al. 2012, LFM—Lyon et al. 2004, OpenGGCM—Raeder et al. 2001 and GUMICS—Janhunen et al. 2012) that can model the 3D magnetosphere with roughly real time speed on a moderate number of CPU cores (50–100) and with a reasonable grid resolution ($\approx 1/4$RE) near the inner boundary that is placed at 2 to 3 planet radii from the center of the planet. Although MHD models miss many kinetic effects, they can still show good agreement with some in situ observations: the location of the bow shock, the location of the magnetopause, the plasma density, temperature and velocity in the magnetosheath, and the magnetic field in most of the magnetosphere. The global MHD models are less successful in reproducing the measurements in the inner magnetosphere and the magnetotail especially during disturbed times.

Coupling multiple models can provide an efficient yet reasonably accurate way to represent the structure and dynamics of the magnetosphere, as illustrated in Fig. 10. Almost all global MHD models are coupled to an ionosphere electrodynamics model that provides the inner boundary conditions for the MHD code. Although this is an improvement over a simpler boundary condition (like a perfectly conducting sphere) it does not address the major shortcomings of the fluid description.

A major improvement over a pure MHD model is using a kinetic code for the inner magnetosphere defined by the closed field lines that are attached to the planet at both ends. These kinetic codes use a bounce-averaged description of the electrons and ions, which is valid as long as the bounce period of the charged particles between the magnetic mirror points is significantly shorter than the dynamic time scales of the system. Bounce-averaging reduces the spatial dimensionality of the problem from three to two dimensions: the location of a particle is given by magnetic field line it is gyrating around, and the field line is identified by its foot point on the surface of the planet, or the intersection with the magnetic equatorial plane. The velocity distribution is also described with reduced dimensionality. Some inner magnetosphere models, like the Rice Convection Model (RCM—Wolf et al. 1982; Toffoletto et al. 2003), assume an isotropic velocity distribution and solve for the energy distribution, while other models (e.g., CRCM–Fok et al. 2001, HEIDI—Liemohn et al. 2001, RAM-SCB—Jordanova et al. 1994; Zaharia et al. 2006) solve for energy and pitch angle distribution. These models treat the electrons as a separate species and also allow for multiple ion species. This means that the MHD model should also allow for multiple ion species

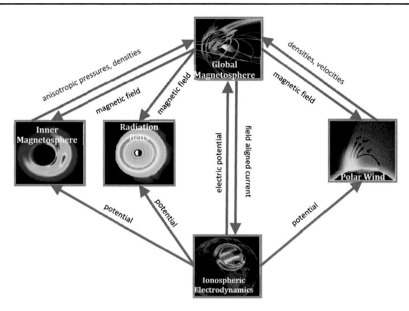

Fig. 10 It is necessary to couple multiple models to simulate the magnetosphere. The *text next to green arrows* show the most important quantities passed from one model to another. The *red text* indicates potential difficulties due to differences between the physics of different models

to supply consistent boundary conditions for the inner magnetosphere model. The major source of heavy ions for the Earth comes from the ionosphere along the open field lines. One can use an empirical model (Strangeway et al. 2000) or a first-principles model (e.g., PWOM—Glocer et al. 2009b) that solves for the advection and acceleration of individual fluids along the magnetic field lines as well as the motion of the field lines due to the ionospheric $E \times B$ drift. Radiation belt models (e.g., RBE—Fok et al. 2008) are essentially the same as inner magnetosphere models but they deal with higher energy electrons only.

Coupling these models is a non-trivial problem. There are challenges due to the different physics and numerical schemes in the different models, as well as the requirement of making the coupling accurate and efficient. Coupling a single fluid MHD code to an inner magnetosphere model provides an excellent example for both problems. The MHD code has to provide magnetic field information for the inner magnetosphere model as well as the boundary conditions at the edge of the closed field line region. Ideally these boundary conditions should provide the pitch angle and energy distribution function of the electrons and the ion species used in the inner magnetosphere model. Clearly, a single fluid ideal MHD code cannot provide that. One has to assume an isotropic Maxwellian velocity distribution and pick some ratio between ion densities and temperatures as well some fraction for the electron pressure. Using a more advanced extended MHD (XMHD) model that allows for multiple ion fluids, anisotropic pressure tensor, and separate electron pressure can provide a more consistent boundary condition, although it is still incomplete. Providing the magnetic field to the inner magnetosphere model is physically straightforward, but computationally challenging. One needs to trace tens of thousands of field lines through the MHD solution every time the codes are coupled. Doing this accurately and efficiently, so that the overall simulation is not slowed down, is quite challenging, and in fact, currently only the Space Weather Modeling Framework (SWMF—Toth et al. 2005, 2012) can do coupled global-inner magnetosphere simulations with real time speed (De Zeeuw et al. 2004;

Glocer et al. 2009a). In the other direction the inner magnetosphere provides full distribution functions for the MHD code inside the closed field line region. Simply overwriting the pressure and density (or pressures and densities for XMHD) can lead to numerical instabilities and artefacts. It was found that it works better to 'nudge' the MHD pressure(s) and density (densities) towards the inner magnetosphere values with a short but finite relaxation time (De Zeeuw et al. 2004).

The SWMF attained reasonable success modelling the magnetosphere of the Earth using coupled models, e.g. (Ridley et al. 2002; Zhang et al. 2007; Welling and Ridley 2010). Recently Glocer et al. (2013) integrated and coupled the CRCM inner magnetosphere model that solves for multiple ions with an arbitrary pitch angle distribution into the SWMF. This, as well as other applications, required the extension of the capabilities of the BATS-R-US code towards XMHD. Currently it can solve for Hall MHD (Toth et al. 2008) with separate electron pressure, multiple ion and neutral fluids (Glocer et al. 2009c), and anisotropic ion pressure (Meng et al. 2012a, 2012b). BATS-R-US can also solve for a non-ideal equation of state, heat conduction (isotropic or field aligned), resistivity (constant or anomalous) and viscosity. Not all combinations of these extensions are allowed at this time, but there is steady progress towards that goal. It is also important to realize that replacing ideal MHD with XMHD does not automatically lead to a more realistic model. A pure multi-fluid MHD model can develop unrealistic large relative velocities of the fluids along field lines, while the pure anisotropic MHD code can produce unreasonably large anisotropies. The reason for this unphysical behaviour is that XMHD does not contain the kinetic instabilities that limit the relative speed or the anisotropy. To solve these issues, one can use source terms based on the theory of kinetic instabilities that mimic such kinetic effects on the XMHD scales (Glocer et al. 2009c; Meng et al. 2012a).

To illustrate the above points, we mention our simulations of the July 22, 2008 geomagnetic storm by Meng et al. (2013). We performed four simulations with identical grids and solvers in BATS-R-US and the ionosphere electrodynamics code RIM, but we used two different inner magnetosphere models (RCM and CRCM) and two different MHD models (isotropic and anisotropic ion pressure). The simulations were then compared to magnetic field measurements from several satellites. Although none of the four models are perfect, the RMS errors are the smallest for the anisotropic MHD model coupled with CRCM that also allows for anisotropy. This single example shows (but does not prove in general) that a more sophisticated model can provide better agreement with measurements. In a recent study, requested by the U.S. national Space Weather Prediction Center (SWPC), Pulkkinen et al. (2013) compared several empirical and physics based magnetosphere models for several geomagnetic events and found that the SWMF using the coupled BATS-R-US, RCM and RIM models consistently outperformed all the empirical models as well as the global MHD models that were not coupled with an inner magnetosphere model, and in fact the SWMF has been selected to be transitioned to operations at NOAA's Space Weather Prediction Center (SWPC).

The required sophistication of the computational model is very problem dependent. Up to this point we have focused on the modelling of Earth's magnetosphere and emphasized space weather type applications. While the coupled global MHD and kinetic inner magnetosphere models of the SWMF are quite efficient and reasonably accurate for this application, they do not address all aspects of the dynamics and structure of the magnetosphere. For example, the fine structure of the collisionless bow shock and the detailed dynamics of magnetic reconnection are not captured by these codes. One would need to use a local kinetic code to properly model these phenomena. In some cases, using a more sophisticated local model can have a significant effect on the global dynamics. For example, the rate of reconnection in a small volume can influence the magnetic topology and the overall dynamics of the whole system.

One approach is to couple the global MHD code with a test particle model to study the acceleration and propagation of energetic particles (e.g., Peroomian et al. 2011). A test particle code does not provide feedback to the MHD code so it is only valid if the charge carried by the energetic particles is small compared to the thermal plasma represented by the MHD model. A more self-consistent approach is to couple the MHD model with an embedded kinetic model in regions where the kinetic effects are important. Daldorff et al. (2014) have recently succeeded in two-way coupling the implicit Particle-in-Cell code iPIC3D with the global MHD code BATS-R-US. In their 2D magnetospheric simulation the embedded PIC model solves for the dayside reconnection while the coupled MHD code provides the global solution. Future extensions of this new MHD with Embedded PIC (MHD-EPIC) method to 3D will allow for realistic representation of the reconnection and other local kinetic processes in a global 3D magnetospheric simulation. The upcoming Magnetospheric Multiscale (MMS) mission is aimed at studying the reconnection process at the electron scales. This will provide an excellent opportunity to validate local, embedded or possibly global kinetic magnetosphere models if they become feasible (Pokhotelov et al. 2013).

Once one moves away from the Earth, the requirements for a good model can vary drastically. For unmagnetized planets, like Mars or Venus, the solar wind directly interacts with the ionosphere, so the computational grid has to extend down to the surface and one needs to model a large number of ion (and neutral) species with corresponding chemical reactions, charge exchange, photoionization and recombination processes. The neutral atmosphere may be represented by a separate atmosphere/thermosphere model that is coupled to the multispecies/multifluid MHD solver. If the ion gyroradius becomes comparable with the typical length scales due to the weak magnetic field, one may need to switch to Hall MHD (Ma et al. 2007), or in some cases fluid solvers may not be appropriate at all, and global hybrid/kinetic models are needed (e.g., Jarvinen et al. 2009; Kallio et al. 2010; Brecht and Ledvina 2012), for example to model heavy ion escape. Mercury represents a unique challenge, being physically small but magnetised and without a significant atmosphere (Omidi et al. 2006). Other magnetized planets, like Jupiter and Saturn, have moons that load the plasma with heavy ions that play a major role in the dynamics of these magnetospheres (e.g., Jia et al. 2012). Near the polar regions of these giant planets the classical Alfvén speed may become larger than the speed of light so one has to use a semi-relativistic MHD model (Gombosi et al. 2002).

In general, the required complexity of an appropriate computational model is determined both by the system studied and by the quantities of interest, but it is often limited by the lack of observations and computational resources. Computational modelling, inevitably, is an art of compromise.

6 Conclusions

In this review we have attempted to summarize some aspects concerning the structure and dynamics of planetary magnetospheres, mainly concentrating on the Earth's magnetosphere. We have examined three important regions—the bow shock, the magnetopause and the magnetotail. Whilst there are many important controlling parameters, perhaps the most important from a simple qualitative point of view is the orientation of the interplanetary magnetic field. At the bow shock, this controls the regions of quasi-parallel and quasi-perpendicular geometry, and therefore the structure of the foreshock, the locations of particle acceleration, and the occurrence of magnetosheath high speed jets. At the magnetopause, rapid transport of solar wind energy via dayside reconnection occurs under southward IMF conditions. Under northward IMF, the KHI at the flanks, together with dual-lobe reconnection, means that mass transport (of cold dense plasma) is enhanced. Finally, in the Earth's magnetotail, the storage and release of energy in magnetic substorms and storms is strongly controlled by the IMF orientation and the manner in which plasma enters at the magnetopause. As such, for cosmic plasma structures in general, ascertaining the orientation of the magnetic field is of key importance.

The future prospects for gaining an even deeper and more detailed understanding of magnetospheric structure and dynamics are very bright. The demand for further improvements in our understanding is driven in part by an increasing societal need to understand the physics of space weather, so as to better predict and mitigate against its effects (Eastwood 2008). In this context our experimental knowledge of the key processes at work will continue to grow as new missions such as Magnetospheric Multi-Scale are launched, and novel concepts (e.g. using X-rays created by solar wind charge exchange to perform magnetospheric imaging) offer intriguing opportunities. Finally, continuing developments in computing power and processing speeds mean that ever more complex and detailed simulations will be available to explore the underlying physics.

Turning to the connection between magnetospheric observations and more general studies of cosmic plasma structure, we make the following observations. Firstly, it is clear that very precise studies of specific plasma physics processes can be performed in the magnetosphere. Examples include the structure of the bow shock, wave-particle interactions that occur in the foreshock, the physics of magnetic reconnection under asymmetric boundary conditions, plasma transport rates by the KHI, the onset of explosive energy release in current sheets, the structure of the reconnection diffusion region, the generation of energetic particles during storms and substorms, etc.

However, care must be taken in generalising these observations. For example, the Earth's bow shock is highly curved, and rather small in an astrophysical sense, so there is significant cross-talk between the quasi-perpendicular and -parallel sides, which may not occur in general. As a second example, studies based on Earth alone may lead to the conclusion that magnetospheres are driven by stellar winds, but the comparison of Mercury, Earth, Jupiter and Saturn shows that this is not always the case. In applying our knowledge of these magnetospheres to other astrophysical objects, it is very important to determine all the plasma conditions described in Sect. 1 so as to make the correct comparison.

We would like to finish by clarifying that this connection is not a one-way street. The parameter regime experienced in the solar system is inevitably limited, but the plasma conditions that cause infrequent, yet very important, extreme events (e.g. a Carrington-class geomagnetic storm) may occur more commonly in other astrophysical environments. Thus studying similar plasma phenomena elsewhere under different, more unusual conditions will help us to understand how Earth's magnetosphere would behave under extreme conditions. As such, an improved dialogue and cross-fertilization of ideas between all those studying different cosmic plasma structures can only help to further improve our understanding of magnetospheric physics.

Acknowledgements JPE is supported by an STFC Advanced Fellowship, and research at Imperial is supported by STFC grants ST/K001051/1 and ST/G00725X/1. HH also thanks the Alfred Kordelin foundation for financial support.

References

M.H. Acuña, K.W. Ogilvie, D.N. Baker, S.A. Curtis, D.H. Fairfield, W.H. Mish, The global geospace program and its investigations. Space Sci. Rev. **71**, 5–21 (1995)

S.-I. Akasofu, The development of the auroral substorm. Planet. Space Sci. **12**, 273–282 (1964)

E. Amata, S.P. Savin, D. Ambrosino, Y.V. Bogdanova, M.F. Marcucci, S. Romanov, A. Skalsky, High kinetic energy density jets in the Earth's magnetosheath: a case study. Planet. Space Sci. **59**, 482–494 (2011)

B.J. Anderson, C.L. Johnson, H. Korth, M.E. Purucker, R.M. Winslow, J.A. Slavin, S.C. Solomon, R.L. McNutt, J.M. Raines, T.H. Zurbuchen, The global magnetic field of mercury from messenger orbital observations. Science **333**(6051), 1859–1862 (2011). doi:10.1126/science.1211001

V. Angelopoulos, The THEMIS mission. Space Sci. Rev. **141**, 5–34 (2008)

V. Angelopoulos et al., Tail reconnection triggering substorm onset. Science **321**, 931–935 (2008)

V. Angelopoulos, A. Runov, X.Z. Zhou, D.L. Turner, S.A. Kiehas, S.S. Li, I. Shinohara, Electromagnetic energy conversion at reconnection fronts. Science **341**(6153), 1478–1482 (2013). doi:10.1126/science.1236992

M.O. Archer, T.S. Horbury, Magnetosheath dynamic pressure enhancements: occurrence and typical properties. Ann. Geophys. **31**(2), 319–331 (2013). doi:10.5194/angeo-31-319-2013

M.O. Archer, T.S. Horbury, J.P. Eastwood, J.M. Weygand, T.K. Yeoman, Magnetospheric response to magnetosheath pressure pulses: a low pass filter effect. J. Geophys. Res. Space Phys. **118**, 5454–5466 (2013). doi:10.1002/jgra.50519

W.I. Axford, C.O. Hines, A unifying theory of high-latitude geophysical phenomena and geomagnetic storms. Can. J. Phys. **39**(10), 1433–1464 (1961)

F. Bagenal, T.E. Dowling, W.B. McKinnon, W.B. McKinnon, *Jupiter: The Planet, Satellites and Magnetosphere* (Cambridge University Press, Cambridge, 2007)

D.N. Baker, T.I. Pulkkinen, V. Angelopoulos, W. Baumjohann, R.L. McPherron, Neutral line model of substorms: past results and present view. J. Geophys. Res. **101**, 12975–13010 (1996)

S.D. Bale et al., Quasi-perpendicular shock structure and processes. Space Sci. Rev. **118**, 161–203 (2005)

W. Baumjohann, R.A. Treumann, *Basic Space Plasma Physics* (Imperial College Press, London, 1996)

W. Baumjohann, G. Paschmann, C.A. Cattell, Average plasma properties in the central plasma sheet. J. Geophys. Res. **94**(A6), 6597–6606 (1989)

K.W. Behannon, Intrinsic magnetic properties of the lunar body. J. Geophys. Res. **73**, 7257 (1968)

X. Blanco-Cano, N. Omidi, C.T. Russell, Global hybrid simulations: foreshock waves and cavitons under radial interplanetary magnetic field geometry. J. Geophys. Res. **114**, A01216 (2009). doi:10.1029/2008JA013406

J.E. Borovsky, R.J. Nemzek, R.D. Belian, The occurrence rate of magnetospheric-substorm onsets: random and periodic substorms. J. Geophys. Res. **98**(A3), 3807–3813 (1993)

J.E. Borovsky, M.F. Thomsen, R.C. Elphic, The driving of the plasma sheet by the solar wind. J. Geophys. Res. **103**(A8), 17617–17639 (1998)

G. Branduardi-Raymont et al., AXIOM: advanced X-ray imaging of the magnetosphere. Exp. Astron. **33**(2–3), 403–443 (2012). doi:10.1007/s10686-011-9239-0

S.H. Brecht, S.A. Ledvina, Control of ion loss from Mars during solar minimum. Earth Planets Space **64**(2), 165–178 (2012). doi:10.5047/eps.2011.05.037

E.J. Bunce, S.W.H. Cowley, D.M. Wright, A.J. Coates, M.K. Dougherty, N. Krupp, W.S. Kurth, A.M. Rymer, In situ observations of a solar wind compression-induced hot plasma injection in Saturn's tail. Geophys. Res. Lett. **32**, L20S04 (2005). doi:10.1029/2005GL022888

D. Burgess, On the effect of a tangential discontinuity on ions specularly reflected at an oblique shock. J. Geophys. Res. **94**(A1), 472–478 (1989)

D. Burgess, Foreshock-shock interaction at collisionless quasi-parallel shocks. Adv. Space Res. **15**(8/9), 159–169 (1995)

D. Burgess, M. Scholer, Microphysics of quasi-parallel shocks in collisionless plasmas. Space Sci. Rev. **178**(2–4), 513–533 (2013). doi:10.1007/s11214-013-9969-6

D. Burgess, E. Lucek, M. Scholer, S. Bale, M. Balikhin, A. Balogh, T. Horbury, V. Krasnoselskikh, H. Kucharek, B. Lembège, Quasi-parallel shock structure and processes, in *Outer Magnetospheric Boundaries: Cluster Results* (Springer, Berlin, 2005), pp. 205–222

D. Burgess, E. Mobius, M. Scholer, Ion acceleration at the Earth's bow shock. Space Sci. Rev. **173**(1–4), 5–47 (2012). doi:10.1007/s11214-012-9901-5

P.A. Cassak, M.A. Shay, Scaling of asymmetric magnetic reconnection: general theory and collisional simulations. Phys. Plasmas **14**(10), 102114 (2007). doi:10.1063/1.2795630

C. Chaston, J. Bonnell, J. McFadden, C. Carlson, C. Cully, O. Le Contel, A. Roux, H. Auster, K. Glassmeier, V. Angelopoulos, Turbulent heating and cross-field transport near the magnetopause from THEMIS. Geophys. Res. Lett. **35**(17), L17S08 (2008)

S.P. Christon et al., Magnetospheric plasma regimes identified using Geotail measurements—2. Statistics, spatial distribution, and geomagnetic dependence. J. Geophys. Res. **103**(A10), 23521–23542 (1998). doi:10.1029/98ja01914

S.W.H. Cowley, Comments on the merging of nonantiparallel magnetic fields. J. Geophys. Res. **81**, 3455–3458 (1976)

S.W.H. Cowley, Magnetospheric asymmetries associated with the y-component of the IMF. Planet. Space Sci. **29**, 79–96 (1981)

S.W.H. Cowley, C.J. Owen, A simple illustrative model of open flux tube motion over the dayside magnetopause. Planet. Space Sci. **37**, 1461–1475 (1989)

S.W.H. Cowley, A. Balogh, M.K. Dougherty, T.M. Edwards, R.J. Forsyth, R.J. Hynds, K. Staines, Ulysses observations of anti-sunward flow on Jovian polar-cap field lines. Planet. Space Sci. **41**(11–12), 987–998 (1993). doi:10.1016/0032-0633(93)90103-9

S.W.H. Cowley, E.J. Bunce, T.S. Stallard, S. Miller, Jupiter's polar ionospheric flows: theoretical interpretation. Geophys. Res. Lett. **30**(5), 1220 (2003). doi:10.1029/2002gl016030

S.W.H. Cowley, S.V. Badman, E.J. Bunce, J.T. Clarke, J.-C. Gérard, D. Grodent, C.M. Jackman, S.E. Milan, T.K. Yeoman, Reconnection in a rotation-dominated magnetosphere and its relation to Saturn's auroral dynamics. J. Geophys. Res. **110**, A02201 (2005). doi:10.1029/2004JA010796

T.E. Cravens, *Physics of Solar System Plasmas* (Cambridge University Press, Cambridge, 1997)

N.U. Crooker, Dayside merging and cusp geometry. J. Geophys. Res. **84**, 951–959 (1979)

L.K.S. Daldorff, G. Toth, T.I. Gombosi, G. Lapenta, J. Amaya, S. Markidis, J.U. Brackbill, Two-way coupling of a global Hall magnetohydrodynamic model with a local implicit Particle-in-Cell model. J. Comp. Physiol. **268**, 236 (2014)

J. De Keyser, M.W. Dunlop, C.J. Owen, B.U.O. Sonnerup, S.E. Haaland, A. Vaivads, G. Paschmann, R. Lundin, L. Rezeau, Magnetopause and boundary layer. Space Sci. Rev. **118**(1–4), 231–320 (2005). doi:10.1007/s11214-005-3834-1

D.L. De Zeeuw, S. Sazykin, R.A. Wolf, T.I. Gombosi, A.J. Ridley, G. Toth, Coupling of a global MHD code and an inner magnetospheric model: initial results. J. Geophys. Res. **109**(A12), 14 (2004). doi:10.1029/2003ja010366

P.A. Delamere, F. Bagenal, Solar wind interaction with Jupiter's magnetosphere. J. Geophys. Res., Space Phys. **115**, A10201 (2010). doi:10.1029/2010ja015347

G.A. DiBraccio, J.A. Slavin, S.A. Boardsen, B.J. Anderson, H. Korth, T.H. Zurbuchen, J.M. Raines, D.N. Baker, R.L. McNutt, S.C. Solomon, MESSENGER observations of magnetopause structure and dynamics at Mercury. J. Geophys. Res., Space Phys. **118**(3), 997–1008 (2013). doi:10.1002/jgra.50123

A.V. Dmitriev, A.V. Suvorova, Traveling magnetopause distortion related to a large-scale magnetosheath plasma jet: THEMIS and ground-based observations. J. Geophys. Res. **117**, A08217 (2012). doi:10.1029/2011JA016861

M. Dougherty, L.W. Esposito, S.S.M. Krimigis, *Saturn from Cassini-Huygens* (Springer, Berlin, 2009)

J.W. Dungey, Interplanetary magnetic field and the auroral zones. Phys. Rev. Lett. **6**(2), 47–48 (1961)

J.W. Dungey, The structure of the exosphere or adventures in velocity space, in *Geophysique Exterieure/Geophysics the Earth's Environment*, ed. by C. DeWitt, J. Hieblot, A. Lebeau (Gordon & Breach, New York, 1963), pp. 504–550

M.W. Dunlop et al., Extended magnetic reconnection across the dayside magnetopause. Phys. Rev. Lett. **107**, 025004 (2011a)

M.W. Dunlop et al., Magnetopause reconnection accross wide local time. Ann. Geophys. **29**, 1683–1697 (2011b)

T.E. Eastman, B. Popielawska, L.A. Frank, Three-dimensional plasma observations near the outer magnetospheric boundary. J. Geophys. Res. **90**(A10), 9519–9539 (1985)

J.P. Eastwood, The science of space weather. Philos. Trans. R. Soc. A **366**, 4489–4500 (2008)

J.P. Eastwood, S. Kiehas, Origin and evolution of plasmoids and flux ropes in the magnetotails of Earth and Mars, in *Magnetotails in the Solar System*, ed. by A. Keiling, C.M. Jackman, P.T. Delamere (AGU, Washington, 2014)

J.P. Eastwood, A. Balogh, E.A. Lucek, C. Mazelle, I. Dandouras, Quasi-monochromatic ULF foreshock waves as observed by the four-spacecraft Cluster mission 1. Statistical properties. J. Geophys. Res. **110**, A11219 (2005a). doi:10.1029/2004JA010617

J.P. Eastwood, E.A. Lucek, C. Mazelle, K. Meziane, Y. Narita, J.S. Pickett, R.A. Treumann, The foreshock. Space Sci. Rev. **118**, 41–94 (2005b)

J.P. Eastwood et al., THEMIS observations of a hot flow anomaly: solar wind, magnetosheath and ground-based measurements. Geophys. Res. Lett. **35**, L17S03 (2008). doi:10.1029/2008GL033475

J.P. Eastwood, S.J. Schwartz, T.S. Horbury, C.M. Carr, K.-H. Glassmeier, I. Richter, C. Koenders, F. Plaschke, J.A. Wild, Transient Pc3 wave activity generated by a hot flow anomaly: cluster, Rosetta, and ground-based observations. J. Geophys. Res. **116**, A08224 (2011). doi:10.1029/2011JA016467

J.P. Eastwood, T.D. Phan, R.C. Fear, D.G. Sibeck, V. Angelopoulos, M. Oieroset, M.A. Shay, Survival of flux transfer event (FTE) flux ropes far along the tail magnetopause. J. Geophys. Res., Space Phys. **117**, A08222 (2012). doi:10.1029/2012ja017722

J.P. Eastwood, T.D. Phan, J.F. Drake, M.A. Shay, A.L. Borg, B. Lavraud, M. Taylor, Energy partition in magnetic reconnection in Earth's magnetotail. Phys. Rev. Lett. **110**(22), 225001 (2013a). doi:10.1103/PhysRevLett.110.225001

J.P. Eastwood, T.D. Phan, M. Oieroset, M.A. Shay, K. Malakit, M. Swisdak, J.F. Drake, A. Masters, Influence of asymmetries and guide fields on the magnetic reconnection diffusion region in collisionless space plasmas. Plasma Phys. Control. Fusion **55**, 124001 (2013b)

D.H. Fairfield, On the structure of the distant magnetotail: ISEE 3. J. Geophys. Res. **97**, 1403 (1992)

D.H. Fairfield, A. Otto, T. Mukai, S. Kokubun, R.P. Lepping, J.T. Steinberg, A.J. Lazarus, T. Yamamoto, Geotail observations of the Kelvin-Helmholtz instability at the equatorial magnetotail boundary for parallel northward fields. J. Geophys. Res., Space Phys. **105**(A9), 21159–21173 (2000). doi:10.1029/1999ja000316

C.J. Farrugia, F.T. Gratton, R.B. Torbert, Viscous-type processes in the solar wind-magnetosphere interaction. Space Sci. Rev. **95**(1–2), 443–456 (2001). doi:10.1023/a:1005288703357

R.C. Fear, S.E. Milan, A.N. Fazakerley, E.A. Lucek, S.W.H. Cowley, I. Dandouras, The azimuthal extent of three flux transfer events. Ann. Geophys. **26**, 2353–2369 (2008)

R.C. Fear, S.E. Milan, K. Oksavik, Determining the axial direction of high-shear flux transfer events: implications for models of FTE structure. J. Geophys. Res., Space Phys. **117**, A09220 (2012). doi:10.1029/2012ja017831

E. Fermi, On the origin of the cosmic radiation. Phys. Rev. **75**(8), 1169–1174 (1949)

M.O. Fillingim, J.P. Eastwood, G.K. Parks, V. Angelopoulos, I.R. Mann, S.B. Mende, A.T. Weatherwax, Polar UVI and THEMIS GMAG observations of the ionospheric response to a hot flow anomaly. J. Atmos. Sol.-Terr. Phys. **73**, 137–145 (2011)

M.C. Fok, R.A. Wolf, R.W. Spiro, T.E. Moore, Comprehensive computational model of Earth's ring current. J. Geophys. Res., Space Phys. **106**(A5), 8417–8424 (2001). doi:10.1029/2000ja000235

M.C. Fok, R.B. Horne, N.P. Meredith, S.A. Glauert, Radiation belt environment model: application to space weather nowcasting. J. Geophys. Res., Space Phys. **113**, A03S08 (2008). doi:10.1029/2007ja012558

T.G. Forbes, A review on the genesis of coronal mass ejections. J. Geophys. Res. **105**(A10), 23153–23165 (2000)

H.U. Frey, T.D. Phan, S.A. Fuselier, S.B. Mende, Continuous magnetic reconnection at Earth's magnetopause. Nature **426**(6966), 533–537 (2003). doi:10.1038/nature02084

M. Fujimoto, A. Nishida, T. Mukai, Y. Saito, T. Yamamoto, S. Kokubun, Plasma entry from the flanks of the near-earth magnetotail: GEOTAIL observations in the dawnside LLBL and the plasma sheet. J. Geomagn. Geoelectr. **48**(5–6), 711–727 (1996)

M. Fujimoto, T. Terasawa, T. Mukai, Y. Saito, T. Yamamoto, S. Kokubun, Plasma entry from the flanks of the near-Earth magnetotail: Geotail observations. J. Geophys. Res. **103**, 4391–4408 (1998)

M. Fujimoto, T. Mukai, S. Kokubun, *The Structure of the Plasma Sheet Under Northward IMF* (Elsevier, Amsterdam, 2005), pp. 19–27

S.A. Fuselier, Suprathermal Ions Upstream and Downstream From the Earth's Bow Shock, in *Solar Wind Sources of Magnetospheric Ultra-Low-Frequency Waves*, ed. by M.J. Engebretson, K. Takahashi, M. Scholer. Geophysical Monograph, vol. 81 (American Geophysical, Union, Washington, 1994), pp. 107–119

S.A. Fuselier, Ion distributions in the Earth's foreshock upstream from the bow shock. Adv. Space Res. **15**(8/9), 43–52 (1995)

S.A. Fuselier, M.F. Thomsen, J.T. Gosling, S.J. Bame, C.T. Russell, M.M. Mellott, Fast shocks at the edges of hot diamagnetic cavities upstream from the Earth's bow shock. J. Geophys. Res. **92**(A4), 3187–3194 (1987)

A. Glocer, G. Toth, M. Fok, T. Gombosi, M. Liemohn, Integration of the radiation belt environment model into the space weather modeling framework. J. Atmos. Sol.-Terr. Phys. **71**(16), 1653–1663 (2009a). doi:10.1016/j.jastp.2009.01.003

A. Glocer, G. Toth, T. Gombosi, D. Welling, Modeling ionospheric outflows and their impact on the magnetosphere, initial results. J. Geophys. Res., Space Phys. **114**, A05216 (2009b). doi:10.1029/2009ja014053

A. Glocer, G. Toth, Y. Ma, T. Gombosi, J.C. Zhang, L.M. Kistler, Multifluid block-adaptive-tree solarwind Roe-type upwind scheme: magnetospheric composition and dynamics during geomagnetic storms initial results. J. Geophys. Res., Space Phys. **114**, A12203 (2009c). doi:10.1029/2009ja014418

A. Glocer, M. Fok, X. Meng, G. Toth, N. Buzulukova, S. Chen, K. Lin, CRCM+BATS-R-US two-way coupling. J. Geophys. Res., Space Phys. **118**(4), 1635–1650 (2013). doi:10.1002/jgra.50221

T.I. Gombosi, G. Toth, D.L. De Zeeuw, K.C. Hansen, K. Kabin, K.G. Powell, Relativistic magnetohydrodynamics and physics-based convergence acceleration. J. Comput. Phys. **177**(1), 176–205 (2002). doi:10.1006/jcph.2002.7009

W.D. Gonzalez, F.S. Mozer, A quantitative model for the potential resulting from reconnection with an arbitrary interplanetary magnetic field. J. Geophys. Res. **79**, 4186–4194 (1974)

W.D. Gonzalez, J.A. Joselyn, Y. Kamide, H.W. Kroehl, G. Rostoker, B.T. Tsurutani, V.M. Vasyliunas, What is geomagnetic storm? J. Geophys. Res. **99**(A4), 5771–5792 (1994)

W.D. Gonzalez, B.T. Tsurutani, A.L. Clua de Gonzalez, Interplanetary origin of geomagnetic storms. Space Sci. Rev. **88**, 529–562 (1999)

J.T. Gosling, M.F. Thomsen, S.J. Bame, C.T. Russell, Accelerated plasma flows at the near-tail magnetopause. J. Geophys. Res., Space Phys. **91**(A3), 3029–3041 (1986). doi:10.1029/JA091iA03p03029

J.T. Gosling, M.F. Thomsen, S.J. Bame, T.G. Onsager, C.T. Russell, The electron edge of the low latitude boundary-layer during accelerated flow events. Geophys. Res. Lett. **17**(11), 1833–1836 (1990). doi:10.1029/GL017i011p01833

J.T. Gosling, D.J. McComas, J.L. Phillips, S.J. Bame, Geomagnetic activity associated with earth passage of interplanetary shock disturbances and coronal mass ejections. J. Geophys. Res. **96**(5), 7831–7839 (1991)

J.T. Gosling, S. Eriksson, T.D. Phan, D.E. Larson, R.M. Skoug, D.J. McComas, Direct evidence for prolonged magnetic reconnection at a continuous x-line within the heliospheric current sheet. Geophys. Res. Lett. **34**, L06102 (2007). doi:10.1029/2006gl029033

G. Gustafsson et al., First results of electric field and density observations by cluster EFW based on initial months of operation. Ann. Geophys. **19**(10–12), 1219–1240 (2001)

S. Haaland, G. Paschmann, M. Forster, J.M. Quinn, R.B. Torbert, C.E. McIlwain, H. Vaith, P.A. Puhl-Quinn, C.A. Kletzing, High-latitude plasma convection from cluster EDI measurements: method and IMF-dependence. Ann. Geophys. **25**(1), 239–253 (2007)

S. Haaland, E.A. Kronberg, P.W. Daly, M. Franz, L. Degener, E. Georgescu, I. Dandouras, Spectral characteristics of protons in the Earth's plasmasheet: statistical results from cluster CIS and RAPID. Ann. Geophys. **28**(8), 1483–1498 (2010). doi:10.5194/angeo-28-1483-2010

A. Hasegawa, *Plasma Instabilities and Nonlinear Effects*. Series on Physics Chemistry Space, vol. 8 (Springer, Berlin, 1975)

H. Hasegawa, Structure and dynamics of the magnetopause and its boundary layers. Monogr. Environ. Earth Planets **1**(2), 71–119 (2012)

H. Hasegawa, M. Fujimoto, T.-D. Phan, H. Rème, A. Balogh, M.W. Dunlop, C. Hashimoto, R. TanDokoro, Transport of solar wind into Earth's magnetosphere through rolled-up Kelvin-Helmholtz vortices. Nature **430**, 755–758 (2004)

H. Hasegawa et al., Evidence for a flux transfer event generated by multiple X-line reconnection at the magnetopause. Geophys. Res. Lett. **37**, L16101 (2010). doi:10.1029/2010GL044219

P. Hellinger, P. Travnicek, H. Matsumoto, Reformation of perpendicular shocks: hybrid simulations. Geophys. Res. Lett. **29**(24), 2234 (2002). doi:10.1029/2002gl015915

H. Hietala, F. Plaschke, On the generation of magnetosheath high speed jets by bow shock ripples. J. Geophys. Res. **118**, 7237–7245 (2013). doi:10.1002/2013JA019172

H. Hietala, T.V. Laitinen, K. Andreeova, R. Vainio, A. Vaivads, M. Palmroth, T.I. Pulkkinen, H.E.J. Koskinen, E.A. Lucek, H. Reme, Supermagnetosonic jets behind a collisionless quasiparallel shock. Phys. Rev. Lett. **103**, 245001 (2009)

H. Hietala, N. Agueda, K. Andreeova, R. Vainio, S. Nylund, E.K.J. Kilpua, H.E.J. Koskinen, In situ observations of particle acceleration in shock-shock interaction. J. Geophys. Res. **116**, A10105 (2011). doi:10.1029/2011ja016669

H. Hietala, N. Partamies, T.V. Laitinen, L.B.N. Clausen, G. Facsko, A. Vaivads, H.E.J. Koskinen, I. Dandouras, H. Reme, E.A. Lucek, Supermagnetosonic subsolar magnetosheath jets and their effects: from the solar wind to the ionospheric convection. Ann. Geophys. **30**, 33–48 (2012a). doi:10.5194/angeo-30-33-2012

H. Hietala, A. Sandroos, R. Vainio, Particle acceleration in shock-shock interaction: model to data comparison. Astrophys. J. Lett. **751**(1), 6 (2012b). doi:10.1088/2041-8205/751/1/l14

E.W. Hones Jr., Substorm processes in the magnetotail: comments on 'On hot tenuous plasmas, fireballs, and boundary layers in the Earth's magnetotail' by L.A. Frank, K.L. Ackerson and R.P. Lepping. J. Geophys. Res. **82**(35), 5633–5640 (1977)

K.-J. Hwang, M.M. Kuznetsova, F. Sahraoui, M.L. Goldstein, E. Lee, G.K. Parks, Kelvin-Helmholtz waves under southward IMF. J. Geophys. Res. **116**, A08210 (2011). doi:10.1029/2011JA016596

D.S. Intriligator, H.R. Collard, J.D. Mihalov, O.L. Vaisberg, J.H. Wolfe, Evidence for Earth magnetospheric tail associated phenomena at 3100 RE. Geophys. Res. Lett. **6**(7), 585–588 (1979)

C.M. Jackman, C.T. Russell, D.J. Southwood, C.S. Arridge, N. Achilleos, M.K. Dougherty, Strong rapid dipolarizations in Saturn's magnetotail: in situ evidence of reconnection. Geophys. Res. Lett. **34**, L11203 (2007). doi:10.1029/2007GL029764

C.M. Jackman, J.A. Slavin, S.W.H. Cowley, Cassini observations of plasmoid structure and dynamics: implications for the role of magnetic reconnection in magnetospheric circulation at Saturn. J. Geophys. Res. **116**, A10212 (2011). doi:10.1029/2011ja016682

K.S. Jacobsen et al., THEMIS observations of extreme magnetopause motion caused by a hot flow anomaly. J. Geophys. Res. **114**, A08210 (2009). doi:10.1029/2008JA013873

P. Janhunen, M. Palmroth, T. Laitinen, I. Honkonen, L. Juusola, G. Facsko, T.I. Pulkkinen, The GUMICS-4 global MHD magnetosphere-ionosphere coupling simulation. J. Atmos. Sol.-Terr. Phys. **80**, 48–59 (2012)

R. Jarvinen, E. Kallio, P. Janhunen, S. Barabash, T.L. Zhang, V. Pohjola, I. Sillanpaa, Oxygen ion escape from Venus in a global hybrid simulation: role of the ionospheric O+ ions. Ann. Geophys. **27**(11), 4333–4348 (2009)

X.Z. Jia, K.C. Hansen, T.I. Gombosi, M.G. Kivelson, G. Toth, D.L. DeZeeuw, A.J. Ridley, Magnetospheric configuration and dynamics of Saturn's magnetosphere: a global MHD simulation. J. Geophys. Res., Space Phys. **117**, A05225 (2012). doi:10.1029/2012ja017575

J.R. Johnson, C.Z. Cheng, Kinetic Alfvén waves and plasma transport at the magnetopause. Geophys. Res. Lett. **24**(11), 1423–1426 (1997)

C.L. Johnson et al., MESSENGER observations of Mercury's magnetic field structure. J. Geophys. Res. **117**, E00L14 (2012). doi:10.1029/2012je004217

V.K. Jordanova, J.U. Kozyra, G.V. Khazanov, A.F. Nagy, C.E. Rasmussen, M.C. Fok, A bounce-averaged kinetic-model of the ring current ion population. Geophys. Res. Lett. **21**(25), 2785–2788 (1994). doi:10.1029/94gl02695

E. Kallio, K.J. Liu, R. Jarvinen, V. Pohjola, P. Janhunen, Oxygen ion escape at Mars in a hybrid model: high energy and low energy ions. Icarus **206**(1), 152–163 (2010). doi:10.1016/j.icarus.2009.05.015

Y. Kamide et al., Current understanding of magnetic storms: storm-substorm relationships. J. Geophys. Res. **103**(A8), 17705–17728 (1998)

H. Karimabadi, H.X. Vu, D. Krauss-Varban, Y. Omelchenko, *Global Hybrid Simulations of the Earth's Magnetosphere* (Astronomical Soc Pacific, San Francisco, 2006), pp. 257–263

H. Karimabadi, B. Loring, H.X. Vu, Y. Omelchenko, M. Tatineni, A. Majumdar, U. Ayachit, B. Geveci, *Petascale Global Kinetic Simulations of the Magnetosphere and Visualization Strategies for Analysis of Very Large Multi-Variate Data Sets* (Astronomical Soc Pacific, San Francisco, 2011), pp. 281–291

H. Karimabadi, V. Roytershteyn, W. Daughton, Y.H. Liu, Recent evolution in the theory of magnetic reconnection and its connection with turbulence. Space Sci. Rev. **178**(2–4), 307–323 (2013). doi:10.1007/s11214-013-0021-7

A. Keiling, C.M. Jackman, P.T. Delamere, *Magnetotails in the Solar System* (AGU, Washington, 2014)

Y.V. Khotyaintsev, C.M. Cully, A. Vaivads, M. Andre, C.J. Owen, Plasma jet braking: energy dissipation and nonadiabatic electrons. Phys. Rev. Lett. **106**(16), 165001 (2011). doi:10.1103/PhysRevLett.106.165001

M.G. Kivelson, C.T. Russell, *Introduction to Space Physics* (Cambridge University Press, Cambridge, 1995)

M.G. Kivelson, D.J. Southwood, Dynamical consequences of two modes of centrifugal instability in Jupiter's outer magnetosphere. J. Geophys. Res. **110**, A12209 (2005). doi:10.1029/2005ja011176

S. Krimigis, Voyager energetic particle observations at interplanetary shocks and upstream of planetary bow shocks: 1977–1990. Space Sci. Rev. **59**(1–2), 167–201 (1992)

B. Lavraud, M.F. Thomsen, M.G.G.T. Taylor, Y.L. Wang, T.-D. Phan, S.J. Schwartz, R.C. Elphic, A.N. Fazakerley, H. Rème, A. Balogh, Characteristics of the magnetosheath electron boundary layer under northward interplanetary magnetic field: implications for high-latitude reconnection. J. Geophys. Res. **110**, A06209 (2005). doi:10.1029/2004JA010808

B. Lavraud, M.F. Thomsen, B. Lefebvre, S.J. Schwartz, K. Seki, T.D. Phan, Y.L. Wang, A. Fazakerley, H. Reme, A. Balogh, Evidence for newly closed magnetosheath field lines at the dayside magnetopause under northward IMF. J. Geophys. Res. **111**, A05211 (2006). doi:10.1029/2005ja011266

B. Lavraud, C. Foullon, C.J. Farrugia, J.P. Eastwood, The magnetopause, its boundary layers and pathways to the magnetotail, in *The Dynamic Magnetosphere*, ed. by W. Liu, M. Fujimoto (Springer, Dordrecht, 2011)

G. Le, C.T. Russell, The Morphology of ULF Waves in the Earth's Foreshock, in *Solar Wind Sources of Magnetospheric Ultra-Low-Frequency Waves*, ed. by M.J. Engebretson, K. Takahashi, M. Scholer. Geophysical Monograph, vol. 81 (American Geophysical, Union, Washington, 1994), pp. 87–98

L.C. Lee, Z.F. Fu, A theory of magnetic flux transfer at the Earth's magnetopause. Geophys. Res. Lett. **12**, 105–108 (1985)

W.H. Li, J. Raeder, J. Dorelli, M. Oieroset, T.D. Phan, Plasma sheet formation during long period of northward IMF. Geophys. Res. Lett. **32**, L12S08 (2005). doi:10.1029/2004gl021524

M.W. Liemohn, J.U. Kozyra, C.R. Clauer, A.J. Ridley, Computational analysis of the near-Earth magnetospheric current system during two-phase decay storms. J. Geophys. Res., Space Phys. **106**(A12), 29531–29542 (2001). doi:10.1029/2001ja000045

J. Liu, C. Gabrielse, V. Angelopoulos, N.A. Frissell, L.R. Lyons, J.P. McFadden, J. Bonnell, K.H. Glassmeier, Superposed epoch analysis of magnetotail flux transport during substorms observed by THEMIS. J. Geophys. Res., Space Phys. **116**(A5), A00I29 (2011). doi:10.1029/2010JA015886

W. Lotko, B.U.O. Sonnerup, The low-latitude boundary layer on closed field lines, in *Physics of the Magnetopause*. Geophysical Monograph Series, vol. 90 (1995)

E.A. Lucek, T.S. Horbury, A. Balogh, I. Dandouras, H. Rème, Cluster observations of hot flow anomalies. J. Geophys. Res. **109**, A06207 (2004). doi:10.1029/2003JA010016

R. Lundin, On the magnetospheric boundary-layer and solar-wind energy-transfer into the magnetosphere. Space Sci. Rev. **48**(3–4), 263–320 (1988)

J.G. Lyon, J.A. Fedder, C.M. Mobarry, The Lyon-Fedder-Mobarry (LFM) global MHD magnetospheric simulation code. J. Atmos. Sol.-Terr. Phys. **66**(15–16), 1333–1350 (2004). doi:10.1016/j.jastp.2004.03.020

Y.J. Ma et al., 3D global multi-species Hall-MHD simulation of the Cassini T9 flyby. Geophys. Res. Lett. **34**, L24S10 (2007). doi:10.1029/2007gl031627

K. Maezawa, T. Hori, T. Mukai, Y. Saito, T. Yamamoto, S. Kokubun, A. Nishida, Structure of the distant magnetotail and its dependence on the IMF B-y component: Geotail observations, in *Results of the Iastp Program*, ed. by C.T. Russell (Pergamon Press, Oxford, 1997), pp. 949–959. doi:10.1016/s0273-1177(97)00503-6

A. Masters, J.P. Eastwood, M. Swisdak, M.F. Thomsen, C.T. Russell, N. Sergis, F.J. Crary, M.K. Dougherty, A.J. Coates, S.M. Krimigis, The importance of plasma beta conditions for magnetic reconnection at Saturn's magnetopause. Geophys. Res. Lett. **39**, L08103 (2012). doi:10.1029/2012gl051372

A. Masters, L. Stawarz, M. Fujimoto, S.J. Schwartz, N. Sergis, M.F. Thomsen, A. Retino, H. Hasegawa, B. Zieger, G.R. Lewis, Electron acceleration to relativistic energies at a strong quasi-parallel shock wave. Nat. Phys. **9**(3), 164–167 (2013)

X. Meng, G. Toth, M.W. Liemohn, T.I. Gombosi, A. Runov, Pressure anisotropy in global magnetospheric simulations: a magnetohydrodynamics model. J. Geophys. Res. **117**, A08216 (2012a). doi:10.1029/2012ja017791

X. Meng, G. Toth, I.V. Sokolov, T.I. Gombosi, Classical and semirelativistic magnetohydrodynamics with anisotropic ion pressure. J. Comput. Phys. **231**(9), 3610–3622 (2012b). doi:10.1016/j.jcp.2011.12.042

X. Meng, G. Toth, A. Glocer, M.C. Fok, T.I. Gombosi, Pressure anisotropy in global magnetospheric simulations: coupling with ring current models. J. Geophys. Res., Space Phys. **118**(9), 5639–5658 (2013). doi:10.1002/jgra.50539

J. Merka, A. Szabo, J.A. Slavin, M. Peredo, Three-dimensional position and shape of the bow shock and their variation with upstream Mach numbers and interplanetary magnetic field orientation. J. Geophys. Res. **110**, A04202 (2005). doi:10.1029/2004ja010944

S.E. Milan, Both solar wind-magnetosphere coupling and ring current intensity control the size of the auroral oval. Geophys. Res. Lett. **36**, L18101 (2009). doi:10.1029/2009GL039997

S.E. Milan, Sun et Lumiere: solar wind-magnetosphere coupling as deduced from ionospheric flows and polar auroras, in *Festschrift Jim Dungey at 90*, ed. by D.J. Southwood, S.W.H. Cowley (Cambridge University Press, Cambridge, 2014)

D.G. Mitchell, F. Kutchko, D.J. Williams, T.E. Eastman, L.A. Frank, C.T. Russell, An extended study of the low-latitude boundary layer on the dawn and dusk flanks of the magnetosphere. J. Geophys. Res. **92**(A7), 7394–7404 (1987)

A. Miura, Dependence of the magnetopause Kelvin-Helmholtz instability on the orientation of the magnetosheath magnetic field. Geophys. Res. Lett. **22**(21), 2993–2996 (1995)

Y. Miyashita et al., A state-of-the-art picture of substorm-associated evolution of the near-Earth magnetotail obtained from superposed epoch analysis. J. Geophys. Res., Space Phys. **114**(A1), A01211 (2009). doi:10.1029/2008JA013225

T.E. Moore, J.L. Horwitz, Stellar ablation of planetary atmospheres. Rev. Geophys. **45**, RG3002 (2007). doi:10.1029/2005rg000194

T. Nagai, M. Fujimoto, Y. Saito, S. Machida, T. Terasawa, R. Nakamura, T. Yamamoto, T. Mukai, A. Nishida, S. Kokubun, Structure and dynamics of magnetic reconnection for substorm onsets with Geotail observations. J. Geophys. Res. **103**(A3), 4419–4440 (1998)

Z. Nemecek, J. Safrankova, L. Prech, D.G. Sibeck, S. Kokubun, T. Mukai, Transient flux enhancements in the magnetosheath. Geophys. Res. Lett. **25**, 1273–1276 (1998)

N.F. Ness, The Earth's magnetic tail. J. Geophys. Res. **70**, 2989 (1965)

A. Nishida, The Earth's dynamic magnetotail. Space Sci. Rev. **91**(3–4), 507–577 (2000). doi:10.1023/a:1005223124330

Y. Nishimura, L. Lyons, S. Zou, V. Angelopoulos, S. Mende, Substorm triggering by new plasma intrusion: THEMIS all-sky imager observations. J. Geophys. Res., Space Phys. **115**(A7), A07222 (2010). doi:10.1029/2009JA015166

K. Nykyri, A. Otto, Plasma transport at the magnetospheric boundary due to reconnection in Kelvin-Helmholtz vortices. Geophys. Res. Lett. **28**(18), 3565–3568 (2001). doi:10.1029/2001gl013239

M. Øieroset, J. Raeder, T.-D. Phan, S. Wing, J.P. McFadden, W. Li, M. Fujimoto, H. Reme, A. Balogh, Global cooling and densification of the plasma sheet during an extended period of purely northward IMF on October 22–24, 2003. Geophys. Res. Lett. **32**, L12S07 (2005). doi:10.1029/2004GL021523

M. Øieroset, T.D. Phan, V. Angelopoulos, J.P. Eastwood, J. McFadden, D. Larson, C.W. Carlson, K.-H. Glassmeier, M. Fujimoto, J. Raeder, THEMIS multi-spacecraft observations of magnetosheath plasma penetration deep into the dayside low-latitude magnetosphere for northward and strong B_y IMF. Geophys. Res. Lett. **35**, L17S11 (2008). doi:10.1029/2008GL033661

M. Øieroset et al., Direct evidence for a three-dimensional magnetic flux rope flanked by two active magnetic reconnection X lines at Earth's magnetopause. Phys. Rev. Lett. **107**, 165007 (2011)

M. Oka, T. Terasawa, Y. Seki, M. Fujimoto, Y. Kasaba, H. Kojima, I. Shinohara, H. Matsui, H. Matsumoto, Y. Saito, Whistler critical Mach number and electron acceleration at the bow shock: Geotail observation. Geophys. Res. Lett. **33**(24), L24104 (2006)

N. Omidi, D.G. Sibeck, Flux transfer events in the cusp. Geophys. Res. Lett. **34**, L04106 (2007a). doi:10.1029/2006GL028698

N. Omidi, D.G. Sibeck, Formation of hot flow anomalies and solitary shocks. J. Geophys. Res. **112**, A01203 (2007b). doi:10.1029/2006JA011663

N. Omidi, X. Blanco-Cano, C.T. Russell, Macrostructure of collisionless bow shocks: 1. Scale lengths. J. Geophys. Res. **110**, A12212 (2005). doi:10.1029/2005ja011169

N. Omidi, X. Blanco-Cano, C. Russell, H. Karimabadi, Global hybrid simulations of solar wind interaction with magnetospheric boundaries. Adv. Space Res. **38**(4), 632–638 (2006)

C.J. Owen, M. Taylor, I.C. Krauklis, A.N. Fazakerley, M.W. Dunlop, J.M. Bosqued, Cluster observations of surface waves on the dawn flank magnetopause. Ann. Geophys. **22**(3), 971–983 (2004)

M. Palmroth, T.I. Pulkkinen, P. Janhunen, C.C. Wu, Stormtime energy transfer in global MHD simulation. J. Geophys. Res. **108**(A1), 1048 (2003). doi:10.1029/2002ja009446

M. Palmroth, T.V. Laitinen, T.I. Pulkkinen, Magnetopause energy and mass transfer: results from a global MHD simulation. Ann. Geophys. **24**, 3467–3480 (2006)

M. Palmroth, I. Honkonen, A. Sandroos, Y. Kempf, S. von Alfthan, D. Pokhotelov, Preliminary testing of global hybrid-Vlasov simulation: magnetosheath and cusps under northward interplanetary magnetic field. J. Atmos. Sol.-Terr. Phys. **99**, 41–46 (2013). doi:10.1016/j.jastp.2012.09.013

G. Paschmann, Observational evidence for transfer of plasma across the magnetopause. Space Sci. Rev. **80**, 217–234 (1997)

G. Paschmann, B.U.Ö. Sonnerup, I. Papamastorakis, N. Sckopke, G. Haerendel, S.J. Bame, J.R. Asbridge, J.T. Gosling, C.T. Russell, R.C. Elphic, Plasma acceleration at the Earth's magnetopause: evidence for reconnection. Nature **282**, 243–246 (1979)

G. Paschmann, G. Haerendel, N. Sckopke, E. Möbius, H. Luhr, C.W. Carlson, Three-dimensional plasma structures with anomalous flow directions near the Earth's bow shock. J. Geophys. Res. **93**(A10), 11279–11294 (1988)

G. Paschmann, M. Øieroset, T. Phan, In-situ observations of reconnection in space. Space Sci. Rev. **178**, 385–417 (2013). doi:10.1007/s11214-012-9957-2 (online)

V. Peroomian, M. El-Alaoui, P.C. Brandt, The ion population of the magnetotail during the 17 April 2002 magnetic storm: large-scale kinetic simulations and IMAGE/HENA observations. J. Geophys. Res. **116**, A05214 (2011). doi:10.1029/2010JA016253

S.M. Petrinec, C.T. Russell, Near-Earth magnetotail shape and size as determined from the magnetopause flaring angle. J. Geophys. Res., Space Phys. **101**(A1), 137–152 (1996). doi:10.1029/95ja02834

T.D. Phan et al., Low-latitude dusk flank magnetosheath, magnetopause, and boundary layer for low magnetic shear: wind observations. J. Geophys. Res. **102**(A9), 19883–19895 (1997)

T.D. Phan et al., Extended magnetic reconnection at the Earth's magnetopause from detection of bi-directional jets. Nature **404**, 848–850 (2000)

T. Phan et al., Simultaneous Cluster and IMAGE observations of cusp reconnection and auroral proton spot for northward IMF. Geophys. Res. Lett. **30**(10), 1509 (2003). doi:10.1029/2003gl016885

T.D. Phan et al., Cluster observations of continuous reconnection at the magnetopause under steady interplanetary magnetic field conditions. Ann. Geophys. **22**(7), 2355–2367 (2004)

T. Phan, C.P. Escoubet, L. Rezeau, R.A. Treumann, A. Vaivads, G. Paschmann, S.A. Fuselier, D. Attié, B. Rogers, B.U.Ö. Sonnerup, Magnetopause processes. Space Sci. Rev. **118**, 367–424 (2005a)

T.D. Phan, M. Oieroset, M. Fujimoto, Reconnection at the dayside low-latitude magnetopause and its nonrole in low-latitude boundary layer formation during northward interplanetary magnetic field. Geophysical research letters **32**(17), L17101 (2005b)

T.D. Phan, J.T. Gosling, G. Paschmann, C. Pasma, J.F. Drake, M. Oieroset, D. Larson, R.P. Lin, M.S. Davis, The dependence of magnetic reconnection on plasma beta and magnetic shear: evidence from solar wind observations. Astrophys. J. Lett. **719**(2), L199–L203 (2010). doi:10.1088/2041-8205/719/2/l199

T.D. Phan, G. Paschmann, J.T. Gosling, M. Oieroset, M. Fujimoto, J.F. Drake, V. Angelopoulos, The dependence of magnetic reconnection on plasma beta and magnetic shear: evidence from magnetopause observations. Geophys. Res. Lett. **40**(1), 11–16 (2013a). doi:10.1029/2012gl054528

T.D. Phan, M.A. Shay, J.T. Gosling, M. Fujimoto, J.F. Drake, G. Paschmann, M. Oieroset, J.P. Eastwood, V. Angelopoulos, Electron bulk heating in magnetic reconnection at Earth's magnetopause: dependence on the inflow Alfven speed and magnetic shear. Geophys. Res. Lett. **40**(17), 4475–4480 (2013b). doi:10.1002/grl.50917

F. Plaschke, K.H. Glassmeier, D.G. Sibeck, H.U. Auster, D. Constantinescu, V. Angelopoulos, W. Magnes, Magnetopause surface oscillation frequencies at different solar wind conditions. Ann. Geophys. **27**, 4521–4532 (2009). doi:10.5194/angeo-27-4521-2009

F. Plaschke, H. Hietala, V. Angelopoulos, Anti-sunward high-speed jets in the subsolar magnetosheath. Ann. Geophys. **31**, 1877–1889 (2013). doi:10.5194/angeo-31-1877-2013

D. Pokhotelov, S. Von Alfthan, Y. Kempf, R. Vainio, H.E.J. Koskinen, M. Palmroth, Ion distributions upstream and downstream of the Earth's bow shock: first results from Vlasiator. Ann. Geophys. **31**, 2207–2212 (2013)

K.G. Powell, P.L. Roe, T.J. Linde, T.I. Gombosi, D.L. De Zeeuw, A solution-adaptive upwind scheme for ideal magnetohydrodynamics. J. Comput. Phys. **154**(2), 284–309 (1999). doi:10.1006/jcph.1999.6299

A. Pulkkinen et al., Community-wide validation of geospace model ground magnetic field perturbation predictions to support model transition to operations. Space Weather, Int. J. Res. Appl. **11**(6), 369–385 (2013). doi:10.1002/swe.20056

J. Raeder, Flux transfer events: 1. Generation mechanism for strong southward IMF. Ann. Geophys. **24**, 381–392 (2006)

J. Raeder, R.J. Walker, M. Ashour-Abdalla, The structure of the distant geomagnetic tail during long periods of northward IMF. Geophys. Res. Lett. **22**, 349–352 (1995)

J. Raeder, J. Berchem, M. AshourAbdalla, L.A. Frank, W.R. Paterson, K.L. Ackerson, S. Kokubun, T. Yamamoto, J.A. Slavin, Boundary layer formation in the magnetotail: Geotail observations and comparisons with a global MHD simulation. Geophys. Res. Lett. **24**(8), 951–954 (1997). doi:10.1029/97gl00218

J. Raeder, R.L. McPherron, L.A. Frank, S. Kokubun, G. Lu, T. Mukai, W.R. Paterson, J.B. Sigwarth, H.J. Singer, J.A. Slavin, Global simulation of the Geospace Environment Modeling substorm challenge event. J. Geophys. Res., Space Phys. **106**(A1), 381–395 (2001). doi:10.1029/2000ja000605

H. Rème et al., First multispacecraft ion measurements in and near the Earth's magnetosphere with the identical Cluster ion spectrometry (CIS) experiment. Ann. Geophys. **19**, 1303–1354 (2001)

I.G. Richardson, E.W. Cliver, H.V. Cane, Sources of geomagnetic storms for solar minimum and maximum conditions during 1972–2000. Geophys. Res. Lett. **28**(13), 2569–2572 (2001)

A.J. Ridley, K.C. Hansen, G. Toth, D.L. De Zeeuw, T.I. Gombosi, K.G. Powell, University of Michigan MHD results of the geospace global circulation model metrics challenge. J. Geophys. Res. **107**(A10), 1290 (2002). doi:10.1029/2001ja000253

G. Rostoker, S.-I. Akasofu, J. Foster, R.A. Greenwald, Y. Kamide, A.T.Y. Lui, R.L. McPherron, C.T. Russell, Magnetospheric substorms—definition and signatures. J. Geophys. Res. **85**(A4), 1663–1668 (1980)

A. Runov, V. Angelopoulos, M.I. Sitnov, V.A. Sergeev, J. Bonnell, J. McFadden, D. Larson, K.-H. Glassmeier, U. Auster, THEMIS observations of an earthward-propagation dipolarization front. Geophys. Res. Lett. **36**, L14106 (2009). doi:10.1029/2009GL038980

C.T. Russell, The dynamics of planetary magnetospheres. Planet. Space Sci. **49**(10–11), 1005–1030 (2001). doi:10.1016/s0032-0633(01)00017-4

C.T. Russell, R.C. Elphic, Initial ISEE magnetometer results: magnetopause observations. Space Sci. Rev. **22**, 681–715 (1978)

S. Savin et al., High kinetic energy jets in the Earth's magnetosheath: implications for plasma dynamics and anomalous transport. JETP Lett. **87**, 593–599 (2008)

M. Scholer, Magnetic flux transfer at the magnetopause based on single X-line bursty reconnection. Geophys. Res. Lett. **15**, 291–294 (1988)

S.J. Schwartz et al., An active current sheet in the solar wind. Nature **318**(21), 269–271 (1985)

S.J. Schwartz, R.L. Kessel, C.C. Brown, L.J.C. Woolliscroft, M.W. Dunlop, C.J. Farrugia, D.S. Hall, Active current sheets near the Earth's bow shock. J. Geophys. Res. **93**(A10), 11295–11310 (1988). doi:10.1029/JA093iA10p11295

V.A. Sergeev, V. Angelopoulos, R. Nakamura, Recent advances in understanding substorm dynamics. Geophys. Res. Lett. **39**, L05101 (2012). doi:10.1029/2012gl050859

M.A. Shay, J.F. Drake, J.P. Eastwood, T.D. Phan, Super-Alfvénic propagation of substorm reconnection signatures and poynting flux. Phys. Rev. Lett. **107**(6), 065001 (2011)

J.H. Shue, J.K. Chao, P. Song, J.P. McFadden, A. Suvorova, V. Angelopoulos, K.H. Glassmeier, F. Plaschke, Anomalous magnetosheath flows and distorted subsolar magnetopause for radial interplanetary fields. Geophys. Res. Lett. **36**, L18112 (2009). doi:10.1029/2009GL039842

D.G. Sibeck, G.L. Siscoe, J.A. Slavin, C.W. Smith, B.T. Tsurutani, R.P. Lepping, The distant magnetotail's response to a strong interplanetary magnetic field by: twisting, flattening, and field line bending. J. Geophys. Res. **90**(A5), 4011–4019 (1985)

D.G. Sibeck, N.L. Borodkova, G.N. Zastenker, S.A. Romanov, J.A. Sauvaud, Gross deformation of the dayside magnetopause. Geophys. Res. Lett. **25**(4), 453–456 (1998)

D.G. Sibeck et al., Comprehensive study of the magnetospheric response to a hot flow anomaly. J. Geophys. Res. **104**(A3), 4577–4593 (1999a)

D.G. Sibeck et al., Plasma transfer processes at the magnetopause. Space Sci. Rev. **88**(1–2), 207–283 (1999b). doi:10.1023/a:1005255801425

M.I. Sitnov, M. Swisdak, Onset of collisionless magnetic reconnection in two-dimensional current sheets and formation of dipolarization fronts. J. Geophys. Res. **116**, A12216 (2011). doi:10.1029/2011ja016920

M.I. Sitnov, M. Swisdak, A.V. Divin, Dipolarization fronts as a signature of transient reconnection in the magnetotail. J. Geophys. Res. **114**, A04202 (2009). doi:10.1029/2008JA013980

J.A. Slavin et al., Dual spacecraft observations of lobe magnetic field perturbations before, during and after plasmoid release. Geophys. Res. Lett. **26**(19), 2897–2900 (1999)

J.A. Slavin et al., Simultaneous observations of earthward flow bursts and plasmoid ejection during magnetospheric substorms. J. Geophys. Res. **107**(A7), 13-1–13-23 (2002). doi:10.1029/2000JA003501

J.A. Slavin et al., MESSENGER and Mariner 10 flyby observations of magnetotail structure and dynamics at Mercury. J. Geophys. Res. **117**, A01215 (2012a). doi:10.1029/2011ja016900

J.A. Slavin et al., Messenger observations of a flux-transfer-event shower at Mercury. J. Geophys. Res. **117**, A00M06 (2012b). doi:10.1029/2012ja017926

P. Song, C.T. Russell, Model of the formation of the low-latitude boundary layer for strongly northward interplanetary magnetic field. J. Geophys. Res. **97**(A2), 1411–1420 (1992)

P. Song, B.U.Ö. Sonnerup, M.F. Thomsen, *Physics of the Magnetopause* (American Geophysical Union, Washington, 1995)

B. Sonnerup, Magnetopause reconnection rate. J. Geophys. Res. **79**(10), 1546–1549 (1974)

B.U.O. Sonnerup, Theory of the low-latitude boundary layer. J. Geophys. Res. **85**(A5), 2017–2026 (1980)

B.U.Ö. Sonnerup, G. Paschmann, I. Papamastorakis, N. Sckopke, G. Haerendel, S.J. Bame, J.R. Asbridge, J.T. Gosling, C.T. Russell, Evidence for magnetic field reconnection at the Earth's magnetopause. J. Geophys. Res. **86**(A12), 10049–10067 (1981)

D.J. Southwood, The hydrodynamic stability of the magnetospheric boundary. Planet. Space Sci. **16**, 587–605 (1968)

D.J. Southwood, C.J. Farrugia, M.A. Saunders, What are flux transfer events? Planet. Space Sci. **36**, 503–508 (1988)

R.J. Strangeway, C.T. Russell, C.W. Carlson, J.P. McFadden, R.E. Ergun, M. Temerin, D.M. Klumpar, W.K. Peterson, T.E. Moore, Cusp field-aligned currents and ion outflows. J. Geophys. Res., Space Phys. **105**(A9), 21129–21142 (2000). doi:10.1029/2000ja900032

T. Sugiyama, Time sequence of energetic particle spectra in quasiparallel shocks in large simulation systems. Phys. Plasmas **18**, 022302 (2011). doi:10.1063/1.3552026

T. Sugiyama, M. Fujimoto, T. Mukai, Quick ion injection and acceleration at quasi-parallel shocks. J. Geophys. Res. **106**(A10), 21657–21673 (2001)

A.V. Suvorova, J.H. Shue, A.V. Dmitriev, D.G. Sibeck, J.P. McFadden, H. Hasegawa, K. Ackerson, K. Jelinek, J. Safrankova, Z. Nemecek, Magnetopause expansions for quasi-radial interplanetary magnetic field: THEMIS and Geotail observations. J. Geophys. Res. **115**, A10216 (2010). doi:10.1029/2010ja015404

M. Swisdak, B.N. Rogers, J.F. Drake, M.A. Shay, Diamagnetic suppression of component magnetic reconnection at the magnetopause. J. Geophys. Res., Space Phys. **108**(A5), 1218 (2003). doi:10.1029/2002ja009726

M. Swisdak, M. Opher, J.F. Drake, F.A. Bibi, The vector direction of the interstellar magnetic field outside the heliosphere. Astrophys. J. **710**(2), 1769–1775 (2010). doi:10.1088/0004-637x/710/2/1769

T. Terasawa et al., Solar wind control of density and temperature in the near-Earth plasma sheet: WIND/GEOTAIL collaboration. Geophys. Res. Lett. **24**(8), 935–938 (1997)

V.A. Thomas, D. Winske, Kinetic simulations of the Kelvin-Helmholtz instability at the magnetopause. J. Geophys. Res., Space Phys. **98**(A7), 11425–11438 (1993). doi:10.1029/93ja00604

V.A. Thomas, D. Winske, M.F. Thomsen, T.G. Onsager, Hybrid simulation of the formation of a hot flow anomaly. J. Geophys. Res. **96**(A7), 11625–11632 (1991)

M.F. Thomsen, J.T. Gosling, S.A. Fuselier, S.J. Bame, C.T. Russell, Hot, diamagnetic cavities upstream from the Earth's bow shock. J. Geophys. Res. **91**(A3), 2961–2973 (1986)

M.F. Thomsen, J.T. Gosling, S.J. Bame, K.B. Quest, C.T. Russell, S.A. Fuselier, On the origin of hot diamagnetic cavities near the Earth's bow shock. J. Geophys. Res. **93**(A10), 11311–11325 (1988)

M.F. Thomsen, V.A. Thomas, D. Winske, J.T. Gosling, M.H. Farris, C.T. Russell, Observational test of hot flow anomaly formation by the interaction of a magnetic discontinuity with the bow shock. J. Geophys. Res. **98**(A9), 15319–15330 (1993)

F. Toffoletto, S. Sazykin, R. Spiro, R. Wolf, Inner magnetospheric modeling with the rice convection model. Space Sci. Rev. **107**(1–2), 175–196 (2003). doi:10.1023/a:1025532008047

G. Toth et al., Space weather modeling framework: a new tool for the space science community. J. Geophys. Res. **110**, A12226 (2005). doi:10.1029/2005ja011126

G. Toth, Y. Ma, T.I. Gombosi, Hall magnetohydrodynamics on block-adaptive grids. J. Comput. Phys. **227**(14), 6967–6984 (2008). doi:10.1016/j.jcp.2008.04.010

G. Toth et al., Adaptive numerical algorithms in space weather modeling. J. Comput. Phys. **231**(3), 870–903 (2012). doi:10.1016/j.jcp.2011.02.006

K.J. Trattner, J.S. Mulcock, S.M. Petrinec, S.A. Fuselier, Location of the reconnection line at the magnetopause during southward IMF conditions. Geophys. Res. Lett. **34**(3), 5 (2007a). doi:10.1029/2006gl028397

K.J. Trattner, J.S. Mulcock, S.M. Petrinec, S.A. Fuselier, Probing the boundary between antiparallel and component reconnection during southward interplanetary magnetic field conditions. J. Geophys. Res. **112**, A08210 (2007b). doi:10.1029/2007ja012270

K.J. Trattner, S.M. Petrinec, S.A. Fuselier, T.D. Phan, The location of reconnection at the magnetopause: testing the maximum magnetic shear model with THEMIS observations. J. Geophys. Res. **117**, A01201 (2012). doi:10.1029/2011ja016959

R.A. Treumann, J. LaBelle, T.M. Bauer, *Diffusion Processes: An Observational Perspective*. Geophysical Monograph Series, vol. 90 (1995), pp. 331–341

B.T. Tsurutani, R.G. Stone, *Collisionless Shocks in the Heliosphere: Reviews of Current Research* (American Geophysical Union, Washington, 1985)

B.T. Tsurutani et al., Corotating solar wind streams and recurrent geomagnetic activity: a review. J. Geophys. Res. **111**, A07S01 (2006). doi:10.1029/2005JA011273

V. Vasyliunas, Plasma distribution and flow. Phys. Jovian Magnetosph. **1**, 395–453 (1983)

R. Walker et al., Source and loss processes in the magnetotail. Space Sci. Rev. **88**(1–2), 285–353 (1999). doi:10.1023/a:1005207918263

Y.L. Wang, R.C. Elphic, B. Lavraud, M.G.G.T. Taylor, J. Birn, C.T. Russell, J. Raeder, H. Kawano, X.X. Zhang, Dependence of flux transfer events on solar wind conditions from 3 years of Cluster observations. J. Geophys. Res. **111**, A04224 (2006). doi:10.1029/2005JA011342

D.T. Welling, A.J. Ridley, Validation of SWMF magnetic field and plasma. Space Weather **8**, S03002 (2010). doi:10.1029/2009sw000494

S. Wing, P.T. Newell, 2D plasma sheet ion density and temperature profiles for northward and southward IMF. Geophys. Res. Lett. **29**, 21-1–21-4 (2002). doi:10.1029/2001gl013950

R.A. Wolf, M. Harel, R.W. Spiro, G.H. Voigt, P.H. Reiff, C.K. Chen, Computer-simulation of inner magnetospheric dynamics for the magnetic storm of July 29, 1977. J. Geophys. Res., Space Phys. **87**(A8), 5949–5962 (1982). doi:10.1029/JA087iA08p05949

S. Zaharia, V.K. Jordanova, M.F. Thomsen, G.D. Reeves, Self-consistent modeling of magnetic fields and plasmas in the inner magnetosphere: application to a geomagnetic storm. J. Geophys. Res. **111**, A11S14 (2006). doi:10.1029/2006ja011619

J.C. Zhang et al., Understanding storm-time ring current development through data-model comparisons of a moderate storm. J. Geophys. Res. **112**, A04208 (2007). doi:10.1029/2006ja011846

J. Zhong et al., Three-dimensional magnetic flux rope structure formed by multiple sequential X-line reconnection at the magnetopause. J. Geophys. Res., Space Phys. **118**(5), 1904–1911 (2013). doi:10.1002/jgra.50281

DOI 10.1007/978-1-4939-3547-5_10
Reprinted from *Space Science Reviews* Journal, DOI 10.1007/s11214-014-0037-7

Properties of Magnetic Field Fluctuations in the Earth's Magnetotail and Implications for the General Problem of Structure Formation in Hot Plasmas

Lev Zelenyi · Anton Artemyev · Anatoli Petrukovich

Received: 24 December 2013 / Accepted: 5 February 2014 / Published online: 18 March 2014
© Springer Science+Business Media Dordrecht 2014

Abstract In this review we discuss the formation of plasma structures in hot plasmas with large β. We use spacecraft observations of magnetic field fluctuations in the Earth magnetotail to reveal the main multiscale plasma properties. Fourier spectra of magnetic field fluctuations observed by various spacecraft in the different domains of the magnetotail at quiet or moderately disturbed times demonstrated a number of practically universal features: (1) the presence of two kinks at low ($\sim 5 \times 10^{-2}$ Hz) and high ($\sim 5 \times 10^{-1}$ Hz) frequencies, (2) in the frequency interval between the kinks the power law shape with an index ~ 2.5, and (3) the significant intensification of fluctuations with the increase of plasma flow velocities. To describe these spectra we consider properties of a principal structure supporting the equilibrium of a hot plasma configuration—the magnetotail current sheet. The observed spatial scales of current sheets are often very small and almost reach the ion Larmor radius. Convection of such mesoscale plasma structures by plasma flows can be responsible for the formation of a certain part of the spectrum of magnetic field fluctuations. Lower- and higher- frequency domains in Fourier spectra can be generated respectively by large-scale MHD oscillations of current sheets and by kinetic small-scale instabilities excited by strong local magnetic field gradients existing in active current sheets.

Keywords Current sheets · Magnetic field turbulence

1 Introduction

Direct spacecraft observations in the near-Earth space give an unique opportunity for detailed investigation of plasma structure formation and dynamics in a system which could be considered as a proxy for many astrophysical and laboratory plasma configurations. There are three types of plasmas with substantially different conditions in the near-Earth environment. One can characterize these types by two parameters: plasma beta β (ratio of plasma and magnetic pressures) and Alfven Mach number $M_A = v_{flow}/v_A$ (ratio of

L. Zelenyi (✉) · A. Artemyev · A. Petrukovich
Space Research Institute, RAS, Moscow, Russia
e-mail: lzelenyi@iki.rssi.ru

plasma flow velocity and Alfven velocity). Parameters of the upstream solar wind before its braking at the collisionless bow shock correspond to $\beta \sim 1$ and $M_A \sim 5$–10. Thus, the main energy supply to the magnetosphere is provided by the weakly magnetized fast plasma flow, while the magnetic field energy and the plasma thermal energy are relatively small. In this case, one of the most important processes is the transformation of the plasma flow energy into the plasma thermal energy via shock waves (Burgess and Scholer 2013; Krasnoselskikh et al. 2013). Outer regions of planetary magnetospheres are filled by the hot shocked plasmas with β up to ~ 100–1000 and $M_A \leq 1$ (in the central region) and by cold rarefied plasma with $\beta < 0.1$ (in the magnetospheric lobes). Thus, the energy supply stored at lobes is provided by magnetic fields, while the magnetic energy transformation into particle kinetic and thermal energies is supported by the magnetic reconnection triggered by currents flowing in a hot plasma (Priest and Forbes 2000). Finally, the inner regions of planetary magnetospheres are characterized by the absence of plasma flows (i.e. $M_A \to 0$) and strong magnetic fields with $\beta \ll 1$. In this case, we deal with the redistribution of energy between particle populations via excitation of various plasma instabilities and resonant wave-particle interaction (Millan and Baker 2012, and references therein).

Each of these plasma regimes has some analogues in astrophysics and laboratory plasmas (see discussion in Yamada et al. 2010; Frank 2010; Zelenyi and Artemyev 2013). In this review we concentrate mainly on the outer region of planetary magnetospheres where currents carried by hot plasma particles play the dominant role. For such systems the process of the magnetic field line reconnection supports the plasma heating (and/or acceleration) and dissipation of the magnetic energy. Such energy transformation is the essential element in numerous scenarios of particle acceleration in both astrophysical (e.g. Kirk 2004; Birn and Priest 2007; Somov 2010; Arons 2012) and laboratory plasmas (Wesson 1990; Porcelli et al. 2002; Kyrie et al. 2012). Moreover, the magnetic reconnection is responsible for current filamentation, destruction of magnetic field surfaces (Galeev et al. 1986) and formation of multiscale structure in plasma turbulence (e.g., Milovanov et al. 1996).

There are two general approaches for the investigations of distributions of plasma currents and magnetic fields obtained in spacecraft observations. First of all, one can concentrate on the examination of Fourier spectra of electromagnetic field fluctuations (see, e.g., reviews by Zimbardo et al. 2010; Dudok de Wit et al. 2013). In this case the comparison of the observed dispersion (relation between a frequency and a spatial scale of fluctuations) and theoretical models can give some information about wave modes responsible for the formation of plasma structures (Eastwood et al. 2009; Huang and et al. 2012; Chaston et al. 2012; Viberg et al. 2013). An example of the magnetic field spectrum observed in the Earth magnetotail in the vicinity of the reconnection region is shown in Fig. 1. One can distinguish particular interval in this spectrum in the frequency range $f \in [5 \times 10^{-2}, 5 \times 10^{-1}]$ Hz, corresponding roughly to fluctuations of plasma structures with the spatial sales around the ion Larmor radius. However, the spectral analysis cannot separate temporal and spatial fluctuations. The same spectrum can be formed by (1) monoscale fluctuations (having a single spatial scale), but with corresponding frequency distribution of magnetic power density, or (2) by transformation of some spectrum of spatial scales of magnetic fluctuations frozen into the plasma to the frequency spectrum due to the plasma convective motion.

Analysis of spectra of magnetic field fluctuations could be accomplished by several methods (Wooliscroft et al. 1996; Dudok de Wit 2003; Petrukovich 2005; Dudok de Wit et al. 2013). On one side, the determination of a power-law exponent α of the spectrum $P(f) \sim f^\alpha$ for various frequency ranges can give a certain information about processes contributing to the formation of magnetic field fluctuations (see discussion in Bauer et al. 1995a; Consolini and Lui 1999; Weygand et al. 2005; Huang and et al. 2012). On the other side, the

Fig. 1 Spectrum of magnetic field fluctuations collected by Cluster in the vicinity of the reconnection region. Figure is adapted from Huang and et al. (2012)

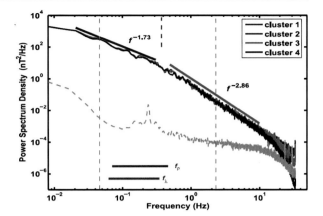

measured time series of magnetic field a fluctuations can be considered as a one-dimensional fractal curve with corresponding fractal dimension δ (Mandelbrot 1977). The fractal dimension δ depends on the frequency range of magnetic field fluctuations (Ohtani et al. 1998; Petrukovich 2005). The theoretical relation between δ and α (so called Berry relation) is $\alpha + 2\delta = 5$ (Mandelbrot 1977) and have been tested with magnetic field fluctuations measured in the magnetotail (Abry et al. 2000; Petrukovich and Malakhov 2009, and references therein).

Another approach to the investigation of plasma structures could be realized with the help of multispacecraft observations, when simultaneous measurements of magnetic fields in several points allow to separate spatial and temporal characteristics of plasma structures. The investigation of a self-consistent mixture of small-scale currents and magnetic field perturbations requires the determination of correlations between magnetic field measurements at different points. For example, Cluster mission includes four spacecraft, while the distance between satellites varies form 100 km up to 10000 km (Escoubet et al. 2013). In the magnetotail region hot ions have Larmor radii around $\sim$100–1000 km. Therefore, magnetic field measurements by Cluster system are capable to distinguish plasma structures with the kinetic scales of the order of the ion Larmor length. Moreover, correlation of magnetic field measurements at different spacecraft can give an estimate of the velocities of these structures. Spatial scales around the ion Larmor radius correspond to various mesoscale structures discovered by Cluster and THEMIS multispacecraft missions during the last decade (Sharma et al. 2008; Runov et al. 2011). One of the most important plasma structures in this range of spatial scales is the thin magnetotail current sheet (Baumjohann et al. 2007).

The range of frequencies of magnetic field fluctuations in the magnetotail is bounded by the time-scale of an entire magnetotail reconfiguration ($\sim$tens of minutes, or 10^{-3} Hz) and by the sensitivity of modern fluxgate magnetometers ($\sim$10 Hz). Fluctuations with higher frequencies can be measured by search-coil devices, but matching of magnetic measurements of both types of instruments is not a simple problem. Moreover, magnetic fluctuations with frequencies larger than 10 Hz could probably be controlled by electrostatic effects with less energy input into magnetic fields (Baumjohann et al. 1989). Typical amplitudes of a plasma bulk velocity in the Earth's magnetotail are about 10 km/s–1000 km/s (depending on the geomagnetic conditions), thus, in a framework of Taylor hypothesis the frequency range $\sim$$10^{-3}$–1 Hz directly corresponds to the spatial range 100–1000 km. Therefore, one has two different approaches to the investigation of plasma structures: analysis of a Fourier spectra of electromagnetic fluctuations within the frequency range $\sim$$10^{-3}$–1 Hz and the study of magnetic field structures within the range of spatial scales ~100–1000 km. Combination of these two methods should give us answers to the following important questions: (1) what is the nature of plasma structures contributing to the spectrum of electromagnetic field fluctuations (2) what mechanisms are responsible for the formation of the spectrum of fluctuations obtained by *in situ* spacecraft observations.

2 Spacecraft Observations of Fourier Spectra of Magnetic Field Fluctuations in the Earth's Magnetotail

Early observational information about magnetic field fluctuations in the magnetotail was obtained and analyzed by Russell (1972) and Garrett (1973). Using OGO-5 measurements Russell (1972) demonstrated that for B_x-component (hereinafter GSM coordinate system is used) the oscillations of the power density reaches maximum in the central region of the magnetotail (see Fig. 2). This power density decreases with a distance from the neutral plane $B_x \sim 0$ (in the agreement with previous estimates based on IMP-1 observations, see Hruška and Hrušková 1969). The spectrum slope of the power density (nT^2/Hz) observed by Russell (1972) for the frequency range below 1 Hz is about $\alpha \sim -2.5$. Similar characteristics of magnetic field fluctuations were found by Garrett (1973) with the help of Explorer-34 measurements for frequencies below 0.2 Hz. Intensification of high-frequency (10 Hz $< f < 10^3$ Hz) magnetic field fluctuations were detected by Imp-7 and Imp-8 spacecraft in the vicinity of the neutral plane of the magnetotail CS (Scarf et al. 1974; Gurnett et al. 1976). Moreover, Gurnett et al. (1976) have shown that the enhancement of high-frequency magnetic field fluctuations correlates with observations of strong plasma flows and is localized around midnight. Similar correlation between high-frequency magnetic field fluctuations ($f \sim 0.1$–1 Hz) and plasma flows was reported by Coroniti et al. (1980), who studied the dynamics of the magnetotail CS with IMP-7 measurements during several substorms.

Fig. 2 *Left panel* shows magnetic field spectra for time interval before entering into the plasma sheet. *Right panel* shows magnetic field spectra in the plasma sheet. Figure is adapted from Russell (1972)

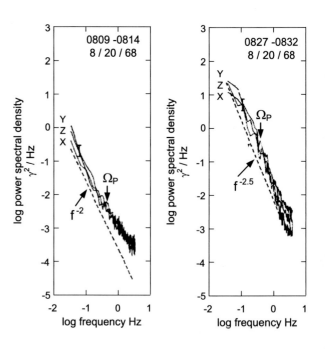

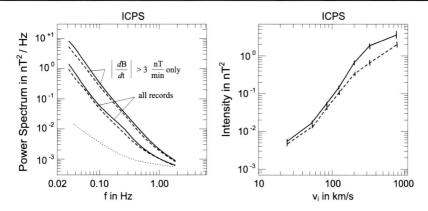

Fig. 3 *Left panel* shows average 30-second spectra of right-hand-polarized transverse fluctuations S_r (*dotted line*), left-hand-polarized transverse fluctuations S_l (*dashed line*), and parallel fluctuations $S_\parallel$ (*solid line*). Spectra obtained for data with $|d\mathbf{B}/dt| > 3$ nT/min are shown separately. The mean value and the standard deviation of ion gyrofrequency are shown by circle and cross (for $|d\mathbf{B}/dt| > 3$ nT/min data). *Right panel* shows averaged intensities $I_{r,l,\parallel} = \int_{0.03\,\text{Hz}}^{0.13\,\text{Hz}} S_{r,l,\parallel} df$ as a function of the plasma bulk velocity. Figure is adapted from Bauer et al. (1995b)

Measurements of quasi-stationary and turbulent electric fields by ISEE-1 spacecraft confirmed the strong relation between the enhancement of magnetic field fluctuations in the vicinity of the neutral plane and plasma flows (Cattell and Mozer 1982; Cattell et al. 1986). In this case, the quasi-stationary electric fields were used to estimate the plasma drift velocity $\sim \mathbf{E} \times \mathbf{B}$ as an independent evidence of the intensification of plasma flows. Bauer et al. (1995a) used AMPTE/IRM measurements of the magnetic field in the inner region of the plasma sheet ($\sqrt{B_x^2 + B_y^2} < 15$ nT) to plot the spectra of the transverse and parallel magnetic field fluctuations. Figure 3 shows that parallel fluctuations are more intense in comparison with the transverse ones. Moreover, the intensity of magnetic field fluctuations substantially increases with the increase of the plasma bulk velocity (see right panel of Fig. 3). This result was confirmed by ISEE-2 spacecraft observations (see comprehensive statistics in Borovsky and Funsten 2003; Neagu et al. 2005). Particularly, Neagu et al. (2005) found the correlation between amplitudes of magnetic field fluctuations δB and an amplitude of the plasma bulk velocity v_{flow}: $(\delta B[\text{nT}])^2 \sim 10^{-4} \cdot (v_{flow}[\text{km/s}])^2$ (see Fig. 4, right panel). The increase of δB with the spacecraft approach to the neutral plane (in other words with the increase of the plasma pressure) was also found in these experiments (see Fig. 4, left panel). Not only fluctuations of the magnetic field, but also plasma velocity fluctuations intensified with the increase of the velocity of plasma flows (Neagu et al. 2002, 2005).

Measurements of the next generation of spacecraft at 90-ies (Geotail, Interball-Tail) gave the opportunity to investigate the fine structure of magnetic field spectra (Hoshino et al. 1994; Petrukovich 2005). Examples of Geotail observations in a distant tail are shown in Fig. 5. The dominance of B_x fluctuations is well documented. Spectra were approximated by the double power-law function with the break point around ~ 0.04 Hz. The similar double power-law spectrum was measured by Interball-Tail in the magnetotail during the geomagnetically active period (see Fig. 6, left panel). The breaking point is located around 0.02 Hz (compare with Figs. 5, 1). Geotail statistics presented by Petrukovich and Malakhov (2009) have shown the same breaking of spectra (see Fig. 6, right panel). Moreover, Geotail data have shown clearly the dependence of the power density of fluctuations with frequencies

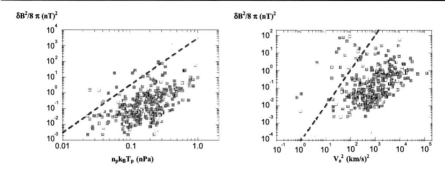

Fig. 4 *Left panel* shows correlation of amplitudes of magnetic field fluctuations and the plasma pressure. *Right panel* shows correlation of amplitudes of magnetic field fluctuations and the plasma bulk velocity. Figure is adapted from Neagu et al. (2005)

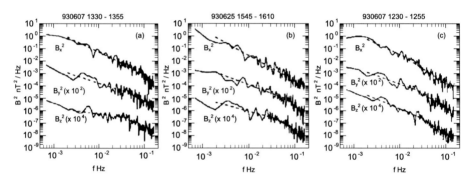

Fig. 5 Three examples of magnetic field spectra collected by Geotail spacecraft. Figure is adapted from Hoshino et al. (1994)

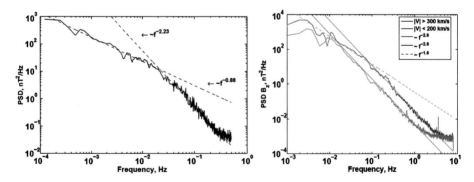

Fig. 6 *Left panel* shows an example of magnetic field spectra collected by Interball-tail spacecraft. Figure is adapted from Petrukovich (2005). *Right panel* shows two spectra collected by Geotail for different ranges of the plasma flow velocity. Figure is adapted from Petrukovich and Malakhov (2009)

$f > 10^{-2}$ Hz on the plasma flow velocity v_{flow}. These observations demonstrated also that for $f > 10^{-1}$ Hz the slope of the fluctuation spectrum does not change with the increase of v_{flow}, but amplitudes of magnetic field fluctuations exhibit very strong ($\sim$5 times) growth.

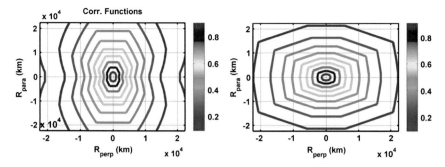

Fig. 7 Correlation coefficients (shown by color) as functions of spacecraft separation along and across the magnetic field direction are calculated for quiet (*left panel*) and disturbed (*right panel*) conditions. Figure is adapted from Weygand et al. (2010)

V The increase of v_{flow} results also in appearance of a new slope of the power density of magnetic field fluctuations within the frequency range $f \in [10^{-2}, 10^{-1}]$ Hz.

Cluster observations within the magnetotail confirmed the presence of a double slope of the power density in the frequency range $f < 10^{-1}$ Hz—Weygand et al. (2005) found two maxima in the distribution of slope values (around -2 and -1.6). Weygand et al. (2005) also have shown that the distribution function of magnetic field fluctuations in the vicinity of the neutral plane at the time scales 10–100 s is very different from the Gaussian one.

Multispacecraft epoch coming with the Cluster mission opens new opportunities for the investigation of magnetic field structures and fluctuations. Simultaneous measurements of the magnetic field at four points allowed to calculate the correlation function of fluctuations (Weygand et al. 2009, 2010). The correlation coefficient was obtained as a function of a spacecraft separation along and across the magnetic field direction. Examples of such correlation coefficients calculated for disturbed and quiet conditions are shown in Fig. 7. For the quiet conditions the correlation length is the longest along the mean magnetic field direction and the shortest perpendicular to it. The situation changes substantially for the disturbed conditions—the correlation length becomes the shortest along the direction of the mean magnetic field and the longest perpendicular to it. These results can be interpreted in terms of plasma flows: one can assume that during quiet conditions the magnetic field structures are transported mainly along the field lines, while disturbed conditions are characterized by the intensification of transverse (relative to the magnetic field) plasma transport (e.g., Petrukovich et al. 2001).

Two distinct methods to study possible relations between magnetic field fluctuations and plasma flows have been applied to process Cluster measurements. The first method corresponds to the intensification of wave modes contributing to magnetic field fluctuations. It was suggested that the increase of the plasma flow velocity results in intensification of compressional waves propagating together with plasma towards the Earth (Volwerk et al. 2003, 2004c). Analyzing Cluster measurements Volwerk et al. (2004b) suggested that eigenoscillations of the magnetotail CS can be initiated by fast flows. Volwerk et al. (2004a) found good correlation between the intensification of ULF waves (magnetic field fluctuations in the frequency range $\sim$0.06 Hz) and plasma flows and showed that these fluctuations propagate with the plasma flow. The spatial scales of plasma flows in the magnetotail are usually $\sim$2–4R_E in the dawn-dusk direction (Nakamura et al. 2004). Thus, as was speculated by Volwerk et al. (2004a) these flows can be considered as a large-scale structures with internal instabilities producing magnetic field fluctuations. For example, Volwerk et al. (2007)

Fig. 8 Spectra of magnetic and electric fields inside the strong plasma flow originated in the reconnection region (*solid curves*) and outside this flow (*dotted curves*). Figure is adapted from Eastwood et al. (2009)

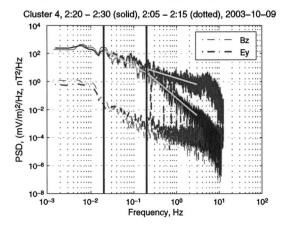

found the enhanced wave activity at the boundary of fast flows and interpreted these waves as a manifestation of the Kelvin-Helmholtz instability. Further investigations showed that plasma flows are often accompanied by plasma vortices at the boundaries (e.g., Panov et al. 2010). Vörös et al. (2009) investigated generation of kink-like fluctuations in the magnetotail CS due to fast plasma flows originated in the reconnection region. All these investigations deal with the low-frequency ($f < 1$ Hz) perturbations of magnetic field.

The second approach deals with the high-frequency fluctuations ($f > 1$ Hz). According to this approach, multispacecraft measurements can be used to reveal the correlation between various properties of magnetic field fluctuations (like multi-scale magnetic field intermittence, see Vörös et al. 2003 measured in different points. Vörös et al. (2004) have shown that magnetic fluctuations with short temporal scales (0.08–0.3 s) having the same scaling index $\alpha \sim -2.6$ as magnetic fluctuations with longer temporal scales (~ 0.75 s) occurring during periods with fast plasma flows. During quiet periods (when fast flows are absent) the energy transfer to the short scales is suppressed, and the scaling index in a long scale range is $\alpha \sim -1.7$. Vörös et al. (2005) estimated the dissipation time-scale of the magnetic field and showed that it is inversely proportional to the flow velocity amplitude. Vörös et al. (2008) performed the systematic comparisons of spectral characteristics, including the variance anisotropy and the scale-dependent spectral anisotropy in a wave vector space for magnetic field fluctuations observed under quiet and disturbed conditions.

Multispacecraft measurements of the magnetic field fluctuations gave a chance to separate spatial and temporal variations (Paschmann and Schwartz 2000). Thus, the position of spacecraft relative to the reconnection region can be clarified (see review by Paschmann et al. 2013). In this case one can separate measurements of magnetic field fluctuations in the outflow region (where strong plasma flows are transporting small-scale magnetic structures from the reconnection region) and far from the outflow region (Eastwood et al. 2009, 2010; Huang and et al. 2012). Examples of such spectra are shown in Figs. 1 and 8. Figure 8 clearly demonstrates the presence of a double-slope of the power-law distribution of magnetic field fluctuations. Moreover, in the outflow region with fast plasma flows the strong increase of the power density of magnetic field fluctuations is well seen (for frequency range $f < 5$ Hz). This increase of the power density is more substantial for electric field fluctuations in comparison with magnetic field fluctuations. Therefore, there are different slopes of magnetic and electric field fluctuations for the frequency range $f > 0.02$ Hz.

3 Current Sheets in Hot Plasmas

The Earth magnetotail with its magnetic field reversal supported by the current sheet (CS) immersed in a hot $\beta \gg 1$ collisionless plasma sheet was discovered by Ness (1965). In the plane of magnetic field reversal (sometimes called the *neutral plane*) the main component of the magnetic field $B_x(z)$ changes its sign and only a weak B_z component does not vanish. The vertical scale of the magnetotail domain is about few Earth radii R_E (much larger than the ion Larmor radius). Thus, this global configuration can be considered as a large-scale MHD structure. However, recent multispacecraft observations show that the 'heart' of the magnetotail CS is some relatively small-scale plasma structure where a very intense cross-tail (in the dawn-dusk direction) current is localized. This is so-called thin CS (TCS) with the transverse spatial scale about the ion Larmor radius $\sim \rho_i$ (Runov et al. 2006; Petrukovich et al. 2011).

Strong transverse gradients of magnetic field (i.e. intense currents) characteristic for TCS assume that the motion of ions is not anymore adiabatic, i.e. the corresponding adiabatic invariant (magnetic moment of particles) is not conserved (Northrop 1963; Sivukhin 1965). Moreover, due to a weakness of the nonvanishing B_z component the curvature radius of field lines is much smaller than ρ_i. Ion dynamics in such a regime could be described as quasi-adiabatic (Büchner and Zelenyi 1989). In this case, the new mechanism of current generation starts playing an important role (besides the slow diamagnetic drifts due to plasma inhomogeneity are also operating in a quasi-adiabatic case). This new mechanism corresponds to the openness of particle orbits (similar to paramagnetic currents in the systems with sharp boundaries, see Frank-Kamenetskii 1968; Zelenyi et al. 2011b). If one considers the particle motion averaged over fast transverse oscillations, then particles with the trajectories crossing the TCS boundaries carry the strong current in the central region of TCS due to the incomplete Larmor rotation around B_z field (Speiser 1965; Eastwood 1972). The spatial scale of this current layer across the neutral plane $B_x = 0$ is about the ion Larmor radius (Francfort and Pellat 1976). There are several self-consistent models of TCS based on this mechanism of the cross-tail current generation (Sitnov et al. 2000; Zelenyi et al. 2000).

Many predictions of TCS models were confirmed by recent Cluster observations (see review Artemyev and Zelenyi 2013, and references therein). TCS models including currents carried by ions at open orbits approximate experimental profiles of the transverse current density surprisingly well (Sitnov et al. 2006; Artemyev et al. 2008). Moreover, some features of the particle distribution in the velocity space predicted by theoretical TCS models were revealed in spacecraft observations (Zhou et al. 2009; Artemyev et al. 2010). The most intense TCS were observed in the vicinity of the reconnection region (Nakamura et al. 2006a; Nakamura et al. 2006b) where the current density exceeds 50 nA/m^2 (Fig. 9(a)). Quite abundant statistics have shown that observed TCS indeed has the predicted spatial scales $\sim \rho_i$ (Fig. 9(b)). The comparison of direct spacecraft observations with the laboratory modelling of thin and strong CSs demonstrated also a number of common features (see details in Artemyev et al. 2013).

The equilibrium of the TCS plasma configuration assumes that the transverse (vertical relative to the plane $z = 0$) stress balance is supported by the variation of the plasma pressure across the plasma sheet: $B_{ext}^2 \approx 8\pi P_i$ where B_{ext} is the B_x amplitude, $P_i = n_i T_i$ is the ion thermal pressure (the electron temperature in the magnetotail is substantially smaller than the ion one, see Baumjohann et al. 1989; Artemyev et al. 2011b) and n_i, T_i are ion density

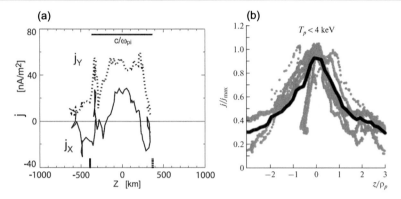

Fig. 9 Panel (**a**) shows an example of intense TCS observed in the vicinity of the reconnection region. Figure is adapted from (Nakamura ct al. 2006a). Panel (**b**) shows normalized profiles of the current density in several TCS (in all TCS the proton temperature is less than 4 keV) and the averaged profile. Spatial scale is normalized on the Larmor radius of protons with the temperature 3 keV. Figure is adapted from Zelenyi et al. (2011b)

and temperature, respectively. Thus, the ion Larmor radius calculated with B_{ext} is equal to the ion inertial length: $\sqrt{2T_i m_i} c/e B_{ext} = \sqrt{m_i/4\pi n_i} c/e = c/\omega_{pi}$ where m_i is the ion mass.

Value of B_{ext} is determined at the boundary of the plasma sheet where the plasma pressure vanishes. However, the thickness of TCS could be substantially smaller than the thickness of the plasma sheet, because at the TCS boundary the B_x magnetic field could reach for example only an amplitude $B_0 \sim (1/2) B_{ext}$ (Artemyev et al. 2010; Petrukovich et al. 2011). This property is called TCS embedding into the plasma sheet. Thus, the TCS thickness is of the order of ion Larmor radius ρ_i calculated with the field B_0 at the TCS boundary and $\rho_i = (B_{ext}/B_0)(c/\omega_{pi})$. The effect of embedding is important for the generation of various plasma instabilities (Burkhart et al. 1992; Sitnov et al. 2004; Zelenyi et al. 2008, 2010). Moreover, this effect allows to create the strong maximum of the cross-tail current density without a comparable maximum in the plasma density profile, i.e. TCS can be supported by a relatively small population of particles at open orbits (the density of this population could be about only 10 % of the total plasma density, see Artemyev et al. 2010).

Ions are demagnetized in TCS due to the small transverse scale of such sheets, while electrons still magnetized (the Larmor radius of thermal electrons is even much smaller than the small curvature radius of magnetic field lines). Thus, we need to deal with the separation of ion and electron motions in TCS. This separation results in the appearance of electric fields, which are required to support the quasi-neutrality in the system (Zelenyi et al. 2011a; Schindler et al. 2012). This transverse electric field creates the cross-field drift $\sim \mathbf{E} \times \mathbf{B}$, which diminishes the ion drift velocity and increases the electron drift velocity. Roughly speaking, ions create the current due to the openness of their orbits, but the current is redistributed to electrons via the generation of a cross-field drift $\mathbf{E} \times \mathbf{B}$. These ambipolar effects are controlled by electron temperature T_e and are vanishing completely for $T_e \to 0$. The total current density is not influenced by $\mathbf{E} \times \mathbf{B}$ drifts, but these drifts can substantially increase the amplitudes of partial electron currents registered *in situ* (Artemyev et al. 2011a). This consideration explains some puzzling observations when spacecraft measurements demonstrate the dominance of electron cross-tail currents (Asano et al. 2004; Israelevich et al. 2008; Artemyev et al. 2009).

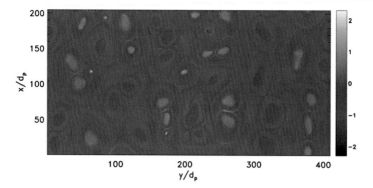

Fig. 10 Figure shows intensity of the current density (out of plane) in the system with developed tearing mode. Figure is adapted from Drake et al. (2010)

4 Dynamics of Current Sheets

TCS in hot plasma represents a reservoir of a free energy for various dynamic processes including a multitude of plasma instabilities (e.g., Lui 2004). The nonlinear processes of saturation and interaction of these modes give rise to the generation of turbulent electromagnetic fields (Zelenyi and Milovanov 2004). This concept of turbulence generation allows to describe some principal features of observed magnetic field fluctuations. On the basis of their macroscopic consequences for the CS structure, all modes of instabilities can be separated into two main branches: (1) various drift instabilities resulting in global oscillations and deformations of TCS structure as a whole and (2) tearing modes corresponding to the destruction of magnetic surfaces and modification of the inner TCS structure. We postpone the discussion of the large-scale TCS oscillations to the end of this section and concentrate now on properties of perturbations resulting in the violation of the magnetic field topology.

For the simple one-dimensional TCS without a B_z component of the magnetic field (or with a very weak one) the tearing mode growth is provided by the resonant interaction of electromagnetic field perturbations with electrons (Laval et al. 1966) or ions (Schindler 1974; Galeev and Zelenyi 1976). This instability results in the growth of magnetic field perturbations producing the reconnection of magnetic field lines due to the emerging B_z component. If the initial TCS configuration includes finite B_y (shear) component, then the reconnection of field lines is supplemented by the overlapping and destruction of magnetic surfaces at periphery (Galeev and Zelenyi 1977; Galeev et al. 1986). The same effects have been obtained for oblique tearing modes (Wiegelmann and Büchner 2000; Lapenta and Brackbill 2000; Zelenyi et al. 2009). Finally, at the nonlinear stage of the tearing mode the neutral plane of TCS looses its regular structure and becomes filled by small-scale ($\sim \rho_i$) magnetic islands (see results of numerical modelling in Fig. 10).

Important components of nonlinear CS dynamics correspond to the magnetic island merging and the current filament disruption (Pritchett and Wu 1979; Biskamp and Welter 1980; Zelenyi and Taktakishvili 1988). For example, double-slope of spectra (see Figs. 1, 5, 8) with the kink around ~ 0.04 Hz can be obtained from the basic geometrical model of the multiple tearing mode excitation and evolution in TCS (Milovanov et al. 1996). This self-consistent model is based on the following assumptions: magnetic field fluctuations δB_z result in particle scattering and corresponding diffusion in the neutral plane; this diffusion produces the effective conductivity σ; due to the presence of dawn-dusk electric field E_y the presence of this effective conductivity allows to estimate the net current density $\langle j_y \rangle \sim \sigma E_y$; the current density $\langle j_y \rangle$ supports the main magnetic field component $\langle j_y \rangle \sim (c/4\pi)B_0/L$ where L is the thickness of TCS. At the same time fluctuations of current δj_y are self-consistently related (by means of Maxwell equations) to magnetic fluctuations δB_z, which closed the logical loop and provided the qualitative estimates of the basic properties of fluctuations. All these assumptions allow to estimate the level of magnetic field fluctuations $\delta B_z/B_0 \sim \sqrt{T_i \rho_0/e E_y L^2}$ within the range of spatial scales $\in [\sqrt{\rho_0 L}, 2\pi L]$ where ρ_0 is the Larmor radius of thermal ions (with the temperature T_i) in the magnetic field B_0. For reasonable values of system parameters one can get $\delta B_z/B_0 \sim 0.1$ for the spatial scale range $\in [400\text{ km}, 8000\text{ km}]$ (Milovanov et al. 1996). Thus, the spectrum of magnetic field fluctuations should have two kinks at $f \sim v_{flow}/8000\text{ km} \sim 0.03$ Hz and $f \sim v_{flow}/400\text{ km} \sim 0.5$ Hz. The first kink (~ 0.03 Hz) is well seen in Figs. 1, 5, 8. The investigation of the second kink requires long enough intervals with sufficiently high-level of magnetic field fluctuations. Thus, this problem is still under consideration. For a dynamic system where variations of magnetic field produce the inductive electric fields $\sim \partial \delta B_z/\partial t \neq 0$, corresponding particle acceleration results in the formation of a power-law particle energy distribution. Slope of the energy distribution can be obtained from the same assumptions of the self-consistency of the time-variable magnetic field fluctuations and currents of accelerated particles scattered and accelerated by these fluctuations (Milovanov and Zelenyi 2001). Comparison of slopes estimated from such model with the spacecraft observations demonstrates quite reasonable agreement between them (Milovanov and Zelenyi 2002).

Besides the nonlinear evolution of tearing modes resulting in formation of magnetic islands, various secondary instabilities can develop in the system with reconnected TCS. The tearing mode growth gives rise to the gradients of the B_z component of the magnetic field along the x direction. For the transient spontaneous reconnection such gradients could become quite strong (see, e.g., Sitnov et al. 2009). Several electromagnetic modes can be excited in the system with $\partial B_z/\partial x \neq 0$. First of all, the kinetic ballooning/interchange mode becomes unstable in the region with the strong longitudinal gradients (Hurricane et al. 1999; Pritchett and Coroniti 2011). This instability perturbs B_z field and results in appearance of wave-like B_z variations with the time-scale $\sim 1/\Omega_i$ (Pritchett and Coroniti 2010; Vapirev et al. 2013) where Ω_i is the ion Larmor frequency (this mode can contribute to magnetic field fluctuations with $f \in [10^{-1}, 1]$ Hz). Moreover, strong transverse gradients of the magnetic field represent also a free energy source for the lower-hybrid drift instability (Huba et al. 1977; Daughton 2003). This mode can excite parallel magnetic field fluctuations with the time-scales $\sim \sqrt{m_e/m_i}/\Omega_i$ ($f \sim 1$–10 Hz).

For the frequency range $f > v_{flow}/\rho_i$ the wave activity in the Earth magnetotail is often described in terms of the kinetic Alfven wave model (e.g., Chaston et al. 2009). This mode for hot plasma with $T_i \gg T_e$ has simple relation between perturbations of electric and magnetic fields: $E_1/B_1 \approx (v_A/c)k_\perp \rho_i$ (Stasiewicz et al. 2000). Thus, observations of the growth of the ratio E_1/B_1 with $k_\perp$ indicate on the substantial contribution of this mode to a turbulent spectrum (e.g., Chaston et al. 2012). However, we can underline here the analogy between the polarization of the tearing mode in TCS and the polarization of kinetic Alfven waves. Perturbations of the electric field for the tearing mode correspond to the separation of motions of unmagnetized ions and magnetized electrons (the so-called effect of the electron compressibility, see Schindler 2006, and references therein) and can be described by estimates $E_{1x} \approx k_x \phi_1 \approx k_x n_{i1} T_i/n_i e \approx k_x B_{z1} T_i/B_z e$ where $n_{i1} \approx n_{e1}$ are perturbations of ion and electron densities, B_{1z} is perturbation of B_z component of the magnetic field (e.g., Galeev and Zelenyi 1976; Coroniti 1980; Lembege and Pellat 1982). Finally, we get $E_{1x}/B_{1z} \approx k_x \tilde{\rho}_i v_A^2/c v_{Ti}$ where v_{Ti} is the thermal velocity of ions and $\tilde{\rho}_i$ is their Larmor radius in the vicinity of the neutral plane $z = 0$.

The tearing mode is unstable for $k_x L > 1$ (Galeev and Sudan 1985), i.e. the corresponding frequency range is $\omega = k_x v_{flow} > v_{flow}/\rho_i$ where $L \sim \rho_i$ for TCS. Therefore, the tearing mode can describe the growth of the ratio E_1/B_1 with the increase of $k_\perp$ for the same frequency range as kinetic Alfven waves. Thus, the interpretation of observed electromagnetic field fluctuations given by Chaston et al. (2009) can be substituted by the alternative interpretation which takes into account the tearing mode instead of kinetic Alfven waves.

TCS can be considered as a plasma layer with the decoupling of ion and electron perpendicular flows moving in the dawn-dusk directions. Moreover, TCS is always embedded into the background plasma sheet. Thus, there is the relative motion of ions belonging to the background and current carrying components. Both these effects excite the large-scale kink mode, which develops as a wavy deformation of the current layer (Lapenta and Brackbill 1997; Daughton 1999; Karimabadi et al. 2003). The time-scale of magnetic field perturbations produced by the kink mode is larger than 100 s (for magnetotail conditions). These perturbations are usually detected in the magnetotail as the flapping motions of TCS (Zhang et al. 2002). Corresponding vertical quasi-periodic oscillations of TCS (Sergeev et al. 2006; Petrukovich et al. 2006) are contributing to the lower-frequency range of spectra (Bauer et al. 1995a). In addition to the drift kink modes, such oscillations can be exited by so-called double-gradient instability of TCS produced by a weak gradient $\partial B_z/\partial x$ (Erkaev et al. 2007; Korovinskiy et al. 2013). Corresponding timescales of TCS motion due to this mode are also about $\sim$100 s.

5 Plasma Convection

Current sheets (CSs) could be formed only in sufficiently hot plasmas where the thermal pressure of particles can counterbalance the excessive magnetic field pressure. Therefore, some mechanism is required for heating (or acceleration) of the originally cold solar wind plasma. This problem cannot be solved solely by the plasma energization at the planetary bow shock. In planetary magnetotails there exists another powerful mechanism of plasma heating controlled by a global magnetotail convection. The ultimate source of energy for this process is the relative motion of the solar wind flow and planetary magnetosphere (some analog of homopolar generator) producing charge separation at magnetospheric flanks and correspondingly global convective electric field between these flanks (Dungey 1961, 1963). Such interaction forms two large-scale null-points where magnetic field lines of the solar wind reconnect with planetary magnetic field lines (see Fig. 11). Solar wind plasmas penetrate into the magnetosphere through the distant null-point (X-line). Presence of the convection electric field results in the plasma motion inside the magnetosphere towards the planet (the motion, opposite to the solar wind flow outside the magnetosphere). Therefore, presence of a finite planetward plasma flow velocity is the intrinsic property of the interaction of the planetary magnetic field with the supersonic solar wind plasma flow.

Although, the energy of plasma flows ($\sim v_{flow}^2 m n_i/2$) in hot plasmas (inside the magnetosphere) is smaller than the thermal plasma energy, existence of the plasma bulk motion is very important for the formation of turbulent spectra. In planetary magnetotails the bulk flow is either originated from a local magnetic reconnection (sporadic activity) or supported by the global convection even at geomagnetically quiet times (Kan 1990; Angelopoulos et al. 1993). Velocities of plasma flows v_{flow} are usually within the wide range 10 km/s–1000 km/s. Larger values of $v_{flow} > 100$ km/s correspond to short-time intervals of bursty bulk flows originated from the local reconnection regions (Baumjohann et al. 1990), while the steady convection is characterized by smaller amplitudes of v_{flow} (Hori

Fig. 11 Schematic picture of the solar wind interaction with the planetary magnetic field. Figure is adapted from Dungey (1961)

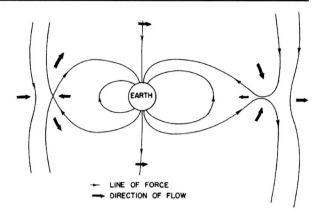

- LINE OF FORCE
- DIRECTION OF FLOW

et al. 2000; Wang et al. 2009; Kissinger et al. 2012). Therefore, the v_{flow} value criterion is often used to demarcate the active and quiet conditions in the magnetotail: $v_{flow} < 100$ km/s for the quiet time magnetotail and $v_{flow} > 100$ km/s for the disturbed one.

The intensification of the wave-activity with the increase of the plasma flow amplitude v_{flow} was revealed for electrostatic fluctuations by Baumjohann et al. (1989) and for magnetic fluctuations by Bauer et al. (1995b). Using statistics collected by AMPTE/IRM, Bauer et al. (1995b) have shown that in the central region of the plasma sheet the intensity of magnetic field fluctuations (within the frequency range $f > 0.03$ Hz) grows rapidly with the increase of the variation levels of the plasma density and temperature. Moreover, the fluctuation intensity is higher in the region with higher plasma β. The similar increase of the intensity of magnetic field fluctuations (two orders of magnitude) inside the strong plasma flows was found by Eastwood et al. (2009). The dispersion relation derived from four spacecraft measurements indicates that the frequency of fluctuations is about $v_A k_x$, where $v_A \sim v_{flow}$ and k_x is a wavenumber in the direction along the plasma flow (Eastwood et al. 2009).

The presence of plasma flows results in a transformation of the spatial spectrum of magnetic field structures into the frequency spectrum due to the Doppler effect $f = v_{flow}/\ell$ where ℓ is spatial scale of plasma structures. Therefore, for larger v_{flow} we should observe the increase of fluctuations in the frequency range corresponding to spatial scales $\ell \sim \rho_i$ (typical TCS transverse scale). This range can be defined as $f \in [10^{-2}, 1]$ Hz for $v_{flow} \sim 10$–1000 km/s and $\rho_i \sim 100$–1000 km. Magnetic field fluctuations within such frequency range are provided by the convection of current sheet structures frozen into the convecting plasma flow. Indeed, Figs. 1, 3, 6, 8 obtained in different experiments remarkably well demonstrate the increase of spectra density for $f \in [10^{-2}, 1]$ Hz with the increase of v_{flow}.

To support our conclusion concerning the convective nature of the increase of the fluctuation intensity with the growth of the flow velocity we refer to the study by Volwerk et al. (2003). In this paper the authors indicated the distinct increase of the power spectra density of magnetic field fluctuations with the increase of the plasma bulk velocity (see Fig. 12(a)). Moreover, it was shown that variations of the spectral index in the chosen frequency range (0.08–1.0 Hz) are very weak, and there is no clear dependence of the spectral index on the plasma bulk velocity(see Fig. 12(b)). Thus, we can conclude that the entire spectrum of magnetic field fluctuations is simply shifted by convection towards the higher frequency range. In this case, for the fixed frequency range we should observe the increase of the power density effectively coming from the lower frequencies, but the spectral index should remain more or less the same, because there is not real modification of the spectrum except its shift.

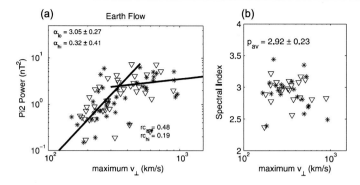

Fig. 12 Panel (**a**) shows the total power in the Pi2 band (8–25 mHz) for Cluster 1 (*stars*) and Cluster 3 (*triangles*) as a function of the amplitude value of the plasma velocity in the Earthward flow. Here α_{lo} and α_{hi} are the slopes of the fits for the low and high velocity ranges, respectively, and accordingly rc_{lo} and rc_{hi} are goodness of fit coefficients for these ranges. Panel (**b**) shows the spectral index for the frequency range 0.08–1.0 Hz as a function of $v_{\perp \max}$. The average spectral index p with its standard deviation is shown. Figure is adapted from Volwerk et al. (2003)

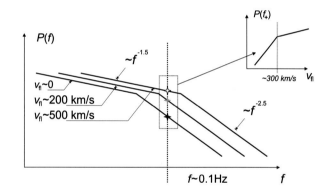

Fig. 13 Schematic view of the spectrum modification with the plasma bulk velocity

We illustrate schematically this process of the spectrum modification with the plasma bulk velocity in Fig. 13. The increase of fluctuation intensity with v_{flow} should have two scales—for low v_{flow} and for large v_{flow}. Such double dependence comes from the existence of the knee in the shape of fluctuation spectra. For low enough values of v_{flow} we observe the same part of the spectrum. Thus, the dependence of the corresponding increase of fluctuation intensity with v_{flow} should have a certain slope in a double logarithmic scales. For larger v_{flow} we already meet the low-frequency part of the spectrum and this change produces the change of the slope in the dependence of the fluctuation intensity on v_{flow} (see Fig. 13). The same behavior can be recognized in experimental data—see Fig. 12(a). This explanation assumes the direct link between the kink in the frequency spectra $P(f)$ and the kink in the dependence of $P(f_*)$ on v_{flow} for some chosen f_*. At the same time, we do not expect to observe a strong variation of the spectral index with v_{flow}. To calculate the spectral index one should consider certain frequency range. Thus, the increase of v_{flow} should result eventually in the mixing of two spectral slopes and one should obtain the value of spectral index in the range between indices typical for low and high-frequency parts of spectrum. This new index should be smaller than one calculated in absence of plasma flow, but still larger than the one typical for the low-frequency range. Indeed, we observe in experimental

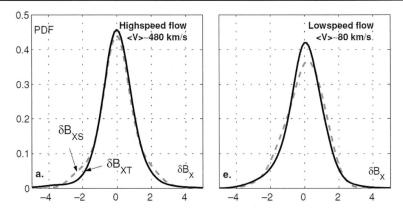

Fig. 14 Probability density functions estimated from spatial fluctuations δB_{XS} between Cluster spacecraft 1 and 4 (the distance between spacecraft is $\sim$1700 km), and from temporal time-delayed fluctuations, δB_{XT}. Figure is adapted from Vörös et al. (2006)

data the decrease of the spectral index with the increase of v_{flow}, but this variation is rather weak (Fig. 12(b)).

An additional test of the convective nature of magnetic field fluctuations in specific frequency range was presented by Vörös et al. (2006). In this paper the probability density distribution of magnetic field fluctuations was determined by two methods: as a difference of the magnetic field measured by two spacecraft δB_{XT} and as fluctuations of the magnetic field observed at one spacecraft with time delay. This time delay was calculated as a ratio of spacecraft separation (along the direction of the plasma flow) and the plasma flow velocity. Figure 14 shows that two distributions are similar for the high-speed flow when fluctuations measured by one spacecraft are convected to another one without substantial intrinsic dynamics. In contrast, for the low-speed flow two spacecraft measure local (different) distributions of fluctuations.

6 Discussion and Conclusions

In this paper we discuss the formation of the spectrum of magnetic field fluctuations in a hot ($\beta \geq 1$) plasma. The proposed synthetic scenario is based on the key element of such system—thin current sheet (TCS). This scenario assumes that (1) TCSs with ion scales appear in hot plasma as self-consistent structures, (2) the drift kink modes and/or double-gradient modes contribute to a low-frequency part of spectra $f < 10^{-2}$ Hz, (3) the tearing and/or ballooning instabilities in TCS result in the filamentation of currents and the magnetic islands formation with the corresponding contribution to fluctuations at $f \sim 10^{-1}$ Hz, (4) the current filamentation results in appearance of a smaller-scale strong magnetic gradients which initiate secondary instabilities contributing to the high-frequency part of spectra $f > 10^{-1}$ Hz, (5) finally, the motion of TCS structures due to the convection shifts all spectral profiles towards higher frequencies and contributes to spectra of magnetic field fluctuations within $f \sim 10^{-2}$–5×10^{-1} Hz (mainly where Doppler shift is most effective). The schematic view of this scenario is shown in Fig. 15.

It is interesting to note that the slope of spectra of magnetic field fluctuations for low frequency range is often about -1. This is so-called flicker noise spectrum (or pink spectrum)

Fig. 15 Schematic view of the formation of the spectrum of magnetic field fluctuations

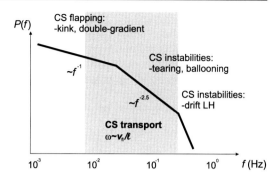

(e.g, Bochkov and Kuzovlev 1983). The same spectrum of fluctuations is found in interplanetary space (e.g., Matthaeus and Goldstein 1986). The formation of such fluctuations can be described by the model of uncorrelated processes contributing to the spectrum (e.g., Zelenyi and Milovanov 2004). For the Earth magnetotail Vörös et al. (2007) suggested that magnetic fluctuations with $1/f$ spectrum are associated with multiple bursty flows which represent completely uncorrelated events. The same spectrum could be provided by uncorrelated drivers of CS flapping oscillations: kink/double-gradient instabilities (Erkaev et al. 2007; Karimabadi et al. 2003; Zelenyi et al. 2009), plasma flows (Gubchenko 1985; Vörös et al. 2009; Erkaev et al. 2009), variations of the solar wind pressure (Sergeev et al. 2008; Gordeev et al. 2012). The nature of this universality of the low-frequency part of fluctuation spectra is still not fully understood and requires further theoretical efforts.

Although properties of the cold solar wind plasma with the strong flow ($M_A > 5$) are quite different from the properties of the hot plasma within the magnetosphere, some similarities of the patterns of the formation of magnetic fluctuation spectra can be found for these two plasma regimes (e.g., Vörös 2011). The solar wind carries with it magnetic field fluctuations with spectra similar to the one shown in Fig. 1: there is an evident difference of spectrum power law indices (around the local ion Larmor frequency) for low-frequency and high-frequency parts (e.g., Sahraoui et al. 2009; Alexandrova et al. 2009). However, due to smallness of particle temperatures in the solar wind (while plasma beta is $\beta \sim 1$ due to a weak magnetic field) the fast transient reconnection of current layers with the corresponding formation of fast flows seems to be not so important in the solar wind plasma. Thus, the convection of small-scale structures (like CSs) is dominant process there. To investigate fluctuations of magnetic field in the solar wind one can study density fluctuations (Chen et al. 2012) and plasma anisotropy (Bale et al. 2009). Magnetic field fluctuations in solar wind are often accompanied by strong density fluctuations (Šafránková et al. 2013, and reference therein). Therefore, the incompressible whistler turbulence cannot be responsible for the formation of fluctuation spectra alone (Chen et al. 2013). As an alternative source of the turbulence the Alfven waves are considered (Šafránková et al. 2013; Chen et al. 2013). However, the same correlation of magnetic field fluctuations and plasma density variations is the intrinsic property of any mode of TCS dynamics where the pressure balance is established. Thus, correlation of the plasma density and magnetic field fluctuations can be associated with TCS dynamics. To support our conclusions about the important role played by TCS in the formation of magnetic field fluctuation spectra in the solar wind we can mention several recent publications showing that TCSs are intrinsic structures of the classical MHD turbulence believed to be observed in the solar wind (Servidio et al. 2011a, 2011b; Zhdankin et al. 2013).

We described in this review peculiarities of the spectra of magnetic field fluctuations in hot plasmas of the Earth magnetotail obtained by numerous *in situ* spacecraft measurements.

We have shown that an important modification of these spectra comes from the convection motion of current sheets with bulk velocities up to ~1000 km/s. This effect explains the observed increase of a power-density of magnetic field fluctuations with the intensification of plasma flows. We speculate how small-scale kinetic instabilities of TCS can produce magnetic field structures with spatial scales smaller than 1000 km/s, while large-scale MHD modes of sheet oscillations can contribute to a lower frequency part of spectra. Thus, dynamics of TCSs are responsible for many observed features of magnetic field fluctuations.

Acknowledgements   Work was partially supported by Programm 22 of the Presidium of the Russian Academy of Sciences.

References

P. Abry, P. Flandrin, M.S. Taqqu, D. Veitch, Wavelets for the analysis, estimation and synthesis of scaling data, in *Self-Similar Network Traffic and Performance Evaluation*, ed. by K. Park, W. Willinger, V.S. Semenov (Wiley (Interscience Division), New York, 2000), pp. 39–88

O. Alexandrova, J. Saur, C. Lacombe, A. Mangeney, J. Mitchell, S.J. Schwartz, P. Robert, Universality of solar-wind turbulent spectrum from MHD to electron scales. Phys. Rev. Lett. **103**(16), 165003 (2009). doi:10.1103/PhysRevLett.103.165003

V. Angelopoulos, C.F. Kennel, F.V. Coroniti, R. Pellat, H.E. Spence, M.G. Kivelson, R.J. Walker, W. Baumjohann, W.C. Feldman, J.T. Gosling, Characteristics of ion flow in the quiet state of the inner plasma sheet. Geophys. Res. Lett. **20**, 1711–1714 (1993). doi:10.1029/93GL00847

J. Arons, Pulsar wind nebulae as cosmic pevatrons: a current sheet's tale. Space Sci. Rev. **173**, 341–367 (2012). doi:10.1007/s11214-012-9885-1

A.V. Artemyev, L.M. Zelenyi, Kinetic structure of current sheets in the Earth magnetotail. Space Sci. Rev. (2013). doi:10.1007/s11214-012-9954-5

A.V. Artemyev, A.A. Petrukovich, L.M. Zelenyi, H.V. Malova, V.Y. Popov, R. Nakamura, A. Runov, S. Apatenkov, Comparison of multi-point measurements of current sheet structure and analytical models. Ann. Geophys. **26**, 2749–2758 (2008)

A.V. Artemyev, A.A. Petrukovich, L.M. Zelenyi, R. Nakamura, H.V. Malova, V.Y. Popov, Thin embedded current sheets: cluster observations of ion kinetic structure and analytical models. Ann. Geophys. **27**, 4075–4087 (2009)

A.V. Artemyev, A.A. Petrukovich, R. Nakamura, L.M. Zelenyi, Proton velocity distribution in thin current sheets: cluster observations and theory of transient trajectories. J. Geophys. Res. **115**, 12255 (2010). doi:10.1029/2010JA015702

A.V. Artemyev, A.A. Petrukovich, R. Nakamura, L.M. Zelenyi, Cluster statistics of thin current sheets in the Earth magnetotail: specifics of the dawn flank, proton temperature profiles and electrostatic effects. J. Geophys. Res. **116**, 0923 (2011a). doi:10.1029/2011JA016801

A.V. Artemyev, W. Baumjohann, A.A. Petrukovich, R. Nakamura, I. Dandouras, A. Fazakerley, Proton/electron temperature ratio in the magnetotail. Ann. Geophys. **29**, 2253–2257 (2011b). doi:10.5194/angeo-29-2253-2011

A.V. Artemyev, A.A. Petrukovich, A.G. Frank, R. Nakamura, L.M. Zelenyi, Intense current sheets in the magnetotail: peculiarities of electron physics. J. Geophys. Res. **118**, 2789–2799 (2013). doi:10.1002/jgra.50297

Y. Asano, T. Mukai, M. Hoshino, Y. Saito, H. Hayakawa, T. Nagai, Statistical study of thin current sheet evolution around substorm onset. J. Geophys. Res. **109**, 5213 (2004). doi:10.1029/2004JA010413

S.D. Bale, J.C. Kasper, G.G. Howes, E. Quataert, C. Salem, D. Sundkvist, Magnetic fluctuation power near proton temperature anisotropy instability thresholds in the solar wind. Phys. Rev. Lett. **103**(21), 211101 (2009). doi:10.1103/PhysRevLett.103.211101

T.M. Bauer, W. Baumjohann, R.A. Treumann, Neutral sheet oscillations at substorm onset. J. Geophys. Res. **100**, 23737–23742 (1995a). doi:10.1029/95JA02448

T.M. Bauer, W. Baumjohann, R.A. Treumann, N. Sckopke, H. Lühr, Low-frequency waves in the near-Earth plasma sheet. J. Geophys. Res. **100**, 9605–9618 (1995b). doi:10.1029/95JA00136

W. Baumjohann, G. Paschmann, C.A. Cattell, Average plasma properties in the central plasma sheet. J. Geophys. Res. **94**, 6597–6606 (1989). doi:10.1029/JA094iA06p06597

W. Baumjohann, G. Paschmann, H. Luehr, Characteristics of high-speed ion flows in the plasma sheet. J. Geophys. Res. **95**, 3801–3809 (1990). doi:10.1029/JA095iA04p03801

W. Baumjohann, R.A. Treumann, J. Labelle, R.R. Anderson, Average electric wave spectra across the plasma sheet and their relation to ion bulk speed. J. Geophys. Res. **94**, 15221–15230 (1989). doi:10.1029/JA094iA11p15221

W. Baumjohann, A. Roux, O. Le Contel, R. Nakamura, J. Birn, M. Hoshino, A.T.Y. Lui, C.J. Owen, J. Sauvaud, A. Vaivads, D. Fontaine, A. Runov, Dynamics of thin current sheets: cluster observations. Ann. Geophys. **25**, 1365–1389 (2007)

J. Birn, E.R. Priest, *Reconnection of Magnetic Fields: Magnetohydrodynamics and Collisionless Theory and Observations* (Cambridge University Press, Cambridge, 2007)

D. Biskamp, H. Welter, Coalescence of magnetic islands. Phys. Rev. Lett. **44**, 1069–1072 (1980). doi:10.1103/PhysRevLett.44.1069

G.N. Bochkov, Y.E. Kuzovlev, New aspects in $1/f$ noise studies. Sov. Phys. Usp. **26**, 829–844 (1983). doi:10.1070/PU1983v026n09ABEH004497

J.E. Borovsky, H.O. Funsten, MHD turbulence in the Earth's plasma sheet: dynamics, dissipation, and driving. J. Geophys. Res. **108**, 1284 (2003). doi:10.1029/2002JA009625

J. Büchner, L.M. Zelenyi, Regular and chaotic charged particle motion in magnetotaillike field reversals. I—Basic theory of trapped motion. J. Geophys. Res. **94**, 11821–11842 (1989). doi:10.1029/JA094iA09p11821

D. Burgess, M. Scholer, Microphysics of quasi-parallel shocks in collisionless plasmas. Space Sci. Rev. (2013). doi:10.1007/s11214-013-9969-6

G.R. Burkhart, J.F. Drake, P.B. Dusenbery, T.W. Speiser, Ion tearing in a magnetotail configuration with an embedded thin current sheet. J. Geophys. Res. **97**, 16749–16756 (1992). doi:10.1029/92JA01523

C.A. Cattell, F.S. Mozer, Electric fields measured by ISEE-1 within and near the neutral sheet during quiet and active times. Geophys. Res. Lett. **9**, 1041–1044 (1982). doi:10.1029/GL009i009p01041

C.A. Cattell, F.S. Mozer, E.W. Hones Jr., R.R. Anderson, R.D. Sharp, ISEE observations of the plasma sheet boundary, plasma sheet, and neutral sheet. I—Electric field, magnetic field, plasma, and ion composition. J. Geophys. Res. **91**, 5663–5688 (1986). doi:10.1029/JA091iA05p05663

C.C. Chaston, J.R. Johnson, M. Wilber, M. Acuna, M.L. Goldstein, H. Reme, Kinetic Alfvén wave turbulence and transport through a reconnection diffusion region. Phys. Rev. Lett. **102**(1), 015001 (2009). doi:10.1103/PhysRevLett.102.015001

C.C. Chaston, J.W. Bonnell, L. Clausen, V. Angelopoulos, Energy transport by kinetic-scale electromagnetic waves in fast plasma sheet flows. J. Geophys. Res. **117**, 9202 (2012). doi:10.1029/2012JA017863

C.H.K. Chen, C.S. Salem, J.W. Bonnell, F.S. Mozer, S.D. Bale, Density fluctuation spectrum of solar wind turbulence between ion and electron scales. Phys. Rev. Lett. **109**(3), 035001 (2012). doi:10.1103/PhysRevLett.109.035001

C.H.K. Chen, S. Boldyrev, Q. Xia, J.C. Perez, Nature of subproton scale turbulence in the solar wind. Phys. Rev. Lett. **110**(22), 225002 (2013). doi:10.1103/PhysRevLett.110.225002

G. Consolini, A.T.Y. Lui, Sign-singularity analysis of current disruption. Geophys. Res. Lett. **26**, 1673–1676 (1999). doi:10.1029/1999GL900355

F.V. Coroniti, On the tearing mode in quasi-neutral sheets. J. Geophys. Res. **85**, 6719–6728 (1980). doi:10.1029/JA085iA12p06719

F.V. Coroniti, F.L. Scarf, L.A. Frank, D.J. Williams, R.P. Lepping, S.M. Krimigis, G. Gloeckler, Variability of plasma sheet dynamics. J. Geophys. Res. **85**, 2957–2977 (1980). doi:10.1029/JA085iA06p02957

W. Daughton, The unstable eigenmodes of a neutral sheet. Phys. Plasmas **6**, 1329–1343 (1999). doi:10.1063/1.873374

W. Daughton, Electromagnetic properties of the lower-hybrid drift instability in a thin current sheet. Phys. Plasmas **10**, 3103–3119 (2003). doi:10.1063/1.1594724

J.F. Drake, M. Opher, M. Swisdak, J.N. Chamoun, A magnetic reconnection mechanism for the generation of anomalous cosmic rays. Astrophys. J. **709**, 963–974 (2010). doi:10.1088/0004-637X/709/2/963

T. Dudok de Wit, Numerical schemes for the analysis of turbulence—a tutorial, in *Space Plasma Simulation*, ed. by J. Büchner, C. Dum, M. Scholer. Lecture Notes in Physics, vol. 615 (Springer, Berlin, 2003), pp. 315–343

T. Dudok de Wit, O. Alexandrova, I. Furno, L. Sorriso-Valvo, G. Zimbardo, Methods for characterising microphysical processes in plasmas. Space Sci. Rev. (2013). doi:10.1007/s11214-013-9974-9

J.W. Dungey, Interplanetary magnetic field and the auroral zones. Phys. Rev. Lett. **6**, 47–48 (1961). doi:10.1103/PhysRevLett.6.47

J.W. Dungey, Interactions of solar plasma with the geomagnetic field. Planet. Space Sci. **10**, 233–237 (1963). doi:10.1016/0032-0633(63)90020-5

J.P. Eastwood, T.D. Phan, S.D. Bale, A. Tjulin, Observations of turbulence generated by magnetic reconnection. Phys. Rev. Lett. **102**(3), 035001 (2009). doi:10.1103/PhysRevLett.102.035001

J.P. Eastwood, T.D. Phan, M. Øieroset, M.A. Shay, Average properties of the magnetic reconnection ion diffusion region in the Earth's magnetotail: the 2001–2005 cluster observations and comparison with simulations. J. Geophys. Res. **115**, 8215 (2010). doi:10.1029/2009JA014962

J.W. Eastwood, Consistency of fields and particle motion in the 'Speiser' model of the current sheet. Planet. Space Sci. **20**, 1555–1568 (1972). doi:10.1016/0032-0633(72)90182-1

N.V. Erkaev, V.S. Semenov, H.K. Biernat, Magnetic double-gradient instability and flapping waves in a current sheet. Phys. Rev. Lett. **99**(23), 235003 (2007). doi:10.1103/PhysRevLett.99.235003

N.V. Erkaev, V.S. Semenov, I.V. Kubyshkin, M.V. Kubyshkina, H.K. Biernat, MHD model of the flapping motions in the magnetotail current sheet. J. Geophys. Res. **114**, 3206 (2009). doi:10.1029/2008JA013728

C.P. Escoubet, M.G.G.T. Taylor, A. Masson, H. Laakso, J. Volpp, M. Hapgood, M.L. Goldstein, Dynamical processes in space: cluster results. Ann. Geophys. **31**, 1045–1059 (2013). doi:10.5194/angeo-31-1045-2013

P. Francfort, R. Pellat, Magnetic merging in collisionless plasmas. Geophys. Res. Lett. **3**, 433–436 (1976). doi:10.1029/GL003i008p00433

A.G. Frank, Dynamics of current sheets underlying flare-type events in magnetized plasmas. Phys. Usp. **53**, 941–947 (2010). doi:10.3367/UFNe.0180.201009h.0982

D.A. Frank-Kamenetskii, *Course on Plasma Physics* (Atomizdat, Moscow, 1968)

A.A. Galeev, R.N. Sudan, *Handbook of Plasma Physics. Vol. 2: Basic Plasma Physics Ii* (North-Holland, Amsterdam, 1985)

A.A. Galeev, L.M. Zelenyi, Tearing instability in plasma configurations. Sov. Phys. JETP **43**, 1113 (1976)

A.A. Galeev, L.M. Zelenyi, Model of magnetic-field reconnection in a plane layer of collisionless plasma. JETP Lett. **25**, 380 (1977)

A.A. Galeev, M.M. Kuznetsova, L.M. Zelenyi, Magnetopause stability threshold for patchy reconnection. Space Sci. Rev. **44**, 1–41 (1986). doi:10.1007/BF00227227

H.B. Garrett, ULF magnetic fluctuations in the plasma sheet as recorded by the Explorer 34 satellite. J. Geophys. Res. **78**, 3799 (1973). doi:10.1029/JA078i019p03799

E. Gordeev, M. Amosova, V. Sergeev, IMF Bx effect on the magnetotail neutral sheet geometry and dynamics, in *EGU General Assembly Conference Abstracts*, ed. by A. Abbasi, N. Giesen. EGU General Assembly Conference Abstracts, vol. 14 (2012), p. 14437

V.M. Gubchenko, Stratification in a neutral current sheet with counterstreaming plasma. Sov. J. Plasma Phys. **11**, 467–476 (1985)

D.A. Gurnett, L.A. Frank, R.P. Lepping, Plasma waves in the distant magnetotail. J. Geophys. Res. **81**, 6059–6071 (1976). doi:10.1029/JA081i034p06059

T. Hori, K. Maezawa, Y. Saito, T. Mukai, Average profile of ion flow and convection electric field in the near-Earth plasma sheet. Geophys. Res. Lett. **27**, 1623–1626 (2000). doi:10.1029/1999GL003737

M. Hoshino, A. Nishida, T. Yamamoto, S. Kokubun, Turbulent magnetic field in the distant magnetotail: bottom-up process of plasmoid formation? Geophys. Res. Lett. **21**, 2935–2938 (1994). doi:10.1029/94GL02094

A. Hruška, J. Hrušková, Long time-scale magnetodynamic noise in the geomagnetic tail. Planet. Space Sci. **17**, 1497–1504 (1969). doi:10.1016/0032-0633(69)90170-6

S.Y. Huang, et al., Observations of turbulence within reconnection jet in the presence of guide field. Geophys. Res. Lett. **39**, 11104 (2012). doi:10.1029/2012GL052210

J.D. Huba, N.T. Gladd, K. Papadopoulos, The lower-hybrid-drift instability as a source of anomalous resistivity for magnetic field line reconnection. Geophys. Res. Lett. **4**, 125–126 (1977). doi:10.1029/GL004i003p00125

O.A. Hurricane, B.H. Fong, S.C. Cowley, F.V. Coroniti, C.F. Kennel, R. Pellat, Substorm detonation. J. Geophys. Res. **104**, 10221–10232 (1999). doi:10.1029/1999JA900012

P.L. Israelevich, A.I. Ershkovich, R. Oran, Current carriers in the bifurcated tail current sheet: ions or electrons? J. Geophys. Res. **113**, 4215 (2008). doi:10.1029/2007JA012541

J.R. Kan, Tail-like reconfiguration of the plasma sheet during the substorm growth phase. Geophys. Res. Lett. **17**, 2309–2312 (1990). doi:10.1029/GL017i013p02309

H. Karimabadi, P.L. Pritchett, W. Daughton, D. Krauss-Varban, Ion-ion kink instability in the magnetotail: 2. Three-dimensional full particle and hybrid simulations and comparison with observations. J. Geophys. Res. **108**, 1401 (2003). doi:10.1029/2003JA010109

J.G. Kirk, Particle acceleration in relativistic current sheets. Phys. Rev. Lett. **92**(18), 181101 (2004). doi:10.1103/PhysRevLett.92.181101

J. Kissinger, R.L. McPherron, T.-S. Hsu, V. Angelopoulos, Diversion of plasma due to high pressure in the inner magnetosphere during steady magnetospheric convection. J. Geophys. Res. **117**, 5206 (2012). doi:10.1029/2012JA017579

D.B. Korovinskiy, A. Divin, N.V. Erkaev, V.V. Ivanova, I.B. Ivanov, V.S. Semenov, G. Lapenta, S. Markidis, H.K. Biernat, M. Zellinger, MHD modeling of the double-gradient (kink) magnetic instability. J. Geophys. Res. **118**, 1146–1158 (2013). doi:10.1002/jgra.50206

V. Krasnoselskikh, M. Balikhin, S.N. Walker, S. Schwartz, D. Sundkvist, V. Lobzin, M. Gedalin, S.D. Bale, F. Mozer, J. Soucek, Y. Hobara, H. Comisel, The dynamic quasiperpendicular shock: cluster discoveries. Space Sci. Rev. (2013). doi:10.1007/s11214-013-9972-y

N.P. Kyrie, V.S. Markov, A.G. Frank, Generation of superthermal plasma flows in current sheets. JETP Lett. **95**, 14–19 (2012). doi:10.1134/S0021364012010067

G. Lapenta, J.U. Brackbill, A kinetic theory for the drift-kink instability. J. Geophys. Res. **102**, 27099–27108 (1997). doi:10.1029/97JA02140

G. Lapenta, J.U. Brackbill, 3D reconnection due to oblique modes: a simulation of Harris current sheets. Nonlinear Process. Geophys. **7**, 151–158 (2000)

G. Laval, R. Pellat, M. Vuillemin, Instabilités Électromagnétiques des Plasmas Sans Collisions (cn-21/71), in *Plasma Physics and Controlled Nuclear Fusion Research*, vol. II (1966), pp. 259–277

B. Lembege, R. Pellat, Stability of a thick two-dimensional quasineutral sheet. Phys. Fluids **25**, 1995–2004 (1982). doi:10.1063/1.863677

A.T.Y. Lui, Potential plasma instabilities for substorm expansion onsets. Space Sci. Rev. **113**, 127–206 (2004). doi:10.1023/B:SPAC.0000042942.00362.4e

B.B. Mandelbrot, *The Fractal Geometry of Nature* (1977)

W.H. Matthaeus, M.L. Goldstein, Low-frequency $1/f$ noise in the interplanetary magnetic field. Phys. Rev. Lett. **57**, 495–498 (1986). doi:10.1103/PhysRevLett.57.495

R.M. Millan, D.N. Baker, Acceleration of particles to high energies in Earth's radiation belts. Space Sci. Rev. **173**, 103–131 (2012). doi:10.1007/s11214-012-9941-x

A.V. Milovanov, L.M. Zelenyi, "Strange" Fermi processes and power-law nonthermal tails from a self-consistent fractional kinetic equation. Phys. Rev. E **64**(5), 052101 (2001). doi:10.1103/PhysRevE.64.052101

A.V. Milovanov, L.M. Zelenyi, Nonequilibrium stationary states in the earth's magnetotail: stochastic acceleration processes and nonthermal distribution functions. Adv. Space Res. **30**, 2667–2674 (2002). doi:10.1016/S0273-1177(02)80378-7

A.V. Milovanov, L.M. Zelenyi, G. Zimbardo, Fractal structures and power law spectra in the distant Earth's magnetotail. J. Geophys. Res. **101**, 19903–19910 (1996). doi:10.1029/96JA01562

R. Nakamura, W. Baumjohann, C. Mouikis, L.M. Kistler, A. Runov, M. Volwerk, Y. Asano, Z. Vörös, T.L. Zhang, B. Klecker, H. Rème, A. Balogh, Spatial scale of high-speed flows in the plasma sheet observed by cluster. Geophys. Res. Lett. **31**, 9804 (2004). doi:10.1029/2004GL019558

R. Nakamura, W. Baumjohann, Y. Asano, A. Runov, A. Balogh, C.J. Owen, A.N. Fazakerley, M. Fujimoto, B. Klecker, H. RèMe, Dynamics of thin current sheets associated with magnetotail reconnection. J. Geophys. Res. **111**, 11206 (2006a). doi:10.1029/2006JA011706

R. Nakamura, W. Baumjohann, A. Runov, Y. Asano, Thin current sheets in the magnetotail observed by cluster. Space Sci. Rev. **122**, 29–38 (2006b). doi:10.1007/s11214-006-6219-1

E. Neagu, J.E. Borovsky, M.F. Thomsen, S.P. Gary, W. Baumjohann, R.A. Treumann, Statistical survey of magnetic field and ion velocity fluctuations in the near-Earth plasma sheet: active magnetospheric particle trace Explorers/Ion release module (AMPTE/IRM) measurements. J. Geophys. Res. **107**, 1098 (2002). doi:10.1029/2001JA000318

E. Neagu, J.E. Borovsky, S.P. Gary, A.M. Jorgensen, W. Baumjohann, R.A. Treumann, Statistical survey of magnetic and velocity fluctuations in the near-Earth plasma sheet: international Sun Earth Explorer (ISEE-2) measurements. J. Geophys. Res. **110**, 5203 (2004). doi:10.1029/2004JA010448

N.F. Ness, The Earth's magnetic tail. J. Geophys. Res. **70**, 2989–3005 (1965). doi:10.1029/JZ070i013p02989

T.G. Northrop, *The Adiabatic Motion of Charged Particles* (Wiley, New York-London-Sydney, 1963)

S. Ohtani, K. Takahashi, T. Higuchi, A.T.Y. Lui, H.E. Spence, J.F. Fennell, AMPTE/CCE-SCATHA simultaneous observations of substorm-associated magnetic fluctuations. J. Geophys. Res. **103**, 4671–4682 (1998). doi:10.1029/97JA03239

E.V. Panov, R. Nakamura, W. Baumjohann, V. Angelopoulos, A.A. Petrukovich, A. Retinò, M. Volwerk, T. Takada, K.-H. Glassmeier, J.P. McFadden, D. Larson, Multiple overshoot and rebound of a bursty bulk flow. Geophys. Res. Lett. **37**, 8103 (2010). doi:10.1029/2009GL041971

G. Paschmann, S.J. Schwartz, *Issi Book on Analysis Methods for Multi-Spacecraft Data*. ESA Special Publication, vol. 449 (ESA, Noordwijk, 2000)

G. Paschmann, M. Øieroset, T. Phan, In-situ observations of reconnection in space. Space Sci. Rev. (2013). doi:10.1007/s11214-012-9957-2

A.A. Petrukovich, Low frequency magnetic fluctuations in the Earth's plasma sheet, in *Astrophysics and Space Science Library*, ed. by A.S. Sharma, P.K. Kaw. Astrophysics and Space Science Library, vol. 321 (2005), p. 145

A.A. Petrukovich, D.V. Malakhov, Variability of magnetic field spectra in the Earth's magnetotail. Nonlinear Process. Geophys. **16**, 691–698 (2009)

A.A. Petrukovich, W. Baumjohann, R. Nakamura, R. Schödel, T. Mukai, Are earthward bursty bulk flows convective or field-aligned? J. Geophys. Res. **106**, 21211–21216 (2001). doi:10.1029/2001JA900019

A.A. Petrukovich, T.L. Zhang, W. Baumjohann, R. Nakamura, A. Runov, A. Balogh, C. Carr, Oscillatory magnetic flux tube slippage in the plasma sheet. Ann. Geophys. **24**, 1695–1704 (2006)

A.A. Petrukovich, A.V. Artemyev, H.V. Malova, V.Y. Popov, R. Nakamura, L.M. Zelenyi, Embedded current sheets in the Earth magnetotail. J. Geophys. Res. **116**, 1–25 (2011). doi:10.1029/2010JA015749

F. Porcelli, D. Borgogno, F. Califano, D. Grasso, M. Ottaviani, F. Pegoraro, Recent advances in collisionless magnetic reconnection. Plasma Phys. Control. Fusion **44**, 389 (2002)

E. Priest, T. Forbes, *Magnetic Reconnection* (Cambridge University Press, Cambridge, 2000)

P.L. Pritchett, F.V. Coroniti, A kinetic ballooning/interchange instability in the magnetotail. J. Geophys. Res. **115**, 06301 (2010). doi:10.1029/2009JA014752

P.L. Pritchett, F.V. Coroniti, Plasma sheet disruption by interchange-generated flow intrusions. Geophys. Res. Lett. **381**, 10102 (2011). doi:10.1029/2011GL047527

P.L. Pritchett, C.C. Wu, Coalescence of magnetic islands. Phys. Fluids **22**, 2140–2146 (1979). doi:10.1063/1.862507

A. Runov, V.A. Sergeev, R. Nakamura, W. Baumjohann, S. Apatenkov, Y. Asano, T. Takada, M. Volwerk, Z. Vörös, T.L. Zhang, J. Sauvaud, H. Rème, A. Balogh, Local structure of the magnetotail current sheet: 2001 cluster observations. Ann. Geophys. **24**, 247–262 (2006)

A. Runov, V. Angelopoulos, X.-Z. Zhou, X.-J. Zhang, S. Li, F. Plaschke, J. Bonnell, A THEMIS multi-case study of dipolarization fronts in the magnetotail plasma sheet. J. Geophys. Res. **116**, 5216 (2011). doi:10.1029/2010JA016316

C.T. Russell, Noise in the geomagnetic tail. Planet. Space Sci. **20**, 1541 (1972). doi:10.1016/0032-0633(72)90055-4

F. Sahraoui, M.L. Goldstein, P. Robert, Y.V. Khotyaintsev, Evidence of a cascade and dissipation of solar-wind turbulence at the electron gyroscale. Phys. Rev. Lett. **102**(23), 231102 (2009). doi:10.1103/PhysRevLett.102.231102

F.L. Scarf, L.A. Frank, K.L. Ackerson, R.P. Lepping, Plasma wave turbulence at distant crossings of the plasma sheet boundaries and the neutral sheet. Geophys. Res. Lett. **1**, 189–192 (1974). doi:10.1029/GL001i005p00189

K. Schindler, A theory of the substorm mechanism. J. Geophys. Res. **79**, 2803–2810 (1974). doi:10.1029/JA079i019p02803

K. Schindler, *Physics of Space Plasma Activity* (Cambridge University Press, Cambridge, 2006). doi:10.2277/0521858976

K. Schindler, J. Birn, M. Hesse, Kinetic model of electric potentials in localized collisionless plasma structures under steady quasi-gyrotropic conditions. Phys. Plasmas **19**(8), 082904 (2012). doi:10.1063/1.4747162

V.A. Sergeev, D.A. Sormakov, S.V. Apatenkov, W. Baumjohann, R. Nakamura, A.V. Runov, T. Mukai, T. Nagai, Survey of large-amplitude flapping motions in the midtail current sheet. Ann. Geophys. **24**, 2015–2024 (2006)

V.A. Sergeev, N.A. Tsyganenko, V. Angelopoulos, Dynamical response of the magnetotail to changes of the solar wind direction: an MHD modeling perspective. Ann. Geophys. **26**, 2395–2402 (2008). doi:10.5194/angeo-26-2395-2008

S. Servidio, P. Dmitruk, A. Greco, M. Wan, S. Donato, P.A. Cassak, M.A. Shay, V. Carbone, W.H. Matthaeus, Magnetic reconnection as an element of turbulence. Nonlinear Process. Geophys. **18**, 675–695 (2011a). doi:10.5194/npg-18-675-2011

S. Servidio, A. Greco, W.H. Matthaeus, K.T. Osman, P. Dmitruk, Statistical association of discontinuities and reconnection in magnetohydrodynamic turbulence. J. Geophys. Res. **116**, 9102 (2011b). doi:10.1029/2011JA016569

A.S. Sharma, R. Nakamura, A. Runov, E.E. Grigorenko, H. Hasegawa, M. Hoshino, P. Louarn, C.J. Owen, A. Petrukovich, J. Sauvaud, V.S. Semenov, V.A. Sergeev, J.A. Slavin, B.U.-A. Sonnerup, L.M. Zelenyi, G. Fruit, S. Haaland, H. Malova, K. Snekvik, Transient and localized processes in the magnetotail: a review. Ann. Geophys. **26**, 955–1006 (2008)

M.I. Sitnov, L.M. Zelenyi, H.V. Malova, A.S. Sharma, Thin current sheet embedded within a thicker plasma sheet: self-consistent kinetic theory. J. Geophys. Res. **105**, 13029–13044 (2000). doi:10.1029/1999JA000431

M.I. Sitnov, A.T.Y. Lui, P.N. Guzdar, P.H. Yoon, Current-driven instabilities in forced current sheets. J. Geophys. Res. **109**, 3205 (2004). doi:10.1029/2003JA010123

M.I. Sitnov, M. Swisdak, P.N. Guzdar, A. Runov, Structure and dynamics of a new class of thin current sheets. J. Geophys. Res. **111**, 8204 (2006). doi:10.1029/2005JA011517

M.I. Sitnov, M. Swisdak, A.V. Divin, Dipolarization fronts as a signature of transient reconnection in the magnetotail. J. Geophys. Res. **114**, 04202 (2009). doi:10.1029/2008JA013980

D.V. Sivukhin, Motion of charged particles, in *Electromagnetic Fields in the Drift Approximation*, ed. by M.A. Leontovich, vol. 1 (Consultants Bureau, New York, 1965), pp. 1–104

B.V. Somov, Magnetic reconnection in solar flares. Phys. Usp. **53**, 954–958 (2010). doi:10.3367/UFNe.0180.201009j.0997

T.W. Speiser, Particle trajectories in model current sheets, 1, analytical solutions. J. Geophys. Res. **70**, 4219–4226 (1965). doi:10.1029/JZ070i017p04219

K. Stasiewicz, P. Bellan, C. Chaston, C. Kletzing, R. Lysak, J. Maggs, O. Pokhotelov, C. Seyler, P. Shukla, L. Stenflo, A. Streltsov, J.-E. Wahlund, Small scale Alfvénic structure in the aurora. Space Sci. Rev. **92**, 423–533 (2000)

J. Šafránková, Z. Němeček, L. Přech, G.N. Zastenker, Ion kinetic scale in the solar wind observed. Phys. Rev. Lett. **110**(2), 025004 (2013). doi:10.1103/PhysRevLett.110.025004

A.E. Vapirev, G. Lapenta, A. Divin, S. Markidis, P. Henri, M. Goldman, D. Newman, Formation of a transient front structure near reconnection point in 3-D PIC simulations. J. Geophys. Res. **118**, 1435–1449 (2013). doi:10.1002/jgra.50136

H. Viberg, Y.V. Khotyaintsev, A. Vaivads, M. André, J.S. Pickett, Mapping HF waves in the reconnection diffusion region. Geophys. Res. Lett. **40**, 1032–1037 (2013). doi:10.1002/grl.50227

M. Volwerk, R. Nakamura, W. Baumjohann, R.A. Treumann, A. Runov, Z. Vörös, T.L. Zhang, Y. Asano, B. Klecker, I. Richter, A. Balogh, H. Rème, A statistical study of compressional waves in the tail current sheet. J. Geophys. Res. **108**, 1429 (2003). doi:10.1029/2003JA010155

M. Volwerk, W. Baumjohann, K. Glassmeier, R. Nakamura, T. Zhang, A. Runov, Z. Vörös, B. Klecker, R. Treumann, Y. Bogdanova, H. Eichelberger, A. Balogh, H. Rème, Compressional waves in the Earth's neutral sheet. Ann. Geophys. **22**, 303–315 (2004a). doi:10.5194/angeo-22-303-2004

M. Volwerk, K.-H. Glassmeier, A. Runov, R. Nakamura, W. Baumjohann, B. Klecker, I. Richter, A. Balogh, H. RèMe, K. Yumoto, Flow burst-induced large-scale plasma sheet oscillation. J. Geophys. Res. **109**, 11208 (2004b). doi:10.1029/2004JA010533

M. Volwerk, Z. Vörös, W. Baumjohann, R. Nakamura, A. Runov, T. Zhang, K. Glassmeier, R. Treumann, B. Klecker, A. Balogh, H. Rème, Multi-scale analysis of turbulence in the Earth's current sheet. Ann. Geophys. **22**, 2525–2533 (2004c). doi:10.5194/angeo-22-2525-2004

M. Volwerk, K.-H. Glassmeier, R. Nakamura, T. Takada, W. Baumjohann, B. Klecker, H. Rème, T.L. Zhang, E. Lucek, C.M. Carr, Flow burst-induced Kelvin-Helmholtz waves in the terrestrial magnetotail. Geophys. Res. Lett. **34**, 10102 (2007). doi:10.1029/2007GL029459

Z. Vörös, Magnetic reconnection associated fluctuations in the deep magnetotail: ARTEMIS results. Nonlinear Process. Geophys. **18**, 861–869 (2011). doi:10.5194/npg-18-861-2011

Z. Vörös, W. Baumjohann, R. Nakamura, A. Runov, T.L. Zhang, M. Volwerk, H.U. Eichelberger, A. Balogh, T.S. Horbury, K.-H. Glaßmeier, B. Klecker, H. Rème, Multi-scale magnetic field intermittence in the plasma sheet. Ann. Geophys. **21**, 1955–1964 (2003). doi:10.5194/angeo-21-1955-2003

Z. Vörös, W. Baumjohann, R. Nakamura, M. Volwerk, A. Runov, T.L. Zhang, H.U. Eichelberger, R. Treumann, E. Georgescu, A. Balogh, B. Klecker, H. Réme, Magnetic turbulence in the plasma sheet. J. Geophys. Res. **109**, 11215 (2004). doi:10.1029/2004JA010404

Z. Vörös, W. Baumjohann, R. Nakamura, A. Runov, M. Volwerk, H. Schwarzl, A. Balogh, H. Rème, Dissipation scales in the Earth's plasma sheet estimated from cluster measurements. Nonlinear Process. Geophys. **12**, 725–732 (2005)

Z. Vörös, W. Baumjohann, R. Nakamura, M. Volwerk, A. Runov, Bursty bulk flow driven turbulence in the Earth's plasma sheet. Space Sci. Rev. **122**, 301–311 (2006). doi:10.1007/s11214-006-6987-7

Z. Vörös, W. Baumjohann, R. Nakamura, A. Runov, M. Volwerk, Y. Asano, D. Jankovičová, E.A. Lucek, H. Rème, Spectral scaling in the turbulent Earth's plasma sheet revisited. Nonlinear Process. Geophys. **14**, 535–541 (2007)

Z. Vörös, R. Nakamura, V. Sergeev, W. Baumjohann, A. Runov, T.L. Zhang, M. Volwerk, T. Takada, D. Jankovičová, E. Lucek, H. Rème, Study of reconnection-associated multiscale fluctuations with cluster and double star. J. Geophys. Res. **113**, 7 (2008). doi:10.1029/2007JA012688

Z. Vörös, M.P. Leubner, A. Runov, V. Angelopoulos, W. Baumjohann, Evolution of kinklike fluctuations associated with ion pickup within reconnection outflows in the Earth's magnetotail. Phys. Plasmas **16**(12), 120701 (2009). doi:10.1063/1.3271410

C. Wang, L.R. Lyons, R.A. Wolf, T. Nagai, J.M. Weygand, A.T.Y. Lui, Plasma sheet $PV^{5/3}$ and nV and associated plasma and energy transport for different convection strengths and AE levels. J. Geophys. Res. **114**, 0 (2009). doi:10.1029/2008JA013849

J.A. Wesson, Sawtooth reconnection. Nucl. Fusion **30**, 2545–2549 (1990)

J.M. Weygand, M.G. Kivelson, K.K. Khurana, H.K. Schwarzl, S.M. Thompson, R.L. McPherron, A. Balogh, L.M. Kistler, M.L. Goldstein, J. Borovsky, D.A. Roberts, Plasma sheet turbulence observed by cluster II. J. Geophys. Res. **110**, 1205 (2005). doi:10.1029/2004JA010581

J.M. Weygand, W.H. Matthaeus, S. Dasso, M.G. Kivelson, L.M. Kistler, C. Mouikis, Anisotropy of the Taylor scale and the correlation scale in plasma sheet and solar wind magnetic field fluctuations. J. Geophys. Res. **114**, 7213 (2009). doi:10.1029/2008JA013766

J.M. Weygand, W.H. Matthaeus, M. El-Alaoui, S. Dasso, M.G. Kivelson, Anisotropy of the Taylor scale and the correlation scale in plasma sheet magnetic field fluctuations as a function of auroral electrojet activity. J. Geophys. Res. **115**, 12250 (2010). doi:10.1029/2010JA015499

T. Wiegelmann, J. Büchner, Kinetic simulations of the coupling between current instabilities and reconnection in thin current sheets. Nonlinear Process. Geophys. **7**, 141–150 (2000)

L.J.C. Wooliscroft, T. Dudok de Wit, V.V. Krasnosel'skikh, M.A. Balikhin, On aspects of the measurement of non-linear turbulence processes using the cluster wave experiments. Nonlinear Process. Geophys. **3**, 58–65 (1996)

M. Yamada, R. Kulsrud, H. Ji, Magnetic reconnection. Rev. Mod. Phys. **82**, 603–664 (2010). doi:10.1103/RevModPhys.82.603

L. Zelenyi, A. Artemyev, Mechanisms of spontaneous reconnection: from magnetospheric to fusion plasma. Space Sci. Rev. (2013). doi:10.1007/s11214-013-9959-8

L.M. Zelenyi, A.V. Milovanov, Fractal topology and strange kinetics: from percolation theory to problems in cosmic electrodynamics. Phys. Usp. **47**, 749–788 (2004). doi:10.1070/PU2004v047n08ABEH001705

L.M. Zelenyi, A.L. Taktakishvili, A kinetic theory of the magnetic islands merging instability. Plasma Phys. Control. Fusion **30**, 663–679 (1988). doi:10.1088/0741-3335/30/6/003

L.M. Zelenyi, M.I. Sitnov, H.V. Malova, A.S. Sharma, Thin and superthin ion current sheets. Quasi-adiabatic and nonadiabatic models. Nonlinear Process. Geophys. **7**, 127–139 (2000)

L.M. Zelenyi, A.V. Artemyev, V.Y. Popov, Marginal stability of thin current sheets in the Earth's magnetotail. J. Atmos. Sol.-Terr. Phys. **70**, 325–333 (2008). doi:10.1016/j.jastp.2007.08.019

L.M. Zelenyi, A.V. Artemyev, A.A. Petrukovich, R. Nakamura, H.V. Malova, V.Y. Popov, Low frequency eigenmodes of thin anisotropic current sheets and cluster observations. Ann. Geophys. **27**, 861–868 (2009)

L.M. Zelenyi, A.V. Artemyev, K.V. Malova, A.A. Petrukovich, R. Nakamura, Metastability of current sheets. Phys. Usp. **53**, 933–941 (2010). doi:10.3367/UFNe.0180.201009g.0973

L.M. Zelenyi, S.D. Rybalko, A.V. Artemyev, A.A. Petrukovich, G. Zimbardo, Charged particle acceleration by intermittent electromagnetic turbulence. Geophys. Res. Lett. **381**, 17110 (2011a). doi:10.1029/2011GL048983

L.M. Zelenyi, H.V. Malova, A.V. Artemyev, V.Y. Popov, A.A. Petrukovich, Thin current sheets in collisionless plasma: equilibrium structure, plasma instabilities, and particle acceleration. Plasma Phys. Rep. **37**, 118–160 (2011b). doi:10.1134/S1063780X1102005X

T.L. Zhang, W. Baumjohann, R. Nakamura, A. Balogh, K. Glassmeier, A wavy twisted neutral sheet observed by CLUSTER. Geophys. Res. Lett. **29**(19), 190000 (2002). doi:10.1029/2002GL015544

V. Zhdankin, D.A. Uzdensky, J.C. Perez, S. Boldyrev, Statistical analysis of current sheets in three-dimensional magnetohydrodynamic turbulence. Astrophys. J. **771**, 124 (2013). doi:10.1088/0004-637X/771/2/124

X. Zhou, V. Angelopoulos, A. Runov, M.I. Sitnov, F. Coroniti, P. Pritchett, Z.Y. Pu, J. Zong, J.P. McFadden, D. Larson, K. Glassmeier, Thin current sheet in the substorm late growth phase: modeling of THEMIS observations. J. Geophys. Res. **114**, 3223 (2009). doi:10.1029/2008JA013777

G. Zimbardo, A. Greco, L. Sorriso-Valvo, S. Perri, Z. Vörös, G. Aburjania, K. Chargazia, O. Alexandrova, Magnetic turbulence in the geospace environment. Space Sci. Rev. **156**, 89–134 (2010). doi:10.1007/s11214-010-9692-5

DOI 10.1007/978-1-4939-3547-5_11
Reprinted from *Space Science Reviews* Journal, DOI 10.1007/s11214-014-0126-7

Current Sheets in the Earth Magnetotail: Plasma and Magnetic Field Structure with Cluster Project Observations

**Anatoli Petrukovich · Anton Artemyev · Ivan Vasko ·
Rumi Nakamura · Lev Zelenyi**

Received: 17 October 2014 / Accepted: 8 December 2014 / Published online: 7 February 2015
© Springer Science+Business Media Dordrecht 2015

Abstract Thin current sheets having kinetic scales are an important plasma structure, where the magnetic energy dissipation and charged particle acceleration are the most effective. It is believed that such current sheets are self-consistently formed by the specific nonadiabatic dynamics of charged particles and play a critical role in many space plasma and astrophysical objects. Current sheets in the near-Earth plasma environment, e.g., the magnetotail current sheet, are readily available for *in-situ* investigations. The dedicated multi-spacecraft Cluster mission have revealed basic properties of this current sheet, which are presented in this review: typical spatial profiles of magnetic field and current density, distributions of plasma temperature and density, role of heavy ions and electron currents, etc. Being important for the Earth magnetosphere physics, the new knowledge also could provide the basis for advancement in general plasma physics as well as in plasma astrophysics.

Keywords Current sheet · Planetary magnetospheres · Kinetic plasma structures

1 Introduction

Current sheets (CSs) represent the typical magnetic field configuration in plasmas where the thermal pressure dominates in comparison with the magnetic one (i.e. plasma systems with a large value of plasma β-parameter). CSs are often introduced qualitatively as spatially localized regions, where magnetic field substantially changes the direction or/and amplitude.

A. Petrukovich · A. Artemyev (✉) · I. Vasko · L. Zelenyi
Space Research Institute (IKI), RAS, Moscow, Russia
e-mail: ante0226@gmail.com

A. Petrukovich
e-mail: apetruko@iki.rssi.ru

L. Zelenyi
e-mail: lzelenyi@iki.rssi.ru

R. Nakamura
Space Research Institute (IWF), OAW, Graz, Austria

One can distinguish forced CS formed due to external driving at the boundary of two plasma domains with different properties (e.g. at the magnetopause of the Earth's magnetosphere) and spontaneous CS formed in the course of self-consistent evolution of plasma systems without direct external driving (e.g. magnetotail). CSs play the fundamentally important role because the magnetic energy dissipates within these structures to kinetic energy of accelerated charged particles and thermal plasma energy (Parker 1994; Priest and Forbes 2000; Birn and Priest 2007).

In-situ spacecraft observations have demonstrated that CSs are formed in practically all very different magnetospheres (Jackman et al. 2014): the dynamic and compact Mercury magnetosphere (Slavin et al. 2012), the induced magnetosphere of Venus (Vaisberg and Zeleny 1984; McComas et al. 1986a; Vasko et al. 2014b), the Earth magnetosphere (Ness 1965), the disk-like magnetospheres of Saturn (Dougherty et al. 2009) and Jupiter (Smith et al. 1974; Behannon et al. 1981; Artemyev et al. 2014b), the cylindrically symmetric magnetospheres of Neptune and Uranus (Ness et al. 1989), magnetospheres of comets (McComas et al. 1987), etc. CSs in solar corona represent the critical element of numerous modern solar-flare theories (Aschwanden 2002). In solar wind CSs can also be formed due to the inhomogeneity of solar wind flows (Gosling 2012). Detected radiation of accelerated electrons from pulsar nebulae also points to the possible important role of CS in this exotic system (Arons 2012).

The Earth magnetotail CS is for the obvious reasons the most investigated one. CS with magnetic field reversal is a core part of the plasma sheet—the region with the closed (connected to the Earth ionosphere) magnetic field lines, filled with hot plasmas. According to the modern scenario of magnetospheric disturbances (substorms), the CS structure and stability determines dynamics of the whole magnetotail (Baker et al. 1996).

It is commonly (and erroneously) assumed that the scales of variations of the plasma parameters in space and astrophysical objects are comparable with the scales of these objects. In particular magnetic field and plasma density variations in the magnetotail for many years were supposed to occur on the scale comparable with that of the magnetosphere cross-section, i.e. few tens of 10^3 km (e.g., Bird and Beard 1972; Schindler 1974; Kan 1973; Schindler 1979). On the other hand the classic theory predicts that the most efficient dissipation of magnetic energy in collisionless plasma occurs on kinetic plasma scales within such structures as shock waves, magnetic discontinuities, kinetic current sheets, etc. Kinetic scales (in magnetized plasma controlled by the Larmor radius of the heaviest plasma species) are much smaller than scales of the majority of space plasma objects (e.g., ion Larmor radius in the outer Earth's magnetosphere is one hundred times smaller than the magnetotail diameter).

General plasma characteristics (magnetic field, density, temperature) as well as some geometrical properties of the magnetotail have been successfully investigated with the help of a number of single and double spacecraft missions (OGO, IMP, ISEE, AMPTE, Geotail, Interball-tail) of twentieth century (McComas et al. 1986b; Baumjohann et al. 1989; Mitchell et al. 1990; Sergeev et al. 1993; Hoshino et al. 1996; Petrukovich et al. 1999; Asano et al. 2004b). These observations clearly pointed out that dynamics of the tail CS indeed should be driven by charged particle kinetics and thus strongly depends on the CS spatial profile (e.g., thickness scale). However, determination of spatial scales of magnetic field and plasma structures in the near-Earth domain cannot be solved reliably by the measurements on a single spacecraft.

A four-spacecraft European Cluster mission (since 2000) was specially designed for determination on a regular basis of spatial gradients in the magnetotail on both fluid (10 000 km) and ion kinetic scales (100–1000 km) (Credland et al. 1997; Escoubet et al.

2001). Indeed, many structures with kinetic scales have been found (Sharma et al. 2008). These new observations revealed the multiscale nature of the Earth's magnetotail, in which very thin (~100–1000 km) current sheets can be responsible for the generation of 50 % of the magnetic field total amplitude. Although, these results are important for the modeling of magnetotail stability and dynamics, they also give lessons on the possible configurations of self-consistent plasma systems. Namely, independently of the general system scale, significant processes (electric current generation, magnetic reconnection, charged particle acceleration, etc.) can occur at kinetic scales where dynamics of relatively small populations of charged particles may drive the whole magnetic field configuration. These properties should be (and can be) taken into account in the CS theoretical models and simulations.

There are several aspects of magnetotail CS structure, which we would like to discuss below in this paper:

1. Statistics of current sheet thicknesses, i.e. scales of magnetic field inhomogeneities in the vicinity of the magnetotail central region and its relation to plasma spatial scales (ion gyroradious, ion inertial length).
2. Relation between thicknesses of the plasma sheet and CS, spatial gradients of density and temperature.
3. The role of heavy ions in the sheet structure (in addition to protons and electrons).
4. Underlying physics of the electric current (diamagnetic drift of hot ions, non-adiabatic ions, Hall electric field, etc.).
5. Vertical current sheets as newly discovered dynamical structures.
6. Formation of thin current sheets and their two-dimensional configuration (including gradients along the tail)

2 Models of Current Sheets

In this section we give a brief review of CS models. Although, there is a large number of analytical and numerical MHD models of CS (Bird and Beard 1972; Birn et al. 1975; Birn 1979; Schindler and Birn 1986; Whipple et al. 1991; Steinhauer et al. 2008) we concentrate here on kinetic models, which allow to describe structures, comparable with ion scales and which represent self-consistent solutions of Vlasov–Maxwell equations (Schindler 2006). We use the GSM coordinate system: X axis is directed from the Earth to the Sun, Z axis is in the plane, containing X-axis and Earth magnetic moment, Y axis completes the right-handed coordinate system. For a typical (horizontal) sheet the normal as well as the main inhomogeneity scale are directed almost along Z (hereafter called vertical structure), electric current is along Y. Weaker gradients in the magnetotail are possible along X (tail axis).

The simplest CS model which has been widely used for description of spacecraft observations was proposed by Harris (1962). This self-consistent model describes 1D CS with the single component of the magnetic field: $B_x = B_0 \tanh(z/L)$, where B_0 is the magnetic field amplitude, L is CS thickness. Corresponding single component of the current density $j_y = (cB_0/4\pi L) \cosh^{-2}(z/L)$ is supported by diamagnetic drift of the entire populations of ions and electrons. Thus, the profile of current density coincides with the profile of plasma density, while the drift velocity and plasma temperature are constant across CS. Magnetic field pressure in such CS is balanced (along the normal) by variation of perpendicular plasma thermal pressure $p_\perp(z)$: $8\pi p_\perp(z) + B_x^2(z) = $ const.

Although Harris (1962) CS model is very popular and easy for applications, it cannot describe the realistic magnetotail CS due to absence of the B_z magnetic field component. This component, in a natural configuration connects magnetic field lines in southern and northern magnetotail lobes through the neutral plane $z = 0$ (at least for the near-Earth region, $X > -30\ R_E$). Even small, but finite B_z component destroys the self-consistency of Harris (1962) model, since the tension of magnetic field lines $\sim j_y B_z$ remains unbalanced. This force can be compensated by the gradients of a plasma pressure $\partial p / \partial x$ in 2D models. Such a scenario is realized in several 2D CS models with isotropic pressure tensor (Schindler 1972; Kan 1973; Schindler and Birn 2002; Yoon and Lui 2005; Vasko et al. 2013). In case of 2D models with isotropic pressure the L_x-scale of system inhomogeneity along the tail is coupled to the CS thickness L_z and amplitudes of magnetic field components: $L_x = L_z B_0 / 2 B_z$ (Burkhart and Chen 1993).

The alternative approach to compensation of $\sim j_y B_z$ force takes into account the anisotropy of plasma pressure (Rich et al. 1972; Francfort and Pellat 1976; Cowley and Pellat 1979), which can be generated by transient nonadiabatic ions (Speiser 1965, 1967). This approach allows to construct a 1D CS model with a finite B_z component and ion beams at system edges (see, e.g., Sitnov et al. 2006 and review by Zelenyi et al. 2011). In 1D models L_x is formally infinite (much larger than the estimate made above for 2D models). The main current in such 1D models is carried by transient hot ions. The bulk velocity of such ions is comparable with their thermal velocity and, as a result, rather small ion population can support almost the entire cross tail current required for the magnetotail field reversal Eastwood (1972, 1974).

Modern models are capable to describe almost all details of CS observations obtained thus far (Sitnov et al. 2006; Artemyev et al. 2008; Zhou et al. 2009). To fit the vertical (along the normal) profile of current density, we can pick up any model with a necessary number of free parameters (Yoon and Lui 2004; Birn et al. 2004; Zelenyi et al. 2004; Artemyev et al. 2009). Moreover, even simple Harris (1962) CS model can be fitted to observations after some modifications (Petrukovich et al. 2011). The primary difference between CS models is in their configurations along the magnetotail (e.g., the principal choice is between 2D isotropic and 1D anisotropic models). Another way is to compare the kinetic structure of plasma distributions in observed CSs and the one predicted by different models (see details in Artemyev and Zelenyi 2013).

3 Dataset and Methods of Processing of Multispacecraft Measurements

To investigate CS structure we use measurements of multispacecraft Cluster mission (Escoubet et al. 2001). Magnetic fields are measured by FGM experiment (Balogh et al. 2001), electric fields are measured by EFW experiment (Gustafsson et al. 2001), electron distribution functions are measured by PEACE experiment (Johnstone et al. 1997), while ion velocity distributions are measured by CIS/HIA/CODIF experiment (Rème et al. 2001). The time resolution of measurements of electron and ion distribution functions is 4–16 seconds (1–4 spacecraft spins). We use electric and magnetic field data with time resolution corresponding to one point per 4s or five points per second. The main statistics of CS crossings is based on the first four years (2001–2004) of measurements, when distances between spacecraft were of the order **of ion scale**: 1000 km, 3000 km, 300 km, 600 km. During this period Cluster crossed magnetotail CS at 14–20 R_E from Earth. The data collected in later years (2005–2009) with the larger distances between spacecraft are used to investigate the magnetic field gradients along the tail.

The main advantage of multispacecraft missions is a possibility of simultaneous measurements at four positions. Using magnetic field data at four points (properly spaced, e.g., in a form of tetrahedron) one can estimate the magnetic field gradient. Corresponding assumptions and estimations of errors are described in details in several papers (Dunlop et al. 2002; Runov et al. 2005b; Paschmann and Schwartz 2000). Gradient of B_x component can be used to determine the direction of normal to CS for typical magnetotail configuration. Gradients of all three magnetic field components give an estimate of the curl of magnetic field curl $\mathbf{B}$, i.e. the electric currents. This method is called "curlometer" technique. For the majority of observed CSs the main component of curl $\mathbf{B}$ is dB_x/dz.

The alternative method to determine the normal direction and the magnetic field gradient is the timing technique. In this case, time delays between observations of the same magnetic field features at different spacecraft provide the estimates of the direction and velocity of CS motion along the normal. This velocity is later used to recalculate the time delays to spatial scale.

To unify the consideration of CS, the local coordinate system $(\mathbf{l}, \mathbf{m}, \mathbf{n})$ could be introduced. For a model 1D current sheet it is in fact cartesian, the gradient of B_l component of magnetic field is supported by the current density j_m. For real data obtained vectors are not exactly orthogonal and a certain procedure is necessary (Runov et al. 2005b). $\mathbf{l}$ vector is defined as a direction of a maximum variation of magnetic field. The current density vector $\mathbf{j}$ is used to determine the vector $\mathbf{m} = \mathbf{j}/j - (\mathbf{lj})/j$. The normal vector completes the system to right-handed one: $\mathbf{n} = [\mathbf{l} \times \mathbf{m}]$.

The profiles across the current sheet (along the normal) are taken during episodes of fast relative motion of a spacecraft and current sheet (usually within 10 min for a thin sheet) and require assumption of stationarity of CS. The combination of current density j_m profiles and magnetic field B_l as a nonlinear coordinate across CS allows us to restore the spatial gradient of any plasma parameter A measured by a single spacecraft (Zelenyi et al. 2010; Artemyev and Zelenyi 2013): $dA/dz = (4\pi/c)j_m(dA/dB_l)$ (e.g. A can be an ion or electron temperature, or plasma pressure).

The determination of CS thickness L (scale along the normal) requires additional assumptions, because L definition depends on the particular profile of j_m. In this review we estimate $L = cB_0/4\pi j_0$, B_0 and j_0 are the maximal value of magnetic field and current density. The determination of B_0 is not straightforward, because B_0 (maximal field in the thin current sheet) can differ substantially from the magnetic field B_{ext} at lobes. In general case the magnetic field B_{ext} can be readily estimated using the vertical pressure balance:

$$B_{ext}^2/8\pi = B_l^2/8\pi + P_{th}$$

where P_{th} is a plasma (ions and electrons) thermal pressure. In the next section we describe possible approaches for the determination of B_0.

4 Embedding of CS

To illustrate the main properties of CSs observed by Cluster in the Earth magnetotail we use the figure published in one of the first statistical papers by Runov et al. (2005a, 2006). Figure 1 demonstrates the profiles of current and plasma densities across the sheet (as a function of B_l magnetic field component normalized on the lobe field B_{ext}). In contrast to Harris (1962) model, in which profiles of current density and plasma density coincide exactly, in observed CS both these profiles are very different: the current density profile is much narrower than the plasma density one. The average profile of plasma temperature is similar to the density profile. Thus, the current sheet is localized deeply inside the plasma density profile and such a configuration is called embedded sheet.

Fig. 1 Profiles of current density (**a**), plasma density (**b**) and temperature (**c**) across the CS. The plasma and current densities are normalized to their value at the neutral sheet ($B_l = 0$). The profiles averaged over the statistics are shown by solid curves. Figure is adopted from Runov et al. (2006)

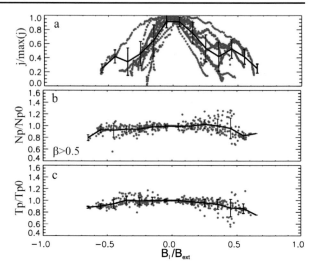

Fig. 2 Schematic view of current density profile in the embedded CS (*green curve*) and in simple CS described by Harris (1962) model. Figure is adopted from Petrukovich et al. (2011)

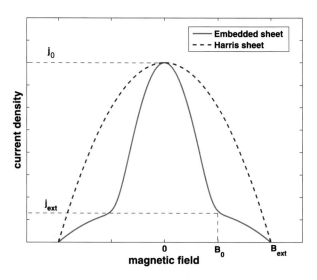

In absence of CS embedding we can characterize the CS configuration by two parameters: magnetic field amplitude at lobes B_{ext} and CS thickness L_{ext} (or maximal current density $j_0 = cB_{ext}/4\pi L_{ext}$). The current density amplitude j_0 can be derived from four-point measurements of magnetic field. B_{ext} can be estimated using the vertical pressure balance.

To describe the embedded CS one should introduce at least two additional parameters (Fig. 2): B_0 is magnetic field amplitude at the boundary of thin embedded CS and j_{ext} is the current density amplitude at the boundary of thin embedded CS. In the most of cases two parameters j_{ext} and j_0 cannot be estimated for the same CS, because Cluster spacecraft can probe reliably only one scale corresponding to the distance between spacecraft, while we often have $j_{ext} \ll j_0$ and, as a result, $L_{ext} \gg L = cB_0/4\pi j_0$. To measure weaker magnetic field gradients related to j_{ext} one needs larger spacecraft separation. Thus, only B_0 can be considered as a practical parameter characterizing embedded CS.

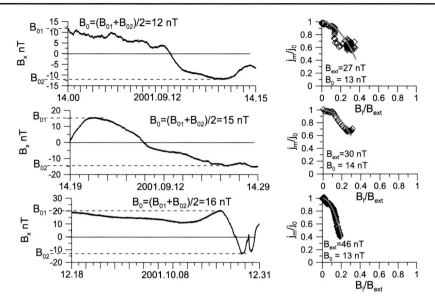

Fig. 3 Three examples of B_0 determination. On the *left panels* B_0 is estimated as the average of magnetic fields observed at the CS boundaries, $B_0 = (B_{01} + B_{02})/2$. On the *right panels* B_0 is estimated using the profile of the current density (see text for details). Figure is adopted from Artemyev et al. (2010)

Although, magnetic field measurements are reliable nowadays, the determination of B_0 is not a straightforward task. The main problem is the determination of the boundary of thin embedded CS. This boundary can be accurately defined only in case of substantial difference between thicknesses of embedded and enveloping CSs. Artemyev et al. (2010) introduced two approaches to the B_0 determination. The first one assumes that B_l variation is due to motion of the CS relative to motion of the spacecraft. Then B_0 can be defined as a B_l value measured at the moment when variation rate of B_l is substantially reduced, when the spacecraft escapes the thin CS with strong electric current (Fig. 3, left panels). The second approach approximates current density profile by a function of B_l field, $j_m = j_0(1 - sB_l^2)$, with s as a free parameter, so that $B_0 \approx s^{-1/2}$ (Fig. 3, right panels). The accuracy of both methods is about 10–20 %. Thus, it is more reliable to analyze the statistics of B_0 values rather than B_0 for individual crossings.

The statistics of CS crossings by Cluster spacecraft (42 examples, see Artemyev et al. 2010) allows us to consider the distribution of B_0/B_{ext} ratio. For the most of cases B_0/B_{ext} is about 0.3–0.5 with the average value ≈ 0.4 (Fig. 4). The ratio of the current densities measured in the center and at the boundary of thin CS (where $0.9 < |B_l/B_0| < 1$) is about ~ 5. The variation of plasma density across thin CS does not exceed 25 % in agreement with an estimate given by the pressure balance $(B_0/B_{ext})^2 \sim 0.16$.

The strong variation of the current density across CS without corresponding variation of the plasma density suggests that the main current in this thin embedded CS is carried by a relatively small (10–15 %) ion population with a large bulk velocity. An estimate of this velocity can be obtained as a ratio of measured (curlometer) current density and a variation of the plasma density. Such estimate gives a bulk velocity (along the direction of the electric current) about the thermal ion velocity (Artemyev et al. 2010). Presence of such ion population is predicted by the models with transient ions (Ashour-Abdalla et al. 1991, 1993; Burkhart et al. 1992a; Kropotkin et al. 1997; Sitnov et al.

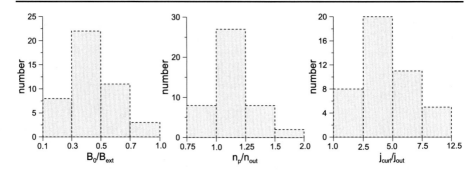

Fig. 4 Histograms of ratio B_0/B_{ext}, ratios of plasma and current densities at CS center ($B_l = 0$) and at CS boundaries ($0.9 < |B_l/B_0| < 1$). Figure is adopted from Artemyev et al. (2010)

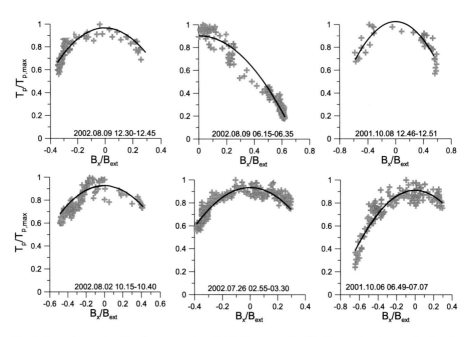

Fig. 5 Several examples of proton temperature distribution across the CS. Temperature is normalized on its value at CS central region. Figure is adopted from Artemyev et al. (2011a)

2000; Zelenyi et al. 2000). Figure 1c and previous investigations (Hoshino et al. 1996; Runov et al. 2006) have shown that proton temperature also vary substantially across the tail. A detailed Cluster analysis (Fig. 5) results in an approximation by a simple function $T_p = T_{p,\max}(1 - \alpha_T(B/B_{ext})^2)$, $\alpha_T \approx 0.8$. If this temperature increase in the CS center is also due to current-carrying ions, their temperature is three-five times larger than the background temperature (Artemyev et al. 2011a).

A similar to Fig. 5 variation is found for the electron temperature (see Zelenyi et al. 2010; Artemyev et al. 2012). These variations of ion (proton) and electron temperatures are often more significant than the variation of the plasma density. The scale of T_p variation can be estimated as $L_T \sim |d \ln T_p/dz|_{B_x \approx 0}^{-1} \approx (B_{ext}/\alpha_T B_0)L \approx (2-10)L$ (L is the CS thickness).

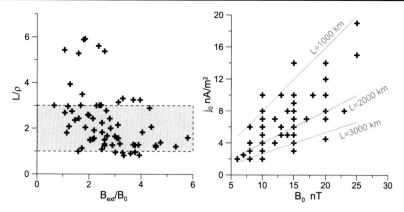

Fig. 6 *Left panel* shows the dependence of ratio L/ρ on B_{ext}/B_0 value (both CS thickness and proton Larmor radius are calculated with B_0 field). *Right panel* shows the current density amplitude j_0 versus B_0 field. *Three lines* correspond to characteristic CS thicknesses

Finally one can test the relation between the CS thickness and typical plasma spatial scales (Larmor radius ρ and/or ion inertial length). For the most of CSs it was shown that the ratio L_{ext}/ρ_{ext} (where proton Larmor radius ρ_{ext} and L_{ext} are calculated with B_{ext}) varies in a wide range with the mean value ~ 10 (Runov et al. 2005a, 2006). However, if we recalculate the Larmor radius and CS thickness, using the magnetic field B_0 at the boundary of thin embedded CS this ratio should substantially decrease $L/\rho = (L_{ext}/\rho_{ext})(B_0/B_{ext})^2 \sim 0.2(L_{ext}/\rho_{ext})$. To verify this estimate we plot L/ρ as a function of B_{ext}/B_0 for statistics of 72 CS crossings (Artemyev et al. 2011a) in Fig. 6. For almost all CSs L/ρ is within the range [3, 1]. Thus, for observed CSs we can use the proton Larmor radius calculated in B_0 field as a very good approximation for CS thickness. In such a case the current density amplitude j_0 should grow with B_0. Indeed, this growth can be found (right panel in Fig. 6).

As ρ also depends on the proton temperature T_p, the CS thickness should increase with T_p. We select two groups of CSs with different temperatures ($T_p < 4$ keV and $T_p > 8$ keV) and normalize CS thicknesses on the Larmor radius calculated for $T_p = 3$ keV. Resulting current density profiles are in Fig. 7. One can clearly see the increase of CS thickness with T_p.

Another test of relation $L/\rho \approx$ const can be done using the variable content of oxygen ions O^+ (up to 50 % of total plasma density). In CSs with larger oxygen content the corresponding thickness should be larger because for the same temperature the Larmor radius of oxygen ion is four times larger than the proton Larmor radius. We select three groups of CSs with different populations of oxygen ions and normalize CS thicknesses on the proton Larmor radius calculated with the measured temperature (Fig. 8). The increase of oxygen content clearly results in the increase of CS thickness.

Although, Fig. 6 shows that for the majority of observed CSs the thickness is about 1–3 proton Larmor radii (or 1000–3000 km), there is one specific class of CSs with smaller thickness. These CSs are observed in the vicinity of X-lines and often have current density amplitudes higher than 30 nA/m² (Nakamura et al. 2006, 2008). Corresponding thickness of such intense CSs is about ion inertial length $d_i \approx \rho(B_0/B_{ext})$ and is smaller than the proton Larmor radius. However, it is still much larger than the electron Larmor radius, $L/\rho_e \in [5, 30]$ (Artemyev et al. 2013a). The example of such intense CS is shown in Fig. 9. Comparison of the curlometer current density and electron moments shows that almost all

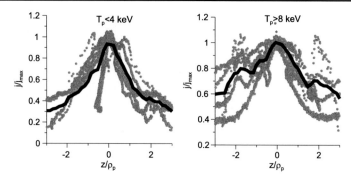

Fig. 7 Profiles of normalized current density for two groups of CSs with different proton temperatures. *Black curves* show the average profile. Figure is adopted from Zelenyi et al. (2011)

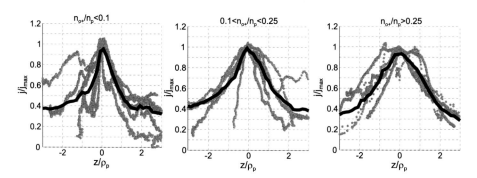

Fig. 8 Profiles of normalized current density across the CS for three groups of CSs with different densities of oxygen ions with respect to total plasma density. *Black curves* show the average profile. Figure is adopted from Zelenyi et al. (2011)

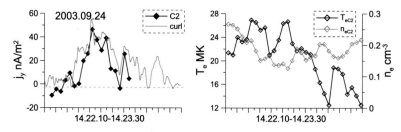

Fig. 9 *Right panel* shows the profile of current density in the intense CS: *grey color* is for current density calculated by curlometer technique, while *black color* is for electron current density measured by PEACE experiment onboard Cluster 2 spacecraft. *Left panel* shows profiles of electron temperature and electron density across CS. Figure is adopted from Artemyev et al. (2013a)

observed current density is due to the electron current (for this CS $L/\rho_e \approx 12$). There is almost no variation of electron density (and plasma density as well) across the sheet, while the electron temperature has a significant maximum in center of CS. Due to small CS thickness the ion temperature does not vary across the sheet and the pressure balance is supported by specific configuration of parallel currents creating a local maximum of B_m field in the central region of CS (see details in Artemyev et al. 2013a) (these curves are not shown here).

Investigations of such intense thin CS is possible only for one season of Cluster observations when spacecraft separation was small enough (2003). However, even for this season the time resolution of measurements of plasma parameters does not allow us to study well an internal structure of these CSs. Thus, the detailed investigation of these important class of CS could be performed only in future missions (like MMS).

5 Ion Kinetics

Vertical profiles of plasma and current densities indicate that the main cross-tail current should be carried by a relatively small part of ion population contributing only about $\sim$10–15 % to the total plasma density with large bulk velocity along the dawn-dusk direction. Thus, it is interesting to reveal its velocity distribution. However, only in a few cases the observed ion current density is close to the value of current density estimated using the curlometer (magnetic field gradient) technique (see next section for more details). Thus, to study ion kinetics in embedded CSs we identify several crossings having strong ion currents (Artemyev et al. 2010). One example is shown in Fig. 10. There is a strong increase of ion bulk velocity u_y up to 100 km/s in the central region of CS, while the plasma density varies very weakly.

To reveal reliably the current-carrying ion population we divide the whole time interval of CS crossing into five subintervals (see numbers in Fig. 10). For each subinterval we calculate the average ion distribution $f(v_x, v_y)$. Then, we calculate the differences $\Delta f_i(v_x, v_y) = f_i(v_x, v_y) - f_0(v_x, v_y)$ for $i = 1..4$ (interval "0" is out of the current sheet). Distributions $\Delta f_i(v_x, v_y)$ should contain the ion population which is present in the center and absent at the CS boundary, and thus is mainly responsible for the support of the transverse current. Right panels of Fig. 10 demonstrate that positive $\Delta f_i(v_x, v_y)$ (grey color) corresponds to an ion population mostly located in the $v_y > 0$ domain. Distribution of this population in velocity space has a local maximum at $v_y > 0$ (there are almost no particles for $v_y < 0$) and, thus, can be considered as a beam. However, it should be mentioned that the same distribution is often called bean distribution due to the half-ring shape in (v_x, v_y) plane. Its shape resembles in

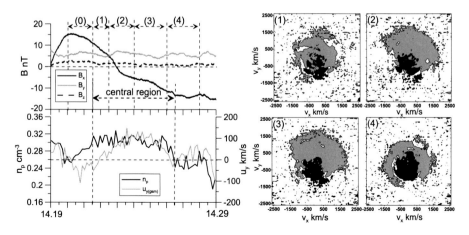

Fig. 10 CS crossing at 12 September 2001. *Left panels* show profiles of magnetic field, plasma density and ion bulk velocity. *Right panels* show $\Delta f_i(v_x, v_y)$ distributions. *Grey color* corresponds to $\Delta f_i > 0$ and *black color* to $\Delta f_i < 0$ (see text for details). Figure is adopted from Artemyev et al. (2010)

many aspects the theoretical distribution of transient ions (see corresponding comparison and discussion in Artemyev and Zelenyi 2013). Artemyev et al. (2010) demonstrated that all observed (measured by the curlometer technique) current density can be almost fully supported by these ions if their density is only about 10–15 % of the total plasma density which is in agreement with the measured variation of the plasma density (see Fig. 4).

An additional interesting question concerns a possible contribution of high-energy ions (energy larger than 35 keV) to the cross-tail current. The Cluster CIS/CODIF experiment (Rème et al. 2001) measures ion velocity distributions for energies below 35 keV, while for high-energy ions one should consider data obtained by the RAPID experiment (Wilken et al. 2001). However, the accuracy of the ion flow data obtained from high-energy ion measurement is not sufficient for estimations of the corresponding current density. Nonetheless, by extrapolating the velocity distribution profile to high energy range gives an upper limit of a possible current density carried by ions with energies above 35 keV. For several CSs with relatively high ion temperature (7–9 keV) Artemyev et al. (2009) estimated this upper limit to be less than 30 % of the total ion current. However, for the total plasma pressure the contribution of high energy ions can be larger (see discussion in Runov et al. 2006).

6 Current Carriers and Sources of Electron Currents

Accurate and regular measurements of electron velocity distribution by Cluster spacecraft allow for the first time to perform a comprehensive statistical investigation of electron parameters in CS. We compare electron currents with currents estimated by curlometer technique. First of all, it is important to demonstrate that the sum of ion and electron currents is equivalent to curlometer current density. Figure 11 shows two examples of CSs, in which the sum of ion and electron currents approximate curlometer data quite well (see details in Artemyev et al. 2009).

To analyze the relation between particle currents and curlometer data over a broader statistics we consider averaged values of currents calculated for $|B_l| < 5$ nT. For each crossing we identify three values: average ion current j_i, average electron current j_e, and average curlometer current j_{curl}. Comparison of these three values shows (see Fig. 12 and publications (Asano et al. 2003; Israelevich et al. 2008; Artemyev et al. 2009)): 1. for the most of observed CSs j_i is significantly smaller than j_{curl}, while the sum $j_i + j_e$ describes j_{curl} on average quite well; 2. there are CSs with negative and positive j_i, while j_e is almost always positive; 3. for some CSs the sum $j_i + j_e$ significantly differs from j_{curl}. The last effect can be explained in several ways. First of all, there could be a technical problem related to the

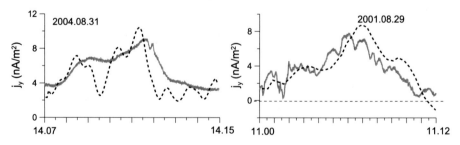

Fig. 11 Two CS crossings: *grey solid curve* shows curlometer estimates of the current density, while *dashed black curve* shows the smoothed profiles of the sum of electron and ion currents. Figure is adopted from Artemyev et al. (2009)

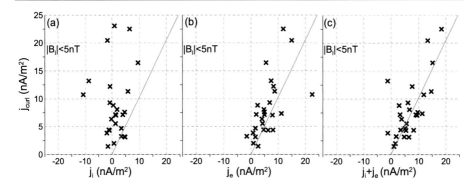

Fig. 12 Comparison of curlometer estimate of the current density j_{curl} with (**a**) ion current density j_i, (**b**) electron current density j_e, (**c**) amplitudes of sum of particle currents $j_i + j_e$. Figure is adopted from Artemyev et al. (2009)

reliability of the measurements of particle currents. These measurements are performed with the time resolution 4–16 s. For some fast crossings of CSs this resolution can be insufficient for reliable determination of maximal current. Second, the finite energy range of particle measurements can result in underestimation of particle currents. Third, one cannot exclude the possibility of fast variations of particle currents. For example, Asano et al. (2004a, 2003) demonstrated the presence of fast substantial variations of electron currents in CSs. These variations can be considered as a manifestation of natural small-scale electron substructures in CS. However, detailed analysis of this problem requires more accurate measurements of particle velocity distributions, e.g. in a future MMS project (Burch et al. 2014).

To explain surprisingly large contribution of electrons to the total cross-tail current we estimate three main components of the electron current density: diamagnetic drift current, curvature drift current, and cross-field drift current. The electron diamagnetic drift current is proportional to the ratio of particle temperatures $T_e/(T_e + T_i)$ (Artemyev and Zelenyi 2013). Cluster statistics gives $T_i/T_e \sim 3.5$–5 (Artemyev et al. 2011b), and electron diamagnetic effects can be responsible only for 10–20 % of the total electric current. Amplitude of electron drift curvature current density is about $j_{curl}(T_{e\parallel}/T_{e\perp} - 1)4\pi T_{e\perp}/B_z^2$ (Artemyev and Zelenyi 2013). Temperature anisotropy of electron population in CSs is about $T_{e\parallel}/T_{e\perp} \sim 1.1$–1.2 (Artemyev et al. 2012) and for some CSs the curvature currents indeed can be comparable with the total curlometer current density. However, for the most of CSs the curvature drift current gives less than 50 % of j_{curl} (Zelenyi et al. 2010). Thus, the remaining 50 % should be provided by electron cross-field drift $v_{E \times B} = c\mathbf{E} \times \mathbf{B}/B^2$. This drift does not produce any total current but reduces ion contribution and increases the electron one (Artemyev et al. 2011a). Estimate of $v_{E \times B}$ gives ~ -40 km/s for the abundant statistics of observed CSs (Zelenyi et al. 2010).

For the classical CS configuration $v_{E \times B} = c(E_z B_x - E_x B_z)/B^2 \approx cE_x/B_z$ in the CS central region where $B_x \sim 0$. To get $v_{E \times B} \sim -40$ km/s one needs an electric field $E_x \sim 0.15$ mV/m. So small electric field cannot be measured in the magnetotail and was never broadly discussed. However, positive E_x was obtained in laboratory modelling of the magnetotail CS (Minami et al. 1993). The model of E_x field as a part of ambipolar (polarization) electrostatic field can be developed for weakly 2D CSs (Zelenyi et al. 2010). Estimates based on Cluster observations show that the drift $v_{E \times B}$ can indeed significantly reduce current of transient ions or even result in negative values of observed ion current density (Artemyev et al. 2011a; Artemyev and Zelenyi 2013). Taking into account the role of this *hidden* electric field (which, by itself, does not produce any cross-tail current) enables to explain many paradoxical features of current observations in the magnetotail (see Fig. 12).

7 Tilted Current Sheets

In the classical configuration of the Earth magnetotail the current sheet is horizontal: the main gradient of magnetic field B_l is directed along the north-south direction (Z-axis), the main component of the normal vector $\mathbf{n}$ in the GSM is n_z. The current is in equatorial plane, the main component of vector $\mathbf{m}$ is m_y. Cluster spacecraft discovered a new type of CSs with vector $\mathbf{n}$ directed almost along the dawn-dusk direction and an electric current flowing almost along the Z-axis (Zhang et al. 2002; Petrukovich et al. 2003; Sergeev et al. 2006). We call CSs with $n_y > 0.85$ strongly-tilted—the angle between vector $\mathbf{n}$ and the Y-axis is smaller than $30°$. Formation of strongly-tilted CSs and their structure are considered also in several later publications (Petrukovich et al. 2006, 2008; Rong et al. 2010; Vasko et al. 2014a).

One example of the observation of a strongly-tilted CS is presented in Fig. 13. This example demonstrates the main properties of the strongly-tilted CSs in the statistical study by Vasko et al. (2014a): The magnetic field has a significant B_m (B_z in the GSM) component, on average B_m is about ~ 7 nT (see also Petrukovich et al. 2003), $B_m/B_0 \sim 0.45$. B_n component (B_y in the GSM) is much smaller than B_m, on average B_n is about ~ 1.5 nT, $B_n/B_m \sim 0.35$. Almost all current is along the Z-axis so that a significant portion of this current is field-aligned with respect to the magnetic field, $j_\parallel/j_\perp \sim 2.1$. Almost all current is carried by electrons (Vasko et al. 2014a).

Strongly-tilted CSs represent locally 1D structures with the magnetic field configuration resembling the horizontal CSs with a strong shear magnetic field B_m (or B_z in the GSM system) (Petrukovich et al. 2007; Nakamura et al. 2008; Shen et al. 2008; Rong et al. 2012). The presence of a strong shear component drastically changes charged

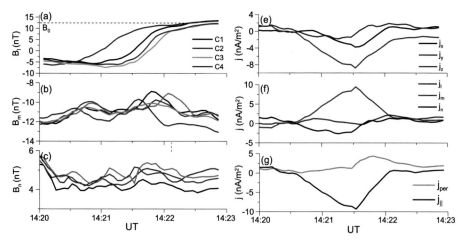

Fig. 13 An example of a strongly-tilted CS crossed by Cluster on 7 September 2004: (**a**), (**b**), (**c**) the magnetic field observed by four Cluster spacecraft in the local coordinate system; (**e**), (**f**) the current density determined by the curlometer technique in the GSM system and in the local coordinate system, respectively; (**g**) the components of the curlometer current density parallel and perpendicular to the magnetic field. Figure is adopted from Vasko et al. (2014a)

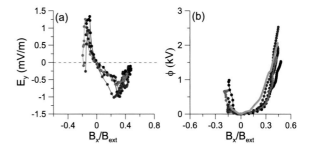

Fig. 14 Profiles of the polarization electric field (**a**) and the corresponding scalar potential (**b**) across the CS for the event presented in Fig. 13. *Grey curve* presents the theoretical profile $\phi = T_p (B_x/B_{ext})^2$. Figure is adopted from Vasko et al. (2014a)

particle dynamics (Büchner and Zelenyi 1989; Artemyev et al. 2013a). The particle dynamics is determined by the adiabaticity parameter κ that is the ratio of the field line curvature to the particle Larmor radius at the neutral sheet (where $B_l \approx 0$). For horizontal CSs with a small shear component the curvature radius in the neutral sheet is about $(B_n/B_0)L \sim L/10 \sim 100$–$200$ km, while for strongly-tilted CSs (with large shear) it is about $(B_n/B_0)(1 + (B_m/B_n)^2)L \sim 1.5L \sim 3000$ km. Thus, in horizontal CSs the curvature radius is smaller than the ion Larmor radius (Runov et al. 2006) and the adiabaticity parameter κ_p is generally smaller than 0.5. For tilted CS the curvature radius is larger than ion Larmor radius and generally $\kappa_p > 0.5$ (Vasko et al. 2014a). As a result in horizontal CSs the ion dynamics is quasi-adiabatic (Zelenyi et al. 2013), while in strongly-tilted CSs the ion dynamics is likely stochastic. In both horizontal and strongly-tilted CSs the curvature radius is significantly larger than electron Larmor radius so that $\kappa_e > 3$ and hence electrons are magnetized.

Strongly-tilted CSs provide one additional opportunity for investigation of general CS structure. The decoupling of motions of chaotic (or quasi-adiabatic) ions and magnetized electrons should result in generation of (polarization) electric fields if CSs are sufficiently thin (Schindler et al. 2012). This field has two components—E_n and E_l (i.e. E_z and E_x in horizontal CSs) (Zelenyi et al. 2004; Birn et al. 2004). As has been mentioned in the previous section, E_x component plays an important role in the redistribution of intensities of ion and electron currents and is coupled with the E_z component (Schindler and Birn 2002; Zelenyi et al. 2010). However, the configuration of the probe measurement (perpendicular to spacecraft rotation axis) at Cluster spacecraft does not permit to measure $E_n \approx E_z$ in horizontal CSs. In contrast, in strongly tilted CSs $E_n \approx E_y$ can be already reliably measured. E_y is the sum of convection and polarization electric fields. The former one corresponds to the large-scale earthward plasma convection and is about 0.1–0.3 mV/m (Kennel 1973; Angelopoulos et al. 1993). This field does not vanish at the neutral sheet, while the polarization field equals zero therein. Thus, one can determine and subtract convection field from the observed electric field E_n.

Vasko et al. (2014a) have determined the profiles of the polarization electric field E_y for statistics of strongly-tilted CSs (Fig. 14(a)): E_y has an asymmetrical profile across the CS ($E_y B_x > 0$) and the amplitude of about 0.5–1 mV/m. Integration of E_y across the CS gives the profile of the scalar potential ϕ (Fig. 14(b)). For statistics of tilted CS the amplitude of ϕ is about few kV, while ϕ-profile has a minimum at the neutral plane and can be well described by following equation in accordance with the model expectations (Vasko et al. 2014a)

$$\phi = T_p \frac{B_x^2}{B_{ext}^2}$$

where T_p is ion temperature.

8 Current Sheet Dynamics During Substorm Growth Phase

Dynamics of the magnetotail current sheet during growth phase of geomagnetic substorm is of particular interest (Petrukovich et al. 2007, 2013; Snekvik et al. 2012; Davey et al. 2012) because CSs are assumed to serve as accumulators of magnetic field energy stored in the magnetotail. Being driven by the global magnetospheric convection the tail plasma sheet thins and the entire magnetic configuration stretches from more dipolar to more tail-like geometry. As a result an intense CS with a thickness of the order of several thousand kilometers forms in the plasma sheet. Prior to Cluster mission this process was characterized mainly by the decrease of B_z magnetic field. Cluster provided regular measurements of the cross-tail current density and, in some cases, of horizontal magnetic gradient $\partial B_z / \partial x$. It should be noted, however, that the task of prolonged (tens of minutes) monitoring of CS is in a certain contradiction with the basic method of its study (using the fast crossings as described in previous sections). Thus, some sheet parameters are determined less reliably (in particular, B_0).

There were 39 Cluster CS observations during growth phase in 2001, 2002, 2004 at distances 17–20 R_E downtail (Petrukovich et al. 2007). An example is in Fig. 15. The key moment is the determination of an instant of substorm onset. Properties of CS directly before an onset may provide important information on the development of an expected magnetotail instability. In this case the onset was defined as the beginning of tailward flow (negative V_x) at 13:01 UT.

Before the onset the classical isolated growth phase was observed for about 40 min. During the growth phase the total magnetic field in the tail lobes B_{ext} increased from 29 to 32 nT (calculated using the full pressure), denoting the accumulation of open magnetic flux. The current density increased from 2 to 8 nA/m^2, but the maximum current density of 15 nA/m^2 was registered just after onset. B_z decreased from 5 to 2 nT. The small up-and-down CS oscillations allowed to recover the current density profile and estimate the

Fig. 15 Current sheet, observed September 12, 2001: (**a**)–(**c**) magnetic field; (**d**) current density, (**e**) proton bulk velocity V_x. *Colors* denote the satellite (C1)–(C4), respectively: *black, red, green, blue*. Figure is adopted from Petrukovich et al. (2009)

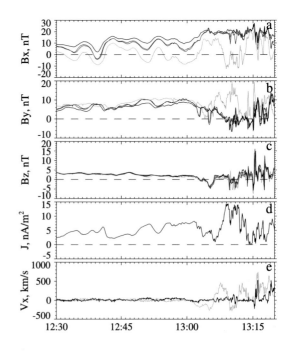

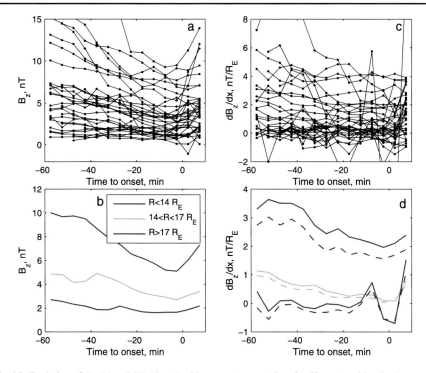

Fig. 16 Evolution of B_z (**a**) and $\partial B_z/\partial x$ (**c**) with respect to onset time for 59 events with a 5 minute averaging. Averaged profiles of B_z (**b**) and $\partial B_z/\partial x$ (**d**) in the three ranges of downtail distances. At the panel (d) *solid lines* correspond to the gradient $\partial B_z/\partial x$, *dashed lines*—to $\partial B_z/\partial r$. Figure is adopted from Petrukovich et al. (2013)

increase of B_0/B_{ext} from 0.3 to 0.5. The estimate of sheet thickness gives several thousand km and could not be accurate because of the uncertainty in B_0. These parameters are close to average for the whole dataset of 39 crossings.

Cluster orbit after 2005 allowed to probe also the near-Earth magnetotail (up to 9 R_E downtail), where one expects a significant influence of two-dimensionality on the formation of the current sheet. For the period 2005–2009 59 observations of growth phase have been identified, in which small longitudinal gradients of magnetic field $\partial B_z/\partial x$ were measured. These measurements become possible due to substantially increased spacecraft separation after 2005 (Petrukovich et al. 2013).

General statistics of gradient $\partial B_z/\partial x$ and B_z for three ranges of downtail distances is shown in Fig. 16. At a distance more than 17 R_E a negligible gradient was registered (fluctuating around zero, but the error in our data set is not less than 0.5 nT/R_E). B_z average also changes a little and is about 2 nT. This is the zone of elongated almost one-dimensional sheet. Sharp changes in the gradient near onset are produced by the earthward motion of dipolarization fronts and thus essentially represent an unavoidable uncertainty of onset time definition. Near the Earth at a distance less than 14 R_E there is a clear decrease of the average gradient from 3–4 to 2 nT/R_E, and B_z from 10 to 6 nT. This is the zone where the effects of second dimension play an essential role. At the range of distances 14–17 R_E properties of magnetic field are intermediate. In the first part of the growth phase gradient decreases to 1 nT/R_E, and B_z from 5 to 3–4 nT. During the second part gradient is close to zero, and B_z is stable.

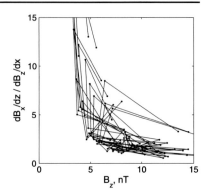

Fig. 17 Evolution of sheets in coordinates $(\partial B_x/\partial z)/(\partial B_z/\partial x)$, B_z. Figure is adopted from Petrukovich et al. (2013)

The sheet evolution from predominantly two-dimensional to predominantly one-dimensional structure can be followed in Fig. 17 (Petrukovich et al. 2013). For 10 cases with the stable positive gradient $\partial B_z/\partial x$ we show the dependence of the ratio $(\partial B_x/\partial z)/(\partial B_z/\partial x)$ on B_z. All events are close to the same hyperbolic-type curve. At the initial stage of thinning at $B_z \sim 5$–10 nT the ratio of the gradients is of the order 2–3. Upon further decrease of B_z below 5 nT vertical gradient increases sharply (CS intensifies).

Changes of magnetic field at the CS boundary, B_0, cannot be reliably monitored for such type of observations, however some estimates of dynamics of B_0/B_{ext} ratio can be obtained basing on general assumptions. Magnetic flux in a thin embedded CS with a thickness of the order of ion Larmor radius is $F_0 \sim 2$–$5B_0\rho = 0.006$–0.03 Wb/m (per unit width in y and proton temperature 4 keV), which is at least ten times less than the estimate of the total magnetic flux in the plasma sheet $F_{ext} \sim 0.4$–0.5 Wb/m at a distance from the Earth about 15 R_E (note, flux F_0 does not depend on B_0 because $\rho \sim 1/B_0$). Such a large difference in flux values imposes significant restrictions on possible B_0/B_{ext}. First of all, the minimal possible B_0/B_{ext} (minimal identifiable embedding) is set if current density in embedded and background sheets are equal: $B_0/L = B_{ext}/L_{ext}$. Converting this to magnetic flux units and neglecting all coefficients of the order of unity under the square root one could obtain:

$$(B_0/B_{ext})_{min} \sim \sqrt{F_0/F_{ext}} \qquad (1)$$

Thus, for our statistics, the estimate $(B_0/B_{ext})_{min}$ is 0.2–0.3, which is consistent with observations (Fig. 6a, for example).

On the other hand, let us consider the opposite case with $B_0 \approx B_{ext}$, $F_0 \ll F_{ext}$. Then embedded sheet is a very thin plasma layer, in a thick background sheet, dominated by the magnetic pressure. Since the development of an embedded sheet begins during the growth phase within a sufficiently thick plasma sheet filled by hot and relatively dense plasma, such a configuration is unrealistic in the quiet magnetotail. During growth phases B_0/B_{ext} will be likely closer to $\sqrt{F_0/F_{ext}}$. After the onset a substantial part of plasma and magnetic flux are lost from the plasma sheet and $B_0 \approx B_{ext}$ is more feasible (Petrukovich et al. 2011). Configuration with $B_0 \approx B_{ext}$ is more likely also in the distant magnetotail, where F_{ext} is small.

9 Discussion and Conclusions

The availability of Cluster mission four-point data allowed for the first time to obtain information on spatial scales in the magnetotail on a regular basis and proceed with quantitative analysis of current sheet plasma physics as well as comparison with theoretical models.

Embedded configurations with thin intense CS inside a much wider plasma sheet were found to be typical for the magnetotail. Thickness of an embedded current density profile is about 1000–3000 km, which is only about 1–3 proton Larmor radii. Moreover, in the vicinity of the reconnection region one can expect to observe intense CSs with the thickness about a fraction of ρ (Nakamura et al. 2006, 2008; Artemyev et al. 2013a).

Effect of CS embedding manifests in ion kinetics. The strong increase of the current density without a comparable increase of the plasma density indicates that there is a small population of ions carrying significant cross-tail current. Such a population should have specific velocity distribution with the substantial asymmetry along the v_y direction. Indeed, detailed investigation of embedded CS reveals the beam-like ion distribution responsible for almost entire current density in embedded CS. Similar ion distributions were observed by THEMIS spacecraft (Zhou et al. 2009). The remaining plasma ($\sim$80–90 % of plasma density) can be considered as a background with a the uniform density profile across thin CS.

Oxygen ions (or high-temperature protons) influence CS structure. The increase of O^+ results in the increase of the CS thickness in agreement with modeling (Zelenyi et al. 2006). This effect confirms kinetic nature of CS and can play the important role in CS stability (Shay and Swisdak 2004; Markidis et al. 2011; Karimabadi et al. 2011; Liu et al. 2014).

The variations of ion (proton) and electron temperatures across CS are often more significant than the variations of plasma density. The scale of T_p variation is 2–10 times larger than the CS thickness. The development of some plasma instabilities (e.g., lower hybrid drift instability, drift kink/sausage instability) strongly depends on spatial gradients of the plasma density and/or temperature (Büchner and Kuska 1999; Daughton 1999a, 1999b, 2003; Yoon and Lui 2001; Tummel et al. 2014). Thus, in a realistic configuration these instabilities can be driven by the temperature gradients rather than by plasma density inhomogeneities.

Due to the small thickness of embedded CSs, ions are significantly demagnetized. The decoupling of ion and electron motions results in generation of polarization electrostatic field responsible for relatively strong cross-field electric drifts. Such drifts in weakly 2D configuration redistribute ion and electron contributions to the cross-tail current density in CSs with weak B_y and small $B_z \ll B_0$, making electron current comparable to the ion one. We conclude that electrostatic fields responsible for these drifts although very weak and hardly measurable are very important and should be included in any model pretending on realistic CS description.

Presence of the intense tilted sheets requires to reconsider criteria of general classification in the magnetotail. For example, the presence of intense currents $\sim$10–20 nA/m^2 cannot be considered as an direct indication of magnetotail reconfiguration before a substorm (Lui 2004), because these values of current density are typical for tilted CSs even in a quiet time.

Formation of thin CS during substorms reveals itself in three variables B_z, j_y, $\partial B_z/\partial x$. Substorm onsets (sheet instability) are associated with rather moderately intense current sheet (current density below 10 nA/m^2) and relatively small ratio B_0/B_{ext}. Much more intense current sheets are observed after onsets. This is a significant observational constraint for instability models. Predominantly two-dimensional structure with substantial gradients in X direction is gradually changing to predominantly one-dimensional with the negligible gradient along x-axis. Gradients along X play the critical role in many recent models of CS stability (Erkaev et al. 2009; Pritchett and Coroniti 2010; Sitnov and Schindler 2010, 2013; Pritchett and Coroniti 2013; Korovinskiy et al. 2013).

In summary, data collected by Cluster mission demonstrate that any accurate modelling of the magnetotail CS requires taking into account many specific properties: embedding, temperature gradients, polarization electrostatic field, redistribution of currents and the specific shape of ion distribution function containing beam-like features. All these effects can influence CS stability and determine rate of magnetic reconnection (Burkhart et al. 1992b; Krallmann et al. 1994; Sitnov et al. 2002; Zelenyi et al. 2008, 2010; Karimabadi et al. 2003a, 2003b). Modern theories of stability concentrate on specific $B_z(x)$ profiles (Pritchett and Coroniti 2010, 2011; Sitnov and Schindler 2010; Sitnov et al. 2013), while many other CS properties are waiting for their comprehensive numerical modeling.

Some aspects of CS physics remain beyond this review and potential of Cluster data. First of all, we did not consider here B_y (shear) magnetic field component in CSs. For the majority of observed CSs this component is almost absent or very small $B_y \leq B_z$, but cases with $B_y \geq B_z$ are still abundant (Nakamura et al. 2008; Petrukovich 2009, 2011). Cluster observations have shown that the B_y component can be significantly enhanced by local field-aligned currents (Rong et al. 2012; Grigorenko et al. 2013). In this case, we may observe a parabolic profile of $B_y \sim 1 - (B_x/B_0)^2$ well described in theoretical models (Artemyev 2011; Mingalev et al. 2012; Vasko et al. 2014). However, the origin of such specific CS configuration with enhanced B_y is still unknown. In the events with a large magnetic shear component, field-aligned (rather than transverse) currents dominate, which are poorly addressed by existing theories. Moreover the theory of charged particle motion in CSs with B_y and a finite convection electric field E_y predicts the strong chaotization of ion trajectories and reduction of the cross-tail current (Artemyev et al. 2013b, 2014a), which is not readily observed. Thus, further experimental and theoretical investigation of CSs with intermediate B_y values ($|B_y| < B_0$) seems to be necessary.

While the vertical structure of the magnetotail CS is well investigated, the horizontal (along the tail) configuration is much less known. Cluster measurements of local B_z gradients in the near-Earth tail are not quite sufficient to restore the instantaneous profiles of crucial sheet characteristics (B_0, B_z, j_y) along X. Knowledge of these profiles is important for the identification of the most effective instability or select between specific sheet models. Only several statistical studies (Kan and Baumjohann 1990; Wang et al. 2009; Kissinger et al. 2012; Rong et al. 2014), are available now which considered averaged data collected during several years of spacecraft measurements. Potentially data necessary for reconstruction of the instantaneous CS configuration can be provided by five THEMIS spacecraft distributed along the magnetotail (Artemyev et al. 2013b).

The third important deficiency of our knowledge is insufficient understanding of electron current variability and structure of very thin and very intense CSs with domination of electron currents. It can be resolved only with future experiments having substantially better temporal and spatial resolutions such as the MMS, which is planned to be launched in 2015. It includes four spacecraft with separation 10–100 km, specially designed to study the electron physics in reconnecting current sheets. Cluster mission still remaining operative with in-depth investigation of 100–1000 km scales provide an important baseline for future studies.

Acknowledgements   Work of A.A.P., A.V.A., I.Y.V. is supported by Russian Scientific Fund (project # 14-12-00824). R.N. is supported by the Austrian Science Fund (FWF I429-N16, P23862-N16).

References

V. Angelopoulos, C.F. Kennel, F.V. Coroniti, R. Pellat, H.E. Spence, M.G. Kivelson, R.J. Walker, W. Baumjohann, W.C. Feldman, J.T. Gosling, Characteristics of ion flow in the quiet state of the inner plasma sheet. Geophys. Res. Lett. **20**, 1711–1714 (1993). doi:10.1029/93GL00847
J. Arons, Pulsar wind nebulae as cosmic pevatrons: a current sheet's tale. Space Sci. Rev. **173**, 341–367 (2012). doi:10.1007/s11214-012-9885-1

A.V. Artemyev, A model of one-dimensional current sheet with parallel currents and normal component of magnetic field. Phys. Plasmas **18**(2), 022104 (2011). doi:10.1063/1.3552141

A.V. Artemyev, L.M. Zelenyi, Kinetic structure of current sheets in the Earth magnetotail. Space Sci. Rev. **178**, 419–440 (2013). doi:10.1007/s11214-012-9954-5

A.V. Artemyev, A.I. Neishtadt, L.M. Zelenyi, Ion motion in the current sheet with sheared magnetic field—Part 1: Quasi-adiabatic theory. Nonlinear Process. Geophys. **20**(1), 163–178 (2013a). doi:10.5194/npg-20-163-2013. http://www.nonlin-processes-geophys.net/20/163/2013/

A.V. Artemyev, A.I. Neishtadt, L.M. Zelenyi, Ion motion in the current sheet with sheared magnetic field—Part 2: Non-adiabatic effects. Nonlinear Process. Geophys. **20**, 899–919 (2013b). doi:10.5194/npg-20-899-2013

A.V. Artemyev, A.I. Neishtadt, L.M. Zelenyi, Rapid geometrical chaotization in slow-fast Hamiltonian systems. Phys. Rev. E **89**(6), 060902 (2014a). doi:10.1103/PhysRevE.89.060902

A.V. Artemyev, I.Y. Vasko, S. Kasahara, Thin current sheets in the Jovian magnetotail. Planet. Space Sci. **96**, 133–145 (2014b). doi:10.1016/j.pss.2014.03.012

A.V. Artemyev, A.A. Petrukovich, L.M. Zelenyi, H.V. Malova, V.Y. Popov, R. Nakamura, A. Runov, S. Apatenkov, Comparison of multi-point measurements of current sheet structure and analytical models. Ann. Geophys. **26**, 2749–2758 (2008)

A.V. Artemyev, A.A. Petrukovich, L.M. Zelenyi, R. Nakamura, H.V. Malova, V.Y. Popov, Thin embedded current sheets: cluster observations of ion kinetic structure and analytical models. Ann. Geophys. **27**, 4075–4087 (2009)

A.V. Artemyev, A.A. Petrukovich, R. Nakamura, L.M. Zelenyi, Proton velocity distribution in thin current sheets: cluster observations and theory of transient trajectories. J. Geophys. Res. **115**, 12255 (2010). doi:10.1029/2010JA015702

A.V. Artemyev, A.A. Petrukovich, R. Nakamura, L.M. Zelenyi, Cluster statistics of thin current sheets in the Earth magnetotail: specifics of the dawn flank, proton temperature profiles and electrostatic effects. J. Geophys. Res. **116**, 0923 (2011a). doi:10.1029/2011JA016801

A.V. Artemyev, W. Baumjohann, A.A. Petrukovich, R. Nakamura, I. Dandouras, A. Fazakerley, Proton/electron temperature ratio in the magnetotail. Ann. Geophys. **29**, 2253–2257 (2011b). doi:10.5194/angeo-29-2253-2011

A.V. Artemyev, A.A. Petrukovich, R. Nakamura, L.M. Zelenyi, Adiabatic electron heating in the magnetotail current sheet: cluster observations and analytical models. J. Geophys. Res. **117**, 06219 (2012). doi:10.1029/2012JA017513

A.V. Artemyev, A.A. Petrukovich, A.G. Frank, R. Nakamura, L.M. Zelenyi, Intense current sheets in the magnetotail: peculiarities of electron physics. J. Geophys. Res. **118**, 2789–2799 (2013a). doi:10.1002/jgra.50297

A.V. Artemyev, A.A. Petrukovich, R. Nakamura, L.M. Zelenyi, Profiles of electron temperature and B_z along Earth's magnetotail. Ann. Geophys. **31**, 1109–1114 (2013b). doi:10.5194/angeo-31-1109-2013

Y. Asano, T. Mukai, M. Hoshino, Y. Saito, H. Hayakawa, T. Nagai, Evolution of the thin current sheet in a substorm observed by Geotail. J. Geophys. Res. **108**, 1189 (2003). doi:10.1029/2002JA009785

Y. Asano, T. Mukai, M. Hoshino, Y. Saito, H. Hayakawa, T. Nagai, Current sheet structure around the near-Earth neutral line observed by Geotail. J. Geophys. Res. **109**, 2212 (2004a). doi:10.1029/2003JA010114

Y. Asano, T. Mukai, M. Hoshino, Y. Saito, H. Hayakawa, T. Nagai, Statistical study of thin current sheet evolution around substorm onset. J. Geophys. Res. **109**, 5213 (2004b). doi:10.1029/2004JA010413

M.J. Aschwanden, Particle acceleration and kinematics in solar flares—A synthesis of recent observations and theoretical concepts. Space Sci. Rev. **101**, 1–227 (2002). doi:10.1023/A:1019712124366

M. Ashour-Abdalla, J. Buechner, L.M. Zelenyi, The quasi-adiabatic ion distribution in the central plasma sheet and its boundary layer. J. Geophys. Res. **96**, 1601–1609 (1991). doi:10.1029/90JA01921

M. Ashour-Abdalla, J.P. Berchem, J. Buechner, L.M. Zelenyi, Shaping of the magnetotail from the mantle—Global and local structuring. J. Geophys. Res. **98**, 5651–5676 (1993). doi:10.1029/92JA01662

D.N. Baker, T.I. Pulkkinen, V. Angelopoulos, W. Baumjohann, R.L. McPherron, Neutral line model of substorms: past results and present view. J. Geophys. Res. **101**, 12975–13010 (1996). doi:10.1029/95JA03753

A. Balogh, C.M. Carr, M.H. Acuña, M.W. Dunlop, T.J. Beek, P. Brown, K. Fornaçon, E. Georgescu, K. Glassmeier, J. Harris, G. Musmann, T. Oddy, K. Schwingenschuh, The cluster magnetic field investigation: overview of in-flight performance and initial results. Ann. Geophys. **19**, 1207–1217 (2001). doi:10.5194/angeo-19-1207-2001

W. Baumjohann, G. Paschmann, C.A. Cattell, Average plasma properties in the central plasma sheet. J. Geophys. Res. **94**, 6597–6606 (1989). doi:10.1029/JA094iA06p06597

K.W. Behannon, L.F. Burlaga, N.F. Ness, The Jovian magnetotail and its current sheet. J. Geophys. Res. **86**, 8385–8401 (1981). doi:10.1029/JA086iA10p08385

M.K. Bird, D.B. Beard, The self-consistent geomagnetic tail under static conditions. Planet. Space Sci. **20**, 2057–2072 (1972). doi:10.1016/0032-0633(72)90062-1

J. Birn, Self-consistent magnetotail theory—General solution for the quiet tail with vanishing field-aligned currents. J. Geophys. Res. **84**, 5143–5152 (1979). doi:10.1029/JA084iA09p05143

J. Birn, E.R. Priest, *Reconnection of Magnetic Fields: Magnetohydrodynamics and Collisionless Theory and Observations* 2007

J. Birn, K. Schindler, M. Hesse, Thin electron current sheets and their relation to auroral potentials. J. Geophys. Res. **109**, 2217 (2004). doi:10.1029/2003JA010303

J. Birn, R. Sommer, K. Schindler, Open and closed magnetospheric tail configurations and their stability. Astrophys. Space Sci. **35**, 389–402 (1975). doi:10.1007/BF00637005

J. Büchner, J. Kuska, Sausage mode instability of thin current sheets as a cause of magnetospheric substorms. Ann. Geophys. **17**, 604–612 (1999). doi:10.1007/s005850050788

J. Büchner, L.M. Zelenyi, Regular and chaotic charged particle motion in magnetotaillike field reversals. I—Basic theory of trapped motion. J. Geophys. Res. **94**, 11821–11842 (1989). doi:10.1029/JA094iA09p11821

J.L. Burch, T.E. Moore, R.B. Torbert, B. Giles, MMS Overview and Science Objectives. Space Sci. Rev. (2014)

G.R. Burkhart, J. Chen, Particle motion in x-dependent Harris-like magnetotail models. J. Geophys. Res. **98**, 89–97 (1993). doi:10.1029/92JA01528

G.R. Burkhart, J.F. Drake, P.B. Dusenbery, T.W. Speiser, A particle model for magnetotail neutral sheet equilibria. J. Geophys. Res. **97**, 13799–13815 (1992a). doi:10.1029/92JA00495

G.R. Burkhart, J.F. Drake, P.B. Dusenbery, T.W. Speiser, Ion tearing in a magnetotail configuration with an embedded thin current sheet. J. Geophys. Res. **97**, 16749–16756 (1992b). doi:10.1029/92JA01523

S.W.H. Cowley, R. Pellat, A note on adiabatic solutions of the one-dimensional current sheet problem. Planet. Space Sci. **27**, 265–271 (1979). doi:10.1016/0032-0633(79)90069-2

J. Credland, G. Mecke, J. Ellwood, The cluster mission: ESA'S spacefleet to the magnetosphere. Space Sci. Rev. **79**, 33–64 (1997). doi:10.1023/A:1004914822769

W. Daughton, The unstable eigenmodes of a neutral sheet. Phys. Plasmas **6**, 1329–1343 (1999a). doi:10.1063/1.873374

W. Daughton, Two-fluid theory of the drift kink instability. J. Geophys. Res. **104**, 28701–28708 (1999b). doi:10.1029/1999JA900388

W. Daughton, Electromagnetic properties of the lower-hybrid drift instability in a thin current sheet. Phys. Plasmas **10**, 3103–3119 (2003). doi:10.1063/1.1594724

E.A. Davey, M. Lester, S.E. Milan, R.C. Fear, C. Forsyth, The orientation and current density of the magnetotail current sheet: a statistical study of the effect of geomagnetic conditions. J. Geophys. Res. **117**, 7217 (2012). doi:10.1029/2012JA017715

M.K. Dougherty, L.W. Esposito, S.M. Krimigis, *Saturn from Cassini-Huygens* 2009. doi:10.1007/978-1-4020-9217-6

M.W. Dunlop, A. Balogh, K.-H. Glassmeier, P. Robert, Four-point cluster application of magnetic field analysis tools: the curlometer. J. Geophys. Res. **107**, 1384 (2002). doi:10.1029/2001JA005088

J.W. Eastwood, Consistency of fields and particle motion in the 'Speiser' model of the current sheet. Planet. Space Sci. **20**, 1555–1568 (1972). doi:10.1016/0032-0633(72)90182-1

J.W. Eastwood, The warm current sheet model, and its implications on the temporal behaviour of the geomagnetic tail. Planet. Space Sci. **22**, 1641–1668 (1974). doi:10.1016/0032-0633(74)90108-1

N.V. Erkaev, V.S. Semenov, I.V. Kubyshkin, M.V. Kubyshkina, H.K. Biernat, MHD model of the flapping motions in the magnetotail current sheet. J. Geophys. Res. **114**, 3206 (2009). doi:10.1029/2008JA013728

C.P. Escoubet, M. Fehringer, M. Goldstein, Introduction: the cluster mission. Ann. Geophys. **19**, 1197–1200 (2001). doi:10.5194/angeo-19-1197-2001

P. Francfort, R. Pellat, Magnetic merging in collisionless plasmas. Geophys. Res. Lett. **3**, 433–436 (1976). doi:10.1029/GL003i008p00433

J.T. Gosling, Magnetic reconnection in the solar wind. Space Sci. Rev. **172**, 187–200 (2012). doi:10.1007/s11214-011-9747-2

E.E. Grigorenko, H.V. Malova, A.V. Artemyev, O.V. Mingalev, E.A. Kronberg, R. Koleva, P.W. Daly, J.B. Cao, J.-A. Sauvaud, C.J. Owen, L.M. Zelenyi, Current sheet structure and kinetic properties of plasma flows during a near-Earth magnetic reconnection under the presence of a guide field. J. Geophys. Res. **118**, 3265–3287 (2013). doi:10.1002/jgra.50310

G. Gustafsson, M. André, T. Carozzi, A.I. Eriksson, C.-G. Fälthammar, R. Grard, G. Holmgren, J.A. Holtet, N. Ivchenko, T. Karlsson, Y. Khotyaintsev, S. Klimov, H. Laakso, P.-A. Lindqvist, B. Lybekk, G. Marklund, F. Mozer, K. Mursula, A. Pedersen, B. Popielawska, S. Savin, K. Stasiewicz, P. Tanskanen, A. Vaivads, J.-E. Wahlund, First results of electric field and density observations by cluster EFW based on initial months of operation. Ann. Geophys. **19**, 1219–1240 (2001). doi:10.5194/angeo-19-1219-2001

E.G. Harris, On a plasma sheet separating regions of oppositely directed magnetic field. Nuovo Cimento **23**, 115–123 (1962)

M. Hoshino, A. Nishida, T. Mukai, Y. Saito, T. Yamamoto, S. Kokubun, Structure of plasma sheet in magnetotail: double-peaked electric current sheet. J. Geophys. Res. **101**, 24775–24786 (1996). doi:10.1029/96JA02313

P.L. Israelevich, A.I. Ershkovich, R. Oran, Current carriers in the bifurcated tail current sheet: ions or electrons? J. Geophys. Res. **113**, 4215 (2008). doi:10.1029/2007JA012541

C.M. Jackman, C.S. Arridge, N. André, F. Bagenal, J. Birn, M.P. Freeman, X. Jia, A. Kidder, S.E. Milan, A. Radioti, J.A. Slavin, M.F. Vogt, M. Volwerk, A.P. Walsh, Large-scale structure and dynamics of the magnetotails of Mercury, Earth, Jupiter and Saturn. Space Sci. Rev. **182**, 85–154 (2014). doi:10.1007/s11214-014-0060-8

A.D. Johnstone, C. Alsop, S. Burge, P.J. Carter, A.J. Coates, A.J. Coker, A.N. Fazakerley, M. Grande, R.A. Gowen, C. Gurgiolo, B.K. Hancock, B. Narheim, A. Preece, P.H. Sheather, J.D. Winningham, R.D. Woodliffe, Peace: a plasma electron and current experiment. Space Sci. Rev. **79**, 351–398 (1997). doi:10.1023/A:1004938001388

J.R. Kan, On the structure of the magnetotail current sheet. J. Geophys. Res. **78**, 3773–3781 (1973). doi:10.1029/JA078i019p03773

J.R. Kan, W. Baumjohann, Isotropized magnetic-moment equation of state for the central plasma sheet. Geophys. Res. Lett. **17**, 271–274 (1990). doi:10.1029/GL017i003p00271

H. Karimabadi, W. Daughton, P.L. Pritchett, D. Krauss-Varban, Ion-ion kink instability in the magnetotail: 1. Linear theory. J. Geophys. Res. **108**, 1400 (2003a). doi:10.1029/2003JA010026

H. Karimabadi, P.L. Pritchett, W. Daughton, D. Krauss-Varban, Ion-ion kink instability in the magnetotail: 2. Three-dimensional full particle and hybrid simulations and comparison with observations. J. Geophys. Res. **108**, 1401 (2003b). doi:10.1029/2003JA010109

H. Karimabadi, V. Roytershteyn, C.G. Mouikis, L.M. Kistler, W. Daughton, Flushing effect in reconnection: effects of minority species of oxygen ions. Planet. Space Sci. **59**, 526–536 (2011). doi:10.1016/j.pss.2010.07.014

C.F. Kennel, Magnetospheres of the planets. Space Sci. Rev. **14**, 511–533 (1973). doi:10.1007/BF00214759

J. Kissinger, R.L. McPherron, T.-S. Hsu, V. Angelopoulos, Diversion of plasma due to high pressure in the inner magnetosphere during steady magnetospheric convection. J. Geophys. Res. **117**, 5206 (2012). doi:10.1029/2012JA017579

D.B. Korovinskiy, A. Divin, N.V. Erkaev, V.V. Ivanova, I.B. Ivanov, V.S. Semenov, G. Lapenta, S. Markidis, H.K. Biernat, M. Zellinger, MHD modeling of the double-gradient (kink) magnetic instability. J. Geophys. Res. **118**, 1146–1158 (2013). doi:10.1002/jgra.50206

T. Krallmann, J. Dreher, K. Schindler, On the stability of the ion-tearing mode in equilibria with embedded thin current sheets, in *Int. Conf. Substorms*, 1994, pp. 499–503

A.P. Kropotkin, H.V. Malova, M.I. Sitnov, Self-consistent structure of a thin anisotropic current sheet. J. Geophys. Res. **102**, 22099–22106 (1997). doi:10.1029/97JA01316

Y.H. Liu, L.M. Kistler, C.G. Mouikis, V. Roytershteyn, H. Karimabadi, The scale of the magnetotail reconnecting current sheet in the presence of O^+. Geophys. Res. Lett. **41**, 4819–4827 (2014). doi:10.1002/2014GL060440

A.T.Y. Lui, Potential plasma instabilities for substorm expansion onsets. Space Sci. Rev. **113**, 127–206 (2004). doi:10.1023/B:SPAC.0000042942.00362.4e

S. Markidis, G. Lapenta, L. Bettarini, M. Goldman, D. Newman, L. Andersson, Kinetic simulations of magnetic reconnection in presence of a background O^+ population. J. Geophys. Res. (2011). doi:10.1029/2011JA016429

D.J. McComas, H.E. Spence, C.T. Russell, M.A. Saunders, The average magnetic field draping and consistent plasma properties of the Venus magnetotail. J. Geophys. Res. **91**, 7939–7953 (1986a). doi:10.1029/JA091iA07p07939

D.J. McComas, S.J. Bame, C.T. Russell, R.C. Elphic, The near-Earth cross-tail current sheet—detailed ISEE 1 and 2 case studies. J. Geophys. Res. **91**, 4287–4301 (1986b). doi:10.1029/JA091iA04p04287

D.J. McComas, J.T. Gosling, C.T. Russell, J.A. Slavin, Magnetotails at unmagnetized bodies—comparison of comet Giacobini–Zinner and Venus. J. Geophys. Res. **92**, 10111–10117 (1987). doi:10.1029/JA092iA09p10111

S. Minami, A.I. Podgornyi, I.M. Podgornyi, Laboratory evidence of earthward electric field in the magnetotail current sheet. Geophys. Res. Lett. **20**, 9–12 (1993). doi:10.1029/92GL02492

O.V. Mingalev, I.V. Mingalev, M.N. Mel'nik, A.V. Artemyev, H.V. Malova, V.Y. Popov, S. Chao, L.M. Zelenyi, Kinetic models of current sheets with a sheared magnetic field. Plasma Phys. Rep. **38**, 300–314 (2012). doi:10.1134/S1063780X12030063

D.G. Mitchell, D.J. Williams, C.Y. Huang, L.A. Frank, C.T. Russell, Current carriers in the near-Earth cross-tail current sheet during substorm growth phase. Geophys. Res. Lett. **17**, 583–586 (1990). doi:10.1029/GL017i005p00583

R. Nakamura, W. Baumjohann, A. Runov, Y. Asano, Thin current sheets in the magnetotail observed by cluster. Space Sci. Rev. **122**, 29–38 (2006). doi:10.1007/s11214-006-6219-1

R. Nakamura, W. Baumjohann, M. Fujimoto, Y. Asano, A. Runov, C.J. Owen, A.N. Fazakerley, B. Klecker, H. Rème, E.A. Lucek, M. Andre, Y. Khotyaintsev, Cluster observations of an ion-scale current sheet in the magnetotail under the presence of a guide field. J. Geophys. Res. **113**, 7 (2008). doi:10.1029/2007JA012760

N.F. Ness, The Earth's magnetic tail. J. Geophys. Res. **70**, 2989–3005 (1965). doi:10.1029/JZ070i013p02989

N.F. Ness, M.H. Acuna, L.F. Burlaga, J.E.P. Connerney, R.P. Lepping, Magnetic fields at Neptune. Science **246**, 1473–1478 (1989). doi:10.1126/science.246.4936.1473

E.N. Parker, Spontaneous current sheets in magnetic fields: with applications to stellar x-rays, in *Spontaneous Current Sheets in Magnetic Fields: with Applications to Stellar x-Rays*. International Series in Astronomy and Astrophysics, vol. 1 (Oxford University Press, New York, 1994)

G. Paschmann, S.J. Schwartz, *Issi Book on Analysis Methods for Multi-Spacecraft Data*. ESA Special Publication, vol. 449 2000

A.A. Petrukovich, Dipole tilt effects in plasma sheet by: statistical model and extreme values. Ann. Geophys. **27**, 1343–1352 (2009). doi:10.5194/angeo-27-1343-2009

A.A. Petrukovich, Origins of plasma sheet B_y. J. Geophys. Res. **116**, 7217 (2011). doi:10.1029/2010JA016386

A.A. Petrukovich, T. Mukai, S. Kokubun, S.A. Romanov, Y. Saito, T. Yamamoto, L.M. Zelenyi, Substorm-associated pressure variations in the magnetotail plasma sheet and lobe. J. Geophys. Res. **104**, 4501–4514 (1999). doi:10.1029/98JA02418

A.A. Petrukovich, W. Baumjohann, R. Nakamura, A. Balogh, T. Mukai, K.-H. Glassmeier, H. Reme, B. Klecker, Plasma sheet structure during strongly northward IMF. J. Geophys. Res. **108**, 1258 (2003). doi:10.1029/2002JA009738

A.A. Petrukovich, T.L. Zhang, W. Baumjohann, R. Nakamura, A. Runov, A. Balogh, C. Carr, Oscillatory magnetic flux tube slippage in the plasma sheet. Ann. Geophys. **24**, 1695–1704 (2006)

A.A. Petrukovich, W. Baumjohann, R. Nakamura, A. Runov, A. Balogh, H. Rème, Thinning and stretching of the plasma sheet. J. Geophys. Res. **112**, 10213 (2007). doi:10.1029/2007JA012349

A.A. Petrukovich, W. Baumjohann, R. Nakamura, A. Runov, Formation of current density profile in tilted current sheets. Ann. Geophys. **26**, 3669–3676 (2008)

A.A. Petrukovich, W. Baumjohann, R. Nakamura, H. Rème, Tailward and earthward flow onsets observed by cluster in a thin current sheet. J. Geophys. Res. **114**, 9203 (2009). doi:10.1029/2009JA014064

A.A. Petrukovich, A.V. Artemyev, H.V. Malova, V.Y. Popov, R. Nakamura, L.M. Zelenyi, Embedded current sheets in the Earth magnetotail. J. Geophys. Res. **116**, 1–25 (2011). doi:10.1029/2010JA015749

A.A. Petrukovich, A.V. Artemyev, R. Nakamura, E.V. Panov, W. Baumjohann, Cluster observations of dBz/dx during growth phase magnetotail stretching intervals. J. Geophys. Res. **118**, 5720–5730 (2013). doi:10.1002/jgra.50550

E. Priest, T. Forbes, *Magnetic Reconnection* 2000

P.L. Pritchett, F.V. Coroniti, A kinetic ballooning/interchange instability in the magnetotail. J. Geophys. Res. **115**, 06301 (2010). doi:10.1029/2009JA014752

P.L. Pritchett, F.V. Coroniti, Plasma sheet disruption by interchange-generated flow intrusions. Geophys. Res. Lett. **381**, 10102 (2011). doi:10.1029/2011GL047527

P.L. Pritchett, F.V. Coroniti, Structure and consequences of the kinetic ballooning/interchange instability in the magnetotail. J. Geophys. Res. **118**, 146–159 (2013). doi:10.1029/2012JA018143

H. Rème, C. Aoustin, J.M. Bosqued, I. Dandouras, B. Lavraud, J.A. Sauvaud, A. Barthe, J. Bouyssou, T. Camus, O. Coeur-Joly, A. Cros, J. Cuvilo, F. Ducay, Y. Garbarowitz, J.L. Medale, E. Penou, H. Perrier, D. Romefort, J. Rouzaud, C. Vallat, D. Alcaydé, C. Jacquey, C. Mazelle, C. D'Uston, E. Möbius, L.M. Kistler, K. Crocker, M. Granoff, C. Mouikis, M. Popecki, M. Vosbury, B. Klecker, D. Hovestadt, H. Kucharek, E. Kuenneth, G. Paschmann, M. Scholer, N. Sckopke, E. Seidenschwang, C.W. Carlson, D.W. Curtis, C. Ingraham, R.P. Lin, J.P. McFadden, G.K. Parks, T. Phan, V. Formisano, E. Amata, M.B. Bavassano-Cattaneo, P. Baldetti, R. Bruno, G. Chionchio, A. di Lellis, M.F. Marcucci, G. Pallocchia, A. Korth, P.W. Daly, B. Graeve, H. Rosenbauer, V. Vasyliunas, M. McCarthy, M. Wilber, L. Eliasson, R. Lundin, S. Olsen, E.G. Shelley, S. Fuselier, A.G. Ghielmetti, W. Lennartsson, C.P. Escoubet, H. Balsiger, R. Friedel, J. Cao, R.A. Kovrazhkin, I. Papamastorakis, R. Pellat, J. Scudder, B. Sonnerup, First multispacecraft ion measurements in and near the Earth's magnetosphere with the identical cluster ion spectrometry (CIS) experiment. Ann. Geophys. **19**, 1303–1354 (2001). doi:10.5194/angeo-19-1303-2001

F.J. Rich, V.M. Vasyliunas, R.A. Wolf, On the balance of stresses in the plasma sheet. J. Geophys. Res. **77**, 4670–4676 (1972). doi:10.1029/JA077i025p04670

Z.J. Rong, C. Shen, A.A. Petrukovich, W.X. Wan, Z.X. Liu, The analytic properties of the flapping current sheets in the Earth magnetotail. Planet. Space Sci. **58**, 1215–1229 (2010). doi:10.1016/j.pss.2010.04.016

Z.J. Rong, W.X. Wan, C. Shen, X. Li, M.W. Dunlop, A.A. Petrukovich, L.-N. Hau, T.L. Zhang, H. Rème, A.M. Du, E. Lucek, Profile of strong magnetic field B_y component in magnetotail current sheets. J. Geophys. Res. **117**, 6216 (2012). doi:10.1029/2011JA017402

Z.J. Rong, W.X. Wan, C. Shen, A.A. Petrukovich, W. Baumjohann, M.W. Dunlop, Y.C. Zhang, Radial distribution of magnetic field in Earth magnetotail current sheet. Planet. Space Sci. (2014). doi:10.1016/j.pss.2014.07.014

A. Runov, V.A. Sergeev, W. Baumjohann, R. Nakamura, S. Apatenkov, Y. Asano, M. Volwerk, Z. Vörös, T.L. Zhang, A. Petrukovich, A. Balogh, J. Sauvaud, B. Klecker, H. Rème, Electric current and magnetic field geometry in flapping magnetotail current sheets. Ann. Geophys. **23**, 1391–1403 (2005a)

A. Runov, V.A. Sergeev, R. Nakamura, W. Baumjohann, T.L. Zhang, Y. Asano, M. Volwerk, Z. Vörös, A. Balogh, H. Rème, Reconstruction of the magnetotail current sheet structure using multi-point cluster measurements. Planet. Space Sci. **53**, 237–243 (2005b). doi:10.1016/j.pss.2004.09.049

A. Runov, V.A. Sergeev, R. Nakamura, W. Baumjohann, S. Apatenkov, Y. Asano, T. Takada, M. Volwerk, Z. Vörös, T.L. Zhang, J. Sauvaud, H. Rème, A. Balogh, Local structure of the magnetotail current sheet: 2001 cluster observations. Ann. Geophys. **24**, 247–262 (2006)

K. Schindler, A self-consistent theory of the tail of the magnetosphere, in *Earth's Magnetospheric Processes*, ed. by B.M. McCormac Astrophysics and Space Science Library, vol. 32, 1972, p. 200

K. Schindler, A theory of the substorm mechanism. J. Geophys. Res. **79**, 2803–2810 (1974). doi:10.1029/JA079i019p02803

K. Schindler, Theories of tail structures. Space Sci. Rev. **23**, 365–374 (1979). doi:10.1007/BF00172245

K. Schindler, *Physics of Space Plasma Activity* (Cambridge University Press, Cambridge, 2006). doi:10.2277/0521858976

K. Schindler, J. Birn, Magnetotail theory. Space Sci. Rev. **44**, 307–355 (1986). doi:10.1007/BF00200819

K. Schindler, J. Birn, Models of two-dimensional embedded thin current sheets from Vlasov theory. J. Geophys. Res. **107**, 1193 (2002). doi:10.1029/2001JA000304

K. Schindler, J. Birn, M. Hesse, Kinetic model of electric potentials in localized collisionless plasma structures under steady quasi-gyrotropic conditions. Phys. Plasmas **19**(8), 082904 (2012). doi:10.1063/1.4747162

V.A. Sergeev, D.G. Mitchell, C.T. Russell, D.J. Williams, Structure of the tail plasma/current sheet at $\sim 11 R_E$ and its changes in the course of a substorm. J. Geophys. Res. **98**, 17345–17366 (1993). doi:10.1029/93JA01151

V.A. Sergeev, D.A. Sormakov, S.V. Apatenkov, W. Baumjohann, R. Nakamura, A.V. Runov, T. Mukai, T. Nagai, Survey of large-amplitude flapping motions in the midtail current sheet. Ann. Geophys. **24**, 2015–2024 (2006)

A.S. Sharma, R. Nakamura, A. Runov, E.E. Grigorenko, H. Hasegawa, M. Hoshino, P. Louarn, C.J. Owen, A. Petrukovich, J. Sauvaud, V.S. Semenov, V.A. Sergeev, J.A. Slavin, B.U.Ö Sonnerup, L.M. Zelenyi, G. Fruit, S. Haaland, H. Malova, K. Snekvik, Transient and localized processes in the magnetotail: a review. Ann. Geophys. **26**, 955–1006 (2008)

M.A. Shay, M. Swisdak, Three-species collisionless reconnection: effect of O^+ on magnetotail reconnection. Phys. Rev. Lett. **93**(17), 175001 (2004). doi:10.1103/PhysRevLett.93.175001

C. Shen, Z.X. Liu, X. Li, M. Dunlop, E. Lucek, Z.J. Rong, Z.Q. Chen, C.P. Escoubet, H.V. Malova, A.T.Y. Lui, A. Fazakerley, A.P. Walsh, C. Mouikis, Flattened current sheet and its evolution in substorms. J. Geophys. Res. **113**, 7 (2008). doi:10.1029/2007JA012812

M.I. Sitnov, K. Schindler, Tearing stability of a multiscale magnetotail current sheet. Geophys. Res. Lett. **37**, 8102 (2010). doi:10.1029/2010GL042961

M.I. Sitnov, L.M. Zelenyi, H.V. Malova, A.S. Sharma, Thin current sheet embedded within a thicker plasma sheet: self-consistent kinetic theory. J. Geophys. Res. **105**, 13029–13044 (2000). doi:10.1029/1999JA000431

M.I. Sitnov, A.S. Sharma, P.N. Guzdar, P.H. Yoon, Reconnection onset in the tail of Earth's magnetosphere. J. Geophys. Res. **107**, 1256 (2002). doi:10.1029/2001JA009148

M.I. Sitnov, M. Swisdak, P.N. Guzdar, A. Runov, Structure and dynamics of a new class of thin current sheets. J. Geophys. Res. **111**, 8204 (2006). doi:10.1029/2005JA011517

M.I. Sitnov, N. Buzulukova, M. Swisdak, V.G. Merkin, T.E. Moore, Spontaneous formation of dipolarization fronts and reconnection onset in the magnetotail. Geophys. Res. Lett. **40**, 22–27 (2013). doi:10.1029/2012GL054701

J.A. Slavin, B.J. Anderson, D.N. Baker, M. Benna, S.A. Boardsen, R.E. Gold, G.C. Ho, S.M. Imber, H. Korth, S.M. Krimigis, R.L. McNutt Jr., J.M. Raines, M. Sarantos, D. Schriver, S.C. Solomon, P. Trávníček, T.H. Zurbuchen, MESSENGER and Mariner 10 flyby observations of magnetotail structure and dynamics at Mercury. J. Geophys. Res. **117**, 1215 (2012). doi:10.1029/2011JA016900

E.J. Smith, L. Davis Jr., D.E. Jones, P.J. Coleman Jr., D.S. Colburn, P. Dyal, C.P. Sonett, A.M.A. Frandsen, The planetary magnetic field and magnetosphere of Jupiter: pioneer 10. J. Geophys. Res. **79**, 3501 (1974). doi:10.1029/JA079i025p03501

K. Snekvik, E. Tanskanen, N. Østgaard, L. Juusola, K. Laundal, E.I. Gordeev, A.L. Borg, Changes in the magnetotail configuration before near-Earth reconnection. J. Geophys. Res. **117**, 2219 (2012). doi:10.1029/2011JA017040

T.W. Speiser, Particle trajectories in model current sheets, 1, analytical solutions. J. Geophys. Res. **70**, 4219–4226 (1965). doi:10.1029/JZ070i017p04219

T.W. Speiser, Particle trajectories in model current sheets, 2, applications to auroras using a geomagnetic tail model. J. Geophys. Res. **72**, 3919–3932 (1967). doi:10.1029/JZ072i015p03919

L.C. Steinhauer, M.P. McCarthy, E.C. Whipple, Multifluid model of a one-dimensional steady state magnetotail current sheet. J. Geophys. Res. **113**, 4207 (2008). doi:10.1029/2007JA012578

K. Tummel, L. Chen, Z. Wang, X.Y. Wang, Y. Lin, Gyrokinetic theory of electrostatic lower-hybrid drift instabilities in a current sheet with guide field. Phys. Plasmas **21**(5), 052104 (2014). doi:10.1063/1.4875720

O.L. Vaisberg, L.M. Zeleny, Formation of the plasma mantle in the Venusian magnetosphere. Icarus **58**, 412–430 (1984). doi:10.1016/0019-1035(84)90087-3

I.Y. Vasko, A.V. Artemyev, V.Y. Popov, H.V. Malova, Kinetic models of two-dimensional plane and axially symmetric current sheets: group theory approach. Phys. Plasmas **20**(2), 022110 (2013). doi:10.1063/1.4792263

I.Y. Vasko, A.V. Artemyev, A.A. Petrukovich, R. Nakamura, L.M. Zelenyi, The structure of strongly tilted current sheets in the Earth magnetotail. Ann. Geophys. **32**, 133–146 (2014a). doi:10.5194/angeo-32-133-2014

I.Y. Vasko, L.M. Zelenyi, A.V. Artemyev, A.A. Petrukovich, H.V. Malova, T.L. Zhang, A.O. Fedorov, V.Y. Popov, S. Barabash, R. Nakamura, The structure of the Venusian current sheet. Planet. Space Sci. **96**, 81–89 (2014b). doi:10.1016/j.pss.2014.03.013

I.Y. Vasko, A.V. Artemyev, A.A. Petrukovich, H.V. Malova, Thin current sheets with strong bell-shape guide field: cluster observations and models with beams. Ann. Geophys. **32**(10), 1349–1360 (2014). doi:10.5194/angeo-32-1349-2014. http://www.ann-geophys.net/32/1349/2014/

C. Wang, L.R. Lyons, R.A. Wolf, T. Nagai, J.M. Weygand, A.T.Y. Lui, Plasma sheet $PV^{5/3}$ and nV and associated plasma and energy transport for different convection strengths and AE levels. J. Geophys. Res. **114**, 1–2 (2009). doi:10.1029/2008JA013849

E. Whipple, R. Puetter, M. Rosenberg, A two-dimensional, time-dependent, near-Earth magnetotail. Adv. Space Res. **11**, 133–142 (1991). doi:10.1016/0273-1177(91)90024-E

B. Wilken, P.W. Daly, U. Mall, K. Aarsnes, D.N. Baker, R.D. Belian, J.B. Blake, H. Borg, J. Büchner, M. Carter, J.F. Fennell, R. Friedel, T.A. Fritz, F. Gliem, M. Grande, K. Kecskemety, G. Kettmann, A. Korth, S. Livi, S. McKenna-Lawlor, K. Mursula, B. Nikutowski, C.H. Perry, Z.Y. Pu, J. Roeder, G.D. Reeves, E.T. Sarris, I. Sandahl, F. Søraas, J. Woch, Q.-G. Zong, First results from the RAPID imaging energetic particle spectrometer on board cluster. Ann. Geophys. **19**, 1355–1366 (2001). doi:10.5194/angeo-19-1355-2001

P.H. Yoon, A.T.Y. Lui, On the drift-sausage mode in one-dimensional current sheet. J. Geophys. Res. **106**, 1939–1948 (2001). doi:10.1029/2000JA000130

P.H. Yoon, A.T.Y. Lui, Model of ion- or electron-dominated current sheet. J. Geophys. Res. **109**, 11213 (2004). doi:10.1029/2004JA010555

P.H. Yoon, A.T.Y. Lui, A class of exact two-dimensional kinetic current sheet equilibria. J. Geophys. Res. **110**, 1202 (2005). doi:10.1029/2003JA010308

L.M. Zelenyi, A.V. Artemyev, A.A. Petrukovich, Earthward electric field in the magnetotail: cluster observations and theoretical estimates. Geophys. Res. Lett. **37**, 6105 (2010). doi:10.1029/2009GL042099

L.M. Zelenyi, M.I. Sitnov, H.V. Malova, A.S. Sharma, Thin and superthin ion current sheets. quasi-adiabatic and nonadiabatic models. Nonlinear Process. Geophys. **7**, 127–139 (2000)

L.M. Zelenyi, H.V. Malova, V.Y. Popov, D. Delcourt, A.S. Sharma, Nonlinear equilibrium structure of thin currents sheets: influence of electron pressure anisotropy. Nonlinear Process. Geophys. **11**, 579–587 (2004)

L.M. Zelenyi, H.V. Malova, V.Y. Popov, D.C. Delcourt, N.Y. Ganushkina, A.S. Sharma, "Matreshka" model of multilayered current sheet. Geophys. Res. Lett. **33**, 5105 (2006). doi:10.1029/2005GL025117

L.M. Zelenyi, A.V. Artemyev, H.V. Malova, V.Y. Popov, Marginal stability of thin current sheets in the Earth's magnetotail. J. Atmos. Sol.-Terr. Phys. **70**, 325–333 (2008). doi:10.1016/j.jastp.2007.08.019

L.M. Zelenyi, A.V. Artemyev, K.V. Malova, A.A. Petrukovich, R. Nakamura, Metastability of current sheets. Phys. Usp. **53**, 933–941 (2010). doi:10.3367/UFNe.0180.201009g.0973

L.M. Zelenyi, H.V. Malova, A.V. Artemyev, V.Y. Popov, A.A. Petrukovich, Thin current sheets in collisionless plasma: equilibrium structure, plasma instabilities, and particle acceleration. Plasma Phys. Rep. **37**, 118–160 (2011). doi:10.1134/S1063780X1102005X

L.M. Zelenyi, A.I. Neishtadt, A.V. Artemyev, D.L. Vainchtein, H.V. Malova, Quasiadiabatic dynamics of charged particles in a space plasma. Phys. Usp. **56**, 347–394 (2013). doi:10.3367/UFNe.0183.201304b.0365

T.L. Zhang, W. Baumjohann, R. Nakamura, A. Balogh, K. Glassmeier, A wavy twisted neutral sheet observed by CLUSTER. Geophys. Res. Lett. **29**(19), 190000 (2002). doi:10.1029/2002GL015544

X. Zhou, V. Angelopoulos, A. Runov, M.I. Sitnov, F. Coroniti, P. Pritchett, Z.Y. Pu, Q. Zong, J.P. McFadden, D. Larson, K. Glassmeier, Thin current sheet in the substorm late growth phase: modeling of THEMIS observations. J. Geophys. Res. **114**, 3223 (2009). doi:10.1029/2008JA013777